Contributions in Mathematical and Computational Sciences • Volume 8

Editors
Hans Georg Bock
Willi Jäger
Otmar Venjakob

For further volumes:
http://www.springer.com/series/8861

Winfried Kohnen • Rainer Weissauer
Editors

Conformal Field Theory, Automorphic Forms and Related Topics

CFT, Heidelberg,
September 19-23, 2011

MATCH
IWR

Springer

Editors
Winfried Kohnen
Rainer Weissauer
Department of Mathematics
University of Heidelberg
Heidelberg
Germany

ISSN 2191-303X ISSN 2191-3048 (electronic)
Contributions in Mathematical and Computational Sciences
ISBN 978-3-662-43830-5 ISBN 978-3-662-43831-2 (eBook)
DOI 10.1007/978-3-662-43831-2
Springer Heidelberg New York Dordrecht London

Library of Congress Control Number: 2014946622

Mathematics Subject Classification (2010): 11-XX, 81-XX

Printed on acid-free paper

Springer is part of Springer Science+Business Media (www.springer.com)

Preface to the Series

Contributions to Mathematical and Computational Sciences

Mathematical theories and methods and effective computational algorithms are crucial in coping with the challenges arising in the sciences and in many areas of their application. New concepts and approaches are necessary in order to overcome the complexity barriers particularly created by nonlinearity, high-dimensionality, multiple scales and uncertainty. Combining advanced mathematical and computational methods and computer technology is an essential key to achieving progress, often even in purely theoretical research.

The term mathematical sciences refers to mathematics and its genuine sub-fields, as well as to scientific disciplines that are based on mathematical concepts and methods, including sub-fields of the natural and life sciences, the engineering and social sciences and recently also of the humanities. It is a major aim of this series to integrate the different sub-fields within mathematics and the computational sciences, and to build bridges to all academic disciplines, to industry and other fields of society, where mathematical and computational methods are necessary tools for progress. Fundamental and application-oriented research will be covered in proper balance.

The series will further offer contributions on areas at the frontier of research, providing both detailed information on topical research, as well as surveys of the state-of-the-art in a manner not usually possible in standard journal publications. Its volumes are intended to cover themes involving more than just a single "spectral line" of the rich spectrum of mathematical and computational research.

The Mathematics Center Heidelberg (MATCH) and the Interdisciplinary Center for Scientific Computing (IWR) with its Heidelberg Graduate School of Mathematical and Computational Methods for the Sciences (HGS) are in charge of providing and preparing the material for publication. A substantial part of the material will be acquired in workshops and symposia organized by these institutions in topical areas of research. The resulting volumes should be more than just proceedings collecting papers submitted in advance. The exchange of information and the discussions during the meetings should also have a substantial influence on the contributions.

This series is a venture posing challenges to all partners involved. A unique style attracting a larger audience beyond the group of experts in the subject areas of specific volumes will have to be developed.

Springer Verlag deserves our special appreciation for its most efficient support in structuring and initiating this series.

Heidelberg University, Germany

Hans Georg Bock
Willi Jäger
Otmar Venjakob

Preface

This volume reports on recent developments in the theory of vertex operator algebras (VOAs) and their applications to mathematics and physics. Historically the mathematical theory of VOAs originated from the famous monstrous moonshine conjectures of J.H. Conway and S.P. Norton, which predicted a deep relationship between the characters of the largest simple finite sporadic group, the Monster, and the theory of modular forms inspired by the observations of J. MacKay and J. Thompson.

Although perhaps implicitly present earlier in conformal field theory, the precise mathematical notion of vertex algebras first emerged from the work of I. Frenkel, J. Lepowsky and A. Meurman and their purely algebraic construction of a vertex algebra with a natural action of the Monster group, laying the foundations for Borcherds' later proof of the moonshine conjectures. Indeed, by isolating the underlying mechanism from analytical aspects, physical field theories and the explicit examples thus shaped the axiomatic definition of vertex algebras as purely algebraic objects, opening a rich new field. By studying them for their own sake, R. Borcherds not only gave the theory its precise form, but also succeeded in proving the moonshine conjectures with the aid of these new concepts. So, it is quite interesting that unlike other algebraic structures like fields, rings, algebras or Lie algebras, the concept of vertex algebras appeared comparatively late in the mathematical literature. However, looking back, even today the underlying mechanism played by vertex operator algebras connecting representations of certain simple groups and modular forms remains to be mysterious. Altogether, Borcherds' results on the monstrous moonshine conjecture relating the representation theory of the monster group with modular forms using the construction of the vertex algebra $V^{\natural}$, whose graded dimensions are the Fourier coefficients of $j(q) - 744$, are a major landmark in the theory. They have led to many subsequent investigations of the structure of VOAs, some of which are addressed in this volume. Another theoretical milestone was Y. Zhu's finding that for rational VOAs (essentially those vertex operator algebras whose category of admissible modules is semisimple), every irreducible representation of a rational VOA gives rise to an elliptic modular form. Hence rational VOAs and certain generalizations have since been studied intensively.

Independently from Zhu's work it should be mentioned that also more general types of modular forms, defined by their Fourier expansion as counting functions, naturally arise in the theory of vertex algebras via the Kac denominator formula of generalized Kac-Moody Lie algebras derived from certain distinguished VOAs and derived generalized Lie algebras. In fact, thereby not only elliptic modular forms seem to appear naturally in the theory, but also modular forms of several variables. Of course there are other interesting circumstances under which modular forms of higher genus naturally occur, some of which will be addressed in this volume.

In a remarkable development, A. Beilinson succeeded in further generalizing the concept of vertex algebras by stressing the importance of the underlying geometric space. Thanks to the combined achievements of Beilinson and V. Drinfeld we now know that vertex algebras are a special case of chiral algebras, where these chiral algebras are certain sheaves on algebraic varieties. It seems, at least if the underlying geometry comes from a Riemann surface, that this new aspect indicates deep connections between conformal field theories, class field theory and various other branches of mathematics.

Quite recently, the study of representations of vertex algebras has produced surprising new developments for simple groups G other than the Monster group. Here, especially the Mathieu groups play a prominent part, with interesting applications to the theory of black holes, which has since become a very active field.

The contributions to this volume are based on lectures held in September 2011 during a conference on Conformal Field Theory, Automorphic Forms and Related Topics, organized by W. Kohnen and R. Weissauer. The conference was part of a special program offered at Heidelberg University in summer 2011 under the sponsorship of the MAThematics Center Heidelberg (MATCH).

We wish to extend our sincere thanks to all contributors to this volume and all conference participants, with special thanks to Geoffrey Mason and Miranda Cheng for their excellent preparatory courses that were held prior to the conference. Geoffrey's course entitled Vertex Operator Algebras, Modular Forms and Moonshine is included as an appendix to this volume.

We are grateful to Sabine Eulentrop for the perfect handling of numerous logistical problems and her help in preparing the final manuscript. Finally, we would like to express our sincere gratitude to MATCH and especially Otmar Venjakob, whose generous support made the conference and the special *Heidelberg Automorphic Semester* possible.

Heidelberg, Germany — Winfried Kohnen
Heidelberg, Germany — Rainer Weissauer
May 2014

Contents

Characters of Modules of Irrational Vertex Algebras

Antun Milas

Abstract We review several properties of characters of vertex algebra modules in connection to q-series and modular-like objects. Four representatives of conformal vertex algebras: regular, C_2-cofinite, tamely irrational and wild, are discussed from various points of view.

1 Introduction

Unlike many algebraic structures, vertex algebras have for long time enjoyed natural and fruitful connection with modular forms. This connection came first to light through the monstrous moonshine, a fascinating conjecture connecting modular forms (or more precisely the Hauptmoduln) and representations of the Monster, the largest finite sporadic simple group. This mysterious connection was partially explained first in the work of Frenkel et al. [37] who constructed a vertex operator algebra $V^\natural$, called the moonshine module, whose graded dimension is $j(q) - 744$ and whose automorphism group is the Monster. The connection with McKay-Thompson series was later proved by Borcherds [21] thus proving the full Conway-Norton conjecture. What is amazing about the vertex algebra $V^\natural$ is that on one hand it is arguably one of the most complicated objects constructed in algebra, yet it has an extremely simple representations theory (that of a field!).

Another important closely related concept in vertex algebra theory (and two-dimensional conformal field theory) is that of modular invariance of characters. This property, proposed by physicists as a consequence of the axioms of rational conformal field theory, was put on firm ground first in the seminal work of Zhu [62]. Among many applications of Zhu's result we point out its power to "explain"

A. Milas (✉)
Department of Mathematics and Statistics, University at Albany (SUNY),
1400 Washington Avenue, Albany, NY 12222, USA
e-mail: amilas@albany.edu

W. Kohnen and R. Weissauer (eds.), *Conformal Field Theory, Automorphic Forms and Related Topics*, Contributions in Mathematical and Computational Sciences 8,
DOI 10.1007/978-3-662-43831-2_1,

modular invariance of characters of integrable highest weight modules for affine Kac-Moody Lie algebras, discovered previously by Kac and Peterson [48]. As rational vertex algebras (with some extra properties) give rise to Modular Tensor Categories [45], this rich underlying structure can be used for to prove the general Verlinde formula [45] (see [20] for definition). This formula, discovered first by Verlinde [61], gives an important connection between the fusion coefficients in the tenor product coefficients of the S-matrix coming from both categorical and analytic $SL(2,\mathbb{Z})$-action on the "space" of modules. In addition, it also gives a fascinating link between analytic q-dimensions and the coefficients of the S-matrix.

There are other important connections between two subjects such as ADE classification of modular invariant partition functions, vertex superalgebras and mock modular forms, orbifold theory, elliptic genus, generalized moonshine, etc.

Everything that we mentioned so far comes from a very special class of vertex algebras called C_2-cofinite rational vertex algebras [62], the moonshine module being a prominent example. In this note we do not try to say much about rational vertex algebras (although we do give some definition and list known results) and almost nothing about the moonshine. Our modest goal is simply to argue that even non-rational (and sometimes even non C_2-cofinite) vertex algebra seem to enjoy properties analogous to properties of rational VOA, but much more complicated, yet reach enough that exploring them leads to some interesting mathematics related to modular forms and other modular-like objects. Another pedagogical aspect of these notes is to convey some ideas and aspects of the theory rarely considered in the literature on vertex algebras. We focus on four different types of vertex algebras:

- rational C_2-cofinite or *regular* (the category of modules has modular tensor category structure, q-dimensions are closely related to categorical dimensions).
- irrational C_2-cofinite (tensor product theory and a version modular invariance are available, a Verlinde-type formula is still to be formulated and proved)
- non C_2-cofinite, mildly irrational (there is evidence of braided tensor category structure on the category of module, or suitable sub-category. A version of modular invariance holds with continuous part added. Usually involve atypical and typical modules, the latter parametrized by continuous parameters. q-dimensions of irreps are finite and nonzero).
- non C_2-cofinite, badly irrational (not likely to have good categorical structure. For example two modules under fusion can give infinitely many modules. Consequently, q-dimensions may be infinite).

As a working example of rational C_2-cofinite vertex algebra we shall use the lattice vertex algebra V_L, where L is an even positive definite lattice. This is, from many different points of view, the most important source of vertex algebras, and in particular leads to the moonshine module via the Leech lattice and orbifolding.

When we move beyond rational vertex algebras, many difficulties arise, and this transition really has to be done in two steps. The nicest examples worth exploring are of course C_2-cofinite vertex algebras. These vertex algebras admit finitely-many inequivalent irreducible modules. Here the most prominent example is triplet vertex algebra [4, 33, 39, 50] being a conformal vertex subalgebra of the rank one lattice vertex algebra of certain rational central charge. Another prominent example is the symplectic fermion vertex superalgebra [1, 50, 59]. As we shall see the triplet vertex algebra enjoys many interesting properties including a version of modular invariance, even a conjectural version of the Verlinde formula.

If we move one step lower in the hierarchy this leads us to non C_2-cofinite vertex algebras. There are at least several candidates here. One is of course the vertex algebra associated to free bosons, called the Heisenberg vertex algebra [37, 49, 51]. Because this algebra has a fairly simple representation theory [37] we decided to consider another family of irrational vertex algebras—certain subalgebras of the Heisenberg vertex algebra. As we shall see this so-called "singlet" vertex algebras involve two types of irreducible representations: typical and atypical, something that persists for many $\mathcal{W}$-algebras. Quite surprisingly, there is a version of modular invariance for the singlet family, including a Verlinde-type formula inferred from the characters [24].

Finally, at the bottom of the barrel sort of speaking, we are left with badly behaved irrational conformal vertex algebra, namely those that are vacuum modules for the Virasoro algebra (or more general affine $\mathcal{W}$-algebras [18]) or for affine Lie algebras [49, 51]. One reason for this type of vertex algebra not being very interesting is due to lack of modular-like properties. Also, their fusion product is somewhat ill behaved. For example, two irreducible modules can produce infinitely many non-isomorphic modules under the fusion.

Four examples representing four types entering our discussion are connected with a chain of VOA embeddings:

$$L(c_{p,1},0) \hookrightarrow W(2,2p-1) \hookrightarrow W(p) \hookrightarrow V_{\sqrt{2p}}.$$

At the end of the paper we show that this diagram can be extended to an arbitrary ADE type simple Lie algebra, the above diagram being the simplest instance coming from $\mathfrak{sl}_2$.

2 Vertex Algebras and Their Characters

We begin by recalling the definition of a vertex operator algebra following primarily [51] (cf. [37, 49]).

Definition 1. A vertex operator algebra is a quadruple $(V, Y, \mathbf{1}, \omega)$ where V is a $\mathbb{Z}$-graded vector space

$$V = \bigoplus_{n \in \mathbb{Z}} V_{(n)}$$

together with a linear map $Y(\cdot, x) : V \to (\text{End } V)[[x, x^{-1}]]$ and two distinguished elements $\mathbf{1}$ and $\omega \in V$, such that for $u, v \in V$ we have

$$Y(u, x)v \in V((x)),$$

$$Y(\mathbf{1}, x) = 1,$$

$$Y(v, x)\mathbf{1} \in V[[x]] \quad \text{and} \quad \lim_{x \to 0} Y(v, x)\mathbf{1} = v,$$

$$[L(m), L(n)] = (m - n)L(m + n) + \frac{m^3 - m}{12}\delta_{m+n,0}c$$

for $m, n \in \mathbb{Z}$, where

$$Y(\omega, x) = \sum_{n \in \mathbb{Z}} L(n)x^{-n-2},$$

and $c \in \mathbb{C}$ (the so-called *central charge*); we also have

$$L(0)w = nw \qquad \text{for } n \in \mathbb{Z} \text{ and } v \in V_{(n)},$$

and the $L(-1)$-axiom

$$Y(L(-1)u, x) = \frac{d}{dx}Y(u, x)$$

and the following Jacobi identity

$$x_0^{-1}\delta\left(\frac{x_1 - x_2}{x_0}\right)Y(u, x_1)Y(v, x_2) - x_0^{-1}\delta\left(\frac{x_2 - x_1}{-x_0}\right)Y(v, x_2)Y(u, x_1) \tag{1}$$

$$= x_2^{-1}\delta\left(\frac{x_1 - x_0}{x_2}\right)Y\left(Y(u, x_0)v, x_2\right). \tag{2}$$

If we omit the Virasoro axiom and the grading the structure is called vertex algebra, but we have to replace $L(-1)$-axiom with the D-derivative axiom [51]. In some constructions it is useful to have another VOA structure on the same space. This is important when we pass to a different coordinate system on the torus E_τ discussed below. With $Y(u, x)$ as above and u homogeneous, we let

$$Y[u, x] = Y(e^{x \deg(u)}u, e^x - 1),$$

which is well-defined if we expand $1/(e^x - 1)^m$, for $m \geq 0$, in finitely many negative powers of x. Then it can be shown [62] that $(V, Y[\cdot, x], \mathbf{1}, \omega - \frac{c}{24}\mathbf{1})$ is also a vertex operator algebra isomorphic to the original one. We also define bracket modes of vertex operator

$$Y[u, x] = \sum_{n \in \mathbb{Z}} u[n]x^{-n-1}; \quad u[n] \in End(V).$$

Definition 2 (Sketch). We say that a vector space W together with a linear map

$$Y_W(\cdot, x) : V \to (\text{End } W)[[x, x^{-1}]]$$

is a weak V-module if Y_W satisfies the Jacobi identity, and "all other defining properties of a vertex algebra that make sense hold". If in addition the space is graded by $L(0)$-eigenvalues such that the grading is compatible with that of V, we say that M is an ordinary module.

Not all vertex algebra modules are of interest to us right now.

Definition 3. An admissible V-module is a weak V-module M which carries a $\mathbb{Z}_{\geq 0}$-grading

$$M = \bigoplus_{n \in \mathbb{Z}_{\geq 0}} M(n)$$

satisfying the following condition: if $r, m \in \mathbb{Z}, n \in \mathbb{Z}_{\geq 0}$ and $a \in V_r$ then

$$a_m M(n) \subseteq M(r + n - m - 1). \tag{3}$$

We call an admissible V-module M irreducible in case 0 and M are the only submodules. An ordinary module is an admissible module where the above grading is decomposition into finite-dimensional $L(0)$-eigenspaces.

A vertex algebra V is called *rational* if every admissible V-module is a direct sum of simple admissible V-modules. That is, we have complete reducibility of admissible V-modules. Observe that the definition of rationality does not seem to involve any internal characterization or property of vertex algebras. The next definition is analogous to "finite-dimensionality" for associative algebras.

Definition 4 (C_2-cofiniteness). A vertex algebra V is said to be C_2-cofinite if the space generated by vectors $\{a_{-2}b, \ a, b \in V\}$ is of finite codimension (in V).

An important consequence of this definition is that a C_2-cofinite vertex algebra has finitely many irreducible modules up to equivalence, which explains "finite-dimensionality" hinted earlier. It is a conjecture that every rational vertex algebra is C_2-cofinite, but the converse is known not to be true (see below).

2.1 One-Point Functions on Torus

To an admissible V-module M with finite dimensional graded subspaces we can associate its modified *graded dimension* or simply *character* [62]:

$$\text{ch}_M(q) := \text{tr}_M q^{L(0)-c/24}, \tau \in \mathbb{H},$$

where c is the central charge. Strictly speaking this function does not necessarily converge so it should be viewed only formally, but in almost all known examples it is holomorphic in the whole upper half-plane.

We are also interested in related graded traces that can be computed on M:

$$\mathrm{tr}_M o(a) q^{L(0)-c/24},$$

where $o(a) = a(\deg(a) - 1)$ is the zero weight operator and a is homogeneous, in the sense that it preserved graded components.

As usual we denote by

$$G_{2k}(\tau) = \frac{B_{2k}}{(2k)!} + \frac{2}{(2k-1)!} \sum_{n \geq 1} \frac{q^n n^{2k-1}}{1-q^n}$$

($k \geq 1$) slightly normalized Eisenstein series as in [62] given by their q-expansions.

Denote by $O_q(V)$ the $\mathbb{C}[G_4, G_6]$-submodule of $V \otimes \mathbb{C}[G_4, G_6]$ generate by

$$a[0]b$$

$$a[-2]b + \sum_{k=2}^{\infty} (2k-1) a[2k-2]b \otimes G_{2k}(\tau)$$

Definition 5. Let V be a VOA. A map $S(-,-) : V[G_2, G_4] \times \mathbb{H} \to \mathbb{C}$ satisfying the following conditions is called a one-point function on the torus $E_\tau = \mathbb{C}/(\mathbb{Z} + \tau\mathbb{Z})$:

(1) For any $a \in V \otimes \mathbb{C}[G_4, G_6]$ the functions $S(a, \tau)$ is holomorphic in $\tau \in \mathbb{H}$.
(2) $S(\sum_i v_i \otimes f_i(\tau), \tau) = \sum_i f_i(\tau) S(a_i, \tau)$ for all $a_i \in V$ and $f_i \in \mathbb{C}[G_4, G_6]$.
(3) $S(a, \tau) = 0$ for all $a \in O_q(V)$,
(4) $S(L[-2]a, \tau) = (q \frac{d}{dq}) S(a, \tau) + \sum_{k=1}^{\infty} G_{2k}(\tau) S(L[2k-2]a, \tau)$.

We denote the space of one-point functions by $\mathscr{C}(V)$. Then any element of the form $S(\mathbf{1}, \tau)$, where $S \in V$, is called a (virtual) *generalized character*. It is possible to show [62], that graded traces $\mathrm{tr}_M o(a) q^{L(0)-c/24}$ give a one-point function on the torus. So in particular an (ordinary) character can be viewed as a generalized character.

Let us explain results pertaining rational vertex algebras first. We denote by M_i, $i \in I$ irreducible V-modules (so I is finite). Later we shall also assume that $i = 0 \in I$ is reserved for the VOA itself, which is also assumed to be simple. We shall also use $\mathrm{Irrep}(V)$ to denote the set of equivalence classes of irreducible V-module.

Theorem 1 (Zhu). *Let V be a rational C_2-cofinite vertex algebra. Then for every homogeneous $a \in V$ with respect to $L[0]$, the expressions $\{\mathrm{tr}_{M_i} o(a) q^{L(0)-c/24}\}$, $i \in I$ defines a vector valued modular form of weight* $\deg(a)$. *In particular, for $a = \mathbf{1}$ this weight is zero. Moreover, the space of one point functions on torus is $|Irrep(V)|$-dimensional and $a \mapsto \mathrm{tr}_{M_i} o(a) q^{L(0)-c/24}$, $i \in I$, is a basis of $\mathscr{C}(V)$.*

Observe that another consequence of this result is that for rational C_2-cofinite vertex algebras every generalized character is an ordinary character.

Because the category of modules of rational vertex algebras has a semisimple braided tensor category structure [47], we have the fusion product:

$$M_i \boxtimes M_j = \sum_{k \in I} N_{ij}^k M_k,$$

where $N_{ij}^k \in \mathbb{Z}_{\geq 0}$ are the fusion coefficients and $\boxtimes$ is Huang, Lepowsky and Zhang's tensor product [47]. On the other hand, the previous theorem furnishes us with a $|Irrep(V)|$-dimensional representations of $SL(2,\mathbb{Z})$ acting on the space of ordinary characters. In particular, we have the special matrix $S \in SL(2,\mathbb{Z})$, called the S-matrix, corresponding to $\tau \to -\frac{1}{\tau}$. If in addition, the vertex algebra is C_2-cofinite the category of V-Mod is a modular tensor category (an important result of Huang [45]), so it also admits a categorical action of $SL(2,\mathbb{Z})$ on the space generated by the equivalence classes of irreducible modules M_i, $i \in I$. In particular $\tau \to -\frac{1}{\tau}$ induces a matrix called the s-matrix. In turns out that $S = s$ [29], after suitable rescaling of s. One important property of MTCs is the Verlinde formula [20] (first formulated in [61]) that allows us to express fusion coefficients simply from the coefficients of the s (and hence S) matrix. The precise statement is: Denote by N_{ij}^k the fusion coefficients, then we have

$$N_{ij}^k = \sum_r \frac{S_{ir} S_{jr} S_{k^* r}}{S_{0r}}, \tag{4}$$

where $r \mapsto r^*$ is the map on indices induced by taking dual of irreducible modules $M_i \mapsto M_i^*$.

Another related important notion in two-dimensional conformal field theory is that of (analytic) q-dimension. For a V-module M we let

$$q\dim(M) = \lim_{y \to 0+} \frac{\mathrm{ch}_M(iy)}{\mathrm{ch}_V(iy)} \tag{5}$$

Of course, such a quantity may not need even exist. But again, for V rational and C_2-cofinite, it is known to be closely related to categorical q-dimension $\dim_q(M)$, computed as the trace of the identity endofunctor, which also equals $\frac{s_{i0}}{s_{00}}$ [20]. Under some favorable conditions on the vertex algebra, this categorical version of the q-dimension coincide with the analytic (see [29], conditions (V1) and (V2) and formula (3.1)):

Proposition 1. *Let V be a rational C_2-cofinite VOA with lowest conformal weights of irreducible modules positive expect for $i = 0$, then*

$$dim_q(M_i) = \frac{S_{i0}}{S_{00}} = qdim(M_i).$$

This proposition is known to hold

Categorical q-dimensions are known to have good properties with respect to tensor products and direct sums:

$$dim_q(M \boxtimes N) = dim_q(M) \cdot dim_q(N),$$
$$dim_q(M \oplus N) = dim_q(M) + dim_q(N).$$

If V is only C_2-cofinite, we shall see in the next sections that Zhu's modular invariance theorem fails and not every one point function on the torus is an ordinary trace. This is closely related to non-semisimplicity of Zhu's algebra $A(V) = V/O(V)$, where $O(V)$ is spanned by $\mathrm{Res}_x \frac{(1+x)^{\deg(a)}}{x^2} Y(a,x)b$. Rationality implies that the space of one-point functions on the torus is isomorphic to the vector space of symmetric functions on $A(V)$:

$$S^V = (A(V)/[A(V), A(V)])^*.$$

But in general this space does not carry a precise description of one-point functions. Still, there is a satisfactory result essentially due to Miyamoto [56]. Assume for completeness that the central charge of the vertex algebra is non-zero (so finite-dimensional V-modules are excluded—these only appear for $c = 0$). Then there is a connection between one-point functions and the Zhu algebra (Miyamoto).

Theorem 2. *The vector space $\mathscr{C}(V)$ admits a finite basis $\mathscr{B}$ such that each $S \in \mathscr{B}$ admits an expansion*

$$S(a,\tau) = \sum_{j=0}^{d} \sum_{k=0}^{\infty} S_{jk}(a) q^{r - c_c/24 + k} (2\pi i \tau)^j$$

for all $a \in V$, where $r \in \mathbb{C}$ and $S_{00} \in S^V$, a symmetric linear functional on $A(V)$. Moreover, $S \mapsto S_{00}$ is an embedding. In particular, the dimension of $\mathscr{C}(V)$ is bounded by the dimension of S^V.

This version of the theorem is proven in [19], but something similar is implicitly used in [56] (see also [6]). One striking feature of the theorem is the appearance of τ-powers, so no q-expansion of one point functions exists in general. This is closely tied to existence of $L(0)$ non-diagonalizable modules, called logarithmic modules [40,53]. For more about this subject and connection to Logarithmic Conformal Field Theory we refer the reader to another review paper [13,40,44], as we do not discuss this subject here. In the aforementioned paper of Miyamoto, he constructs $S(a,\tau)$ via certain *pseudotraces* maps ϕ expressed as $\mathrm{tr}^{\phi,M} o(a) q^{L(0)-c/24}$ where M is a particular module "interlocked" with ϕ. We should point out that in many examples of interest this object is hard to construct explicitly. A slightly more efficient way of constructing one-point functions was obtained by Arike-Nagatomo's paper [19], although it is not clear whether their construction works in general.

3 Rational VOA: Lattice Vertex Algebras

We review the construction of a vertex operator algebra coming from an even lattice following [51] (see also [37, 49]). Let L be a rank $d \in \mathbb{N}$ even positive definite lattice of rank $d \in \mathbb{N}$ with an integer valued nondegenerate symmetric $\mathbb{Z}$-bilinear form $\langle \cdot, \cdot \rangle : L \times L \to \mathbb{Z}$.

Form the vector space

$$\mathfrak{h} = L \otimes_{\mathbb{Z}} \mathbb{C} \tag{6}$$

so that $\dim(\mathfrak{h}) = d$ and extend the bilinear form from L to $\mathfrak{h}$. Now we shall consider the affinization of $\mathfrak{h}$ viewed as an abelian Lie algebra

$$\hat{\mathfrak{h}} = \bigoplus_{n \in \mathbb{Z}} \mathfrak{h} \otimes t^n \oplus \mathbb{C}\mathbf{k}, \tag{7}$$

with bracket relations

$$\begin{aligned} [\alpha \otimes t^m, \beta \otimes t^n] &= \langle \alpha, \beta \rangle m \delta_{m+n,0} \mathbf{k} \\ [\mathbf{k}, \hat{\mathfrak{h}}] &= 0 \end{aligned} \tag{8}$$

for $\alpha, \beta \in \mathfrak{h}$ and $m, n \in \mathbb{Z}$. Consider

$$\hat{\mathfrak{h}}_+ = \mathfrak{h} \otimes t\mathbb{C}[t] \text{ and } \hat{\mathfrak{h}}_- = \mathfrak{h} \otimes t^{-1}\mathbb{C}[t^{-1}]. \tag{9}$$

We now form a vertex operator algebra associated to $\hat{\mathfrak{h}}$ with central charge 1, $M(1)$, by adding structure to the symmetric algebra of $\hat{\mathfrak{h}}_-$. As vector spaces we have

$$M(1) = U(\hat{\mathfrak{h}}_-) = S(\hat{\mathfrak{h}}_-). \tag{10}$$

If we let $\{u^{(1)}, \ldots, u^{(d)}\}$ be an orthonormal basis of $\mathfrak{h}$ we define the conformal vector

$$\omega = \frac{1}{2} \sum_{i=1}^{d} u^{(i)}(-1) u^{(i)}(-1) \mathbf{1}. \tag{11}$$

So we have the Virasoro algebra operators

$$L(n) = \mathrm{Res}_x x^{n+1} Y(\omega, x) = \frac{1}{2} \sum_{i=1}^{d} \sum_{m \in \mathbb{Z}} {}^{\circ}_{\circ} u^{(i)}(m) u^{(i)}(n-m) {}^{\circ}_{\circ}. \tag{12}$$

It is easy to construct irreducible $M(1)$-modules. Those are simply Fock spaces F_λ where $\lambda \in \mathfrak{h}^*$. This is again just an induced module such that $h \in \mathfrak{h}$ acts on the highest weight vector as multiplication by $\lambda(h)$, so as a vector space $F_\lambda \cong M(1)$ and $F_0 = M(1)$. This space will again become relevant in later sections.

The space $S(\hat{\mathfrak{h}}_-)$ makes up one part of the vertex operator algebra associated with L. The other portion is related to the group algebra $\mathbb{C}[L]$. In order to ensure the Jacobi identity we need to modify the product associated to $\mathbb{C}[L]$ so that

$$e^\alpha e^\beta = (-1)^{\langle\alpha,\beta\rangle} e^\beta e^\alpha, \tag{13}$$

for $\alpha, \beta \in L$. To accomplish this we use a central extension, $(\hat{L}, \bar{\ })$, of L by the cyclic group $\langle\kappa|\kappa^2 = 1\rangle$. For $\alpha, \beta \in L$ define the map

$$c_0 : L \times L \to \mathbb{Z}/2\mathbb{Z} \tag{14}$$

as follows:

$$\begin{aligned} c_0(\alpha, \alpha) &= 0 + 2\mathbb{Z}, \\ c_0(\alpha, \beta) &= \langle\alpha, \beta\rangle + 2\mathbb{Z} \text{ and,} \\ c_0(\beta, \alpha) &= -c_0(\alpha, \beta). \end{aligned} \tag{15}$$

This is indeed the commutator map associated to the central extension of the lattice. It may also be uniquely defined by the condition $ab = \kappa^{c_0(\bar{a},\bar{b})} ba$ for $a, b \in \hat{L}$. Define a section of $\hat{L}$, $e : L \to \hat{L}$, so that $\alpha \mapsto e_\alpha$. So e is such that $\bar{\ } \circ e = id_L$. Let

$$\epsilon_0 : L \times L \to \mathbb{Z}/2\mathbb{Z} \tag{16}$$

be the *corresponding 2-cocyle*, defined by

$$e_\alpha e_\beta = \kappa^{\epsilon_0(\alpha,\beta)} e_{\alpha+\beta} \tag{17}$$

Let $\chi : \langle\kappa\rangle \to \mathbb{C}^\times$ be defined by $\chi(\kappa) = -1$. View $\mathbb{C}$ as a $\langle\kappa\rangle$-module where κ acts as -1 and denoted this module as $\mathbb{C}_\chi$. Define

$$\mathbb{C}\{L\} = \mathrm{Ind}_{\langle\kappa\rangle}^{\hat{L}} \mathbb{C}_\chi = \mathbb{C}[\hat{L}] \otimes_{\mathbb{C}[\langle\kappa\rangle]} \mathbb{C}_\chi = \mathbb{C}[\hat{L}]/(\kappa - (-1))\mathbb{C}[\hat{L}]. \tag{18}$$

Let ι be the inclusion $\hat{L} \hookrightarrow \mathbb{C}\{L\}$ such that $\iota(a) = a \otimes 1$. Notice our section e allows us to view $\mathbb{C}\{L\}$ and $\mathbb{C}[L]$ as isomorphic vector spaces with $\iota(e_\alpha) \mapsto e^\alpha$ for $\alpha \in L$.

Now define maps $c, \epsilon : L \times L \to \mathbb{C}^\times$ by $c(\alpha, \beta) = (-1)^{c_0(\alpha,\beta)}$ and $\epsilon(\alpha, \beta) = (-1)^{\epsilon_0(\alpha,\beta)}$. Now we can see the action of $\hat{L}$ on $\mathbb{C}[L]$, for $\alpha, \beta \in L$

$$\begin{aligned} e_\alpha \cdot e^\beta &= \epsilon(\alpha, \beta) e^{\alpha+\beta} \\ \kappa \cdot e^\beta &= -e^\beta \\ e_\alpha \cdot 1 &= e^\alpha \end{aligned} \tag{19}$$

Now set

$$V_L = M(1) \otimes \mathbb{C}\{L\} \tag{20}$$

and

$$\mathbf{1} = 1 \otimes \iota(1) \in V_L. \tag{21}$$

We now add more structure to the space V_L. First we will view $M(1)$ as a trivial $\hat{L}$-module, so that for $\alpha \in L$, e_α acts as $1 \otimes e_\alpha \in End(V_L)$. Also view $\mathbb{C}\{L\}$ as a trivial $\hat{\mathfrak{h}}_*$-module and for $h \in \mathfrak{h}$, define

$$h(0) : \mathbb{C}\{L\} \to \mathbb{C}\{L\} \text{ so that } \iota(a) \mapsto \langle h, \bar{a} \rangle \, \iota(a) \tag{22}$$

for $a \in \hat{L}$. By making the identification $M(1) \cong S(\hat{\mathfrak{h}}) \otimes e^0$ we can transport the structure of a Virasoro algebra module to V_L with the grading given by the action of $L(0)$

$$L(0) \cdot \iota(e_\alpha) = (\mathrm{wt}\, \iota(e_\alpha)) \, \iota(e_\alpha) = \frac{1}{2} \langle \alpha, \alpha \rangle \, \iota(e_\alpha). \tag{23}$$

We keep the same conformal vector so the central charge of V_L is $\mathrm{rank}(L)$.

In order to define the vertex operator $Y(\iota(e_\alpha), x)$ we need the following operator for $h \in \mathfrak{h}$,

$$E^{\pm}(-h, x) = \exp\left(\sum_{n \in \pm\mathbb{Z}} \frac{-h(n)}{n} x^{-n} \right) \in (\mathrm{End}\, V_L)[[x, x^{-1}]]. \tag{24}$$

and define

$$Y(\iota(e_\alpha), x) = E^{-}(-\alpha, x) E^{+}(-\alpha, x) e_\alpha x^\alpha \in (\mathrm{End}\, V_L)[[x, x^{-1}]]. \tag{25}$$

where x^α acts on V_L as

$$x^\alpha (v \otimes \iota(a)) = x^{\langle \alpha, \bar{a} \rangle} (v \otimes \iota(a)). \tag{26}$$

This explains how to construct lattice vertex algebra structure on V_L. If the lattice is of rank one, no central extension is needed. Thus $\mathbb{C}[L] = \mathbb{C}\{L\}$. Also, to simplify the notation we shall write e^α instead of $\iota(e_\alpha)$, where no confusion arise. Everything about representation theory of lattice vertex algebras can be summarized in the following elegant result by Dong (see [27] and [51] for instance):

Theorem 3. *The vertex algebra V_L is rational. Moreover, the set*

$$\{V_{L+\lambda}; \lambda + L \in L^\circ / L\}$$

(where L^0 is the dual lattice) is a complete set of inequivalent irreducible V_L-modules (strictly speaking, we never defined $V_{L+\lambda}$ but this is easily done by replacing $\mathbb{C}[L]$ in the definition with $\mathbb{C}[L+\lambda]$. For a full account on this see [51]).

Characters of V_L-modules are easily determined (keep in mind $c = \mathrm{rank}(L)$). We have

$$\mathrm{ch}_{V_{L+\lambda}}(q) = \frac{\sum_{\alpha \in L+\lambda} q^{\langle \alpha,\alpha \rangle/2}}{\eta(\tau)^c}.$$

By using a well-known formula for the modular transformation formula for the higher rank theta function, we infer

$$\mathrm{ch}_{V_{L+\lambda}}(-\frac{1}{\tau}) = \sum_{\bar{\nu} \in L^\circ/L} S_{\lambda\nu} \mathrm{ch}_{V_{L+\nu}}(\tau),$$

where $S_{\lambda\nu}$ denote the S-matrix of the transformation. Observe that $S_{0\nu} = \frac{1}{\sqrt{det(S)}}$, where S is the Gram matrix of L. This modular invariance part also follows from Zhu's theorem (the vertex algebra V_L is C_2-cofinite). The fusion product for the lattice vertex algebras is simply

$$V_{L+\lambda} \boxtimes V_{L+\nu} = V_{L+\lambda+\nu}.$$

The q-dimensions are also easy to compute and $dim_q(V_{L+\lambda}) = 1$ for all λ.

4 C_2-Cofinite Irrational Case: The Triplet VOA

In this section we examine properties of a specific irrational C_2-cofinite vertex algebra.

4.1 *The Triplet*

Let V_L be as in the previous section, where L is of rank one. First we construct a subalgebra of V_L called the triplet algebra. We should point out that lattice vertex algebra are rarely mentioned in the physics literature, where triplet is usually treated as an extended conformal algebra with $SO(3)$ symmetry [41,42,50] , or as a part of an extended Felder's complex in which we extract the kernel instead of cohomology. Our approach here is slightly different and it follows [4, 33, 39], where the triplet algebra is constructed as kernel of a screening operator acting (as we shall see) among two V_L-modules.

Let $p \in \mathbb{Z}$, $p \geq 2$, and

$$L = \mathbb{Z}\alpha, \quad \langle \alpha, \alpha \rangle = 2p,$$

or simply $L = \sqrt{2p}\mathbb{Z}$, with the usual multiplication. We are interested in the central charge $c_{p,1} = 1 - \frac{6(p-1)^2}{p}$, so we choose

$$\omega = \frac{1}{4p}\alpha(-1)^2\mathbf{1} + \frac{p-1}{2p}\alpha(-2)\mathbf{1}.$$

We also define conformal weights

$$h_{r,s}^{p,q} = \frac{(ps-rq)^2 - (p-q)^2}{4pq}.$$

With this central charge, the generalized vertex algebra V_{L° [28] admits two screenings:

$$\tilde{Q} = e_0^{-\alpha/p} \text{ and } Q = e_0^{\alpha}.$$

Then we let

$$\mathscr{W}(p) = \mathrm{Ker}_{V_L} e_0^{-\alpha/p} \subset V_L, \tag{27}$$

a subalgebra of V_L called the triplet algebra.

The above construction can be recast in terms of automorphisms of infinite order and *generalized twisted* modules introduced by Huang [46]. Consider $\nu = \exp(e_0^{-\alpha/p})$. This operator does not preserve V_L but it can be viewed as an automorphism of V_{L° of infinite order. Then the triplet is $V_{L^\circ}^{\nu} \cap V_L$, where V_L^{ν} denote the ν-fixed vertex subalgebra. In fact, V_{L° can be also replaced by $V_L \oplus V_{L-\alpha/p}$ (see [7]).

As shown in [4], $\mathscr{W}(p)$ is strongly generated by the conformal vector ω and three *primary* vectors

$$F = e^{-\alpha}, \ H = QF, \ E = Q^2 e^{-\alpha}.$$

There is another useful description of $\mathscr{W}(p)$ [32, 39]. As a module for the Virasoro algebra, V_L is not completely reducible but it has a semisimple filtration whose maximal semisimple part is $\mathscr{W}(p)$. More precisely,

$$\begin{aligned} \mathscr{W}(p) &= \mathrm{soc}_{Vir}(V_L) \\ &= \bigoplus_{n=0}^{\infty}\bigoplus_{j=0}^{2n} U(Vir).Q^j e^{-n\alpha} \\ &\cong \bigoplus_{n=0}^{\infty}(2n+1)L(c_{p,1}, h_{1,2n+1}^{p,1}),, \end{aligned} \tag{28}$$

where $L(c, h)$ denote the highest weight Virasoro module of central charge c and lowest conformal weight h. For other examples of irrational C_2-cofinite vertex (super)algebras see [5, 11, 12, 14–16] .

4.2 Irreducible Modules and Characters

The triplet $\mathscr{W}(p)$ is known to be C_2-cofinite but irrational [4] (see also [23]). It also admits precisely $2p$ inequivalent irreducible modules [4] which are usually denoted by:

$$\Lambda(1), \ldots, \Lambda(p), \Pi(1), \ldots, \Pi(p).$$

These modules were previously studied in [33, 34, 39] was proposed as a complete list of irreducibles. Since irreps are admissible, for $1 \leq i \leq p$, the top component of $\Lambda(i)$ is one-dimensional and has lowest conformal weight $h_{i,1}^{p,1}$, and the top component of $\Pi(i)$ is two-dimensional with conformal weight $h_{3p-i,1}^{p,1}$.

The characters of irreducible $\mathscr{W}(p)$-modules are well-known and computed in many papers on logarithmic conformal field theories starting with [39]. For $1 \leq i \leq p$, the formulas are

$$\begin{aligned} \mathrm{ch}_{\Lambda(i)}(\tau) &= \frac{i\Theta_{p,p-i}(\tau) + 2\partial\Theta_{p,p-i}(\tau)}{p\eta(\tau)}, \\ \mathrm{ch}_{\Pi(i)}(\tau) &= \frac{i\Theta_{p,i}(\tau) - 2\partial\Theta_{p,i}(\tau)}{p\eta(\tau)}, \end{aligned} \tag{29}$$

where

$$\Theta_{i,p}(\tau) = \sum_{n\in\mathbb{Z}} q^{(2np+i)^2/4p}, \quad \partial\Theta_{i,p}(\tau) = \sum_{n\in\mathbb{Z}} (n + \frac{i}{2p}) q^{(2np+i)^2/4p}.$$

From here we infer that the space spanned by characters of irreps is not modular invariant! To understand this better observe that in addition to $\Theta_{p,i}$ and $\partial\Theta_{p,i}$ series we also need $\tau\partial\Theta_{p,i}$ series to preserve modularity. This gives indication that one-point functions on the torus might be bigger than the number of irreps.

The next theorem (essentially taken from [6]) settles the problem of finding the space of one-point functions on the torus for the triplet algebra.

Theorem 4. *The space of one-point functions for the triplet vertex algebra is $3p-1$-dimensional.*

The proof breaks down on studying generalized characters. By using general properties of one-point functions and the triplet vertex algebra we first prove that every generalized character $S(\mathbf{1}, \tau)$ satisfies

$$D^{3p-1}S(\mathbf{1},\tau)+\sum_{i=0}^{3p-2}H_i(q)D^iS(\mathbf{1},\tau)=0, \tag{30}$$

where

$$H_i(q)\in\mathbb{C}[G_4,G_6]_{2h-2i}$$

is a modular form of weight $2h-2i$. and

$$D_h=(q\frac{d}{dq})+hG_2(q)$$

where $h\in\mathbb{Z}_{\geq 0}$ and

$$D^n:=D_{2n-2}\cdots D_2D_0.$$

This fact immediately implies several things. First, because the space of solutions of the differential equations is modular invariant, the space of generalized characters is at most $3p-1$-dimensional. But at the same time each ordinary trace associated to an irrep must be a solution to thus equation. So for modular invariance to be preserved the space is at least $3p-1$-dimensional. Therefore there must be contribution coming from $p-1$ generalized characters. Once we observe that $\dim(\mathscr{C}(V))\leq\dim(S^{\mathscr{W}(p)})$, where the right hand side is known to be $3p-1$-dimensional by [9], we have the proof and observation that $\mathscr{C}(V)$ is as large as it can be. By using a method from [19] we can construct all the missing one-point functions explicitly.

4.3 Verlinde-Type Formula for $\mathscr{W}(p)$-Mod

As there is no general Verlinde formula for C_2-cofinite vertex algebras, in what follows "Verlinde-type formula" refers to the following concepts extracted from the (generalized) characters:

1. A way of constructing a genuine finite-dimensional $SL(2,\mathbb{Z})$ representation on the space of irreducible and possible larger generalized characters.
2. By using the S-matrix from (1), for a fixed triple i,j,k, the standard Verlinde sum,that is, the right hand-side of (4), recovers non-negative integers that agree with the known (or at least conjectural) fusion coefficients N_{ij}^k. Because the category of representation is semisimple these fusion coefficients should be understood as multiplicities in the Grothendieck ring.

We do not claim that there is a unique procedure for extracting the S-matrix here, so there might be more than one Verlinde-type formula giving the same answer.

Next, we outline a Verlinde-type formula for the triplet algebra obtained in [39], with some crucial modifications in [43]. We already listed all irreps earlier with their explicit characters. Form a $2n \times 1$ character vector

$$\chi_p(\tau) := (\mathrm{ch}_{\Lambda(p)}, \mathrm{ch}_{\Pi(p)}, \mathrm{ch}_{\Lambda(1)}, \mathrm{ch}_{\Pi(p-1)}, \cdots, \mathrm{ch}_{\Lambda(p-1)}, \mathrm{ch}_{\Pi(1)})^T,$$

where $(\cdot)^T$ stands for the transpose. Easy computation—by using modular transformation formulas for $\Theta_{p,i}$ and $\partial\Theta_{p,i}$—shows that

$$\chi_p(-\frac{1}{\tau}) = S_p(\tau) \cdot \chi_p(\tau), \quad \chi_p(\tau+1) = T_p(\tau) \cdot \chi_p(\tau),$$

where the entries of the matrix are computed by using the formula

$$\mathrm{ch}_{\Lambda(s)}(-\frac{1}{\tau}) = \frac{1}{\sqrt{2p}}\left\{\frac{s}{p}\left(\mathrm{ch}_{\Lambda(p)}(\tau) + (-1)^{p-s}\chi_{\Pi(p)}(\tau)\right.\right.$$

$$\left. + \sum_{s'=1}^{p-1} 2cos(\frac{2(p-s)s'}{p})(\mathrm{ch}_{\Lambda(p-s')}(\tau) + \mathrm{ch}_{\Pi(s')}(\tau))\right)$$

$$\left. - \sum_{s'=1}^{p-1}(-1)^{p+s+s'} 2sin(\frac{2ss'}{p}) i\tau(\frac{p-s'}{p}\mathrm{ch}_{\Lambda(s')}(\tau) - \frac{s'}{p}\mathrm{ch}_{\Pi(p-s')}(\tau))\right\}$$

and a similar formula for $\chi_{\Pi(s)}(-\frac{1}{\tau})$. The matrix $T_p(\tau)$ is clearly independent of τ and diagonal (we omit its explicit form here). This way we do not obtain a $2p$-dimensional representation of the modular group due to τ-dependence. To fix this problem it is convenient to introduce a suitable automorphy factor $j(\gamma, \tau)$, $\gamma \in SL(2,\mathbb{Z})$, satisfying the cocycle condition

$$j(\gamma\gamma', \tau) = j(\gamma', \tau) j(\gamma, \gamma'\tau).$$

In addition, we can define $j(\gamma, \tau)$ such that the modified S,T-matrices

$$\mathbf{S}_p := j(S, \tau) S_p(\tau), \quad \mathbf{T}_p := j(T, \tau) T_p(\tau),$$

do not depend on τ, so $\mathbf{S}_p$ and $\mathbf{T}_p$ define a genuine representation of $SL(2,\mathbb{Z})$. This was achieved explicitly in [39, Sect. 3]. Again we omit explicit formulas for $\mathbf{S}_p$ for brevity. Equipped with a right candidate for the S-matrix we are ready to compute the Verlinde sum

$$N_{ij}^k := \sum_{r \in \{0,\ldots,2p-1\}} \frac{\mathbf{S}_p(ir)\mathbf{S}_p(jr)\mathbf{S}_p(k^*r)}{\mathbf{S}_p(0r)}.$$

These numbers turn out to be non-negative integers, so we can form a free $\mathbb{Z}$-module generated by the equivalence classes of irreps, and on it we let

$$X_I \times X_J := \sum_K N_{IJ}^K X_K. \tag{31}$$

Theorem 5. *The previous product defines an associative ring structure. Moreover,*

$$\begin{aligned}
\Lambda(s) \times \Lambda(t) &= \sum_{r=|s-t|+1, by\ 2}^{min(s+t-1, 2p-s-t-1)} \Lambda(r) \oplus \bigoplus_{r=2p-s-t+1; step=2}^{p,p-1} P_r^+ \\
\Lambda(s) \times \Pi(t) &= \sum_{r=|s-t|+1, by\ 2}^{min(s+t-1, 2p-s-t-1)} \Pi(r) \oplus \bigoplus_{r=2p-s-t+1; step=2}^{p,p-1} P_r^- \\
\Pi(s) \times \Pi(t) &= \sum_{r=|s-t|+1, by\ 2}^{min(s+t-1, 2p-s-t-1)} \Lambda(r) \oplus \bigoplus_{r=2p-s-t+1; step=2}^{p,p-1} P_r^+,
\end{aligned} \tag{32}$$

where $P_r^{\pm}$ are given by

$$P_r^+ = 2\Lambda(r) + 2\Pi(p-r), \quad P_r^- = 2\Pi(r) + 2\Lambda(p-r) \tag{33}$$

and where the summation is up to $p-1$ or p depending on whether $r+s+t$ is even or odd, respectively.

Tsuchiya and Wood in [60] (see also [58]) proved that the above product recovers correct multiplication in the Grothendieck ring of the category $\mathscr{W}(p) - \mathrm{Mod}$ (this one exists thanks to [47]). Moreover, the $P_r^{\pm}$ summands in the formulas should be viewed as projective modules. The approach in [60] is based on the notion of fusion expressed as a certain space of coinvariants. Some special cases of the fusion rules are computed in [7] by using intertwining operators.

Observe also that for $X = \Pi$ or Λ and $1 \leq s \leq p$ we have

$$q\mathrm{dim}(X(s)) = s,$$

which can be easily verified by considering asymptotic properties of the given q-series [22]. It is a priori not clear if this agrees with the categorical q-dimension.

We conclude this section with a comment that we believe this pattern persists for other C_2-cofinite vertex algebras or at least those that are of CFT type and where the vertex algebra is simple. Moreover, we conjecture that in a favorable situation when $V - \mathrm{Mod}$ is rigid [57] the analytic q-dimension agrees with the categorical one. Rigidity in general seems to fail for $\mathscr{W}_{p,q}$ triplet vertex algebras studied in [7, 8, 10, 31, 32].

5 Beyond C_2-Cofinite Vertex Algebras

Very little is known about general categories of representations of irrational non C_2-cofinite vertex algebras (let alone any modularity-type properties!). We only focus on those vertex algebras with good categorical properties in the sense that they admit a subcategory where irreducibles and perhaps projective modules can be classified. An obvious candidate here is the Heisenberg vertex algebra $M(1)$ already discussed in the setup of lattice vertex algebras. The category of $\mathfrak{h}$-diagonalizable $M(1)$-modules is known to be semisimple and the irreps are F_λ, $\lambda \in \mathfrak{h}^*$ [37]. In other words, all irreducible modules are "generic". In addition, the formal fusion product is given by $F_\lambda \times F_\nu = F_{\lambda+\nu}$. A better candidate (in terms of richness of representations) for discuss here is the *singlet* vertex algebra [2, 3, 50], a proper subalgebra of the full rank one Fock space $M(1)$, so all Heisenberg algebra modules are already included. In addition the singlet is included inside the triplet algebra $\mathcal{W}(p)$. So in addition to Fock space modules, the singlet admits a special infinite family of representations that do not look like F_λ and come from decomposition of irreducible $\mathcal{W}(p)$-modules.

The setup is as in the previous section. We fix the central charge to be $c_{p,1}$ and choose the same conformal vector in $M(1)$. Following the notation from [2] (see also [3]), we define

$$\mathcal{W}(2, 2p-1) = \mathrm{Ker}_{M(1)}\tilde{Q}.$$

called the *singlet vertex algebra* of central charge $c_{p,1}$. Since $\tilde{Q}$ commutes with the action of the Virasoro algebra, we have

$$L(c_{p,1}, 0) \subset \mathcal{W}(2, 2p-1).$$

The vertex operator algebra $\mathcal{W}(2, 2p-1)$ is completely reducible as a Virasoro algebra module and the following decomposition holds:

$$\mathcal{W}(2, 2p-1) = \bigoplus_{n=0}^{\infty} U(Vir).\, u^{(n)}; \quad u^{(n)} = Q^n e^{-\alpha n} \cong \bigoplus_{n=0}^{\infty} L(c_{p,1}, n^2 p + np - n),$$

As shown in [2] (see also [3]) all irreducible $\mathcal{W}(2, 2p-1)$-modules are constructed as subquotients of the Fock spaces F_λ. What is peculiar about these irreps is that they come in two groups with very distinct features:

- (Typical or generic) Those isomorphic to irreducible Virasoro Fock spaces denoted by F_λ (it simply means that λ does not satisfy a certain integrability condition).
- (Atypical or generic) A certain family $M_{r,s}$ of subquotients of Fock spaces $F_{\frac{r-1}{2}\sqrt{2p}+\frac{s-1}{\sqrt{2p}}}$, $r \in \mathbb{Z}$, and $1 \leq s \leq p$. Each $M_{r,s}$ is isomorphic to an infinite direct sum of Virasoro irreps.

Each irrep $M_{r,s}$ decomposes as an infinite direct sum of irreducible Virasoro algebra (for explicit decomposition formulas see [4, 24]). This is then used to show:

$$\mathrm{ch}[M_{r,s}](\tau) = \frac{P_{pr-s,p}(0,\tau) - P_{pr+s,p}(0,\tau)}{\eta(\tau)},$$

where

$$P_{a,b}(u,\tau) = \sum_{n=0}^{\infty} z^{n+\frac{a}{2b}} q^{b(n+\frac{a}{2b})^2}, \quad z = e(u). \tag{34}$$

The last expression is what is usually called *partial theta function* of and its properties are well-recorded in the literature [17]. In particular, for $M_{1,1} = \mathcal{W}(2, 2p-1)$, we get

$$\mathrm{ch}[\mathcal{W}(2,2p-1)](\tau) = \frac{\sum_{n\in\mathbb{Z}} \mathrm{sgn}(n) q^{p(n+\frac{p-1}{2p})^2}}{\eta(\tau)},$$

which is precisely false theta function of Rogers.

If we try to naively compute

$$P_{a,b}(\frac{u}{\tau}, -\frac{1}{\tau})$$

some divergent integrals appear, so instead we introduce a regularization, a method used in physics to handle divergent quantities.

Now, we define the regularized characters by introducing a parameter ϵ to achieve better modular properties. We let

$$\begin{aligned} \mathrm{ch}[F_\lambda^\epsilon](\tau) &= e^{2\pi\epsilon(\lambda-\alpha_0/2)} \frac{q^{(\lambda-\alpha_0/2)^2/2}}{\eta(\tau)} \\ \mathrm{ch}[M_{r,s}^\epsilon](u;\tau) &= \frac{1}{\eta(\tau)} \sum_{n=0}^{\infty} \mathrm{ch}[F^\epsilon_{\alpha_{r-2n-1,p-s}}](\tau) - \mathrm{ch}[F^\epsilon_{\alpha_{r-2n-2,s}}](\tau), \end{aligned} \tag{35}$$

where $\alpha_0 = \alpha_+ + \alpha_-$, $\alpha_+ = \sqrt{2p}$ and $\alpha_- = -\sqrt{2/p}$.

Observe that typical ϵ-regularized characters are simply $\mathrm{tr}_{F_\lambda} e^{2\pi\epsilon(\frac{\alpha(0)}{\sqrt{2p}} - \alpha_0/2)} q^{L(0)-c/24}$. But atypical regularization is more subtle and it has no obvious interpretation as graded trace. Let $\beta^\pm_{r,s} = ((r-1)\alpha_+ \pm s\alpha_-)/2$, then the atypical characters are

$$\begin{aligned} &\mathrm{ch}[M^\epsilon_{r,s}](\tau) \\ &\quad = \mathrm{ch}[F^\epsilon_{\alpha_0/2-\beta^-_{r,s}}](\tau) P_{\alpha_+\epsilon}(-\alpha_+\beta^-_{r,s}\tau; \alpha_+^2\tau) \\ &\qquad - \mathrm{ch}[F^\epsilon_{\alpha_0/2-\beta^+_{r,s}}](\tau) P_{\alpha_+\epsilon}(-\alpha_+\beta^+_{r,s}\tau; \alpha_+^2\tau). \end{aligned}$$

We can easily show that

$$\mathrm{ch}[F^{\epsilon}_{\lambda+\alpha_0/2}]\Big(\frac{-1}{\tau}\Big) = \int_{\mathbb{R}} S^{\epsilon}_{\lambda+\alpha_0/2,\mu+\alpha_0/2}\mathrm{ch}[F^{\epsilon}_{\mu+\alpha_0/2}](\tau)d\mu,$$

with $S^{\epsilon}_{\lambda+\alpha_0/2,\mu+\alpha_0/2} = e^{2\pi\epsilon(\lambda-\mu)}e^{-2\pi i\lambda\mu}$.

The next result taken from [24] gives S-"matrix" expressed as a kernel.

Theorem 6.

$$ch[M^{\epsilon}_{r,s}]\Big(-\frac{1}{\tau}\Big) = \int_{\mathbb{R}} S^{\epsilon}_{(r,s),\mu+\alpha_0/2}\mathrm{ch}[F^{\epsilon}_{\mu+\alpha_0/2}](\tau)d\mu + X^{\epsilon}_{r,s}(\tau)$$

with

$$S^{\epsilon}_{(r,s),\mu+\alpha_0/2} = -e^{-2\pi\epsilon((r-1)\alpha_+/2+\mu)}e^{\pi i(r-1)\alpha_+\mu}\frac{\sin\big(\pi s\alpha_-(\mu+i\epsilon)\big)}{\sin\big(\pi\alpha_+(\mu+i\epsilon)\big)}$$

and

$$X^{\epsilon}_{r,s}(\tau) = \frac{1}{4i\eta(\tau)}(\mathrm{sgn}(\mathrm{Re}(\epsilon))+1)\sum_{n\in\mathbb{Z}}(-1)^{rn}e^{\pi i\frac{s}{p}n}q^{\frac{1}{2}(\frac{n^2}{\alpha_+}-\epsilon^2)}\big(q^{-i\epsilon n/\alpha_+^2}-q^{i\epsilon n/\alpha_+^2}\big).$$

5.1 Brewing a Verlinde-Type Formula

If we have a continuous type S-matrix as the one above, the right approach for defining fusion coefficients seems to be [25, 26]

$$\int_{\mathbb{R}} \frac{S^{\epsilon}_{a\rho}S^{\epsilon}_{b\rho}\overline{S^{-\bar{\epsilon}}_{\rho\mu}}}{S^{\epsilon}_{(1,1)\rho}}d\rho.$$

But this integral badly diverges, so we either have to pass to heuristic approach as in [25] and [26] where the integrals are interpreted as a sum of the Dirac delta functions, or we can simply change the order of integration so that the fusion coefficients are genuine distributions. Thus, we redefine the product in the Verlinde algebra of characters as

$$\mathrm{ch}[X_a]\times\mathrm{ch}[X_b] := \int_{\mathbb{R}}\left(\int_{\mathbb{R}}\frac{S^{\epsilon}_{a\rho}S^{\epsilon}_{b\rho}\overline{S^{-\bar{\epsilon}}_{\rho\mu}}}{S^{\epsilon}_{(1,1)\rho}}\mathrm{ch}[F^{\epsilon}_{\mu}]d\mu\right)d\rho \tag{36}$$

It is worth to point out that the map $X_a \mapsto \mathrm{ch}[X_a^\epsilon]$ is injective on irreducible modules, so we don't loose any information by working with the characters, and we can even take the approach as in (31) with integrals added.

It can be shown that this product converges for all irreps, and it give rise to a commutative associative algebra. Finally, we have this remarkable formula [24]

Theorem 7. *With $Re(\epsilon) < 0$, the Verlinde-type algebra of regularized characters is given by*

$$\mathrm{ch}[F_\lambda^\epsilon] \times \mathrm{ch}[F_\mu^\epsilon] = \sum_{\ell=0}^{p-1} \mathrm{ch}[F^\epsilon_{\lambda+\mu+\ell\alpha_-}]$$

$$\mathrm{ch}[M^\epsilon_{r,s}] \times \mathrm{ch}[F_\mu^\epsilon] = \sum_{\substack{\ell=-s+2 \\ \ell+s=0 \bmod 2}}^{s} \mathrm{ch}[F^\epsilon_{\mu+\alpha_{r,\ell}}]$$

$$\begin{aligned}\mathrm{ch}[M^\epsilon_{r,s}] \times \mathrm{ch}[M^\epsilon_{r',s'}] = &\sum_{\substack{\ell=|s-s'|+1 \\ \ell+s+s'=1 \bmod 2}}^{min\{s+s'-1,p\}} \mathrm{ch}[M^\epsilon_{r+r'-1,\ell}] \\ &+ \sum_{\substack{\ell=p+1 \\ \ell+s+s'=1 \bmod 2}}^{s+s'-1} \Big(\mathrm{ch}[M^\epsilon_{r+r'-2,\ell-p}] + \mathrm{ch}[M^\epsilon_{r+r'-1,2p-\ell}] \\ &\qquad + \mathrm{ch}[M^\epsilon_{r+r',\ell-p}]\Big).\end{aligned}$$

Remark 1. The previous result is expected to give relations in the Grothendieck ring of a suitable (sub)category of $\mathscr{W}(2, 2p-1)$-modules. It is not clear to us whether any of the current results in VOA theory (including [47]) gives braided category structure on this category. Also, we conjecture equivalence of categories $\mathscr{W}(2, 2p-1) - Mod \cong U_q(?) - Mod$ where $U_q(?)$ is yet-to-be defined quantum group at $2p$-th root of unity.

6 "Bare" Virasoro Vertex Algebra

We are finally left with the lonely Virasoro vertex operator algebra $L(c_{p,1}, 0)$ sitting inside the singlet $\mathscr{W}(2, 2p-1)$. In this section we also allow $p = 1$. Irreducible admissible $L(c_{p,1}, 0)$-modules are simple module of the form $L(c_{p,1}, h)$, where $h \in \mathbb{C}$. There is a distinguished family of highest weight modules which are not Verma modules, that is $V(c_{p,1}, 0) \neq L(c_{p,1}, 0)$. This is if and only if $h = h_{i,s} = \frac{(ip-s)^2-(p-1)^2}{4p}$, $i > 0$, $0 < s \leq p$. We call them atypical modules.

6.1 Modular-Like Transformation Properties?

It is a well-known fact (due to Feigin and Fuchs) that

$$\mathrm{ch}_{L(c_{p,1},h_{i,s})}(\tau) = \frac{(1-q^{is})q^{\frac{(ip-s)^2}{4p}}}{\eta(\tau)}.$$

$$\mathrm{ch}_{L(c_{p,1},h)}(\tau) = \frac{q^{h+(p-1)^2/4p}}{\eta(\tau)}; \quad h \neq h_{i,s}$$

Evaluating $\tau \mapsto \tau + 1$ transformation on the character is trivial as usual. If we consider $\tau \mapsto -\frac{1}{\tau}$, as in the singlet case of Fock modules, we obtain one or two Gauss' integrals. But this answer will lead to new problems when we start computing a Verlinde-type formula. It turns out that two irreducible modules for this vertex algebra can produce infinitely many (more precisely, uncountably many) irreducible modules after the fusion, so we conclude that there cannot be a reasonable fusion algebra for $L(c_{p,1}, 0)$-modules unless of course we allow some kind of completions that we do not dwell into. Similar problem is already evident at the level of q-dimensions. Observe that for $h \neq h_{i,s}$

$$q\mathrm{dim}(L(c_{p,1}, h)) = \lim_{\tau \to 0} \frac{q^h}{1-q} = \infty.$$

Yet, in sharp contrast, we have

$$q\mathrm{dim}(L(c_{p,1}, h_{i,s})) = \lim_{\tau \to 0} \frac{q^{h_{i,s}}(1-q^{is})}{1-q} = is,$$

indicating that these atypical modules ought to behave much better under the fusion. This is also clear because of the following results (cf. [35, 52–54]):

$$L(c_{p,1}, h_{r,s}) \times L(c_{p,1}, h_{r',s'}) = \sum_{r'' \in A(r,r'), s'' \in A(s,s')} L(c_{p,1}, h_{r'',s''}),$$

where we assume that all indices are positive and $A_{i,j} = \{i+j-1, i+j-3, \cdots, |i-j|+1\}$. We should say that this formula only indicates triples of atypical modules whose fusion rules are 1 and not a relation in a hypothetical Grothendieck ring. We also have

$$qdim(L(c_{p,1}, h_{r,s}) \times L(c_{p,1}, h_{r',s'})) = qdim(L(c_{p,1}, h_{r,s})) \cdot qdim L(c_{p,1}, h_{r',s'}))$$

Because of the infinities involved we do not expect the irreducible modules can be organized in a way that the Verlinde formula holds. This is why to vertex algebras with similar properties we refer to as "wild".

7 Generalization and Higher Rank False Theta Functions

The story told in the previous sections can be generalized by considering the sequence of embeddings of vertex algebras:

$$W_p(\mathfrak{g}) \hookrightarrow \mathcal{W}_Q^0(p) \hookrightarrow \mathcal{W}_Q(p) \hookrightarrow V_{\sqrt{p}Q},$$

where Q is a root lattice of ADE type, $\mathfrak{g}$ is the corresponding simple Lie algebra, $W_p(\mathfrak{g})$ is the affine W-algebra of central charge $c_p(\mathfrak{g})$ associated to $\mathfrak{g}$, and where $\mathcal{W}_Q^0(p)$ and $\mathcal{W}(p)_Q$ are vertex algebras defined below. In the special case of $\mathfrak{g} = sl_2$ and $p \geq 2$ we recover the embedding of vertex algebras given in the introduction.

The affine $\mathcal{W}$-algebra associated to $\hat{\mathfrak{g}}$ at level $k \neq -h^\vee$, denoted by $\mathcal{W}_k(\mathfrak{g})$ is usually defined as the cohomology group obtained via a quantized BRST complex for the Drinfeld-Sokolov hamiltonian reduction [38]. As shown by Feigin and Frenkel (cf. [38] and [36] and citations therein) this cohomology is nontrivial only in the degree zero. Moreover, it is known that $\mathcal{W}_k(\mathfrak{g})$ is a quantum $\mathcal{W}$-(vertex) algebra, in the sense that is freely generated by rank($\mathfrak{g}$) primary fields, not counting the conformal vector.

We will be following the notation from Sect. 3. As before denote by L° the dual lattice of L. Now, we specialize $L = \sqrt{p}Q$, where $p \geq 2$ and Q is root lattice of ADE type. We equip V_L with a vertex algebra structure as earlier in Sect. 3 (by choosing an appropriate 2-cocycle). Let α_i denote the simple roots of Q. For the conformal vector we conveniently choose

$$\omega = \omega_{st} + \frac{p-1}{2\sqrt{p}} \sum_{\alpha \in \Delta_+} \alpha(-2)\mathbf{1},$$

where ω_{st} is the standard (quadratic) Virasoro generator [37, 51] . Then V_L is a conformal vertex algebra of central charge[1]

$$\text{rank}(L) + 12(\rho, \rho)(2 - p - \frac{1}{p}),$$

where ρ is the half-sum of positive roots. Consider the operators

$$e_0^{\sqrt{p}\alpha_i}, \quad e_0^{-\alpha_j/\sqrt{p}}, \quad 1 \leq i, j \leq rank(L) \tag{37}$$

acting between V_L and V_L-modules. These are the so-called *screening operators*. More precisely [55]

Lemma 1. *For every i and j the operators $e_0^{\sqrt{p}\alpha_i}$ and $e_0^{-\alpha_j/\sqrt{p}}$ commute with each other, and they both commute with the Virasoro algebra.*

---

[1]Without the linear term the central charge is $rank(L)$.

We shall refer to $e_0^{\sqrt{p}\alpha_i}$ and $e_0^{-\alpha_j/\sqrt{p}}$, as the *long* and *short* screening, respectively. It is well-known that the intersection of the kernels of residues of vertex operators is a vertex subalgebra (cf. [36]), so the next construction seems very natural.

An important theorem of Feigin and Frenkel [36] says that for k generic and $\mathfrak{g}$ is simply-laced, there is an alternative description of $\mathscr{W}_k(\mathfrak{g})$ in terms of free fields. For this purpose, we let $\nu = k + h^\vee$, where k is generic. Then there are (as above) appropriately defined screenings[2]

$$e_0^{-\alpha_i/\sqrt{\nu}} : M(1) \longrightarrow M(1, -\alpha_i/\sqrt{\nu}),$$

such that

$$\mathscr{W}_\nu(\mathfrak{g}) = \bigcap_{i=1}^{l} \mathrm{Ker}_{M(1)}(e_0^{-\alpha_i/\sqrt{\nu}}),$$

where $l = \mathrm{rank}(L)$. If we assume in addition that $\mathfrak{g}$ is simply laced (ADE type) we also have the following important duality [36]

$$\mathscr{W}_\nu(\mathfrak{g}) = \bigcap_{i=1}^{l} \mathrm{Ker}_{M(1)}(e_0^{\sqrt{\nu}\alpha_i}).$$

Generic values of ν do not have integrality property so in particular the screening operators $e_0^{-\alpha_i/\sqrt{\nu}}$ cannot be extended to a lattice vertex algebra. Still this idea can be used to define much larger vertex algebras which we now describe.

Theorem 8. *Let $\mathfrak{g}$ be simply laced. Then $p = k + h^\vee \in \mathbb{N}_{\geq 2}$ is non-generic. More precisely,*

$$\mathscr{W}^0(p)_Q := \bigcap_{i=1}^{l} \mathrm{Ker}_{M(1)} e_0^{-\alpha_i/\sqrt{p}}$$

is a vertex algebra containing $\mathscr{W}_p(\mathfrak{g})$ as a proper subalgebra. In particular, for $Q = A_1$ this algebra is simply the singlet $\mathscr{W}(2, 2p-1)$ discussed earlier.

The previous algebra can be maximally extended leading to

$$\mathscr{W}(p)_Q := \bigcap_{i=1}^{l} \mathrm{Ker}_{V_L} e_0^{-\alpha_j/\sqrt{p}}. \tag{38}$$

Again, if we let $Q = A_1$ this is just the triplet vertex algebra $\mathscr{W}(p)$.

[2]These screenings to do not extend to a lattice vertex algebra in general.

The following conjecture was mentioned in [13].

Conjecture 1. The vertex algebra $\mathcal{W}(p)_Q$ is C_2-cofinite.

Although there are not many rigorous results on the representation theory of $\mathcal{W}^0(p)_Q$ and of $\mathcal{W}(p)_Q$, we again expect that irreps of $\mathcal{W}(p)_Q$ can be understood as subquotients of V_L-modules, while all atypical irreps of $\mathcal{W}^0(p)_Q$ all appear in the decomposition of irreducible $\mathcal{W}(p)_Q$-modules and typical representations. As the structure of Fock spaces in the higher rank is not well-understood well, one can take a different geometric approach to guess the characters of relevant modules (see [30]).

7.1 Characters of $\mathcal{W}(p)_Q$-Modules

Let ρ denote the half-sum of positive roots, by W we denote the Weyl group and by $(\cdot,\cdot)$ the usual inner product in $L \otimes_{\mathbb{Z}} \mathbb{Q}$ normalized such that $(\alpha,\alpha) = 2$ for each root α. We also let $(\beta,\beta) = ||\beta||^2$. Let

$$\prod_{\alpha\in\Delta_-} (1-\mathbf{z}^\alpha)$$

denote the Weyl denominator (here Δ_- is the set of negative roots) and

$$\mathbf{z}^\alpha = z^{(\alpha_1,\alpha)} \cdots z^{(\alpha_n,\alpha)}.$$

There are two expression that we are concerned about here. The first does not give a proper character but only auxiliary expression to compute the proper (conjectural) characters. Assume $\lambda \in L^\circ$ and let [30]

$$\mathrm{ch}_{\mathcal{W}(p,\lambda)_Q}(\tau,\mathbf{z}) = \frac{\eta(\tau)^{-rank(Q)}}{e^\rho \prod_{\alpha\in\Delta_-}(1-\mathbf{z}^\alpha)} \sum_{w\in W}\sum_{\beta\in Q} (-1)^{l(w)} q^{\frac{||p\beta+\lambda+(p-1)(\rho)||^2}{2p}} \mathbf{z}^{w(\beta+\hat{\lambda}+\rho)}.$$

This expression cannot be evaluated at $z_i = 1$, but the limit

$$\mathrm{ch}_{\mathcal{W}(p,\lambda)_Q}(\tau) = \lim_{\mathbf{z}\to 1} \mathrm{ch}_{W_Q(p,\lambda)}(\tau)$$

is conjecturally expected to give the character of $\mathcal{W}(p,\lambda)_Q$, an irreducible $\mathcal{W}(p)_Q$-module. It is not hard to see by using L'hopital rule that the resulting expression is a linear combination of quasi-modular forms of different weight generalizing the formula in (29). A much harder question to ask is to determine its modular closure [30].

7.2 Characters of $\mathscr{W}^0(p)_Q$-Modules

The previous computation can be motivated to compute (conjectural) expressions for the characters of atypical irreducible $\mathscr{W}^0(p)_Q$-modules. Of course, for typical modules we have $\mathrm{ch}_{F_\lambda}(\tau)$ is just a pure power of q divided with the rank(Q)-th power of the Dedekind η-function. Characters of atypical $\mathscr{W}^0(p)_Q$-modules should be parameterized by $\lambda \in L^0$ (cf. with the singlet algebra)

$$\mathrm{ch}_{\mathscr{W}^0(p,\lambda)_Q}(\tau)$$
$$= \mathrm{CT}_{\mathbf{z}} \left\{ \frac{\eta(\tau)^{-rank(Q)}}{e^\rho \prod_{\alpha\in\Delta_-}(1-\mathbf{z}^\alpha)} \sum_{w\in W} \sum_{\beta\in Q} (-1)^{l(w)} q^{\frac{\|p\beta+\lambda+(p-1)(\rho)\|^2}{2p}} \mathbf{z}^{w(\beta+\hat{\lambda}+\rho)}. \right\},$$

where $\mathrm{CT}_{\mathbf{z}}$ denote the constant term w.r.t. $\mathbf{z}$. Observe that this is precisely in the analogy with the singlet modules. The previous definition motivates our proposal of higher rank false theta functions, generalizing the $\mathfrak{sl}_2$ case, of "weight" $|\Delta_+| - \frac{rank(L)}{2}$:

$$F_{p,\lambda}(\tau) = \mathrm{CT}_{\mathbf{z}} \left\{ \frac{\sum_{w\in W} \sum_{\beta\in Q} (-1)^{l(w)} q^{\frac{\|p\beta+\lambda+(p-1)(\rho)\|^2}{2p}} \mathbf{z}^{w(\beta+\hat{\lambda}+\rho)}.}{e^\rho \prod_{\alpha\in\Delta_-}(1-\mathbf{z}^\alpha)} \right\},$$

Remark 2. We expect many properties of generalized false theta functions to follow the pattern observed in the rank one case, including modularity-like properties of regularized false thetas, etc. This will be the subject of our forthcoming joint work with Bringmann [22].

Acknowledgements I would like to express my gratitude to the organizers of the conference "Conformal Field Theory, Automorphic Forms and Related Topics", Heidelberg 2011, for inviting me to give a talk. This note contains no new results. It is mostly based on joint works with D. Adamovic, T. Creutzig and some work in progress with K. Bringmann. I thank to all of them for discussion and for sharing their ideas.

References

1. Abe, T.: A Z_2-orbifold model of the symplectic fermionic vertex operator superalgebra. Mathematische Zeitschrift **255**(4), 755–792 (2007)
2. Adamović, D.: Classification of irreducible modules of certain subalgebras of free boson vertex algebra. J. Algebra **270**, 115–132 (2003)
3. Adamović, D., Milas, A.: Logarithmic intertwining operators and $W(2, 2p-1)$-algebras. J. Math. Phys. **48**, 073503 (2007)

4. Adamović, D., Milas, A.: On the triplet vertex algebra $W(p)$. Adv. Math. **217**, 2664–2699 (2008)
5. Adamović, D., Milas, A.: The $N = 1$ triplet vertex operator superalgebras. Commun. Math. Phys. **288**, 225–270 (2009)
6. Adamović, D., Milas, A.: An analogue of modular BPZ equation in logarithmic conformal field theory. Contemp. Math. **497**, 1–17 (2009)
7. Adamović, D., Milas, A.: Lattice construction of logarithmic modules for certain vertex algebras. Selecta Math. (N.S.) **15**(4), 535–561 (2009)
8. Adamović, D., Milas, A.: On W-algebras associated to $(2, p)$ minimal models and their representations. Int. Math. Res. Not. **20**, 3896–3934 (2010). arXiv:0908.4053
9. Adamović, D., Milas, A.: The structure of Zhu's algebras for certain W-algebras. Adv. Math. **227**, 2425–2456 (2011)
10. Adamovic, D., Milas, A.: On W-algebra extensions of $(2, p)$ minimal models: $p > 3$. J. Algebra **344**, 313–332 (2011)
11. Adamović, D., Milas, A.: An explicit realization of logarithmic modules for the vertex operator algebra $\mathcal{W}_{p,p'}$. J. Math. Phys. **53**, 073511 (2012)
12. Adamović, D., Milas, A.: The doublet vertex operator algebras $\mathcal{A}(p)$ and $\mathcal{A}_{2,p}$. Contemp. Math. **602**, 23–38 (2013)
13. Adamović, D., Milas, A.: Vertex operator (super)algebras and LCFT, special issue on Logarithmic Conformal Field Theory. J. Phys. A **46**(49), 494005, 23 pp. (2013)
14. Adamović, D., Milas, A.: C_2-cofinite $\mathcal{W}$-algebras and their logarithmic modules. In: Proceedings of the Conference "Conformal Field Theories and Tensor Categories", Vol. IX, Mathematical Lectures from Peking University, Beijing, June 2011, pp. 249–270. Springer, Berlin (2014)
15. Adamović, D., Lin, X., Milas, A.: ADE subalgebras of the triplet vertex algebra $\mathcal{W}(p)$; A-series. Commun. Contemp. Math. **15**, 1350028, 30 pp. (2013)
16. Adamović, D., Lin, X., Milas, A.: ADE subalgebras of the triplet vertex algebra $\mathcal{W}(p)$; D-series. Int. J. Math. **25**, 1450001 (34 pp.) (2014)
17. Andrews, G., Berndt, B.: Ramanujan's Lost Notebook, part. II. Springer (2009)
18. Arakawa, T.: Representation theory of W-algebras. Inventiones Math. **169**, 219–320 (2007). arXiv:math/0506056
19. Arike, Y., Nagatomo, K.: Some remarks on the pseudotrace functions for orbifold models associated to symplectic fermions. Adv. Math. **262**, 520–545 (2014). arXiv:1104.0068
20. Bakalov, B., Kirillov, Jr. A.: Tensor Categories and Modular Functors. American Mathematical Society, Providence (2001)
21. Borcherds, R.: Monstrous moonshine and monstrous lie superalgebras. Invent. Math. **109**, 405–444 (1992)
22. Bringmann, K., Milas, A. (preprint)
23. Carqueville, N., Flohr, M.: Nonmeromorphic operator product expansion and C_2-cofiniteness for a family of W-algebras. J. Phys. A: Math. Gen. **39**, 951 (2006)
24. Creutzig, T., Milas, A.: The false theta functions and the Verlinde formula, arXiv: 1309.6037. Adv. Math. 262, 520–545 (2014)
25. Creutzig, T., Ridout, D.: Modular data and Verlinde formulae for fractional level WZW models I. Nucl. Phys. B **865**, 83 (2012)
26. Creutzig, T., Ridout, D.: Modular data and Verlinde formulae for fractional level WZW models II. Nucl. Phys. B **875**, 423 (2013)
27. Dong, C.: Vertex algebras associated with even lattices. J. Algebra **160**, 245–265 (1993)
28. Dong, C., Lepowsky, J.: Generalized Vertex Algebras and Relative Vertex Operators. Progress in Mathematics. Birkhäuser, Boston (1993)
29. Dong, C., Lin, X., Ng, S.: Congruence Property in Conformal Field Theory. arXiv: 1201.6644
30. Feigin, B., Tipunin, I.: Logarithmic CFTs connected with simple Lie algebras. arXiv: 1002.5047

31. Feigin, B.L., Gaĭnutdinov, A.M., Semikhatov, A.M., Tipunin, I I.Yu.: The Kazhdan-Lusztig correspondence for the representation category of the triplet W-algebra in logorithmic conformal field theories. (Russian) Teoret. Mat. Fiz. **148**(3), 398–427 (2006)
32. Feigin, B.L., Gaĭnutdinov, A.M., Semikhatov, A.M., Tipunin, I.Yu.: Logarithmic extensions of minimal models: characters and modular transformations. Nucl. Phys. B **757**, 303–343 (2006)
33. Feigin, B.L., Gaĭnutdinov, A.M., Semikhatov, A.M., Tipunin, I.Yu.: Modular group representations and fusion in logarithmic conformal field theories and in the quantum group center. Commun. Math. Phys. **265**, 47–93 (2006)
34. Feigin, B.L., Gaĭnutdinov, A.M., Semikhatov, A.M., Tipunin, I.Yu.: Kazhdan–Lusztig-dual quantum group for logarithmic extensions of Virasoro minimal models. J. Math. Phys. **48**, 032303 (2007)
35. Flohr, M.: On modular invariant partition functions of conformal field theories with logarithmic operators. Int. J. Mod. Phys. A **11**(22), 4147–4172 (1996)
36. Frenkel, E., Ben-Zvi, D.: Vertex Algebras and Algebraic Curves. Mathematical Surveys and Monographs, vol. 88. American Mathematical Society, Providence (2001)
37. Frenkel, I.B., Lepowsky, J., Meurman, A.: Vertex Operator Algebras and the Monster. Pure and Applied Mathematics, vol. 134. Academic, New York (1988)
38. Frenkel, E., Kac, V., Wakimoto, M.: Characters and fusion rules for W-algebras via quantized Drinfeld-Sokolov reduction. Commun. Math. Phys. **147**, 295–328 (1992)
39. Fuchs, J., Hwang, S., Semikhatov, A.M., Tipunin, I.Yu.: Nonsemisimple fusion algebras and the Verlinde formula. Commun. Math. Phys. **247**(3), 713–742 (2004)
40. Gaberdiel, M.: An algebraic approach to logarithmic conformal field theory. In: Proceedings of the School and Workshop on Logarithmic Conformal Field Theory and its Applications (Tehran, 2001). Int. J. Mod. Phys. A **18**, 4593–4638 (2003)
41. Gaberdiel, M., Kausch, H.G.: A rational logarithmic conformal field theory. Phys. Lett. B **386**, 131–137 (1996). hep-th/9606050
42. Gaberdiel, M., Kausch, H.G.: A local logarithmic conformal field theory. Nucl. Phys. B **538**, 631–658 (1999). hep-th/9807091
43. Gaberdiel, M., Runkel, I.: From boundary to bulk in logarithmic CFT. J. Phys. **A41**, 075402 (2008)
44. Gurarie, V.: Logarithmic operators in conformal field theory. Nucl. Phys. **B410**, 535–549 (1993)
45. Huang, Y.-Z.: Vertex operator algebra and the Verlinde conjecture. Commun. Contemp. Math. **10**, 108 (2008)
46. Huang, Y.-Z.: Generalized twisted modules associated to general automorphisms of a vertex operator algebra. Commun. Math. Phys. **298**(1), 265–292 (2010). `arXiv:0905.0514`
47. Huang, Y.-Z., Lepowsky, J., Zhang, L.: Logarithmic tensor product theory for generalized modules for a conformal vertex algebra, (2010). `arXiv:0710.2687` (also Parts I–VIII: `arXiv:1012.xxxx` was posted in November of 2010
48. Kac, V.: Infinite-Dimensional Lie Algebras. Cambridge University Press, Cambridge (1990)
49. Kac, V.: Vertex Algebras for Beginners, 2nd edn. American Mathematical Society, Providence (1998)
50. Kausch, H.G.: Extended conformal algebras generated by a multiplet of primary fields. Phys. Lett. B **259**, 448 (1991)
51. Lepowsky, J., Li, H.: Introduction to Vertex Operator Algebras and Their Representations. Birkhäuser, Boston (2003)
52. Lin, X.: Fusion rules of Virasoro vertex operator algebras. `arXiv:1204.4855`
53. Milas, A.: Weak modules and logarithmic intertwining operators for vertex operator algebras. In: Recent Developments in Infinite-Dimensional Lie Algebras and Conformal Field Theory (Charlottesville, VA, 2000). Contemporary Mathematics, vol. 297, pp. 201–225. American Mathematical Society, Providence (2002)
54. Milas, A.: Fusion rings associated to degenerate minimal models. J. Algebra **254**, 300–335 (2002)

55. Milas, A., Penn, M.: Lattice vertex algebras and combinatorial bases: general case and $\mathscr{W}$-algebras. N. Y. J. Math. **18**, 621–650 (2012)
56. Miyamoto, M.: Modular invariance of vertex operator algebras satisfying C_2-cofiniteness. Duke Math. J. **122**, 51–91 (2004)
57. Miyamoto, M.: Flatness of tensor products and semi-rigidity for C_2-cofinite vertex operator algebras I. arXiv:0906.1407
58. Nagatomo, K., Tsuchiya, A.: The triplet vertex operator algebra $W(p)$ and the restricted quantum group at root of unity. In: Exploring New Structures and Natural Constructions in Mathematical Physics. Advanced Studies in Pure Mathematics, vol. 61. Mathematical Society of Japan, Tokyo (2011). arXiv:0902.4607
59. Runkel, I.: A braided monoidal category for free super-bosons. arXiv:1209.5554
60. Tsuchiya, A., Wood, S.: The tensor structure on the representation category of the triplet W_p algebra. arXiv:1210.0419
61. Verlinde, E.: Fusion rules and modular transformations in 2D conformal field theory. Nucl. Phys. B **300**, 360–376 (1988)
62. Zhu, Y.: Modular invariance of characters of vertex operator algebras. J. Am. Math. Soc. **9**, 237–302 (1996)

Lattice Subalgebras of Strongly Regular Vertex Operator Algebras

Geoffrey Mason

Abstract We prove a sharpened version of a conjecture of Dong–Mason about lattice subalgebras of a strongly regular vertex operator algebra V, and give some applications. These include the existence of a canonical conformal vertex operator subalgebra $W \otimes G \otimes Z$ of V, and a generalization of the theory of minimal models.

1 Introduction and Statement of Main Results

This paper concerns the algebraic structure of *strongly regular* vertex operator algebras (VOAs). A VOA $V = (V, Y, \mathbf{1}, \omega)$ is called *regular* [14] if it is *rational* (admissible V-modules are semisimple) and C_2-*cofinite* (the span of $u(n)v$ ($u, v \in V, n \leq -2$) has finite codimension in V). It is strongly regular if, in addition, the $L(0)$-grading (or *conformal grading*) given by $L(0)$-weight has the form

$$V = \mathbb{C}\mathbf{1} \oplus V_1 \oplus \dots \tag{1}$$

and all states in V_1 are *quasiprimary* (i.e. annihilated by $L(1)$). Apart from the still-undecided question of the relationship between rationality and C_2-cofiniteness, changing any of the assumptions in the definition of strong regularity will result in VOAs with quite different properties (cf. [10]). Such VOAs are of interest in their own right, but we will not deal with them here.

To describe the main results, we need some basic facts about strongly regular VOAs V that will be assumed here and reviewed in more detail in later sections. V is equipped with an essentially unique nonzero, invariant, bilinear form $\langle\ ,\ \rangle$, and V is simple if, and only if, $\langle\ ,\ \rangle$ is *nondegenerate*. We assume this is the case from now on.

G. Mason (✉)
Department of Mathematics, UC Santa Cruz, Santa Cruz, CA 95064, USA
e-mail: gem@ucsc.edu

W. Kohnen and R. Weissauer (eds.), *Conformal Field Theory, Automorphic Forms and Related Topics*, Contributions in Mathematical and Computational Sciences 8,
DOI 10.1007/978-3-662-43831-2_2,

Then V_1 carries the structure of a *reductive* Lie algebra and all Cartan subalgebras of V_1 (maximal (abelian) toral Lie subalgebras) are conjugate in $\mathrm{Aut}(V)$. We also refer them as *Cartan subalgebras of* V. We say that a subspace $U \subseteq V$ is *nondegenerate* if the restriction of $\langle\ ,\ \rangle$ to $U \times U$ is nondegenerate. For example, the Cartan subalgebras of V and the solvable radical of V_1 are nondegenerate. We refer to the dimension of H as the *Lie rank* of V.

A *subVOA* of V is a subalgebra $W = (W, Y, \mathbf{1}, \omega')$ with a conformal vector ω' that may not coincide with the conformal vector ω of V. If $\omega = \omega'$ we say that W is a *conformal* subVOA. V contains a unique minimal conformal subVOA (with respect to inclusion), namely the *Virasoro subalgebra* generated by ω. A basic example of a subVOA is the Heisenberg theory $(M_U, Y, \mathbf{1}, \omega_U)$ generated by a nondegenerate subspace U of a Cartan subalgebra of V. M_U has rank (or central charge) $\dim U$ and conformal vector

$$\omega_U := 1/2 \sum_i h^i(-1)h^i, \tag{2}$$

where $\{h^i\}$ is any orthonormal basis of U. A *lattice theory* is a VOA V_L corresponding to a positive-definite, even lattice L.

We can now state the main result.

Theorem 1. *Let V be a strongly regular, simple VOA, and suppose that $U \subseteq H \subseteq V$ where H is a Cartan subalgebra of V and U is a nondegenerate subspace. Let ω_U be as in (2). Then the following hold:*

(a) There is a unique maximal subVOA $W \subseteq V$ with conformal vector ω_U. (3)

(b) $W \cong V_\Lambda$ is a lattice theory, where $\Lambda \subseteq U$ is a positive-definite even lattice with $\dim U = rk\Lambda$. (4)

Remark 2. 1. Part (a) is relatively elementary, and follows from the theory of commutants [20] (cf. Sect. 12). The main point of the Theorem is part (b), the *identification* of W as a lattice theory.
2. U is a Cartan subalgebra of W. Thus, every nondegenerate subspace of H is a Cartan subalgebra of a lattice subVOA of V.

Theorem 1 has many consequences. We discuss some of them here, deferring a fuller discussion until later sections. We can apply Theorem 1 with $U = \mathrm{rad}(V_1)$, and this leads to the next result.

Theorem 3. *Suppose that V is a strongly regular, simple VOA. There is a* canonical *conformal subVOA*

$$T = W \otimes G \otimes Z, \tag{5}$$

the tensor product of subVOAs W, G, Z of V with the following properties:

(a) $W \cong V_\Lambda$ *is a lattice theory and* Λ *has minimal length at least* 4*;*
(b) G *is the tensor product of affine Kac–Moody algebras of positive integral level;*
(c) Z *has no nonzero states of weight* 1*:* $Z = \mathbb{C}\mathbf{1} \oplus Z_2 \ldots$

Remark 4. The gradings on W, G and Z are compatible with that on V in the sense that the nth graded piece of each of them is contained in V_n. T has the tensor product grading, and in particular $T_1 = W_1 \oplus G_1 = V_1$. Indeed, $W_1 = \mathrm{rad}(V_1)$ and G_1 is the Levi factor of V_1. Thus the weight 1 piece of V is contained in a rational subVOA of standard type, namely a tensor product of a lattice theory and affine Kac–Moody algebras.

To a certain extent, Theorem 3 reduces the study of strongly regular VOAs to the following: (A) proof that Z is strongly regular; (B) study of strongly regular VOAs with no nonzero weight 1 states; (C) extension problem for strongly regular VOAs, i.e. characterization of the strongly regular VOAs that contain a *given* strongly regular conformal subVOA T. For example, we have the following immediate consequence of Theorem 3 and Remark 4.

Theorem 5. *Suppose that V is a strongly regular, simple VOA such that the conformal vector ω lies in the subVOA $\langle V_1 \rangle$ generated by V_1. Then the canonical conformal subalgebra (5) is a rational subVOA as in the statement of the Theorem 3*

$$T = W \otimes G,$$

where W and G are as in the statement of Theorem 3.

Remark 6. Let $\mathscr{C}$ consist of the (isomorphism classes of) VOAs satisfying the assumptions of the Theorem. $\mathscr{C}$ contains all lattice theories, all simple affine Kac–Moody VOAs of positive integral level (Siegel–Sugawara construction), and it is closed with respect to tensor products and extensions in the sense of (C) above. Theorem 5 says that every VOA in $\mathscr{C}$ arises this way, i.e. an extension of a tensor product of a lattice theory and affine Kac–Moody theories.

There are applications of Theorem 1 to inequalities involving the Lie rank l and the *effective central charge* $\tilde{c}$ of V. These lead to characterizations of some classes of strongly rational VOAs V according to these invariants. For example, we have

Theorem 7. *Let V be a strongly regular, simple VOA of effective central charge $\tilde{c}$ and Lie rank l. The following are equivalent:*

(a) $\tilde{c} < l + 1$,

(b) V *contains a conformal subalgebra isomorphic to a tensor product* $V_\Lambda \otimes L(c_{p,q}, 0)$ *of a lattice theory of rank* l *and a* simple Virasoro VOA *in the* discrete series.

Remark 8. 1. We *always* have $l \leq \tilde{c}$ ([8]).

2. Define a *minimal model* as a strongly regular simple VOA whose Virasoro subalgebra lies in the *discrete series*. The case $l = 0$ of Theorem 7 characterizes minimal models as those strongly regular simple VOAs which have $\tilde{c} < 1$. This is, of course, very similar to the classification of minimal models in physics (cf. [3], Chaps. 7 and 8), where attention is usually restricted to the *unitary* case, where $c = \tilde{c}$, or equivalently $q = p + 1$, in the notation of Theorem 7. Minimal models with $\tilde{c} = c$ were treated rigorously in [12]; our approach allows us to remove any assumptions about c and permits l to be nonzero.

Our results are a natural continuation of the lines of thought in [8] and [9], which have to do with the weight 1 subspace V_1 of V and its embedding in V. These include the invariant bilinear form of V, the nature of the Lie algebra of V_1 and its action on V-modules, automorphisms of V induced by exponentiating weight 1 states, deformations of V-modules using weight 1 states, and (more recently [26]) weak Jacobi form trace functions defined by weight 1 states. These topics constitute a very satisfying chapter in the theory of rational VOAs. The Heidelberg Conference presented itself as a wonderful opportunity to review this set of ideas and describe how the new results that we have been discussing emerge from them. I am grateful to the organizers, Professors Winfried Kohnen and Rainer Weissauer, for giving me the chance to do so.

2 Background

We review some basic notation and facts about vertex operator algebras that we will need. We refer the reader to [27] for further details.

2.1 *Vertex Operators*

A VOA is a quadruple $(V, Y, \mathbf{1}, \omega)$, often denoted simply by V, satisfying the usual axioms. We write vertex operators as

$$Y(v,z) = \sum_{n \in Z} v(n) z^{-n-1} \quad (v \in V),$$

$$Y(\omega, z) = \sum_{n \in Z} L(n) z^{-n-2}.$$

Useful identities that hold for all $u, v \in V$, $p, q \in Z$ include

$$[u(p), v(q)] = \sum_{i=0}^{\infty} \binom{p}{i} (u(i)v)(p+q-i), \tag{6}$$

$$\{u(p)v\}(q) = \sum_{i=0}^{\infty} (-1)^i \binom{p}{i} (u(p-i)v(q+i) - (-1)^p v(q+p-i)u(i)),$$

called the *commutator formula* and *associativity formula* respectively.

We assume throughout that V is a *simple* VOA that is strongly regular as defined in Sect. 1. One of the main consequences of rationality is the fact that, up to isomorphism, there are only *finitely many* ordinary irreducible V-modules [15]. We let $\mathscr{M} := \{(M^1, Y^1), \ldots, (M^r, Y^r)\}$ denote this set, with $(M^1, Y^1) = (V, Y)$. It is conventional to use $u(n)$ to denote the nth mode of $u \in V$ acting on any V-module, the meaning usually being clear from the context. However, it will sometimes be convenient to distinguish these modes from each other. In particular, we often write $Y^j(u, z) := \sum_{n \in Z} u_j(n) z^{-n-1}$ $(u \in V)$, $Y^j(\omega, z) := \sum_{n \in Z} L_j(n) z^{-n-1}$, dropping the index j from the notation when $j = 1$.

2.2 Invariant Bilinear Form

An *invariant bilinear form* on V is a bilinear map $\langle\, , \rangle : V \times V \to C$ satisfying

$$\langle Y(a, z)b, c\rangle = \langle b, Y(e^{zL(1)}(-z^{-2})^{L(0)}a, z^{-1})\rangle \quad (a, b, c \in V). \tag{7}$$

Such a form is necessarily *symmetric* [21, Proposition 5.3.6].

A theorem of Li [28] says that there is a linear isomorphism between $V_0/L(1)V_1$ and the space of invariant bilinear forms on V. Because V is strongly regular then $V_0/L(1)V_1 = \mathbb{C}\mathbf{1}$, so a nonzero invariant bilinear form exists and it is *uniquely determined* up to scalars.

If $a \in V_k$ is quasi-primary then (7) says that

$$\langle a(n)b, c\rangle = (-1)^k \langle b, a(2k-n-2)c\rangle \quad (n \in \mathbb{Z}). \tag{8}$$

In particular, this applies if $a \in V_1$ (because V is assumed to be strongly regular, so that a is primary), or if $a = \omega$ is the conformal vector (ω is *always* quasiprimary). First apply (8) with $a = \omega$ and $n = 1$, noting that $\omega(1) = L(0)$. Then $k = 2$ and we obtain

$$\langle L(0)b, c\rangle = \langle b, L(0)c\rangle.$$

It follows that eigenvectors of $L(0)$ with distinct eigenvalues are necessarily perpendicular with respect to $\langle\ ,\ \rangle$. Thus (1) is an orthogonal direct sum

$$V = \mathbb{C}\mathbf{1} \perp V_1 \perp \dots \tag{9}$$

(Here and below, for subsets $A, B \subseteq V$ we write $A \perp B$ if $\langle a, b\rangle = 0$ for all $a \in A, b \in B$.)

The radical of $\langle\ ,\ \rangle$ is an *ideal*. Because we are assuming that V is simple then it must be zero, whence $\langle\ ,\ \rangle$ is *nondegenerate*. In particular, $\langle \mathbf{1}, \mathbf{1}\rangle \neq 0$. In what follows, we fix the form so that

$$\langle \mathbf{1}, \mathbf{1}\rangle = -1. \tag{10}$$

Note also that by (9), the restriction of $\langle\ ,\ \rangle$ to each $V_n \times V_n$ is also nondegenerate.

2.3 The Lie Algebra on V_1

The bilinear product $[uv] := u(0)v$ $(u, v \in V_1)$ equips V_1 with the structure of a Lie algebra. This is well-known, and follows easily from (6). Applying (8) with $u, v \in V_1$, we obtain $\langle u, v\rangle = \langle u(-1)\mathbf{1}, v\rangle = -\langle \mathbf{1}, u(1)v\rangle$. With the convention (10), it follows that

$$u(1)v = \langle u, v\rangle \mathbf{1} \quad (u, v \in V_1).$$

Because V is strongly regular, a theorem of Dong–Mason [8] says that the Lie algebra on V_1 is *reductive*. (This result is discussed further in Sect. 3.3 below.) So there is a canonical decomposition

$$V_1 = A \perp S$$

where $A = \mathrm{Rad}(V_1)$ is an *abelian* ideal and S is the unique (semisimple) Levi factor.

The decomposition of S into a direct sum of simple Lie algebras $\oplus_i \mathfrak{g}_i$ is also an orthogonal sum with respect to $\langle\ ,\ \rangle$. There is a refinement of this decomposition, established in [9], namely

$$V_1 = A \perp \mathfrak{g}_{1,k_1} \perp \dots \perp \mathfrak{g}_{s,k_s} \tag{11}$$

where each k_i is a positive integer (the *level* of $\mathfrak{g}_i$).

To explain what this means, for $U \subseteq V$ let $\langle U\rangle$ be the *subalgebra* of V *generated* by U. $\langle U\rangle$ is spanned by states $u = u_1(n_1)\dots u_t(n_t)\mathbf{1}$ with $u_1, \dots, u_t \in U$, $n_1, \dots, n_t \in \mathbb{Z}$, and equipped with vertex operators defined as the *restriction* of $Y(u, z)$ to $\langle U\rangle$.

It is proved in [9] that there is an isomorphism of VOAs $\langle \mathfrak{g}_i \rangle \cong L_{\mathfrak{g}_i}(k_i, 0)$, where $L_{\mathfrak{g}_i}(k_i, 0)$ is the simple VOA (or WZW model) corresponding to the affine Lie algebra $\widehat{\mathfrak{g}_i}$ determined by $\mathfrak{g}_i$, of positive integral level k_i. Orthogonal Lie algebras in (11) determine mutually commuting WZW models. Then the meaning of (11) is that the canonical subalgebra G of V generated by S satisfies

$$G \cong L_{\mathfrak{g}_1}(k_1, 0) \otimes \ldots \otimes L_{\mathfrak{g}_s}(k_s, 0). \tag{12}$$

In particular, G is a rational VOA equipped with the canonical conformal vector ω_G arising from the Sugawara construction associated to each tensor factor [20, 27].

Because V_1 is reductive, it has a *Cartan subalgebra*, that is a maximal (abelian) toral subalgebra, and all Cartan subalgebras are conjugate in $\mathrm{Aut}(V_1)$. (See the following section for further discussion.) Let $H \subseteq V_1$ be a Cartan subalgebra of V_1, say of rank l. By Lie theory, the restriction of $\langle\ ,\ \rangle$ to $H \times H$ is nondegenerate. We also call H a *Cartan subalgebra of* V.

3 Automorphisms

In this Section we discuss automorphisms of a VOA. We are mainly interested in the *linear automorphisms*, which arise from the familiar process of exponentiating the operators $a(0)$ for $a \in V_1$.

3.1 The Group of Linear Automorphisms $\mathfrak{G}$

An *automorphism* of V is an invertible linear map $g : V \to V$ such that $g(\omega) = \omega$ and $ga(n)g^{-1} = g(a)(n)$ for all $a \in V, n \in \mathbb{Z}$, i.e.

$$gY(a, z)g^{-1} = Y(g(a), z). \tag{13}$$

The set of all automorphisms is a group $\mathrm{Aut}(V)$. Because $g\omega(n)g^{-1} = g(\omega)(n) = \omega(n)$, it follows in particular that g commutes with $L(0) = \omega(1)$. Therefore, $\mathrm{Aut}(V)$ acts on each V_n. The uniqueness of $\langle\ ,\ \rangle$ implies that

$$\mathrm{Aut}(V) \text{ leaves } \langle\ ,\ \rangle \text{ invariant.} \tag{14}$$

So each V_n affords an *orthogonal* representation of $\mathrm{Aut}(V)$.

One checks (e.g. using induction and (6)) that for $n \geq 0$,

$$(u(0)^n v)(q) = \sum_{i=0}^{n} (-1)^i \binom{n}{i} u(0)^{n-i} v(q) u(0)^i \quad (u, v \in V,\ q \in \mathbb{Z}).$$

Therefore,

$$\begin{aligned}\left(e^{u(0)}v\right)(q) &= \sum_{n=0}^{\infty} \frac{1}{n!}(u(0)^n v)(q) \\ &= \sum_{n=0}^{\infty}\sum_{i=0}^{n} \frac{(-1)^i}{i!(n-i)!} u(0)^{n-i} v(q) u(0)^i \\ &= e^{u(0)} v(q) e^{-u(0)},\end{aligned}$$

showing that (13) holds with $g = e^{u(0)}$. If we further assume that $u \in V_1$ then we obtain using (6) that

$$\begin{aligned}u(0)\omega &= -[\omega(-1), u(0)]\mathbf{1} = -\sum_{i=0}^{\infty}(-1)^i(\omega(i)u)(-1-i)\mathbf{1} \\ &= -((L(-1)u)(-1) - (L(0)u)(-2))\mathbf{1} = 0.\end{aligned}$$

It follows that $\{e^{u(0)} \mid u \in V_1\}$ is a set of automorphisms of V. Let

$$\mathfrak{G} = \langle e^{u(0)} \mid u \in V_1 \rangle$$

be the group they generate. It is clear from the classical relation between Lie groups and Lie algebras that $\mathfrak{G}$ is the adjoint form of the complex Lie group associated with V_1, and there is a containment

$$\mathfrak{G} \trianglelefteq \mathrm{Aut}(V).$$

(Normality holds because if $g \in \mathrm{Aut}(V)$ and $u \in V_1$ then $g(u) \in V_1$ and $ge^{u(0)}g^{-1} = e^{gu(0)g^{-1}} = e^{g(u)(0)}$.)

One consequence of this is the following. Because $\mathfrak{G}$ acts transitively on the set of Cartan subalgebras of V_1, it follows *ipso facto* that $\mathrm{Aut}(V)$ also acts transitively on the set of Cartan subalgebras of V_1 (or of V). Thus the choice of a Cartan subalgebra in V is unique up to automorphisms of V, in parallel with the usual theory of semisimple Lie algebras.

3.2 Projective Action of Aut(V) on V-Modules

There is a natural action of $\mathrm{Aut}(V)$ on the set $\mathcal{M}$ of (isomorphism classes of) irreducible V-modules $\{(M^j, Y^j) \mid 1 \le j \le r\}$ [6]. Briefly, the argument is as follows. For $g \in \mathrm{Aut}(V)$ and an index j, one checks that the pair (M^j, Y^j_g) defined by $Y^j_g(v,z) := Y^j(gv,z)$ $(v \in V)$ is itself an irreducible V-module. The action of $\mathrm{Aut}(V)$ on $\mathcal{M}$ is then defined by $g : (M^j, Y^j) \mapsto (M^j, Y^j_g)$.

Because $\mathcal{M}$ is finite and $\mathfrak{G}$ is connected, the action of $\mathfrak{G}$ is necessarily *trivial*. Hence, if we fix the index j, then for $g \in \mathrm{Aut}(V)$ there is an isomorphism of V-modules $\alpha_g : (M^j, Y^j) \mapsto (M^j, Y_g^j)$, i.e.

$$\alpha_g Y^j(u,z) = Y_g^j(u,z)\alpha_g = Y^j(gu,z)\alpha_g \quad (u \in V). \tag{15}$$

Because M^j is irreducible, α_g is uniquely determined up to an overall nonzero scalar (Schur's Lemma).

When $j = 1$, so that $M^j = V, \alpha_g$ coincides with g itself, the scalar being implicitly determined by the additional condition $g(\omega) = \omega$. Generally, we find from (15) that

$$\alpha_{gh} Y^j(u,z)\alpha_{gh}^{-1} = \alpha_g\alpha_h Y^j(u,z)\alpha_h^{-1}\alpha_g^{-1} \ (g,h \in \mathrm{Aut}(V)),$$

so that by Schur's Lemma once more there are scalars $c_j(g,h)$ satisfying

$$\alpha_{gh} = c_j(g,h)\alpha_g\alpha_h.$$

The map $c_j : \mathfrak{G} \times \mathfrak{G} \to \mathbb{C}^*$, $(g,h) \mapsto c_j(g,h)$, is a 2-cocycle on $\mathfrak{G}$. It defines a projective action $g \mapsto \alpha_g$ of $\mathfrak{G}$ on M^j that satisfies (15).

While the projective action of $\mathfrak{G}$ on $M^1 = V$ reduces to the linear action previously considered, the 2-cocycles c_j are generally *nontrivial*, i.e. they are not 2-coboundaries. A well-known example is the VOA $V := L_{\mathfrak{sl}_2}(1,0)$, i.e. the level 1 WZW model of type sl_2, which is isomorphic to the lattice theory V_{A_1} defined by the A_1 root lattice. In this case we have $V_1 = \mathfrak{sl}_2$, and the linear group is the adjoint form of $\mathfrak{sl}_2$, i.e., $\mathfrak{G} = SO_3(\mathbb{R})$. There are just two irreducible V-modules, corresponding to the two cosets of A_1 in its dual lattice $A_1^* := (1/\sqrt{2})A_1$, and their direct sum is the Fock space for a generalized VOA $V_{A_1^*}$. The automorphism group of this generalized VOA is $SU_2(\mathbb{C})$, and in particular the projective action of $\mathfrak{G}$ on M^2 lifts to a linear action of its proper twofold (universal) covering group.

3.3 *Linear Reducibility of the V_1-Action*

We discuss the following result.

$$\text{The Lie algebra } V_1 \text{ is } \textit{reductive}\text{, and its action on each simple } V\text{-module } (M^j, Y^j) \text{ is } \textit{completely reducible}. \tag{16}$$

One says in this situation that V_1 is *linearly reductive* in its action on M^j. This follows from results in [8] and [5]. We will need some of the details later, so we sketch the proof.

Each irreducible V-module M^j has a direct sum decomposition into finite-dimensional $L_j(0)$-eigenspaces

$$M^j = \bigoplus_{n=0}^{\infty} M^j_{n+\lambda_j}, \tag{17}$$

where λ_j is a constant called the *conformal weight* of M^j. Each $M^j_{n+\lambda_j}$ is a module for the Lie algebra V_1, acting by the zero mode $u_j(0)$ $(u \in V_1)$, and (16) amounts to the assertion that each of these actions is completely reducible. The simple summands $\mathfrak{g}_i$ $(1 \leq i \leq s)$ of V_1 act completely reducibly by Weyl's theorem, so the main issue is to show that the abelian radical A of V_1 [cf. (11)] acts semisimply.

The first step uses a formula of Zhu [33]. The case we need may be stated as follows (cf. [8]):

$$\begin{gathered}\text{Suppose that } u, v \in V_1. \text{ Then for } 1 \leq j \leq r, \\ \mathrm{Tr}_{M^j} u_j(0) v_j(0) q^{L(0)-c/24} = Z_{M^j}(u[-1]v, \tau) - \langle u, v\rangle E_2(\tau) Z_{M^j}(\tau).\end{gathered} \tag{18}$$

The notation, which is standard, is as follows [16,33]: for $w \in V$, $Z_{M^j}(w, \tau) := \mathrm{Tr}_{M^j} o_j(w) q^{L(0)-c/24}$ is the graded trace of the *zero* mode $o_j(w)$ for the action of w on M^j, $u[-1]$ is the -1st *square bracket* mode for u, and $E_2(\tau) = -1/12 + 2\sum_{n=1}^{\infty}\sum_{d|n} dq^n$ is the usual weight 2 Eisenstein series.

Next we show that if $\langle u, v\rangle \neq 0$ then for some index j we have

$$Z_{M^j}(u[-1]v, \tau) \neq \langle u, v\rangle E_2(\tau) Z_{M^j}(\tau).$$

Indeed, if this does not hold, we can obtain a contradiction using Zhu's modular-invariance theorem [33] and the exceptional transformation law for $E_2(\tau)$ (cf. [8], Sect. 4 for details). From (18) we can conclude that if $\langle u, v\rangle \neq 0$ then there is an index j such that

$$\mathrm{Tr}_{M^j} u_j(0) v_j(0) \neq 0. \tag{19}$$

Now suppose that $u \in V_1$ lies in the nil radical of V_1. Then $u(0)$ annihilates every simple V_1-module, and in particular the lhs of (19) necessarily vanishes for each j. Therefore $\langle u, v\rangle = 0$ $(v \in V_1)$, whence $u = 0$. This shows that V_1 is indeed reductive.

It is known that a VOA is *finitely generated* (f.g.) if it is C_2-cofinite [2, 23], or if it is rational [11]. So certainly a regular VOA is f.g. We need this mainly because Griess and Dong proved [5] that the automorphism group of a f.g. VOA is a (complex) *algebraic group*. It follows that the subgroup $\mathfrak{A} \trianglelefteq \mathfrak{G}$ generated by the exponentials $e^{u(0)}$ $(u \in A = \mathrm{rad}(V_1))$ is an abelian algebraic subgroup, and that A itself is the direct sum of two Lie subalgebras corresponding to the unipotent and semisimple parts of $\mathfrak{A}$. By the same argument as above, the Lie subalgebra corresponding to the unipotent part necessarily vanishes, so that $\mathfrak{A}$ is a complex torus and A consists of semisimple operators. In particular, (16) holds.

Equation (16) was first stated in [8], although the proof there is incomplete. It would be interesting to find a proof that does not depend on the theory of algebraic groups.

4 The Tower $L_0 \subseteq L \subseteq E$

In this section we introduce and study the structure of a certain naturally-defined rational subspace E contained in Cartan subalgebra of V_1.

4.1 Definition of E

Fix a Cartan subalgebra $H \subseteq V_1$ of rank l, say. We have seen in Sect. 3.3 that all of the operators $u_j(0)$ $(u \in H, 1 \leq j \leq r)$ are semisimple. We set

$$\begin{aligned} E &= \{u \in H \mid u(0) \text{ has eigenvalues in } \mathbb{Q}\}, \\ L &= \{u \in H \mid u_j(0) \text{ has eigenvalues in } \mathbb{Z},\ 1 \leq j \leq r\}, \\ L_0 &= \{u \in H \mid u(0) \text{ has eigenvalues in } \mathbb{Z}\}. \end{aligned} \tag{20}$$

(Recall that the operators $u(0)$ in the definition of E act on the VOA V.) E is a $\mathbb{Q}$-vector space in H and $L \subseteq L_0 \subseteq E$ are additive subgroups.

Let $\mathfrak{H} \subseteq \mathfrak{G}$ be the group generated by exponentials $e^{2\pi i u(0)}$ $(u \in H)$. Thus $\mathfrak{H}$ is a complex torus isomorphic to $(\mathbb{C}^*)^l$. There is a short exact sequence

$$0 \to L_0/L \to H/L \xrightarrow{\varphi} \mathfrak{H} \to 1$$

where φ arises from the morphism $u \mapsto e^{2\pi i u(0)}$ $(u \in H)$. H/L is the covering group of H/L_0 that acts linearly on each irreducible module M^j as described in Sect. 3.2, and $H/L_0 \cong \mathfrak{H}$.

Because V is f.g. there is an integer n_0 such that $V = \langle \oplus_{n=0}^{n_0} V_n \rangle$. Then $e^{2\pi i u(0)}$ $(u \in H)$ is the identity if, and only if, its restriction to $\oplus_{n=0}^{n_0} V_n$ is the identity. It follows that the eigenvalues of $u(0)$ for $u \in E$ have bounded denominator, whence

$$E/L_0 = \text{Torsion}(H/L_0) \cong (\mathbb{Q}/\mathbb{Z})^l. \tag{21}$$

In particular, E contains a $\mathbb{C}$-basis of H.

4.2 Deformation of V-Modules

We will need to use Li's theory of deformations of (twisted) V-modules [29]. In Proposition 5.4 (loc. cit.) Li showed how to deform V-modules using a certain operator $\Delta(z)$. We describe the special case that we need here. See [26] for further details of the calculations below, and [13] for further development of the theory.

Fix $u \in L_0$ (cf. 20), and set

$$\Delta_u(z) := z^{u(0)} \exp\left\{-\sum_{k\geq 1} \frac{u(k)}{k}(-z)^{-k}\right\}.$$

For an irreducible V-module $(M^{j'}, Y^{j'})$, set

$$Y^{j'}_{\Delta_u(z)}(v, z) := Y^{j'}(\Delta_u(z)v, z) \ \ (v \in V).$$

Because $u(0)$ has eigenvalues in $\mathbb{Z}$ then $e^{2\pi i u(0)}$ is the identity automorphism of V. In this case, Li's result says that there is an isomorphism of V-modules

$$(M^{j'}, Y^{j'}_{\Delta_u(z)}) \cong (M^j, Y^j) \tag{22}$$

for some j. (Technically, Li's results deal with *weak* V-modules. In the case that we are dealing with, when V is regular, the results apply to ordinary irreducible V-modules, as stated.) Thus there is a linear isomorphism $\psi : M^{j'} \xrightarrow{\cong} M^j$ satisfying

$$\psi^{-1} Y^j(v, z)\psi = Y^{j'}(\Delta_u(z)v, z) \ \ (v \in V). \tag{23}$$

In (23) we choose $j' = 1$ (so $(M^{j'}, Y^{j'}) = (V, Y)$), $v = \omega$, and apply both sides to $\mathbf{1}$. We obtain after some calculation that

$$\psi^{-1} L^j(0)\psi(\mathbf{1}) = 1/2\langle u, u\rangle \mathbf{1}. \tag{24}$$

The $L(0)$-grading on M^j is described in (17). If $\psi(\mathbf{1}) = \sum_n a_n$ with $a_n \in M^j_{n+\lambda_j}$ then $1/2\langle u, u\rangle \sum_n a_n = \sum_n (n + \lambda_j)a_n$. This shows that $\psi(\mathbf{1}) \in M^j_{n_0+\lambda_j}$ for some integer n_0, and moreover

$$1/2\langle u, u\rangle = n_0 + \lambda_j. \tag{25}$$

We use (25) in conjunction with another theorem [1,16] that says that (for regular V) the conformal weight λ_j of the irreducible V-module M^j lies in $\mathbb{Q}$. Then it is immediate from (25) that $\langle u, u\rangle \in \mathbb{Q}$. The only condition on u here is that $u \in L_0$. Because E/L_0 is a torsion group (21) we obtain

$$\langle u, u \rangle \in \mathbb{Q} \ \ (u \in E). \tag{26}$$

Arguing along similar lines, we can also prove the following: (i) if $u \in E$, all eigenvalues of the operators $u_j(0)$ lie in $\mathbb{Q}$ $(1 \leq j \leq r)$; (ii) if $u \in L_0$ then the denominators of the eigenvalues of $u_j(0)$ divide the l.c.m. M of the denominators of the conformal weights λ_j. In other words, L_0/L is a torsion abelian group of exponent dividing M. (It is also f.g., as we shall see. So L_0/L is actually a finite abelian group.)

4.3 Weak Jacobi Forms

The paper [26] develops an extension of Zhu's theory of partition functions [33] to the context of *weak Jacobi forms*. It is also closely related to the deformation theory of V-modules as discussed in the previous subsection. We discuss background sufficient for our purposes. For the general theory of Jacobi forms, cf. [18].

We continue with a strongly regular VOA V. Let $h \in L$ [cf. (20)]. For j in the range $1 \leq j \leq r$, define

$$J_{j,h}(\tau, z) := \mathrm{Tr}_{M^j} q^{L_j(0)-c/24} \zeta^{h_j(0)},$$

where c is the central charge of V. (The definition makes sense because we have seen that $h_j(0)$ is a semisimple operator.) Notation is as follows: $q := e^{2\pi i \tau}$, $\zeta := e^{2\pi i z}$, $\tau \in \mathbb{H}$ (complex upper half-plane), $z \in \mathbb{C}$. The main result [26] is that $J_{j,h}(\tau, z)$ is holomorphic in $\mathbb{H} \times \mathbb{C}$ and satisfies the following functional equations for all $\gamma = \begin{pmatrix} a & b \\ c & d \end{pmatrix} \in SL_2(\mathbb{Z})$, $(u, v) \in \mathbb{Z}^2$, $1 \leq i \leq r$:

(i) there are scalars $a_{ij}(\gamma)$ depending only on γ such that

$$J_{i,h}\left(\gamma\tau, \frac{z}{c\tau + d}\right) = e^{\pi i c z^2 \langle h,h \rangle/(c\tau+d)} \sum_{j=1}^{r} a_{ij}(\gamma) J_{j,h}(\tau, z), \tag{27}$$

(ii) there is a permutation $j \mapsto j'$ of $\{1, \ldots, r\}$ such that

$$J_{j,h}(\tau, z + u\tau + v) = e^{-\pi i \langle h,h \rangle (u^2\tau + 2uz)} J_{j',h}(\tau, z). \tag{28}$$

This says that the r-tuple $(J_{1,h}, \ldots, J_{r,h})$ is a *vector-valued weak Jacobi form* of weight 0 and index $1/2\langle h, h \rangle$. (By (26) we have $\langle h, h \rangle \in \mathbb{Q}$.) Part (i), which we do not need here, is proved by making use of a theorem of Miyamoto [31], which itself extends some of the ideas in Zhu's modular-invariance theorem [33]. The proof of (ii) involves applications of the ideas of Sect. 4.2, and in particular the permutation in (28) is the same as the one that arises from (22).

4.4 $(E, \langle\ ,\ \rangle)$ as a Quadratic Space

We will prove the following result.

$$\begin{aligned}&(E, \langle\ ,\ \rangle) \text{ is a \textit{positive-definite} rational quadratic space}\\&\text{of rank } l, \text{ and } L_0 \subseteq E \text{ is an additive subgroup of rank } l.\end{aligned} \tag{29}$$

We have seen that both E/L_0 and L_0/L are torsion groups. Hence E/L is also a torsion group, so in proving that $\langle h, h\rangle > 0$ for $0 \neq h \in E$, it suffices to prove this under the additional assumption that $h \in L$. We assume this from now on, and set $m = \langle h, h\rangle$. Note that the results of Sect. 4.3 apply in this situation.

We will show that $m \leq 0$ leads to a contradiction. From (28) we know that $J_{j,h}(\tau, z + u\tau + v) = J_{j',h}(\tau, z)$ for $1 \leq j \leq r$. In terms of the Fourier series $J_{j,h} := \sum_{n,t} c(n,t)q^n\zeta^t$, $J_{j',h} := \sum_{n,t} c'(n,t)q^n\zeta^t$, this reads

$$q^{\lambda_j - c/24}\sum_{n\geq 0,t} c(n,t)q^{n+mu^2/2+tu}\zeta^{t+mu} = q^{\lambda_{j'}-c/24}\sum_{n\geq 0,t} c'(n,t)q^n\zeta^t$$

for all $u \in \mathbb{Z}$, j' depending on u. Suppose first that $m = 0$. If for some $t \neq 0$ there is $c(n,t) \neq 0$ we let $u \to -\infty$ and obtain a contradiction. Therefore, $c(n,t) = 0$ whenever $t \neq 0$. This says precisely that $h_j(0)$ is the zero operator on M^j. Furthermore, this argument holds for any index j. But now (19) is contradicted. If $m < 0$ the argument is even easier since we just have to let $u \to -\infty$ to get a contradiction.

This proves that $\langle\ ,\ \rangle$ is positive-definite on E, while rationality has already been established (26). Now we prove that E has rank l, using an argument familiar from the theory of root systems (cf. [24, Sect. 8.5]). We have already seen [cf. (21] and the line following) that E contains a basis of H, say $\{\alpha_1, \ldots, \alpha_l\}$. We assert that $\{\alpha_1, \ldots, \alpha_l\}$ is a $\mathbb{Q}$-basis of E.

Let $u \in E$. There are scalars $c_1, \ldots, c_l \in \mathbb{C}$ such that $u = \sum_j c_j\alpha_j$. We have for $1 \leq i \leq l$ that

$$\langle u, \alpha_i\rangle = \sum_j c_j\langle\alpha_i, \alpha_j\rangle. \tag{30}$$

Each $\langle u, \alpha_i\rangle$ and $\langle\alpha_i, \alpha_j\rangle$ are rational, and the nondegeneracy of $\langle\ ,\ \rangle$ implies that $(\langle\alpha_i, \alpha_j\rangle)$ is *nonsingular*. Therefore, $c_j = \langle u, \alpha_j\rangle / \det(\langle\alpha_i, \alpha_j\rangle) \in \mathbb{Q}$, as required.

We have proved that E is a $\mathbb{Q}$-form for H, i.e. $H = \mathbb{C} \otimes_{\mathbb{Q}} E$. That $L_0 \subseteq E$ is a lattice of the same rank follows from (21). All parts of (29) are now established.

Now observe that the analysis that leads to the proof of (29) carries over *verbatim* to any *nondegenerate* subspace $U \subseteq H$, say of rank l'. For such a subspace we set $E' := U \cap E, L' := U \cap L, L_0' := U \cap L_0$. The result can then be stated as follows:

$$(U, \langle\ ,\ \rangle) \text{ is a } \textit{positive-definite} \text{ rational quadratic space of rank } l', \text{ and } L_0' \subseteq E \text{ is an additive subgroup of rank } l'. \quad (31)$$

Another application of weak Jacobi forms allows us to usefully strengthen the statement (19) in some cases:

$$\text{if } 0 \neq h \in E \text{ then } h_j(0) \neq 0 \text{ for each } 1 \leq j \leq r. \quad (32)$$

Suppose false. Because E/L is a torsion group there is $0 \neq h \in L$ with $h_j(0) = 0$ for some index j. Let $m = \langle h, h\rangle$, so that $m \neq 0$. Then $J_{j,h}(\tau, z)$ is a pure q-expansion, i.e. no nonzero powers of ζ occur in the Fourier expansion. Indeed, it is just the partition function for M^j, so it also does not vanish. By (28), $e^{-\pi i \langle h,h\rangle (u^2\tau + 2uz)} J_{j',h}(\tau, z) = q^{-mu^2/2}\zeta^{-mu} \sum_{n\geq 0,t} c'(n,t) q^{n-\lambda_j}\zeta^t$ is also a pure q-expansion. (As usual, j' depends on u.) But because $m > 0$ we can let $u \to \infty$ to see that in fact this power series is *not* a pure q-expansion. This contradiction proves (32).

5 Commutants and Weights

We now turn to the proof of Theorem 1. The main step is to establish (38) below.

5.1 Commutants

We retain previous notation. In particular, from now on we fix a Cartan subalgebra $H \subseteq V_1$ and a nondegenerate subspace $U \subseteq H$ of rank l'. Let $M_U = (\langle U\rangle, Y, \mathbf{1}, \omega_U)$ be the Heisenberg subVOA of rank l' generated by U [cf. (2)]. We set $Y(\omega_U, z) := \sum_{n\in\mathbb{Z}} L_U(n) z^{-n-2}$.

Consider

$$\mathscr{P}_U := \{(A, Y, \mathbf{1}, \omega_U) \mid A \subseteq V\}. \quad (33)$$

In words, $\mathscr{P}_U$ is the set of subVOAs $A \subseteq V$ which have conformal vector ω_U. $\mathscr{P}_U$ is partially ordered by inclusion. It is nonempty since it contains M_U, for example.

One easily checks that $L(1)\omega_U = 0$. Therefore, the theory of commutants ([20, 27], Sect. 3.11) shows that each $A \in \mathscr{P}_U$ has a *compatible grading* with (1). That is

$$A_n := \{v \in A \mid L_U(0)v = n\} = A \cap V_n.$$

Moreover, $\mathscr{P}_U$ has a *unique maximal element*. Indeed, the *commutant* of $A \in \mathscr{P}(U)$, defined by

$$C_V(A) = ker_V L_U(-1),$$

is *independent* of A, and the maximal element of $\mathscr{P}_U$ is the double commutant $C_V(C_V(A))$.

5.2 U-Weights

Thanks to (16) we can use the language of weights to describe the action of $u(0)$ ($u \in U$). For $\beta \in U$ set

$$V(\beta) := \{w \in V \mid u(0)w = \langle \beta, u \rangle w \ (u \in U)\}.$$

β is a U*-weight*, or simply *weight* (of V) if $V(\beta) \neq 0$, $V(\beta)$ is the β-weight space, and a nonzero $w \in V(\beta)$ is a weight vector of weight β.

Using the action of $Y(u,z)$ ($u \in U$) on weight spaces, one shows that the set of U-weights

$$P := \{\beta \in U \mid V(\beta) \neq 0\} \tag{34}$$

is a *subgroup* of U. See [8], Sect. 4 for further details. By the complete reducibility of $u(0)$ ($u \in U$) and the Stone von-Neumann theorem [22, Sect. 1.7] applied to the Heisenberg subVOA M_U, there is a weight space decomposition

$$V = M_U \otimes \Omega = \bigoplus_{\beta \in P} M_U \otimes \Omega(\beta) \tag{35}$$

where $\Omega := \{v \in V \mid u(n)v = 0 \ (u \in U, n \geq 1)\}, \Omega(\beta) := \Omega \cap V(\beta)$, and $V(\beta) = M_U \otimes \Omega(\beta)$.

$\Omega(0)$ is the commutant $C_V(M_U)$, and $M_U \otimes \Omega(0)$ the zero weight space. By arguments in [7] one sees that $\Omega(0)$ is *simple* VOA (the simplicity of M_U is well-known), moreover each $V(\beta)$ is an *irreducible* $M_U \otimes \Omega(0)$-module. So there is a tensor decomposition

$$V(\beta) = M_U(\beta) \otimes \Omega(\beta)$$

where $M_U(\beta), \Omega(\beta)$ are irreducible modules for $M_U, \Omega(0)$ respectively. Furthermore, $V(\beta) \cong V(\beta')$ if, and only if, $\beta = \beta'$. In particular, there is an identification

$$M_U(\beta) = M_U \otimes e^\beta \tag{36}$$

where $e^\beta \in \Omega(\beta)$.

5.3 Lattice Subalgebras of V

We keep previous notation. In particular, P is the group of U-weights (34) and $E' = U \cap E, L' = L \cap U, L'_0 = L_0 \cap U$ are as in Sect. 4.4 [cf. (31)].

Since $(E', \langle\ ,\ \rangle)$ is a rational space (31) and contains a basis of U, it follows that $E' = \{u \in U \mid \langle u, E'\rangle \subseteq \mathbb{Q}\}$. Because E'/L'_0 is a torsion group, we then see that $(L'_0)^0 \subseteq E'$. (Here, and below, we set $F^0 := \{u \in U \mid \langle u, F\rangle \subseteq \mathbb{Z}\}$ for $F \subseteq E'$.) Now $u \in P^0 \Leftrightarrow \langle P, u\rangle \subseteq \mathbb{Z} \Leftrightarrow$ all eigenvalues of $u(0)$ are integral $\Leftrightarrow u \in L'_0$. We conclude that

$$P = (L'_0)^0 \subseteq E' \tag{37}$$

We will establish

$$\begin{aligned}&\text{there is a positive-definite even lattice } \Lambda \subseteq P \text{ such that } |P : \Lambda| \\ &\text{is finite and the maximal element } W \text{ of } \mathscr{P}_U \text{ satisfies } W \cong V_\Lambda.\end{aligned} \tag{38}$$

The argument utilizes ideas in [9]. Recall the isomorphism (22), which holds for all $u \in L_0$. Set

$$\Gamma := \{u \in L'_0 \mid (V, Y_{\Delta_u(z)}) \cong (V, Y)\}. \tag{39}$$

This is a subgroup of L'_0 of finite index. Although not necessary at this stage, we can show immediately that Γ is an even lattice. Indeed, if $u \in \Gamma$ then the proof of (25) shows that we have $\lambda_j = 0$ in that display, whence $\langle u, u\rangle = n_0$ is a (nonnegative) integer. Now the assertion about Γ follows from (29).

There is another approach that gives more information. The isomorphism of V-modules defined for $u \in \Gamma$ by (39) implies the following assertion concerning the weight spaces in (35):

$$\Omega(\beta) \cong \Omega(\beta + u) \ \ (u \in \Gamma, \beta \in P). \tag{40}$$

In particular, taking $\beta = 0$ shows that $\Omega(u) \neq 0$ $(u \in \Gamma)$, whence $\Gamma \subseteq P$. Using (37) we deduce

$$\Gamma \subseteq P,\ \ P^0 \subseteq \Gamma^0,$$

so Γ is necessarily a positive-definite integral lattice of rank l', and $|P : \Gamma| =: d$ is finite. Equation (40) leads to a refinement of (35), namely a decomposition of V into simple $M_U \otimes \Omega(0)$-modules

$$V = \bigoplus_{i=1}^{d} \bigoplus_{\beta\in\Gamma} M_U(\beta + \gamma_i) \otimes \Omega(\gamma_i),$$

where $\{\gamma_i \mid 1 \leq i \leq d\}$ are coset representatives for P/Γ.

Let

$$\Lambda := \{\beta \in P \mid \Omega(\beta) = \Omega(0)\}, \tag{41}$$

$$W := \bigoplus_{\beta \in \Lambda} M_U(\beta). \tag{42}$$

Then $\Gamma \subseteq \Lambda$ and $W = C_V(\Omega(0)) = C_V(C_V(M_U))$. In particular, Λ is an additive subgroup of P of finite index and W is a subVOA of V. Indeed, it is the maximal element of the poset $\mathscr{P}_U$ discussed in Sect. 5.1.

The $L(0)$-weight of $e^\beta \in W(\beta \in \Lambda)$ [cf. (36)] coincides with its $L_U(0)$-weight (cf. Sect. 5.1). Using the associativity formula, we have

$$\begin{aligned} L(0)e^\beta &= L_0(0).e^\beta = 1/2\sum_{t=1}^{l'}(h_t(-1)h_t)(1)e^\beta \\ &= 1/2\sum_{t=1}^{l'}\left\{\sum_{k\geq 0} h_t(-1-k)h_t(1+k) + h_t(-k)h_t(k)\right\}e^\beta \\ &= 1/2\sum_{t=1}^{l'} h_t(0)h_t(0)e^\beta = 1/2\sum_{t=1}^{l'}\langle\beta, h_t\rangle^2 e^\beta = 1/2\langle\beta,\beta\rangle, \end{aligned}$$

showing that $1/2\langle\beta,\beta\rangle \in \mathbb{Z}$ $(\beta \in \Lambda)$.

This shows that Λ is an *even* lattice of rank l'. The isomorphism $W \cong V_\Lambda$ then follows from the uniqueness of simple current extensions [8, Sect. 5]. This completes the proof of (38) and Theorem 1 is established.

6 Applications of Theorem 1

We present several applications of Theorem 1 to the structure of strongly regular VOAs.

6.1 The Tripartite SubVOA of V

The tripartite subVOA arises when U in Theorem 1 is the radical $\mathrm{rad}(V_1)$. (38) is applicable here because A is indeed nondegenerate (cf. Sect. 4). We keep the notation from previous sections.

Observe that in this case, the lattice Λ *contains no roots*, i.e. there is no $\beta \in \Lambda$ satisfying $\langle\beta,\beta\rangle = 2$. For if $\beta \in \Lambda$ is a root then β is contained in an sl_2-subalgebra of V_1 and hence cannot lie in A. The commutant $\Omega(0)$ of W contains the Levi factor $S \subseteq V_1$, hence also the subVOA G that it generates [cf. (12)]. We can then consider the commutant of G in $\Omega(0)$, call it Z. In this way we obtain the canonical conformal subVOA of V that we call the *tripartite subalgebra*

$$T = W \otimes G \otimes Z.$$

By construction, T is a *conformal subalgebra* of V, and the conformal gradings on W, G, Z are compatible with the $L(0)$-grading on V. Because $(W \otimes G)_1 = W_1 \oplus G_1 = V_1$ then $Z_1 = 0$. This completes the proof of Theorem 3.

Conjecture. In the notation of (5), Z is a strongly regular VOA.

This is just a special case of a more general conjecture, namely that the commutant of a rational subVOA (in a strongly regular VOA, say) is itself rational. If the Conjecture is true then the tripartite subalgebra T is strongly regular, and V reduces to a *finite sum* of irreducible T-modules. In this way, the classification of strongly regular VOAs reduces to the classification of strongly regular VOAs Z with $Z_1 = 0$ and the extension problem as discussed in the introduction.

6.2 The Invariants $\tilde{c}$ and l

We give some further applications of Theorem 1 exemplifying the philosophy of the previous paragraph. Let V be a strongly regular VOA of central charge c and $H \subseteq V$ a Cartan subalgebra of rank l. Recall [8] that the *effective central charge* of V is the quantity

$$\tilde{c}_V = \tilde{c} := c - 24\lambda_{min}.$$

Here, λ_{min} is the *minimum* of the conformal weights λ_j $(1 \le j \le r)$ of the irreducible V-modules. It is known (loc. cit.) that $\tilde{c} \ge l$ and $\tilde{c} > 0$ if $\dim V > 1$. Because of these facts, $\tilde{c}$ is often a more useful invariant than c itself. Note that $\tilde{c}$ is defined for any rational VOA.

We now give the proof of Theorem 7. The basic idea, to combine Zhu's modular-invariance [33] together with growth conditions on the Fourier coefficients of components of vector-valued modular forms [25], was first used in [8]. The availability of Theorem 1 brings added clarity.

It follows easily from the definitions that if $W \subseteq V$ is a *conformal* subalgebra then $\tilde{c}_V \le \tilde{c}_W$. Moreover $\tilde{c}$ is multiplicative over tensor products [21, Sect. 4.6]. So if (b) of Theorem 7 holds then $\tilde{c}_V \le \tilde{c}_{V_\Lambda} + \tilde{c}_{L(c_{p,q},0)}$. Since $\operatorname{rk}\Lambda = l$ then $\tilde{c}_{V_\Lambda} = c = l$ because $\lambda_{min} = 0$ for lattice theories [4]. Moreover, for the discrete series Virasoro VOA we have [8, Sect. 4, Example (e)]

$$\tilde{c} = 1 - \frac{6}{pq} \quad ((p,q) = 1, 2 \le p < q), \tag{43}$$

in particular we always have $\tilde{c}_{L(c_{p,q},0)} < 1$. Therefore $\tilde{c}_V < l + 1$. This establishes the implication (b) $\Rightarrow$ (a) in Theorem 7.

Next, taking $U = H$ in Theorem 1, we find that the maximal element of $\mathscr{P}_H$ is a lattice subVOA $W \cong V_\Lambda$ with $\mathrm{rk}\Lambda = \dim H = l$. Let $C = C_V(W)$ be the commutant of W. Then $W \otimes C$ is a conformal subVOA of V. Now suppose that part (b) of the Theorem does *not* hold. Thus the Virasoro subalgebra of C, call it Vir_C, has a central charge c', say, that is *not* in the discrete series. Then the known submodule structure of Verma modules over the Virasoro algebra [19] shows that the *partition function* $Z_{Vir_C}(\tau)$ of Vir_C satisfies

$$Z_{Vir_C}(\tau) := \mathrm{Tr}_{Vir_C} q^{L(0)-c'/24} = q^{-c'/24} \prod_{n=2}^{\infty} (1-q^n)^{-1}.$$

([8], Proposition 6.1 summarizes exactly what we need here.) Therefore,

$$\begin{aligned} Z_{W\otimes Vir_C}(\tau) &:= Z_W(\tau) Z_{Vir_C}(\tau) \\ &= \frac{\theta_\Lambda(\tau)}{\eta(\tau)^l} \frac{q^{-c'/24}}{\prod_{n=2}^{\infty}(1-q^n)} \\ &= \frac{\theta_\Lambda(\tau)}{\eta(\tau)^{l+1}} q^{(1-c')/24}(1-q). \end{aligned}$$

($\theta_\Lambda(\tau)$ and $\eta(\tau)$ are the *theta-function* of Λ and the *eta-function* respectively.) It follows that for any $\epsilon > 0$, the coefficients of the q-expansion of $\eta(\tau)^{l+1-\epsilon} Z_{W\otimes Vir_C}(\tau)$ have *exponential growth*. Therefore, the same statement holds true *ipso facto* if we replace $W \otimes Vir_C$ with V. We state this as

the coefficients of $\eta(\tau)^{l+1-\epsilon} Z_V(\tau)$ have *exponential growth* ($\epsilon > 0$). (44)

On the other hand, consider the column vector

$$F(\tau) := (Z_{M^1}(\tau), \ldots, Z_{M^r}(\tau))^t$$

whose components are the partition functions of the irreducible V-modules M^j. By Zhu's modular-invariance theorem [33], $F(\tau)$ is a vector-valued modular form of weight 0 on the full modular group $SL_2(\mathbb{Z})$ associated with some representation of $SL_2(\mathbb{Z})$. (See [30], Sect. 8 for a discussion of vector-valued modular forms in the context of VOAs.) Moreover, each $Z_{M^j}(\tau)$ is holomorphic in the complex upper half-plane, so that their only poles are at the cusps. The very definition of $\tilde{c}$, and the reason for its importance, is that the maximum order of a pole of *any* of the partition functions $Z_{M^j}(\tau)$ is $\tilde{c}/24$. It follows from this that

$$\eta(\tau)^{\tilde{c}} F(\tau)$$

is a *holomorphic* vector-valued modular form on $SL_2(\mathbb{Z})$. As such, the Fourier coefficients of the component functions have *polynomial growth* [25]. In particular, this applies to $\eta(\tau)^{\tilde{c}} Z_V(\tau)$, which is one of the components.

Comparing the last statement with (44), it follows that $\tilde{c} > l + 1 - \epsilon$ for all $\epsilon > 0$, i.e. $\tilde{c} \geq l + 1$. So we have shown that if part (b) of the Theorem does not hold, neither does part (a). Theorem 7 is thus proved.

The special case $l = 0$ of the Theorem characterizes *minimal models*. We state it as

Theorem 8. *Let V be a strongly regular VOA. Then $\tilde{c} < 1$ if, and only if, the Virasoro subalgebra of V is in the discrete series.*

Corollary 9. *Let V be a strongly regular VOA with* $\dim V > 1$. *Then $\tilde{c} \geq 2/5$, and equality holds if, and only if, $V \cong L(c_{2,5}, 0)$, the (Yang–Lee) discrete series Virasoro VOA with $c = -22/5$.*

Because $\dim V > 1$ then $\tilde{c} > 0$, and if $\tilde{c} < 1$ then V is a minimal model by Theorem 8. Inspection of (43) shows that the least positive value is $2/5$, corresponding to the Yang–Lee model [3]. This theory has only two irreducible modules, of conformal weight 0 and $-1/5$. Therefore the second irreducible cannot be contained in V, so that $V \cong L(c_{2,5}, 0)$, as asserted in Corollary 9. Informally, the Corollary says that the Yang–Lee theory is the *smallest* rational CFT.

We give a final numerical example. Suppose that V is a strongly regular simple VOA such that $1 < \tilde{c} < 7/5$. Since $l \leq c \leq \tilde{c}$ we must have $l = 0$ or 1. In the latter case, by Theorem 1 we see that V contains as a conformal subVOA a tensor product $V_\Lambda \otimes Vir$ where Vir is a Virasoro algebra in the discrete series with $0 < \tilde{c}_{Vir} < 2/5$. This is impossible by Corollary 9. So in fact $l = 0$, i.e. V has Lie rank 0, meaning that $V_1 = 0$. The smallest value of $\tilde{c}$ in the range $(1, 7/5)$ that I know of is a parafermion theory with $\tilde{c} = 8/7$ [17, 32].

Acknowledgements Geoffrey Mason was supported by the NSF.

References

1. Anderson, G., Moore, G.: Rationality in conformal field theory. Commun. Math. Phys. **117**, 441–450 (1988)
2. Buhl, G.: A spanning set for VOA modules. J. Algebra **254**(1), 125–151 (2004)
3. Di Francesco, P., Mathieu, P., Sénéchal, D.: Conformal Field Theory. Springer, New York (1996)
4. Dong, C.: Vertex algebras associated with even lattices. J. Algebra **160**, 245–265 (1993)
5. Dong, C., Griess, R.: Automorphism groups and derivation algebras of finitely generated vertex operator algebras. Mich. Math. J. **50**, 227–239 (2002)
6. Dong, C., Mason, G.: Nonabelian orbifolds and the boson-fermion correspondence. Commun. Math. Phys. **163**, 523–559 (1994)
7. Dong, C., Mason, G.: On quantum Galois theory. Duke J. Math. **86**(2), 305–321 (1997)

8. Dong, C., Mason, G.: Rational vertex operator algebras and the effective central charge. Int. Math. Res. Not. **56**, 2989–3008 (2004)
9. Dong, C., Mason, G.: Integrability of C_2-cofinite vertex operator algebras. Int. Math. Res. Not. **2006**, 1–15. Art. ID 80468
10. Dong, C., Mason, G.: Shifted vertex operator algebras. Math. Proc. Camb. Philos. Soc. **141**, 67–80 (2006)
11. Dong, C., Zhang, W.: Rational vertex operator algebras are finitely generated. J. Algebra **320**(6), 2610–2614 (2008)
12. Dong, C., Zhang, W.: Toward classification of rational vertex operator algebras with central charges less than 1. ArXiv:0711.4625v1
13. Dong, C., Li, H., Mason, G.: Simple currents and extensions of vertex operator algebras. Commun. Math. Phys. **180**, 671–707 (1996)
14. Dong, C., Li, H., Mason, G.: Regularity of rational vertex operator algebras. Adv. Math. **132**, 148–166 (1997)
15. Dong, C., Li, H., Mason, G.: Twisted representations of vertex operator algebras. Math. Ann. **310**, 571–600 (1998)
16. Dong, C., Li, H., Mason, G.: Modular-invariance of trace functions in orbifold theory and generalized moonshine. Commun. Math. Phys. **214**, 1–56 (2000)
17. Dong, C., Lam, C., Wang, Q., Yamada, H.: The structure of parafermion vertex operator algebras. J. Algebra **323**, 371–381 (2010)
18. Eichler, M., Zagier, D.: The Theory of Jacobi Forms. Progress in Mathematics, vol. 55. Birkhäuser, Boston (1985)
19. Feigin, B., Fuchs, D.: Representations of the Virasoro algebra. In: Representations of Lie Groups and Related Topics. Advanced Studies in Contemporary Mathematics, vol. 7, pp. 465–554. Gordon and Breach, New York (1990)
20. Frenkel, I., Zhu, Y.: Vertex operator algebras associated to representations of affine and Virasoro algebras. Duke J. Math. **66**(1), 123–168 (1992)
21. Frenkel, I., Huang, Y.-Z., Lepowsky, J.: On axiomatic approaches to vertex operator algebras and modules, Memoirs Amer. Math. Soc. **104**, 194 (1993)
22. Frenkel, I., Lepowsky, J., Meurman, A.: Vertex Operator Algebras and the Monster. Academic, London (1988)
23. Gaberdiel, M., Nietzke, A.: Rationality, quasirationality and finite W-algebras. Commun. Math. Phys. **238**(1–2), 305–331 (2003)
24. Humphreys, J.: Introduction to Lie Algebras and Representation Theory. Graduate Texts in Mathematics, vol. 9. Springer, New York (1972)
25. Knopp, M., Mason, G.: On vector-valued modular forms and their Fourier coefficients. Acta Arith. **110**, 117–124 (2003)
26. Krauel, M., Mason, G.: Vertex operator algebras and weak Jacobi forms. Int. J. Math. **23**(1), 10 (2012)
27. Lepowsky, J., Li, H.: Introduction to Vertex Operator Algebras. Birkhaüser, Boston (2004)
28. Li, H.: Symmetric invariant bilinear forms on vertex operator algebras. J. Pure Appl. Math. **96**, 279–297 (1994)
29. Li, H.: Local systems of twisted vertex operators, vertex operator superalgebras and twisted modules. In: Moonshine, the Monster, and Related Topics. Contemporary Mathematics, vol. 193, pp. 203–236. American Mathematical Society, Providence (1996)
30. Mason, G., Tuite, M.: Vertex operators and modular forms. In: A Window into Zeta and Modular Physics. Mathematical Sciences Research Institute Publications, pp. 183–278. Cambridge University Press, Cambridge (2010)
31. Miyamoto, M.: A modular invariance on the theta functions defined on vertex operator algebras. Duke Math. J. **101**(2), 221–236 (2000)

32. Zamolodchikov, A., Fateev, V.: Nonlocal (parafermion) currents in two-dimensional conformal quantum field theory and self-dual critical points in Z_N-symmetric statistical systems. Sov. Phys. J. Exp. Theor. Phys. **62**, 215–225 (1985)
33. Zhu, Y.: Modular-invariance of characters of vertex operator algebras. J. Am. Math. Soc. **9**(1), 237–302 (1996)

A Characterization of the Vertex Operator Algebra $V_{L_2}^{A_4}$

Chongying Dong and Cuipo Jiang

Abstract The rational vertex operator algebra $V_{L_2}^{A_4}$ is characterized in terms of weights of primary vectors. This reduces the classification of rational vertex operator algebras with $c = 1$ to the characterizations of $V_{L_2}^{S_4}$ and $V_{L_2}^{A_5}$.

2000MSC:17B69

1 Introduction

Characterizations of vertex operator algebras $V_{L_2}^G$ for root lattice L_2 of $sl(2,\mathbb{C})$ and finite groups $G = A_4, S_4, A_5$ are the remaining part of classification of rational vertex operator algebras with $c = 1$ after the work of [9–11, 13, 27]. Using the structure and representation theory of $V_{L_2}^{A_4}$ obtained in [12] and [19], we give a characterization of rational vertex operator algebra $V_{L_2}^{A_4}$ in this paper.

The main assumption for the characterization of vertex operator algebra $V_{\mathbb{Z}\gamma}^+$ with $(\gamma,\gamma) \geq 6$ being a positive even integer in [10,11] is that the dimension of the weight 4 subspace is at least three dimensional. Knowing the explicit structure of $V_{L_2}^{A_4}$ we have a different assumption in characterizing $V_{L_2}^{A_4}$. That is, there is a primary vector of weight 9 and the weight of any primary vector which is not a multiple of $\mathbf{1}$ is greater than or equal to 9. Due to a recent result in [21] on the modularity of the q-characters of the irreducible modules for rational and C_2-cofinite vertex operator

C. Dong (✉)
Department of Mathematics, University of California, Santa Cruz, CA 95064, USA
e-mail: dong@math.ucsc.edu

C. Jiang
Department of Mathematics, Shanghai Jiaotong University, Shanghai 200240, China
e-mail: cpjiang@sjtu.edu.cn

W. Kohnen and R. Weissauer (eds.), *Conformal Field Theory, Automorphic Forms and Related Topics*, Contributions in Mathematical and Computational Sciences 8,
DOI 10.1007/978-3-662-43831-2_3,

algebras, we can use the classification of q-characters of rational vertex operator algebras with $c = 1$ from [25] to conclude that the q-character of such a vertex operator algebra and that of $V_{L_2}^{A_4}$ are the same.

Two basic facts are used in the characterization. The first one is that both $V_{L_2}^{A_4}$ and an abstract vertex operator algebra V satisfying the required conditions have the same decomposition as modules for the Virasoro algebra. This allows us to use the fusion rules for the irreducible modules for the Virasoro vertex operator algebra $L(1,0)$ obtained in [9] and [26] to understand the structure of both vertex operator algebras in terms of generators. The other one is that $V_{L_2}^{A_4}$ is generated by a weight 9 primary vector x^1 and has a spanning set in terms Virasoro algebra, the component operators of x^1 and the component operators of y^1 which is a primary vector of weight 16. From the q-characters we know that V also has primary vectors x^2, y^2 of weights 9, 16, respectively. The main task is to show how the vertex operator algebra V has a similar spanning set with x^1, y^1 being replaced by x^2, y^2. The fusion rules for the vertex operator algebra $L(1,0)$ play a crucial role here.

We certainly expect that the ideas and methods presented in this paper can also be used to characterize vertex operator algebras $V_{L_2}^G$ for $G = S_4, A_5$ although $V_{L_2}^G$ have not been understood well. It seems that knowing the generators of $V_{L_2}^G$ and a spanning set is good enough for the purpose of characterization. Of course, the rationality is also needed.

The paper is organized as follows. We review the modular invariance results from [28] and [21] in Sect. 2. These results will be used to conclude that $V_{L_2}^{A_4}$ and an abstract vertex operator algebra V satisfying certain conditions have the same graded dimensions. We also review the fusion rules for the vertex operator algebra $L(1,0)$ from [26] and [9] in this section. In Sect. 3 we discuss the structure of $V_{L_2}^{A_4}$ including the generators and spanning set following [12]. Section 4 is devoted to the characterization of $V_{L_2}^{A_4}$. That is, if a rational, C_2-cofinite and self-dual vertex operator algebra V of central charge 1 satisfies (a) V is a completely reducible module for the Virasoro algebra, (b) V has a primary vector of weight 9 and the weight of any primary vector whose weight is greater than 0 is greater than or equal to 9, then V is isomorphic to $V_{L_2}^{A_4}$. The main idea is to use generators and relations to construct a vertex operator algebra isomorphism from $V_{L_2}^{A_4}$ to V.

2 Preliminaries

Let $V = (V, Y, \mathbf{1}, \omega)$ be a vertex operator algebra [5,23]. We review various notions of V-modules (cf. [17,23,28]) and the definition of rational vertex operator algebras. We also discuss some consequences following [10,18,21,25].

Definition 2.1. A weak V module is a vector space M equipped with a linear map

$$\begin{aligned} Y_M &: V \to \mathrm{End}(M)[[z, z^{-1}]] \\ v &\mapsto Y_M(v,z) = \textstyle\sum_{n\in\mathbb{Z}} v_n z^{-n-1}, \quad v_n \in \mathrm{End}(M) \end{aligned}$$

satisfying the following:

(1) $v_n w = 0$ for $n >> 0$ where $v \in V$ and $w \in M$
(2) $Y_M(\mathbf{1}, z) = Id_M$
(3) The Jacobi identity holds:

$$z_0^{-1}\delta\left(\frac{z_1 - z_2}{z_0}\right) Y_M(u, z_1)Y_M(v, z_2) - z_0^{-1}\delta\left(\frac{z_2 - z_1}{-z_0}\right) Y_M(v, z_2)Y_M(u, z_1)$$
$$= z_2^{-1}\delta\left(\frac{z_1 - z_0}{z_2}\right) Y_M(Y(u, z_0)v, z_2). \quad (1)$$

Definition 2.2. An admissible V module is a weak V-module which carries a $\mathbb{Z}_+$-grading $M = \bigoplus_{n\in\mathbb{Z}_+} M(n)$, such that if $v \in V_r$ then $v_m M(n) \subseteq M(n + r - m - 1)$.

Definition 2.3. An ordinary V module is a weak V-module which carries a $\mathbb{C}$-grading $M = \bigoplus_{\lambda\in\mathbb{C}} M_\lambda$, such that:

(1) $dim(M_\lambda) < \infty$,
(2) $M_{\lambda+n} = 0$ for fixed λ and $n << 0$,
(3) $L(0)w = \lambda w = \mathrm{wt}(w)w$ for $w \in M_\lambda$ where $L(0)$ is the component operator of $Y_M(\omega, z) = \sum_{n\in\mathbb{Z}} L(n)z^{-n-2}$.

Remark 2.4. It is easy to see that an ordinary V-module is an admissible one. If W is an ordinary V-module, we simply call W a V-module.

We call a vertex operator algebra rational if the admissible module category is semisimple. We have the following result from [18] (also see [28]).

Theorem 2.5. *If V is a rational vertex operator algebra, then V has finitely many irreducible admissible modules up to isomorphism and every irreducible admissible V-module is ordinary.*

Suppose that V is a rational vertex operator algebra and let $M^1, \ldots, M^k$ be the irreducible modules such that

$$M^i = \oplus_{n\geq 0} M^i_{\lambda_i+n}$$

where $\lambda_i \in \mathbb{Q}$ [18], $M^i_{\lambda_i} \neq 0$ and each $M^i_{\lambda_i+n}$ is finite dimensional. Let λ_{min} be the minimum of λ_i's. The effective central charge $\tilde{c}$ is defined as $c - 24\lambda_{min}$. For each M^i we define the q-character of M^i by

$$Z_i(q) = q^{-c/24} \sum_{n\geq 0} (\dim M^i_{\lambda_i+n}) q^{n+\lambda_i}.$$

A vertex operator algebra V is called C_2-cofinite if $\dim V/C_2(V)$ is finite dimensional where $C_2(V)$ is a subspace of V spanned by $u_{-2}v$ for $u, v \in V$. If V is C_2-cofinite, then $Z_i(q)$ converges to a holomorphic function on $0 < |q| < 1$

[28]. Let $q = e^{2\pi i\tau}$ and we sometimes also write $Z_i(q)$ by $Z_i(\tau)$ to indicate that $Z_i(q)$ is a holomorphic function on the upper half plane.

For a V-module W, let W' denote the graded dual of W. Then W' is also a V-module [24]. A vertex operator algebra V is called self dual if V' [24] is isomorphic to itself. The following result comes from [21]

Theorem 2.6. *Let V be a rational, C_2-cofinite, self dual simple vertex operator algebra.*

(1) Each $Z_i(\tau)$ is a modular function on a congruence subgroup of $SL_2(\mathbb{Z})$ of level n which is the smallest positive integer such that $n(\lambda_i - c/24)$ is an integer for all i.
(2) $\sum_i |Z_i(\tau)|^2$ is $SL_2(\mathbb{Z})$-invariant.

We now recall the construction of vertex operator algebras $M(1)^+$, V_L^+ and related results from [1–4, 14–16, 20, 23].

Let $L = \mathbb{Z}\alpha$ be a positive definite lattice with $(\alpha, \alpha) = 2k$ for some positive integer k. Set $\mathfrak{h} = \mathbb{C} \otimes_{\mathbb{Z}} L$ and extend $(\cdot, \cdot)$ to $\mathfrak{h}$ by $\mathbb{C}$-linearity. Let $\hat{\mathfrak{h}} = \mathbb{C}[t, t^{-1}] \otimes \mathfrak{h} \oplus \mathbb{C}K$ be the corresponding affine Lie algebra so that

$$[\alpha(m), \alpha(n)] = 2km\delta_{m+n,0}K \text{ and } [K, \hat{\mathfrak{h}}] = 0$$

for any $m, n \in \mathbb{Z}$, where $\alpha(m) = \alpha \otimes t^m$. Note that $\hat{\mathfrak{h}}^{\geq 0} = \mathbb{C}[t] \otimes \mathfrak{h} \oplus \mathbb{C}K$ is an abelian subalgebra. Let $\mathbb{C}e^\lambda$ (for any $\lambda \in \mathfrak{h}$) be one-dimensional $\hat{\mathfrak{h}}^{\geq 0}$-module such that $\alpha(m) \cdot e^\lambda = (\lambda, \alpha)\delta_{m,0}e^\lambda$ and $K \cdot e^\lambda = e^\lambda$ for $m \geq 0$. Consider the induced module

$$M(1, \lambda) = U(\hat{\mathfrak{h}}) \otimes_{U(\hat{\mathfrak{h}}^{\geq 0})} \mathbb{C}e^\lambda \cong S(t^{-1}\mathbb{C}[t^{-1}]) \text{ (linearly)}.$$

Set

$$M(1) = M(1, 0).$$

Then there exists a linear map $Y : M(1) \to \mathrm{End}M(1)[[z, z^{-1}]]$ such that $(M(1), Y, \mathbf{1}, \omega)$ is a simple vertex operator algebra and $M(1, \lambda)$ is an irreducible $M(1)$-module for any $\lambda \in \mathfrak{h}$ (see [23]). The vacuum vector and the Virasoro element are given by $\mathbf{1} = e^0$ and $\omega = \frac{1}{4k}\alpha(-1)^2\mathbf{1}$, respectively.

We use $\mathbb{C}[L]$ to denote the group algebra of L with a basis e^β for $\beta \in L$. Then

$$V_L = M(1) \otimes \mathbb{C}[L]$$

is the lattice vertex operator algebra associated to L [5, 23]. Let L° be the dual lattice of L :

$$L^\circ = \{\lambda \in \mathfrak{h} \mid (\alpha, \lambda) \in \mathbb{Z}\} = \frac{1}{2k}L$$

and $L^\circ = \cup_{i=-k+1}^{k}(L+\lambda_i)$ be the coset decomposition with $\lambda_i = \frac{i}{2k}\alpha$. Set $\mathbb{C}[L+\lambda_i] = \bigoplus_{\beta\in L}\mathbb{C}e^{\beta+\lambda_i}$. Then each $\mathbb{C}[L+\lambda_i]$ is an L-submodule in an obvious way. Set $V_{L+\lambda_i} = M(1)\otimes\mathbb{C}[L+\lambda_i]$. Then V_L is a rational vertex operator algebra and $V_{L+\lambda_i}$ for $i=-k+1,\cdots,k$ are the irreducible modules for V_L (see [5,6,23]).

Let $\theta : V_{L+\lambda_i} \to V_{L-\lambda_i}$ be a linear isomorphism for $i \in \{-k+1,\cdots,k\}$ such that

$$\theta(\alpha(-n_1)\alpha(-n_2)\cdots\alpha(-n_s)\otimes e^{\beta+\lambda_i}) = (-1)^s\alpha(-n_1)\alpha(-n_2)\cdots\alpha(-n_s)\otimes e^{-\beta-\lambda_i}$$

where $n_j > 0$ and $\beta\in L$. In particular, θ is an automorphism of V_L which induces an automorphism of $M(1)$. For any θ-stable subspace U of V_{L°, let $U^\pm$ be the ±1-eigenspace of U for θ. Then V_L^+ is a simple vertex operator algebra.

The θ-twisted Heisenberg algebra $\mathfrak{h}[-1]$ and its irreducible module $M(1)(\theta)$ from [23] are also needed. Define a character χ_s of $L/2L$ such that $\chi_s(\alpha) = (-1)^s$ for $s = 0,1$ and let $T_{\chi_s} = \mathbb{C}$ be the corresponding irreducible $L/2L$. Then $V_L^{T_{\chi_s}} = M(1)(\theta)\otimes T_{\chi_s}$ is an irreducible θ-twisted V_L-module (see [7,23]). We also define actions of θ on $M(1)(\theta)$ and $V_L^{T_{\chi_s}}$ by

$$\theta(\alpha(-n_1)\alpha(-n_2)\cdots\alpha(-n_p)) = (-1)^p\alpha(-n_1)\alpha(-n_2)\cdots\alpha(-n_p)$$
$$\theta(\alpha(-n_1)\alpha(-n_2)\cdots\alpha(-n_p)\otimes t) = (-1)^p\alpha(-n_1)\alpha(-n_2)\cdots\alpha(-n_p)\otimes t$$

for $n_j\in\frac{1}{2}+\mathbb{Z}_+$ and $t\in T_{\chi_s}$. We denote the ±1-eigenspaces of $M(1)(\theta)$ and $V_L^{T_{\chi_s}}$ under θ by $M(1)(\theta)^\pm$ and $(V_L^{T_{\chi_s}})^\pm$ respectively.

The classification of irreducible modules for arbitrary $M(1)^+$ and V_L^+ are obtained in [14–16] and [3]. The rationality of V_L^+ is established in [2] for rank one lattice and [20] in general. One can find the following results from these papers.

Theorem 2.7. *(1) Any irreducible module for the vertex operator algebra $M(1)^+$ is isomorphic to one of the following modules:*

$$M(1)^+, M(1)^-, M(1,\lambda)\cong M(1,-\lambda)\ (0\neq\lambda\in\mathfrak{h}), M(1)(\theta)^+, M(1)(\theta)^-.$$

(2) Any irreducible V_L^+-module is isomorphic to one of the following modules:

$$V_L^\pm, V_{\lambda_i+L}(i\neq k), V_{\lambda_k+L}^\pm, (V_L^{T_{\chi_s}})^\pm.$$

(3) V_L^+ is rational.

The following characterization of V_L^+ is given in [10] and [11].

Theorem 2.8. *Let V be a simple, rational and C_2-cofinite self-dual vertex operator algebra such that V is generated by highest vectors of the Virasoro algebra, $\tilde{c} = c = 1$ and*

$$\dim V_2 = 1,\ \dim V_4 \geq 3.$$

Then V is isomorphic to $V_{\mathbb{Z}\alpha}^{+}$ for some rank one positive definite even lattice $L = \mathbb{Z}\alpha$.

We will need the following result from [25].

Theorem 2.9. *Let V be a rational CFT type vertex operator algebra with $c = \tilde{c} = 1$ such that each $Z_i(\tau)$ is a modular function on a congruence subgroup and $\sum_i |Z_i(\tau)|^2$ is $SL_2(\mathbb{Z})$-invariant, then the q-character of V is equal to the character of one of the following vertex operator algebras V_L, V_L^{+} and $V_{\mathbb{Z}\alpha}^{G}$, where L is any positive definite even lattice of rank 1, $\mathbb{Z}\alpha$ is the root lattice of type A_1 and G is a finite subgroup of $SO(3)$ isomorphic to A_4, S_4 or A_5.*

By Theorems 2.6 we know that the assumptions in Theorem 2.9 hold. So the q-character of a rational vertex operator algebra with $c = 1$ is known.

Recall from [24] the fusion rules of vertex operator algebras. Let V be a vertex operator algebra, and W^i $(i = 1,2,3)$ be ordinary V-modules. We denote by $I_V\begin{pmatrix} W^3 \\ W^1\, W^2 \end{pmatrix}$ the vector space of all intertwining operators of type $\begin{pmatrix} W^3 \\ W^1\, W^2 \end{pmatrix}$. It is well known that fusion rules have the following symmetry [24].

Proposition 2.10. *Let W^i $(i = 1,2,3)$ be V-modules. Then*

$$\dim I_V\begin{pmatrix} W^3 \\ W^1\, W^2 \end{pmatrix} = \dim I_V\begin{pmatrix} W^3 \\ W^2\, W^1 \end{pmatrix}, \quad \dim I_V\begin{pmatrix} W^3 \\ W^1\, W^2 \end{pmatrix} = \dim I_V\begin{pmatrix} (W^2)' \\ W^1\, (W^3)' \end{pmatrix}.$$

Here are some results on the fusion rules for the Virasoro vertex operator algebra. Recall that $L(c,h)$ is the irreducible highest weight module for the Virasoro algebra with central charge c and highest weight h for $c,h \in \mathbb{C}$. It is well known that $L(c,0)$ is a vertex operator algebra [22]. The following two results can be found in [26] and [9].

Theorem 2.11. *(1) We have*

$$\dim I_{L(1,0)}\begin{pmatrix} L(1,k^2) \\ L(1,m^2)\, L(1,n^2) \end{pmatrix} = 1, \quad k \in \mathbb{Z}_+,\ |n-m| \le k \le n+m,$$
$$\dim I_{L(1,0)}\begin{pmatrix} L(1,k^2) \\ L(1,m^2)\, L(1,n^2) \end{pmatrix} = 0, \quad k \in \mathbb{Z}_+,\ k < |n-m| \text{ or } k > n+m,$$

where $n,m \in \mathbb{Z}_+$.

(2) For $n \in \mathbb{Z}_+$ such that $n \neq p^2$, for all $p \in \mathbb{Z}_+$, we have

$$\dim I_{L(1,0)}\begin{pmatrix} L(1,n) \\ L(1,m^2)\, L(1,n) \end{pmatrix} = 1,$$
$$\dim I_{L(1,0)}\begin{pmatrix} L(1,k) \\ L(1,m^2)\, L(1,n) \end{pmatrix} = 0,$$

for $k \in \mathbb{Z}_+$ such that $k \neq n$.

3 The Vertex Operator Subalgebra $V_{L_2}^{A_4}$

Let $L_2 = \mathbb{Z}\alpha$ be the rank one positive-definite lattice such that $(\alpha, \alpha) = 2$. Then $(V_{L_2})_1$ is a Lie algebra isomorphic to $sl_2(\mathbb{C})$ and has an orthonormal basis:

$$x^1 = \frac{1}{\sqrt{2}}\alpha(-1)\mathbf{1},\ x^2 = \frac{1}{\sqrt{2}}(e^{\alpha} + e^{-\alpha}),\ x^3 = \frac{i}{\sqrt{2}}(e^{\alpha} - e^{-\alpha}).$$

There are three involutions $\tau_i \in Aut(V_{L_2}), i = 1, 2, 3$ be such that

$$\tau_1(x^1, x^2, x^3) = (x^1, x^2, x^3)\begin{bmatrix} 1 & & \\ & -1 & \\ & & -1 \end{bmatrix},$$

$$\tau_2(x^1, x^2, x^3) = (x^1, x^2, x^3)\begin{bmatrix} -1 & & \\ & 1 & \\ & & -1 \end{bmatrix},$$

$$\tau_3(x^1, x^2, x^3) = (x^1, x^2, x^3)\begin{bmatrix} -1 & & \\ & -1 & \\ & & 1 \end{bmatrix}.$$

There is also an order 3 automorphism $\sigma \in Aut(V_{L_2})$ defined by

$$\sigma(x^1, x^2, x^3) = (x^1, x^2, x^3)\begin{bmatrix} 0 & 1 & 0 \\ 0 & 0 & -1 \\ -1 & 0 & 0 \end{bmatrix}.$$

It is easy to see that σ and $\tau_i, i = 1, 2, 3$ generate a finite subgroup of $Aut(V_{L_2})$ isomorphic to the alternating group A_4. We simply denote this subgroup by A_4. It is easy to check that the subgroup K generated by $\tau_i, i = 1, 2, 3$ is a normal subgroup of A_4 of order 4. Let

$$J = h(-1)^4\mathbf{1} - 2h(-3)h(-1)\mathbf{1} + \frac{3}{2}h(-2)^2\mathbf{1},\ \ E = e^{\beta} + e^{-\beta}$$

where $h = \frac{1}{\sqrt{2}}\alpha$, $\beta = 2\alpha$. The following lemma comes from [8] and [12].

Lemma 3.1. *(1) The vertex operator algebra $V_{L_2}^K$ and $V_{\mathbb{Z}\beta}^+$ are the same and $V_{\mathbb{Z}\beta}^+$ is generated by J and E. Moreover, $(V_{L_2}^K)_4$ is four dimensional with a basis consisting of $L(-2)^2\mathbf{1}, L(-4)\mathbf{1}, J, E$.*

(2) The vertex operator algebra $V_{L_2}^{A_4}$ and $(V_{\mathbb{Z}\beta}^+)^{\langle\sigma\rangle}$ are the same.

(3) The action of σ on J and E are given by

$$\sigma(J) = -\frac{1}{2}J + \frac{9}{2}E,\ \sigma(E) = -\frac{1}{6}J - \frac{1}{2}E.$$

Clearly, σ preserves the subspace of $(V_{\mathbb{Z}\beta}^+)_4$ spanned by J and E. It is easy to check that

$$\sigma(X^1) = \frac{-1+\sqrt{3}i}{2}X^1, \ \sigma(X^2) = \frac{-1-\sqrt{3}i}{2}X^2 \tag{2}$$

where

$$X^1 = J - \sqrt{27}iE, \ X^2 = J + \sqrt{27}iE. \tag{3}$$

This implies that $(V_{\mathbb{Z}\beta}^+)_4^{\langle\sigma\rangle} = L(1,0)_4$ where $L(1,0)$ is the vertex operator subalgebra of $V_{\mathbb{Z}\beta}$ generated by ω. It follows from [8] that

$$(V_{\mathbb{Z}\beta}^+)^{\langle\sigma\rangle} = L(1,0) \bigoplus \sum_{n\geq 3} a_n L(1,n^2)$$

as a module for $L(1,0)$, where a_n is the multiplicity of $L(1,n^2)$ in $(V_{\mathbb{Z}\beta}^+)^{\langle\sigma\rangle}$. Using (2) shows that for any $n \in \mathbb{Z}$,

$$X_n^1 X^2 \in (V_{\mathbb{Z}\beta}^+)^{\langle\sigma\rangle} = V_{L_2}^{A_4}.$$

We sometimes also call a highest weight vector for the Virasoro algebra a primary vector. From [8] we know that $V_{\mathbb{Z}\beta}^+$ contains two linearly independent primary vectors J and E of weight 4 and one linearly independent primary vector of weight 9. Note from [12] that

$$J_3J=-72L(-4)\mathbf{1}+336L(-2)^2\mathbf{1}-60J, \ E_3E=-\frac{8}{3}L(-4)\mathbf{1}+\frac{112}{9}L(-2)^2\mathbf{1}+\frac{20}{9}J$$

(cf. [11]). By Theorem 2.11 and Lemma 3.1, we have for $n \in \mathbb{Z}$

$$X_n^1 X^2 \in L(1,0) \oplus L(1,9) \oplus L(1,16).$$

The following lemma comes from [12].

Lemma 3.2. *The vector*

$$\begin{aligned} u^{(9)} &= -\frac{\sqrt{2}}{4}(J_{-2}E - E_{-2}J) \\ &= -\frac{1}{\sqrt{2}}(15h(-4)h(-1) + 10h(-3)h(-2) + 10h(-2)h(-1)^3) \otimes (e^\beta + e^{-\beta}) \\ &\quad +(6h(-5) + 10h(-3)h(-1)^2 + \frac{15}{2}h(-2)^2h(-1) + h(-1)^5) \otimes (e^\beta - e^{-\beta}) \end{aligned}$$

is a non-zero primary vector of $V_{\mathbb{Z}\beta}^+$ *of weight* 9.

By Lemma 3.1, we have $J_{-9}J + 27E_{-9}E \in (V_{\mathbb{Z}\beta}^{+})^{\langle\sigma\rangle}$. Then

$$J_{-9}J + 27E_{-9}E = x^0 + X^{(16)} + 27(e^{2\beta} + e^{-2\beta}), \tag{4}$$

where $x^0 \in L(1,0)$, and $X^{(16)}$ is a non-zero primary element of weight 16 in $M(1)^+$. Denote

$$u^{(16)} = X^{(16)} + 27(e^{2\beta} + e^{-2\beta}). \tag{5}$$

Then $u^{(16)} \in (V_{\mathbb{Z}\beta}^{+})^{\langle\sigma\rangle}$ is a non-zero primary vector of weight 16. The following results come from [12].

Theorem 3.3. *The following hold: (1) $V_{L_2}^{A_4}$ is generated by $u^{(9)}$.*

(2) $V_{L_2}^{A_4}$ is linearly spanned by

$$L(-m_s)\cdots L(-m_1)u_n^{(9)}u^{(9)},\ L(-m_s)\cdots L(-m_1)w_{-k_p}^p\cdots w_{-k_1}^1 w,$$

where $w, w^1, \cdots, w^p \in \{u^{(9)}, u^{(16)}\}$, $k_p \geq \cdots \geq k_1 \geq 2$, $n \in \mathbb{Z}$, $m_s \geq \cdots \geq m_1 \geq 1$, $s, p \geq 0$.

Theorem 3.4. *$V_{L_2}^{A_4}$ is C_2-cofinite and rational.*

4 Characterization of $V_{L_2}^{A_4}$

In this section, we will give a characterization of the rational vertex operator algebra $V_{L_2}^{A_4}$. For this purpose we assume the following:

(A) V is a simple, C_2-cofinite rational CFT type and self-dual vertex operator algebra of central charge 1;
(B) V is a sum of irreducible modules for the Virasoro algebra;
(C) There is a primary vector of weight 9 and the weight of any primary vector whose weight is greater than 0 is greater than or equal to 9.

Obviously, $V_{L_2}^{A_4}$ satisfies (A)–(C). By Theorems 2.6 and 2.9, if a vertex operator algebra V satisfies (A)–(C), then V and $V_{L_2}^{A_4}$ have the same trace function. So there is only one linearly independent primary vector of weight 9 in V.

For short, let $V^1 = V_{L_2}^{A_4}$ and V^2 be an arbitrary vertex operator algebra satisfying (A)–(C). We will prove that V^1 and V^2 are isomorphic vertex operator algebras. Since V^1 and V^2 have the same q-character, it follows from the assumption that V^1 and V^2 are isomorphic modules for the Virasoro algebra. Let x^i be a non-zero weight 9 primary vectors in V^i, $i = 1, 2$ such that

$$(x^1, x^1) = (x^2, x^2). \tag{6}$$

By [12], there is only one linearly independent primary elements of weight 16 in V^i, $i = 1, 2$. Now let $y^i \in V^i_{16}$ be linearly independent primary vectors in V^i, $i = 1, 2$ such that

$$(y^1, y^1) = (y^2, y^2). \quad i = 1, 2. \tag{7}$$

The following lemma comes from [12].

Lemma 4.1. *There is no non-zero primary vector of weight 25 in both V^1 and V^2.*

Let $V^{(i,9)}$ be the $L(1,0)$-submodule of V^i generated by x^i and $V^{(i,16)}$ the $L(1,0)$-submodule of V^i generated by y^i, $i = 1, 2$. We may identify the Virasoro vertex operator subalgebra $L(1,0)$ both in V^1 and V^2. Let

$$\phi : L(1,0) \oplus V^{(1,9)} \oplus V^{(1,16)} \to L(1,0) \oplus V^{(2,9)} \oplus V^{(2,16)}$$

be an $L(1,0)$-module isomorphism such that

$$\phi\omega = \omega, \ \phi x^1 = x^2, \ \phi y^1 = y^2.$$

Then

$$(u, v) = (\phi u, \phi v),$$

for $u, v \in L(1,0) \oplus V^{(1,9)} \oplus V^{(1,16)}$.

Let

$$\mathcal{I}^0(u, z)v = \mathcal{P}^0 \circ Y(u, z)v$$

for $u, v \in V^{(1,9)}$ be the intertwining operator of type

$$\begin{pmatrix} L(1,0) \\ V^{(1,9)} \ V^{(1,9)} \end{pmatrix},$$

and $\mathcal{I}^0(\phi u, z)\phi v = \mathcal{Q}^0 \circ Y(\phi u, z)\phi v$ for $u, v \in V^1$ be the intertwining operator of type

$$\begin{pmatrix} L(1,0) \\ V^{(2,9)} \ V^{(2,9)} \end{pmatrix},$$

where $\mathcal{P}^0, \mathcal{Q}^0$ are the projections of V^1 and V^2 to $L(1,0)$ respectively. By (6), we have

$$\mathcal{I}^0(u, z)v = \mathcal{I}^0(\phi u, z)\phi v, \tag{8}$$

for $u, v \in V^{(1,9)}$.

Similarly, let

$$\mathcal{I}^1(u, z)v = \mathcal{P}^1 \circ Y(u, z)v$$

for $u, v \in V^{(1,16)}$ be the intertwining operator of type

$$\begin{pmatrix} L(1,0) \\ V^{(1,16)}\ V^{(1,16)} \end{pmatrix},$$

and $\mathcal{I}^1(\phi u, z)\phi v = \mathcal{Q}^1 \circ Y(\phi u, z)\phi v$ for $u, v \in V^1$ be the intertwining operator of type

$$\begin{pmatrix} L(1,0) \\ V^{(2,16)}\ V^{(2,16)} \end{pmatrix},$$

where $\mathcal{P}^1$, $\mathcal{Q}^1$ are the projections of V^1 and V^2 to $L(1,0)$ respectively. By (7), we have

$$\mathcal{I}^1(u, z)v = \mathcal{I}^1(\phi u, z)\phi v, \tag{9}$$

for $u, v \in V^{(1,16)}$.

Let

$$\mathcal{I}^2(u, z)v = \mathcal{P}^2 \circ Y(u, z)v$$

for $u \in V^{(1,16)}, v \in V^{(1,9)}$ be the intertwining operator of type

$$\begin{pmatrix} V^{(1,9)} \\ V^{(1,16)}\ V^{(1,9)} \end{pmatrix},$$

and $\mathcal{I}^2(\phi u, z)\phi v = \mathcal{Q}^2 \circ Y(\phi u, z)\phi v$ for $u \in V^{(1,16)}, v \in V^{(1,9)}$ be the intertwining operator of type

$$\begin{pmatrix} V^{(2,9)} \\ V^{(2,16)}\ V^{(2,9)} \end{pmatrix},$$

where $\mathcal{P}^2$, $\mathcal{Q}^2$ are the projections of V^1 and V^2 to $V^{(1,9)}$ and $V^{(2,9)}$ respectively. Then we have the following lemma.

Lemma 4.2. *Replacing y^2 by $-y^2$ if necessary, we have*

$$\phi(\mathcal{I}^2(u, z)v) = \mathcal{I}^2(\phi u, z)\phi v,$$

for $u \in V^{(16)}, v \in V^{(1,9)}$.

Proof. Since $V^{(1,9)} \cong V^{(2,9)} \cong L(1,9)$, $V^{(1,16)} \cong V^{(2,16)} \cong L(1,16)$, we may identify $V^{(1,9)}$ with $V^{(2,9)}$ and $V^{(1,16)}$ with $V^{(2,16)}$ through ϕ. So both $\mathcal{I}^1(u,z)v$ and $\mathcal{I}^1(\phi u,z)\phi v$ for $u \in V^{(1,16)}$, $v \in V^{(9)}$ are intertwining operators of type

$$\binom{L(1,9)}{L(1,16)\; L(1,9)}.$$

Therefore

$$\phi(\mathcal{I}^2(u,z)v) = \epsilon \mathcal{I}^2(\phi u,z)\phi v, \tag{10}$$

for some $\epsilon \in \mathbb{C}$. By Theorem 2.11 and (6), we have

$$(y^1_{31}y^1)_{-1}x^1 = (y^1,y^1)x^1,\ (y^2_{31}y^2)_{-1}x^2 = (y^2,y^2)x^2.$$

So

$$\phi(y^1_{31}y^1)_{-1}x^1) = (y^2_{31}y^2)_{-1}x^2.$$

On the other hand, we have

$$(y^i_{31}y^i)_{-1}x^i = \sum_{k=0}^{\infty}\binom{31}{k}(-1)^k(y^i_{31-k}y^i_{-1+k} + y^i_{30-k}y^i_k)x^i,\ i = 1,2.$$

Then by Theorem 2.11, Lemma 4.1 and (10),

$$(y^2_{31}y^2)_{-1}x^2 = \epsilon^2((y^2_{31}y^2)_{-1}x^2).$$

So we have $\epsilon^2 = 1$. If $\epsilon = 1$, then the lemma holds. If $\epsilon = -1$, replacing y^2 by $-y^2$, then we get the lemma. □

Let

$$\mathcal{I}^3(u,z)v = \mathcal{P}^3 \circ Y(u,z)v$$

for $u, v \in V^{(1,9)}$ be the intertwining operator of type

$$\binom{V^{(1,16)}}{V^{(1,9)}\; V^{(1,9)}},$$

and $\mathcal{I}^3(\phi u,z)\phi v = \mathcal{Q}^3 \circ Y(\phi u,z)\phi v$ for $u, v \in V^{(1,9)}$ be the intertwining operator of type

$$\binom{V^{(2,16)}}{V^{(2,9)}\; V^{(2,9)}},$$

where $\mathcal{P}^3$, $\mathcal{Q}^3$ are the projections of V^1 and V^2 to $V^{(1,16)}$ and $V^{(2,16)}$ respectively. Then we have the following lemma.

Lemma 4.3.

$$\phi(\mathcal{I}^3(u,z)v) = \mathcal{I}^3(\phi u, z)\phi v,$$

for $u, v \in V^{(1,9)}$.

Proof. Note that both $\mathcal{I}^3(u,z)v$ and $\mathcal{I}^3(\phi u,z)\phi v$ for $u, v \in V^{(1,9)}$ are intertwining operators of type

$$\begin{pmatrix} L(1,16) \\ L(1,9)\; L(1,9) \end{pmatrix}.$$

Therefore

$$\phi(\mathcal{I}^3(u,z)v) = \epsilon \mathcal{I}^3(\phi u, z)\phi v, \tag{11}$$

for some $\epsilon \in \mathbb{C}$. By Theorem 2.11 and (8), we have

$$x_1^1 x^1 = u + a_1 y^1,\; x_1^2 x^2 = u + a_2 y^2 \tag{12}$$

where $u \in L(1,0)$ and $a_1, a_2 \in \mathbb{C}$. By (11),

$$a_1 = \epsilon a_2. \tag{13}$$

Then by Theorem 2.11, we have

$$x_0^1 x^1 = v + a_1 L(-1) y^1,\; x_0^2 x^2 = v + a_2 L(-1) y^2, \tag{14}$$

for some $v \in L(1,0)$. Notice that

$$(x_1^1 x^1, y^1) = a_1(y^1, y^1),\; (x_1^2 x^2, y^2) = a_2(y^2, y^2).$$

By (10)–(14),

$$(x_1^1 x^1, y^1) = \epsilon (x_1^2 x^2, y^2). \tag{15}$$

On the other hand, we have

$$(x_1^1 x^1, y^1) = -(x^1, x_{15}^1 y^1) = -(x^1, y_{15}^1 x^1),$$
$$(x_1^2 x^2, y^2) = -(x^2, x_{15}^2 y^2) = -(x^2, y_{15}^2 x^2).$$

By Lemma 4.3,

$$\phi(y^1_{15}x^1) = y^2_{15}x^2.$$

So

$$(x^1_1x^1, y^1) = (x^2_1x^2, y^2).$$

This together with (15) deduces that $\epsilon = 1$. □

Let

$$\mathcal{I}^4(u,z)v = \mathcal{P}^4 \circ Y(u,z)v$$

for $u, v \in V^{(1,16)}$ be the intertwining operator of type

$$\begin{pmatrix} V^{(1,16)} \\ V^{(1,16)} \; V^{(1,16)} \end{pmatrix},$$

and $\mathcal{I}^4(\phi u, z)\phi v = \mathcal{Q}^4 \circ Y(\phi u, z)\phi v$ for $u, v \in V^{(1,16)}$ be the intertwining operator of type

$$\begin{pmatrix} V^{(2,16)} \\ V^{(2,16)} \; V^{(2,16)} \end{pmatrix},$$

where $\mathcal{P}^4, \mathcal{Q}^4$ are the projections of V^1 and V^2 to $V^{(1,16)}$ and $V^{(2,16)}$ respectively. Then we have the following lemma.

Lemma 4.4.

$$\phi(\mathcal{I}^4(u,z)v) = \mathcal{I}^4(\phi u, z)\phi v,$$

for $u, v \in V^{(1,16)}$.

Proof. It suffices to prove that

$$\phi(y^1_{15}y^1) = y^2_{15}y^2.$$

By Lemma 4.3 and (12), we have

$$\phi(x^1_1x^1) = \phi(u + a_1y^1) = u + a_1y^2, \tag{16}$$

where u and a_1 are as in (12). Notice that

$$(x^i_1x^i)_{15}y^i = \sum_{k=0}^{\infty} \binom{1}{k} (-1)^k (x^i_{1-k}x^i_{15+k} + x^i_{16-k}x^i_k)y^i, \; i = 1, 2.$$

Then by Lemma 4.3, Lemma 4.2, and the skew-symmetry property of vertex operator algebras, we have

$$\phi((x_1^1x^1)_{15}y^1) = (x_1^2x^2)_{15}y^2.$$

This together with (16) deduces that $\phi(y_{15}^1y^1) = y_{15}^2y^2$. The lemma follows. □

Summarizing Lemmas 4.2–4.4, we have the following proposition:

Proposition 4.5. *(1) For any* $u^1, v^1 \in L(1,0) \oplus V^{(1,9)} \oplus V^{(1,16)}$, *we have*

$$\phi(u_n^1v^1) = (\phi u^1)_n(\phi v^1),$$

for $n \in \mathbb{N}$.
(2) For any $u^1, v^1 \in L(1,0) \oplus V^{(1,9)} \oplus V^{(1,16)}$, *we have*

$$(u^1, v^1) = (\phi(u^1), \phi(v^1)).$$

Recall from Theorem 3.3 that V^1 is generated by x^1 and V^1 is linearly spanned by

$$L(-m_s)\cdots L(-m_1)x_n^1x^1,\ L(-m_s)\cdots L(-m_1)u_{-k_p}^p\cdots u_{-k_1}^1v,$$

where x^1, y^1 are the same as above and $v, u^1, \cdots, u^p \in \{x^1, y^1\}, k_p \geq \cdots \geq k_1 \geq 2$, $n \in \mathbb{Z}, m_s \geq \cdots \geq m_1 \geq 1, s, p \geq 0$. Our goal next is to show that V^2 is generated by $\phi(x^1) = x^2$ and has a similar spanning set.

Lemma 4.6. *For any* $k, l \geq 1, s_i, t_i, p_i \geq 0, m_{i1}, \cdots, m_{is_i}, n_{j1}, \cdots, n_{jt_j}, r_{j1}, \cdots, r_{jp_j} \in \mathbb{Z}_+, n_j \in \mathbb{Z}, u^{j1}, \cdots, u^{jp_j}, u^j \in \{x^1, y^1\}, i = 1, 2, \cdots, k, j = 1, 2, \cdots, l,$ *if*

$$u' = \sum_{i=1}^{k} a_i L(-m_{i1})\cdots L(-m_{is_i})x_{n_i}^2x^2$$
$$+ \sum_{j=1}^{l} b_j L(-n_{j1})\cdots L(-n_{jt_j})(\phi u^{j1})_{-r_{j1}}\cdots(\phi u^{jp_j})_{-r_{jp_j}}(\phi u^j) = 0$$

for some $a_i, b_j \in \mathbb{C}$ *then*

$$u = \sum_{i=1}^{k} a_i L(-m_{i1})\cdots L(-m_{is_i})x_{n_i}^1x^1$$
$$+ \sum_{j=1}^{l} b_j L(-n_{j1})\cdots L(-n_{jt_j})u_{-r_{j1}}^{j1}u_{-r_{j2}}^{j2}\cdots u_{-r_{jp_j}}^{jp_j}u^j = 0.$$

Proof. Without loss, we may assume that u is a linear combination of homogeneous elements with same weight. Suppose that $u \neq 0$. Since V^1 is self-dual and generated by x^1, there is $x_{r_1}^1x_{r_2}^1\cdots x_{r_q}^1x^1 \in V^1$ such that

$$(u, x^1_{r_1} x^1_{r_2} \cdots x^1_{r_q} x^1) \neq 0. \tag{17}$$

Claim. For any $v, u^1, \cdots, u^p \in \{x^1, y^1\}$, $k_p \geq \cdots \geq k_1 \geq 2$, $q_1, q_2, \cdots, q_t, n \in \mathbb{Z}$, $m_s \geq \cdots \geq m_1 \geq 1$, $s, p, t \geq 0$,

$$\begin{aligned}&(L(-m_s)\cdots L(-m_1)x^1_n x^1, x^1_{q_1} x^1_{q_2} \cdots x^1_{q_t} x^1)\\ &\quad = (L(-m_s)\cdots L(-m_1)x^2_n x^2, x^2_{q_1} x^2_{q_2} \cdots x^2_{q_t} x^2),\end{aligned} \tag{18}$$

$$\begin{aligned}&(L(-m_s)\cdots L(-m_1)u^p_{-k_p} \cdots u^1_{-k_1} v, x^1_{q_1} x^1_{q_2} \cdots x^1_{q_t} x^1)\\ &\quad = (L(-m_s)\cdots L(-m_1)\phi(u^p)_{-k_p} \cdots \phi(u^1)_{-k_1}\phi(v), x^2_{q_1} x^2_{q_2} \cdots x^2_{q_t} x^2).\end{aligned} \tag{19}$$

We only show (19) as the proof for (18) is similar and simpler. We may assume that

$$\mathrm{wt}(L(-m_s)\cdots L(-m_1)u^p_{-k_p} \cdots u^1_{-k_1} v) = \mathrm{wt}(x^1_{q_1} x^1_{q_2} \cdots x^1_{q_t} x^1).$$

We prove (19) by induction on $\mathrm{wt}(L(-m_s)\cdots L(-m_1)u^p_{-k_p} \cdots u^1_{-k_1} v)$. By Proposition 4.5, (18) holds if $\mathrm{wt}(L(-m_s)\cdots L(-m_1)u^p_{-k_p} \cdots u^1_{-k_1} v) < 36$. If $s \geq 1$, then

$$\begin{aligned}&(L(-m_s)\cdots L(-m_1)u^p_{-k_p} \cdots u^1_{-k_1} v, x^1_{q_1} x^1_{q_2} \cdots x^1_{q_t} x^1)\\ &\quad = (L(-m_{s-1})\cdots L(-m_1)u^p_{-k_p} \cdots u^1_{-k_1} v, L(m_s)x^1_{q_1} x^1_{q_2} \cdots x^1_{q_t} x^1),\\ &(L(-m_s)\cdots L(-m_1)\phi(u^p)_{-k_p} \cdots \phi(u^1)_{-k_1}\phi(v), x^2_{q_1} x^2_{q_2} \cdots x^2_{q_t} x^2)\\ &\quad = (L(-m_{s-1})\cdots L(-m_1)\phi(u^p)_{-k_p} \cdots \phi(u^1)_{-k_1}\phi(v), L(m_s)x^2_{q_1} x^2_{q_2} \cdots x^2_{q_t} x^2).\end{aligned}$$

So by inductive assumption, (19) holds.

If $s = 0$, then

$$\begin{aligned}&(u^p_{-k_p} \cdots u^1_{-k_1} v, x^1_{q_1} x^1_{q_2} \cdots x^1_{q_t} x^1)\\ &\quad = (u^{p-1}_{-k_{p-1}} \cdots u^1_{-k_1} v, u^p_{2\mathrm{wt}(u^p)+k_p-2} x^1_{q_1} x^1_{q_2} \cdots x^1_{q_t} x^1),\\ &(\phi(u^p)_{-k_p} \cdots \phi(u^1)_{-k_1}\phi(v), x^2_{q_1} \cdots x^2_{q_t} x^2)\\ &\quad = (\phi(u^{p-1})_{-k_{p-1}} \cdots \phi(u^1)_{-k_1}\phi(v), \phi(u^p)_{2\mathrm{wt}(u^p)+k_p-2} x^2_{q_1} \cdots x^2_{q_t} x^2).\end{aligned}$$

Since $k_p \geq 2$, by inductive assumption, (19) holds if $u^p = x^1$. If $u^p = y^1$ by (12) $u^p_{2\mathrm{wt}(u^p)+k_p-2}$ is a sum of operators of forms $aL(n_1)\cdots L(n_\mu), bx^1_i x^1_j$ of the same weight where $n_1 \leq \cdots \leq n_\mu$ and all n_t are nonzero. By induction assumption we know that

$$\begin{aligned}&(u^{p-1}_{-k_{p-1}}\cdots u^1_{-k_1}v, x^1_i x^1_j x^1_{q_1}x^1_{q_2}\cdots x^1_{q_t}x^1)\\&=(\phi(u^{p-1})_{-k_{p-1}}\cdots\phi(u^1)_{-k_1}\phi(v), x^2_i x^2_j x^2_{q_1}\cdots x^2_{q_t}x^2).\end{aligned}$$

Also by Proposition 4.5, relation (16), the fact that x^i are highest weight vectors for the Virasoro algebra with the same weight, and the invariant properties of the bilinear forms,

$$\begin{aligned}&(u^{p-1}_{-k_{p-1}}\cdots u^1_{-k_1}v, L(n_1)\cdots L(n_\mu)x^1_{q_1}x^1_{q_2}\cdots x^1_{q_t}x^1)\\&=(\phi(u^{p-1})_{-k_{p-1}}\cdots\phi(u^1)_{-k_1}\phi(v), L(n_1)\cdots L(n_\mu)x^2_{q_1}\cdots x^2_{q_t}x^2).\end{aligned}$$

So the claim is proved.

By the claim and (17), we have

$$(u', x^1_{r_1}x^2_{r_2}\cdots x^2_{r_q}x^2)\neq 0,$$

which contradicts the assumption that $u'=0$. □

Let U^2 be the subalgebra of V^2 generated by x^2 and y^2. By Theorem 3.3 and Lemma 4.6, for every $n\geq 0$, $\dim V^1_n\leq\dim U^2_n$. Since V^1 and V^2 have the same graded dimensions, it follows that $\dim V^1_n=\dim V^2_n$ for $n\geq 0$. So $\dim V^2_n=\dim U^2_n$ for $n\geq 0$ and $V^2=U^2$. So we have the following corollary which is essentially the V^2 version of Theorem 3.3.

Corollary 4.7. *V^2 is linearly spanned by*

$$L(-m_s)\cdots L(-m_1)x^2_n x^2,\ L(-m_s)\cdots L(-m_1)v^p_{-k_p}\cdots v^1_{-k_1}v,$$

where x^2, y^2 are the same as above and $v, v^1,\cdots,v^p\in\{x^2,y^2\}$, $k_p\geq\cdots\geq k_1\geq 2$, $n\in\mathbb{Z}$, $m_s\geq\cdots\geq m_1\geq 1$, $s,p\geq 0$.

Define $\psi(x^2)=x^1, \psi(y^2)=y^1$, and extend ψ to $\psi: V^2\to V^1$ by

$$\psi(L(-m_s)\cdots L(-m_1)x^2_n x^2)=L(-m_s)\cdots L(-m_1)x^1_n x^1$$

and

$$L(-m_1)v^p_{-k_p}\cdots v^1_{-k_1}v=L(-m_1)\psi(v^p)_{-k_p}\cdots\psi(v^1)_{-k_1}\psi(v),$$

where $v, v^1,\cdots,v^p\in\{x^2,y^2\}$, $k_p\geq\cdots\geq k_1\geq 2$, $n\in\mathbb{Z}$, $m_s\geq\cdots\geq m_1\geq 1$, $s,p\geq 0$. Then by the discussion above, ψ is a linear isomorphism from V^2 to V^1. It follows that ϕ can be extended to a linear isomorphism from V^1 to V^2 such that

$$\phi(L(-m_s)\cdots L(-m_1)x^1_n x^1)=L(-m_s)\cdots L(-m_1)x^2_n x^2$$

and

$$L(-m_1)u^p_{-k_p}\cdots u^1_{-k_1}u = L(-m_1)\phi(p^p)_{-k_p}\cdots\phi(u^1)_{-k_1}\phi(u),$$

where $u, u^1, \cdots, u^p \in \{x^1, y^1\}$, $k_p \geq \cdots \geq k_1 \geq 2$, $n \in \mathbb{Z}$, $m_s \geq \cdots \geq m_1 \geq 1$, $s, p \geq 0$.

We are now in a position to state our main result of this paper.

Theorem 4.8. *If a vertex operator algebra* V *satisfies the conditions (A)–(C), then* V *is isomorphic to* $V_{L_2}^{A_4}$.

Proof. Recall that $V^1 \cong V_{L_2}^{A_4}$ satisfying (A)–(C). So it suffices to show that ϕ is a vertex operator algebra automorphism from V^1 to V^2. Let $u = x^1_{m_1}x^1_{m_2}\cdots x^1_{m_s}x^1$, $v = x^1_{q_1}x^1_{q_2}\cdots x^1_{q_t}x^1 \in V^1$, where $m_i, q_j \in \mathbb{Z}, i = 1, 2, \cdots, p_s, j = 1, 2, \cdots, t$. We need to show that for any $n \in \mathbb{Z}$, $\phi(u^1_n u^2) = \phi(u^1)_n\phi(u^2)$. Note from Theorem 2.11 that for $m_1, m_2 \in \mathbb{Z}_+$, $x^1_{m_1}x^1 \in L(1,0) \oplus V^{(1,16)}$, $y^1_{m_2}x^1 \in V^{(1,9)}$. Since for any $p, q \in \mathbb{Z}$,

$$x^i_q x^i_p = x^i_p x^i_q + \sum_{j=0}^{\infty}\binom{q}{j}(x^i_j x^i)p + q - j,\ i = 1, 2,$$

$$y^i_q x^i_p = y^i_p x^i_q + \sum_{j=0}^{\infty}\binom{q}{j}(y^i_j x^i)p + q - j,$$

Then by Lemma 4.5, it is easy to see that for any fixed $n \in \mathbb{Z}$,

$$u^1_n u^2 = \sum_{i=1}^{k} a_i L(-m_{i1})\cdots L(-m_{is_i})x^1_{n_i}x^1$$
$$+\sum_{i=1}^{l} b_i L(-n_{i1})\cdots L(-n_{it_i})u^{i1}_{-r_{i1}}u^{i2}_{-r_{i2}}\cdots u^{ip_i}_{-r_{ip_i}}u^i,$$

for some $k, l \geq 1, s_i, t_i, p_i \geq 0, m_{i1}, \cdots, m_{is_i}, n_{j1}, \cdots, n_{jt_j}, r_{j1}, \cdots, r_{jp_j} \in \mathbb{Z}_+$, $n_i \in \mathbb{Z}$, $u^{i1}, \cdots, u^{ip_i}, u^i \in \{x^1, y^1\}$, $i = 1, 2, \cdots, k$, $j = 1, 2, \cdots, l$, then

$$\phi(u^1)_n\phi(u^2) = \sum_{i=1}^{k} a_i L(-m_{i1})\cdots L(-m_{is_i})x^2_{n_i}x^2$$
$$+\sum_{i=1}^{l} b_i L(-n_{i1})\cdots L(-n_{it_i})(\phi u^{i1})_{-r_{i1}}\cdots(\phi u^{ip_i})_{-r_{ip_i}}(\phi u^i).$$

The proof is complete. □

Acknowledgements The first author acknowledges the support from NSF grants and a Faculty research grant from the University of California at Santa Cruz. The second author acknowledges the support from China NSF grants (10931006 and 11371245), the RFDP grants of China (20100073110052), and the Innovation Program of Shanghai Municipal Education Commission (11ZZ18).

References

1. Abe, T.: Fusion rules for the free bosonic orbifold vertex operator algebra. J. Algebra **229**, 333–374 (2000)
2. Abe, T.: Rationality of the vertex operator algebra V_L^+ for a positive definite even lattice L. Math. Z. **249**, 455–484 (2005)
3. Abe, T., Dong, C.: Classification of irreducible modules for the vertex operator algebra V_L^+: general case. J. Algebra **273**, 657–685 (2004)
4. Abe, T., Dong, C., Li, H.: Fusion rules for the vertex operator $M(1)^+$ and V_L^+. Commun. Math. Phys. **253**, 171–219 (2005)
5. Borchards, R.: Vertex algebras, Kac-Moody algebras, and the Monster. Proc. Natl. Acad. Sci. USA **83**, 3068–3071 (1986)
6. Dong, C.: Vertex algebras associated with even lattices. J. Algebra **160**, 245–265 (1993)
7. Dong, C.: Twisted modules for vertex algebras associated with even lattices. J. Algebra **165**, 91–112 (1994)
8. Dong, C., Griess, R.L.: Rank one lattice type vertex operator algebras and their automorphism groups. J. Algebra **208**, 262–275 (1998)
9. Dong, C., Jiang, C.: A characterization of vertex operator algebra $L(\frac{1}{2},0)\otimes L(\frac{1}{2},0)$. Commun. Math. Phys. **296**, 69–88 (2010)
10. Dong, C., Jiang, C.: A characterization of vertex operator algebra $V_{\mathbb{Z}\alpha}^+$: I. J. Reine Angew. Math. (to appear). arXiv:1110.1882
11. Dong, C., Jiang, C.: A characterization of vertex operator algebra $V_{\mathbb{Z}\alpha}^+$: II. Adv. Math. **247**, 41–70 (2013)
12. Dong, C., Jiang, C.: Representations of the vertex operator algebra $V_{L_2}^{A_4}$. J. Algebra **377**, 76–96 (2013)
13. Dong, C., Mason, G.: Rational vertex operator algebras and the effective central charge. Int. Math. Res. Not. **56**, 2989–3008 (2004)
14. Dong, C., Nagatomo, K.: Classification of irreducible modules for the vertex operator algebra $M(1)^+$. J. Algebra **216**, 384–404 (1999)
15. Dong, C., Nagatomo, K.: Representations of vertex operator algebra V_L^+ for rank one lattice L. Commun. Math. Phys. **202**, 169–195 (1999)
16. Dong, C., Nagatomo, K.: Classification of irreducible modules for the vertex operator algebra $M(1)^+$: higher rank. J. Algebra **240**, 289–325 (2001)
17. Dong, C., Li, H., Mason, G.: Twisted representations of vertex operator algebras. Math. Ann. **310**, 571–600 (1998)
18. Dong, C., Li, H., Mason, G.: Modular invariance of trace functions in orbifold theory and generalized moonshine. Commun. Math. Phys. **214**, 1–56 (2000)
19. Dong, C., Jiang, C., Jiang, Q., Jiao, X., Yu, N.: Fusion rules for the vertex operator algebra $V_{L_2}^{A_4}$. (2013) arXiv:1310.7235
20. Dong, C., Jiang, C., Lin, X.: Rationality of vertex operator algebra V_L^+ : higher rank. Proc. Lond. Math. Soc. **104**, 799–826 (2012)
21. Dong, C., Lin, X., Ng, S.: Congruence property in conformal field theory. (2012) arXiv: 1201.6644v4
22. Frenkel, I.B., Zhu, Y.: Vertex operator algebras associated to representations of affine and Virasoro algebras. Duke Math. J. **66**, 123–168 (1992)
23. Frenkel, I.B., Lepowsky, J., Meurman, A.: Vertex Operator Algebras and the Monster. Pure and Applied Mathematics, vol. 134. Academic, Boston (1988)
24. Frenkel, I.B., Huang, Y., Lepowsky, J.: On Axiomatic Approaches to Vertex Operator Algebras and Modules. Memoirs of the American Mathematical Society, vol. 104. American Mathematical Society, Providence (1993)
25. Kiritsis, E.: Proof of the completeness of the classification of rational conformal field theories with $c = 1$. Phys. Lett. **B217**, 427–430 (1989)

26. Milas, A.: Fusion rings for degenerate minimal models. J. Algebra **254**, 300–335 (2002)
27. Zhang, W., Dong, C.: W-algebra $W(2,2)$ and the vertex operator algebra $L(1/2,0) \otimes L(1/2,0)$. Commun. Math. Phys. **285**, 991–1004 (2009)
28. Zhu, Y.: Modular invariance of characters of vertex operator algebras. J. Am. Math. Soc. **9**, 237–302 (1996)

Extended Griess Algebras and Matsuo-Norton Trace Formulae

Hiroshi Yamauchi

Abstract We introduce the $\mathbb{Z}_2$-extended Griess algebra of a vertex operator superalgebra with an involution and derive the Matsuo-Norton trace formulae for the extended Griess algebra based on conformal design structure. We illustrate an application of our formulae by reformulating the one-to-one correspondence between 2A-elements of the Baby-monster simple group and $N = 1$ $c = 7/10$ Virasoro subalgebras inside the Baby-monster vertex operator superalgebra.

1 Introduction

A mysterious connection is known to exist between vertex operator algebras (VOAs) and finite simple groups. One can explain that the j-invariant is made of the characters of the Monster simple group as a consequence of the modular invariance of characters of vertex operator algebras [5, 31]. Matsuo [22] introduced the notion *of class* $\mathscr{S}^n$ of a VOA and derived the formulae, which we will call the *Matsuo-Norton trace formulae*, describing trace of adjoint actions of the Griess algebra of a vertex operator algebra. A VOA V is called of class $\mathscr{S}^n$ if the invariant subalgebra of its automorphism group coincides with the subalgebra generated by the conformal vector up to degree n subspace. In the derivation of the formulae the non-associativity of products of vertex operator algebras are efficiently used, resulting that the Matsuo-Norton trace formulae strongly encode structures of higher subspaces of vertex operator algebras. Suitably applying the formula it outputs some information of structures related to automorphisms. Here we exhibit an application of the formulae.

H. Yamauchi (✉)
Department of Mathematics, Tokyo Woman's Christian University, 2-6-1 Zempukuji, Suginami, Tokyo 167-8585, Japan
e-mail: yamauchi@lab.twcu.ac.jp

W. Kohnen and R. Weissauer (eds.), *Conformal Field Theory, Automorphic Forms and Related Topics*, Contributions in Mathematical and Computational Sciences 8,
DOI 10.1007/978-3-662-43831-2_4,

Let V be a VOA of OZ-type with central charge c and e a simple $c = 1/2$ Virasoro vector of V. Then V is a direct sum of irreducible $\langle e \rangle$-modules and the zero-mode $\mathrm{o}(e) = e_{(1)}$ acts semisimply on V. It is shown in [23] that $\tau_e = \exp\left(16\pi\sqrt{-1}\,\mathrm{o}(e)\right)$ defines an element in $\mathrm{Aut}(V)$. The possible $\mathrm{o}(e)$-eigenvalues on the Griess algebra are 2, 0, 1/2 and 1/16. (The eigenspace with eigenvalue 2 is one-dimensional spanned by e.) We denote by d_λ the dimension of $\mathrm{o}(e)$-eigensubspace of V_2 with eigenvalue λ. Then

$$\mathrm{tr}_{V_2}\,\mathrm{o}(e)^i = 2^i + 0^i\, d_0 + \left(\frac{1}{2}\right)^i d_{1/2} + \left(\frac{1}{16}\right)^i d_{1/16}, \tag{1}$$

where we have set $0^0 = 1$. If V is of class $\mathscr{S}^4$ then Matsuo-Norton trace formulae give the values $\mathrm{tr}_{V_2}\,\mathrm{o}(e)^i$ for $i = 0, 1, 2$ and if $c(5c+22) \neq 0$ we can solve (1) and obtain

$$\begin{aligned} d_0 &= \frac{(5c^2 - 100c + 1188)\dim V_2 - 545c^2 - 2006c}{c(5c+22)}, \\ d_{1/2} &= \frac{-2((3c - 110)\dim V_2 + 50c^2 + 192c)}{c(5c+22)}, \\ d_{1/16} &= \frac{64((2c - 22)\dim V_2 + 10c^2 + 37c)}{c(5c+22)}. \end{aligned} \tag{2}$$

Therefore, we get an explicit form of the trace $\mathrm{tr}_{V_2}\,\tau_e = 1 + d_0 + d_{1/2} - d_{1/16}$ as

$$\mathrm{tr}_{V_2}\,\tau_e = \frac{(5c^2 - 234c + 2816)\dim V_2 - 1280c^2 - 4736c}{c(5c+22)}. \tag{3}$$

If we apply this formula to the moonshine VOA $V^\natural$ [5] then we obtain $\mathrm{tr}_{V_2^\natural}\,\mathrm{o}(e) = 4,372$ and hence τ_e belongs to the 2A-conjugacy class of the Monster (cf. [1]). This example seems to suggest an existence of a link between the structure theory of VOAs and the character theory of finite groups acting on VOAs.

Partially motivated by Matsuo's work, Höhn introduced the notion of *conformal designs* based on vertex operator algebras in [11], which would be a counterpart of block designs and spherical designs in the theories of codes and lattices, respectively. The defining condition of conformal designs is already used in [22] to derive the trace formulae. Höhn reformulated it and obtained results analogous to known ones in block and spherical designs in [11]. Contrary to the notion of class $\mathscr{S}^n$, the definition of conformal design does not require the action of automorphism groups of VOAs. Instead, it is formulated by the Virasoro algebra. (The Virasoro algebra is the key symmetry in the two-dimensional conformal field theory.) The conformal design is a purely structure theoretical concept in the VOA theory and seems to measure a structural symmetry of VOAs, as one can deduce the Matsuo-Norton trace formulae via conformal designs.

The purpose of this paper is to extend Matsuo's work on trace formulae to vertex operator superalgebras (SVOAs) with an involution based on conformal design structure. If the even part of the invariant subalgebra of an SVOA by an involution is of OZ-type then one can equip its weight 2 subspace with the structure of a commutative (but usually non-associative) algebra called the Griess algebra. We extend this commutative algebra to a larger one by adding the odd part and call it the *$\mathbb{Z}_2$-extended Griess algebra.* Our $\mathbb{Z}_2$-extended Griess algebra is still commutative but not super-commutative. It is known that Virasoro vectors are nothing but idempotents in a Griess algebra. In the odd part of the extended Griess algebra one can consider square roots of idempotents. We will discuss the structure of the subalgebra generated by a square root of an idempotent in the extended Griess algebra when the top weight of the odd part is small.

Then we derive trace formulae of adjoint actions on the odd part of the extended Griess algebra based on conformal design structure. Our formulae are a variation of Matsuo-Norton trace formulae. As a main result of this paper we apply the trace formulae to the Baby-monster SVOA $VB^\natural$ [8] and reformulate the one-to-one correspondence between the 2A-elements of the Baby-monster simple group and certain $c = 7/10$ Virasoro vectors of $VB^\natural$ obtained in [12] in the supersymmetric setting by considering square roots of idempotents in the extended Griess algebra of $VB^\natural$. This result is in a sense suggestive. It is shown in [22] that if a VOA of OZ-type is of class $\mathscr{S}^8$ and its Griess algebra has dimension $d > 1$ then the central charge is 24 and $d = 196,884$, those of the moonshine VOA. For SVOAs it is shown in [11] that if the odd top level of an SVOA with top weight 3/2 forms a conformal 6-design then its central charge is either 16 or 47/2. Using our formulae we will sharpen this result. We will show if the odd top level of an SVOA with top weight 3/2 forms a conformal 6-design and in addition if it has a proper subalgebra isomorphic to the $N = 1$ $c = 7/10$ Virasoro SVOA then its central charge is 47/2 and the odd top level is of dimension 4,371, those of the Baby-monster SVOA. The author naively expects that the Baby-monster SVOA is the unique example subject to this condition.

The organization of this paper is as follows. In Sect. 2 we review the notion of invariant bilinear forms on SVOAs. Our definition of invariant bilinear forms on SVOAs is natural in the sense that if M is a module over an SVOA V then its restricted dual M^* is also a V-module. We also consider $\mathbb{Z}_2$-conjugation of invariant bilinear forms on SVOAs. In Sect. 3 we introduce the $\mathbb{Z}_2$-extended Griess algebra for SVOAs with involutions and consider square roots of idempotents. When the odd top weight is less than 3 we describe possible structures of subalgebra generated by square roots under mild assumptions. In Sect. 4 we derive trace formulae of adjoint actions on the odd part of the extended Griess algebra based on conformal design structure. A relation between conformal design structure and generalized Casimir vectors is already discussed in [11, 22] but we clarify it in our situation to derive the formulae. In Sect. 5 we apply our formulae to the VOAs with the odd top weight 1 and to the Baby-monster SVOA. In the latter case we reformulate the one-to-one correspondence in [12] in the supersymmetric setting. Final section is the appendix and we list data of the trace formulae.

1.1 Notation and Terminology

In this paper we will work over the complex number field $\mathbb{C}$. We use $\mathbb{N}$ to denote the set of non-negative integers. A VOA V is called *of OZ-type* if it has the $L(0)$-grading $V = \oplus_{n\geq 0} V_n$ such that $V_0 = \mathbb{C}$ and $V_1 = 0$. ("OZ" stands for "Zero-One", introduced by Griess.) We denote the $\mathbb{Z}_2$-grading of an SVOA V by $V = V^0 \oplus V^1$. We allow the case $V^1 = 0$. For $a \in V^i$, we define its parity by $|a| := i \in \mathbb{Z}/2\mathbb{Z}$. We assume that every SVOA in this paper has the $L(0)$-grading $V^0 = \oplus_{n\geq 0} V_n$ and $V^1 = \oplus_{n\geq 0} V_{n+k/2}$ with non-negative $k \in \mathbb{Z}$. It is always assumed that $V_0 = \mathbb{C}$. If $a \in V_n$ then we write $\mathrm{wt}(a) = n$. If $V^1 \neq 0$ then the minimum h such that $V_h^1 \neq 0$ is called *the top weight* and the homogeneous subspace V_h^1 is called *the top level* of V^1. A sub VOA W of V is called *full* if the conformal vector of W is the same as that of V. We denote the Verma module over the Virasoro algebra with central charge c and highest weight h by $M(c,h)$, and $L(c,h)$ denotes its simple quotient. A Virasoro vector e of V with central charge c_e is called *simple* if it generates a simple Virasoro sub VOA isomorphic to $L(c_e, 0)$ in V. Let $(M, Y_M(\cdot, z))$ be a V-module and $g \in \mathrm{Aut}(V)$. The *g-conjugate* $M \circ g$ of M is defined as $(M, Y_M^g(\cdot, z))$ where $Y_M^g(a,z) := Y_M(ga, z)$ for $a \in V$. M is called *g-stable* if $M \circ g \simeq M$, and *G-stable* for $G < \mathrm{Aut}(V)$ if M is g-stable for all $g \in G$. We write $Y(a,z) = \sum_{n\in\mathbb{Z}} a_{(n)} z^{-n-1}$ for $a \in V$ and define its *zero-mode* by $\mathrm{o}(a) := a_{(\mathrm{wt}(a)-1)}$ if a is homogeneous and extend linearly. The supercommutator is denoted by $[\cdot,\cdot]_+$.

2 Invariant Bilinear Form

In this section, we denote the complex number $e^{r\pi\sqrt{-1}}$ by ζ^r for a rational number r. In particular, $\zeta^n = (-1)^n$ for an integer n. We denote the space of V-intertwining operators of type $M^1 \times M^2 \to M^3$ by $\binom{M^3}{M^1\ M^2}_V$.

2.1 Self-Dual Module

Let V be a simple self-dual VOA and M an irreducible self-dual V-module. We assume that M has the $L(0)$-decomposition $M = \oplus_{n=0}^{\infty} M_{n+h}$ such that each homogeneous component is finite dimensional and the top weight h of M is either in $\mathbb{Z}$ or in $\mathbb{Z} + 1/2$. Let $(\cdot \mid \cdot)_V$ and $(\cdot \mid \cdot)_M$ be invariant bilinear forms on V and M, respectively. One can define V-intertwining operators $I(\cdot, z)$ and $J(\cdot, z)$ of types $M \times V \to M$ and $M \times M \to V$, respectively, as follows (cf. Theorem 5.5.1 of [6]). For $a \in V$ and $u, v \in M$,

$$\begin{aligned} I(u,z)a &:= e^{zL(-1)} Y_M(a, -z)u, \\ (J(u,z)v \mid a)_V &:= (u \mid I(e^{zL(1)}(\zeta z^{-2})^{L(0)} u, z^{-1})a)_M. \end{aligned} \tag{4}$$

(Note that $I(u,z)a \in M(\!(z)\!)$ and $J(u,z)v \in V(\!(z)\!)$.) Since $\dim \binom{M}{M\ V}_V = \dim \binom{V}{M\ M}_V = 1$, it follows from Proposition 2.8 of [18] and Proposition 5.4.7 of [6] that there exist $\alpha, \beta \in \{\pm 1\}$ such that

$$(u \mid v)_M = \alpha (v \mid u)_M, \quad e^{zL(-1)} J(v,-z)u = \beta J(u,z)v \quad \text{for } u, v \in M. \tag{5}$$

We can sharpen Proposition 5.6.1 of [6] as follows.

Lemma 1. *Let $\alpha, \beta \in \{\pm 1\}$ be defined as above. Then $\alpha\beta = (-1)^{2h}$.*

Proof. Let $a \in V$ and $u, v \in M$.

$$\begin{aligned}
(J(u,z)v \mid a)_V &= \left(v \mid I(e^{zL(1)}(\zeta z^{-2})^{L(0)}u, z^{-1})a\right)_M \\
&= \left(v \,\middle|\, e^{z^{-1}L(-1)} Y_M(a,-z^{-1}) e^{zL(1)}(\zeta z^{-2})^{L(0)}u\right)_M \\
&= \left(e^{z^{-1}L(1)}v \,\middle|\, Y_M(a,-z^{-1}) e^{zL(1)}(\zeta z^{-2})^{L(0)}u\right)_M \\
&= \left(Y_M(e^{-z^{-1}L(1)}(-z^2)^{L(0)}a, -z) e^{z^{-1}L(1)}v \,\middle|\, e^{zL(1)}(\zeta z^{-2})^{L(0)}u\right)_M \\
&= \left(e^{-zL(-1)} I(e^{z^{-1}L(1)}v, z) e^{-z^{-1}L(1)}(-z^2)^{L(0)}a \,\middle|\, e^{zL(1)}(\zeta z^{-2})^{L(0)}u\right)_M \\
&= \left(I(e^{z^{-1}L(1)}v, z) e^{-z^{-1}L(1)}(-z^2)^{L(0)}a \,\middle|\, (\zeta z^{-2})^{L(0)}u\right)_M \\
&= \alpha \left((\zeta z^{-2})^{L(0)}u \,\middle|\, I(e^{z^{-1}L(1)}v, z) e^{-z^{-1}L(1)}(-z^2)^{L(0)}a\right)_M .
\end{aligned}$$

By the definition of the invariance, one has

$$(v \mid I(u,z)a)_M = \left(J(e^{zL(1)}(\zeta^{-1}z^{-2})^{L(0)}u, z^{-1})v \mid a\right)_V .$$

Then we continue

$$\begin{aligned}
&\alpha \left((\zeta z^{-2})^{L(0)}u \,\middle|\, I(e^{z^{-1}L(1)}v, z) e^{-z^{-1}L(1)}(-z^2)^{L(0)}a\right)_M \\
&= \alpha \Big(J(e^{zL(1)} \underbrace{(\zeta^{-1}z^{-2})^{L(0)} e^{z^{-1}L(1)}}_{= e^{-zL(1)}(\zeta^{-1}z^{-2})^{L(0)}} v, z^{-1})(\zeta z^{-2})^{L(0)}u \,\Big|\, e^{-z^{-1}L(1)}(-z^2)^{L(0)}a\Big)_V \\
&= \alpha \left(J((\zeta^{-1}z^{-2})^{L(0)}v, z^{-1})(\zeta z^{-2})^{L(0)}u \,\middle|\, e^{-z^{-1}L(1)}(-z^2)^{L(0)}a\right)_V \\
&= \alpha \left(e^{-z^{-1}L(-1)} J((\zeta^{-1}z^{-2})^{L(0)}v, z^{-1})(\zeta z^{-2})^{L(0)}u \,\middle|\, (-z^2)^{L(0)}a\right)_V \\
&= \alpha\beta \left(J((\zeta z^{-2})^{L(0)}u, -z^{-1})(\zeta^{-1}z^{-2})^{L(0)}v \,\middle|\, (-z^2)^{L(0)}a\right)_V
\end{aligned}$$

$$\begin{aligned}
&= \alpha\beta \underbrace{\zeta^{\mathrm{wt}(u)-\mathrm{wt}(v)}}_{=(-1)^{\mathrm{wt}(u)-\mathrm{wt}(v)}} \left(J(z^{-2L(0)}u, -z^{-1}) z^{-2L(0)}v \mid (-z^2)^{L(0)}a\right)_V \\
&= \alpha\beta(-1)^{\mathrm{wt}(u)-\mathrm{wt}(v)} \left((-z^2)^{L(0)} J(z^{-2L(0)}u, -z^{-1}) z^{-2L(0)}v \mid a\right)_V .
\end{aligned}$$

Since $(-z^2)^{L(0)}J(u,-z^{-1})v = (-1)^{\mathrm{wt}(u)+\mathrm{wt}(v)}J(z^{2L(0)}u,z)z^{2L(0)}v$, we further continue

$$\begin{aligned}&\alpha\beta\zeta^{\mathrm{wt}(u)-\mathrm{wt}(v)}((-z^2)^{L(0)}J(z^{-2L(0)}u,-z^{-1})z^{-2L(0)}v \mid a)_V\\&=\alpha\beta(-1)^{\mathrm{wt}(u)-\mathrm{wt}(v)}\cdot(-1)^{\mathrm{wt}(u)+\mathrm{wt}(v)}(J(u,z)v \mid a)_V\\&=\alpha\beta(-1)^{2\,\mathrm{wt}(u)}(J(u,z)v \mid a)_V\\&=\alpha\beta(-1)^{2h}(J(u,z)v \mid a)_V.\end{aligned}$$

Therefore, we obtain the desired equation $\alpha\beta = (-1)^{2h}$. ∎

2.2 *Invariant Bilinear Forms on SVOA*

Let $V = V^0 \oplus V^1$ be an SVOA. Let M be an untwisted V-module, that is, M has a $\mathbb{Z}_2$-grading $M = M^0 \oplus M^1$ compatible with that of V. We also assume that M^i $(i = 0,1)$ has an $L(0)$-decomposition $M^i = \oplus_{n\geq 0}M^i_{h+n+i/2}$ where h is the top weight of M and each $L(0)$-eigensubspace is finite dimensional. Let M^* be its restricted dual. We can define a vertex operator map $Y_{M^*}(\cdot,z)$ on M^* by means of the adjoint action $\langle Y_{M^*}(a,z)\nu \mid v\rangle = \langle \nu \mid Y^*_M(a,z)v\rangle$ for $a \in V$, $\nu \in M^*$ and $v \in M$, where

$$Y^*_M(a,z) := Y_M(e^{zL(1)}z^{-2L(0)}\zeta^{L(0)+2L(0)^2}a,z^{-1}). \tag{6}$$

Then one can show the structure $(M^*, Y^*_M(\cdot,z))$ forms a V-module (cf. [6]).

Lemma 2 **$(M^*, Y_{M^*}(\cdot,z))$.** *is a V-module.*

Remark 1. The correction term $\zeta^{2L(0)^2}$ in the definition of the adjoint operator $Y^*_M(\cdot,z)$ is necessary for M^* to be a V-module. It follows from our assumption on the $L(0)$-grading that $\zeta^{L(0)+2L(0)^2}a = \pm a$ if $a \in V$ is $\mathbb{Z}_2$-homogeneous. Therefore, $\zeta^{-L(0)-2L(0)^2}a = \zeta^{L(0)+2L(0)^2}a$ and we have $Y^{**}_M(a,z) = Y_M(a,z)$.

If M^* is isomorphic to M as a V-module, then M is called *self-dual* (as a V-module) and there exists an invariant bilinear form $(\cdot \mid \cdot)_M$ on M satisfying

$$(Y(a,z)u \mid v)_M = \left(u \;\middle|\; Y(e^{zL(1)}z^{-2L(0)}\zeta^{L(0)+2L(0)^2}a,z^{-1})v\right)_M \tag{7}$$

for $a \in V$ and $u,v \in M$.

The following is an easy generalization of known results (cf. Lemma 1).

Proposition 1 ([6, 18, 25]). *Let V be an SVOA of CFT-type.*

(1) Any invariant bilinear form on V is symmetric.
(2) The space of invariant bilinear forms on V is linearly isomorphic to the dual space of $V_0/L(1)V_1$, i.e., $\mathrm{Hom}_{\mathbb{C}}(V_0/L(1)V_1,\mathbb{C})$. *In particular, if V^1 is*

irreducible over V^0 then V is a self-dual V-module if and only if V^0 is a self-dual V^0-module.

2.3 Conjugation of Bilinear Forms

Let $\theta = (-1)^{2L(0)}$ be the canonical $\mathbb{Z}_2$-symmetry of a superalgebra, i.e., $\theta = 1$ on V^0 and $\theta = -1$ on V^1, then we have

$$Y_M^*(\theta a, z) = (-1)^{2\,\mathrm{wt}(a)} Y_M^*(a, z) = Y_M(e^{zL(1)} z^{-2L(0)} \zeta^{L(0)-2L(0)^2} a, z^{-1}). \qquad (8)$$

Consider the θ-conjugate $M \circ \theta = (M, Y_M^\theta(\cdot, z))$ of M. If $M = M^0 \oplus M^1$ is also a superspace then M is always θ-stable, for, one can define an isomorphism $\tilde{\theta}$ between M and $M \circ \theta$ by $\tilde{\theta} = 1$ on M^0 and $\tilde{\theta} = -1$ on M^1. Therefore, if M is self-dual then M, M^* and $(M \circ \theta)^* \simeq M^* \circ \theta$ are all isomorphic. This means there is a freedom of choice of the adjoint operator in the case of SVOA. We can choose the right hand side of (8) as well as (6) for the definition of $Y^*(\cdot, z)$. From now on we will freely choose one of (6) or (8) for the adjoint operator.

3 Extended Griess Algebras

In this section we introduce a notion of $\mathbb{Z}_2$-extended Griess algebras.

3.1 Definition

We will consider SVOAs subject to the following condition.

Condition 1. Let $V = V^0 \oplus V^1$ be a vertex operator superalgebra of CFT-type and g an involution of V. Denote $V^\pm := \{a \in V \mid ga = \pm a\}$. We assume the following.

(1) V is self-dual.
(2) V has the $L(0)$-grading $V^0 = \oplus_{n\ge 0} V_n$ and $V^1 = \oplus_{n \ge 0} V_{n+k/2}$ with non-negative $k \in \mathbb{Z}$ (if V^1 is non-zero).
(3) $V^\pm$ has the $L(0)$-decomposition such that $V^+ = V_0 \oplus V_2^+ \oplus \left(\oplus_{n>2} V_n^+\right)$ and $V^- = V_h \oplus \left(\oplus_{n>h} V_n^-\right)$ where $V_2^+ = V_2 \cap V^+$, $V_h \ne 0$ is the top level of V^- and $h \in \frac{1}{2}\mathbb{Z}$ is its top weight.

Note that $V_h \subset V^0$ if $h \in \mathbb{Z}$ and $V_h \subset V^1$ if $h \in \mathbb{Z} + 1/2$. We will denote $V^{0,+} := V^0 \cap V^+$. By assumption, $V^{0,+}$ is of OZ-type and V_2^+ is the Griess algebra of $V^{0,+}$.

If $V^1 \neq 0$ and $h \in \mathbb{Z} + 1/2$ we choose

$$Y^*(a,z) = \begin{cases} Y(e^{zL(1)} z^{-2L(0)} \zeta^{L(0)-2L(0)^2} a, z^{-1}) & \text{if } h \equiv 1/2 \mod 2, \\ Y(e^{zL(1)} z^{-2L(0)} \zeta^{L(0)+2L(0)^2} a, z^{-1}) & \text{if } h \equiv 3/2 \mod 2, \end{cases} \tag{9}$$

so that if we write $Y^*(a,z) = \sum_{n\in\mathbb{Z}} a^*_{(n)} z^{-n-1}$ then

$$a^*_{(n)} = \epsilon_h (-1)^{\mathrm{wt}(a)-h} \sum_{i=0}^{\infty} \frac{1}{i!} \left(L(1)^i a\right)_{(2\,\mathrm{wt}(a)-n-2-i)}, \tag{10}$$

where the signature $\epsilon_h \in \{\pm 1\}$ is defined by

$$\epsilon_h = \begin{cases} (-1)^h & \text{if } h \in \mathbb{Z}, \\ 1 & \text{if } h \in \mathbb{Z} + 1/2. \end{cases} \tag{11}$$

In particular, we have $(u|v) = \epsilon_h u_{(2h-1)} v$ for $u, v \in V_h$.

Consider the subspace $V_2^+ \oplus V_h$ of $V^+ \oplus V^-$.

Proposition 2. *For $a, b \in V_2^+$ and $u, v \in V_h$, define*

$$\begin{aligned} &ab := a_{(1)}b, \quad au := a_{(1)}u, \quad ua := u_{(1)}a, \quad uv := u_{(2h-3)}v, \\ &(a|b) = a_{(3)}b, \quad (a|u) = 0 = (u|a), \quad (u|v) = \epsilon_h u_{(2h-1)}v. \end{aligned} \tag{12}$$

Then the subspace $V_2^+ \oplus V_h$ forms a unital commutative $\mathbb{Z}_2$-graded algebra with invariant bilinear form which extends the Griess algebra structure on V_2^+, where the invariance of the bilinear form is modified as $(xu|y) = \epsilon_h (x|uy)$ for $u \in V_h$.

Proof. The proof follows by a direct verification. For example, by the skew-symmetry one has

$$\begin{aligned} uv = u_{(2h-3)}v &= \sum_{j\geq 0} \frac{(-1)^{2h-3+1+j+|u|\cdot|v|}}{j!} L(-1)^j v_{(2h-3+j)}u \\ &= \underbrace{(-1)^{2h+|u|\cdot|v|}}_{=1} \left(v_{(2h-3)}u + (-1)\cdot L(-1) \underbrace{v_{(2h-2)}u}_{\in V_1^+ = 0} + \frac{(-1)^2}{2} \underbrace{L(-1)^2 v_{(2h-1)}u}_{\in L(-1)^2 V_0 = 0} \right) \\ &= vu. \end{aligned}$$

That $au = ua$ also follows from the skew-symmetry. The bilinear form clearly satisfies the invariant property. ∎

We call $V_2^+ \oplus V_h$ the *$\mathbb{Z}_2$-extended Griess algebra* of V.

3.2 Square Roots of Idempotents

Recall the following fact (cf. Lemma 5.1 of [23] and Proposition 2.6 of [15]).

Proposition 3. *Let V be a VOA of CFT-type. A vector $e \in V_2$ is a Virasoro vector with central charge c if and only if it satisfies $e_{(1)}e = 2e$ and $2e_{(3)}e = c$.*

By this proposition, for a VOA V of OZ-type we see that $e \in V_2$ is a Virasoro vector if and only if $e/2$ is an idempotent in the Griess algebra. So idempotents are important objects to study in the Griess algebra. If we consider the $\mathbb{Z}_2$-extended Griess algebra, it is possible to consider square roots of idempotents inside the odd part.

Let $V = V^0 \oplus V^1$ be an SVOA and $\theta = (-1)^{2L(0)} \in \mathrm{Aut}(V)$ the canonical $\mathbb{Z}_2$-symmetry of V. In this subsection we assume V and $g = \theta$ satisfy Condition 1 and we consider its extended Griess algebra $V_2 \oplus V_h$. Let a be an idempotent of V_2 and suppose $x \in V_h$ is a square root of a, that is, $xx = a$ hold in the extended Griess algebra. We shall consider the subalgebra $\langle x \rangle$ generated by such a root x. The structure of $\langle x \rangle$ depends on the top weight h.

Case $h = 1/2$

In this case $x_{(n)}x \in V_{-n}$ and $x_{(n)}x = 0$ if $n > 0$ as V is of CFT-type. Since $x_{(0)}x = (x|x)$, we have the following commutation relation:

$$[x_{(m)}, x_{(n)}]_+ = \left(x_{(0)}x\right)_{(m+n)} = (x|x) \ _{(m+n)} = (x|x)\delta_{m+n,-1}. \tag{13}$$

Since $a = xx = x_{(-2)}x$ is an idempotent, we have $aa = a_{(1)}a = a$. Then

$$\begin{aligned}
&(x_{(-2)}x)_{(1)}(x_{(-2)}x) \\
&= \sum_{i \geq 0} (-1)^i \binom{-2}{i} \left(x_{(-2-i)}x_{(1+i)} - (-1)^{-2+|x|\cdot|x|} x_{(-1-i)}x_{(i)} \right) x_{(-2)}x \\
&= x_{(-2)}x_{(1)}x_{(-2)}x + x_{(-1)}x_{(0)}x_{(-2)}x + 2x_{(-2)}x_{(1)}x_{(-2)}x \\
&= 4(x|x)x_{(-2)}x
\end{aligned}$$

and therefore we have $4(x|x) = 1$. The central charge of $2a$ is given by

$$8(a|a) = 8(x_{(-2)}x|x_{(-2)}x) = 8(x|x_{(1)}x_{(-2)}x) = 8(x|(x|x) \cdot x) = 1/2.$$

Set $\psi_{n+1/2} := 2x_{(n)}$ and $\psi(z) := Y(x, z)$. Then (13) is expressed as $[\psi_r, \psi_s]_+ = \delta_{r+s,0}$ for $r, s \in \mathbb{Z} + 1/2$ and we have a free fermionic field

$$\psi(z) = \sum_{r \in \mathbb{Z}+1/2} \psi_r z^{-r-1/2}, \quad \psi(z)\psi(w) \sim \frac{1}{z-w}.$$

Therefore, $\langle x\rangle$ is isomorphic to a simple $c = 1/2$ Virasoro SVOA $L(1/2,0) \oplus L(1/2,1/2)$ (cf. [14]).

Case $h = 3/2$

In this case we have $x_{(0)}x = a$, $x_{(1)}x = 0$, $x_{(2)}x = (x|x)$ and $x_{(n)}x = 0$ for $n \geq 3$. Here we further assume that the $c = 8(a|a)$ Virasoro vector $2a$ is the conformal vector of the subalgebra $\langle x\rangle$. This condition is equivalent to $x \in \operatorname{Ker}_V(\omega - 2a)_{(0)}$ (cf. [4]), where ω is the conformal vector of V. (By Theorem 5.1 of (loc. cit), $\omega-2a$ is a Virasoro vector and $\omega = 2a+(\omega-2a)$ is an orthogonal decomposition.) Then x is a highest weight vector for $\mathrm{Vir}(2a)$ with highest weight $3/2$. Since $xx = x_{(0)}x = a$, we have $(a|x_{(0)}x) = (a|a)$. On the other hand,

$$\begin{aligned}(2a|x_{(0)}x) &= 2a_{(3)}x_{(0)}x = [2a_{(3)}, x_{(0)}]x = \sum_{i\geq 0}\binom{3}{i}\left(2a_{(i)}x\right)_{(3-i)}x\\ &= \left(2a_{(0)}x\right)_{(3)}x + 3\left(2a_{(1)}x\right)_{(2)} = -3x_{(2)}x + 3\cdot\frac{3}{2}x_{(2)}x\\ &= \frac{3}{2}(x|x)\end{aligned}$$

and we get $3(x|x) = 4(a|a)$. Now we can compute the commutation relations:

$$\begin{aligned}[2a_{(m)}, x_{(n)}] &= \sum_{i=0}^{\infty}\binom{m}{i}(2a_{(i)}x)_{(m+n-i)}\\ &= (2a_{(0)}x)_{(m+n)} + m(2a_{(1)}x)_{(m+n-1)}\\ &= -(m+n)x_{(m+n-1)} + m\cdot\frac{3}{2}x_{(m+n-1)}\\ &= \frac{1}{2}(m-2n)x_{(m+n-1)},\end{aligned}$$

$$\begin{aligned}[x_{(p)}, x_{(q)}]_+ &= \sum_{i=0}^{\infty}\binom{p}{i}(x_{(i)}x)_{(p+q-i)}\\ &= (x_{(0)}x)_{(p+q)} + \binom{p}{2}(x_{(2)}x)_{(p+q-2)}\\ &= a_{(p+q)} + \frac{p(p-1)}{2}(x|x)_{\ (p+q-2)}\\ &= a_{(p+q)} + \frac{2p(p-1)}{3}(a|a)\delta_{p+q,1}.\end{aligned} \tag{14}$$

For simplicity, set $L^a(m) := 2a_{(m+1)}$, $c_a := 8(a|a)$ and $G^x(r) := 2x_{(r+1/2)}$. Then (14) looks

$$\begin{aligned}[L^a(m), G^x(r)] &= \frac{1}{2}(m-2r)G^x(m+r),\\ [G^x(r), G^x(s)]_+ &= 2L^a(r+s) + \delta_{r+s,0}\frac{4r^2-1}{12}c_a.\end{aligned} \tag{15}$$

This is exactly the defining relations of the Neveu–Schwarz algebra, also known as the $N = 1$ super Virasoro algebra. Therefore, $\langle x\rangle$ is isomorphic to the $N = 1$ $c = 8(a|a)$ Virasoro SVOA.

Case $h = 5/2$

Again we assume that $xx = a$ and $2a$ is the conformal vector of $\langle x \rangle$. But this is still not enough to determine the structure of $\langle x \rangle$. In order to describe $\langle x \rangle$, we need one more assumption that $x_{(n)}x \in \langle a \rangle$ for $n \geq 0$. Zamolodchikov [30] has studied such a subalgebra.

Proposition 4 ([30]). *Suppose $x_{(n)}x \in \langle a \rangle$ for $n \geq 0$ and $2a$ is the conformal vector of $\langle x \rangle$. Then there is a surjection from $\langle x \rangle$ to $L(-{}^{13}\!/_{14}, 0) \oplus L(-{}^{13}\!/_{14}, {}^{5}\!/_{2})$. In particular, the central charge of $\langle x \rangle$ is uniquely determined.*

Remark 2. Recall the central charges and the highest weights of the minimal series of the Virasoro VOAs (cf. [28]).

$$c_{p,q} := 1 - \frac{6(p-q)^2}{pq}, \quad h_{r,s}^{(p,q)} = \frac{(rq-sp)^2 - (p-q)^2}{4pq}, \quad 0 < r < p,\ 0 < s < q. \tag{16}$$

Then $c_{7,4} = -13/14$ and $h_{6,1}^{(7,4)} = 5/2$. Moreover, $L(-{}^{13}\!/_{14}, {}^{5}\!/_{2})$ is the unique non-trivial simple current module over $L(-{}^{13}\!/_{14}, 0)$ so that the simple quotient in Proposition 4 forms a $\mathbb{Z}_2$-graded simple current extension of $L(-{}^{13}\!/_{14}, 0)$.

Remark 3. The (extended) Griess algebra is a part of the structure of the vertex Lie algebra [24] (or that equivalently known as the conformal algebra [13]). As seen in this subsection, for small h one can determine OPE of elements in $V_2 \oplus V_h$ by the extended Griess algebra. However, for higher h, the extended Griess algebra is insufficient to determine full OPE of $Y(x,z)$ and the subalgebra $\langle x \rangle$ for $x \in V_2 \oplus V_h$.

4 Matsuo-Norton Trace Formulae for Extended Griess Algebras

In this section we derive trace formulae for the extended Griess algebras, which is a variation of Matsuo-Norton trace formulae [22]. In this section we assume V satisfies the following condition.

Condition 2. V is an SVOA satisfying Condition 1 and in addition the following.

(1) The invariant bilinear form is non-degenerate on V.
(2) The restriction of the bilinear form on $\mathrm{Vir}(\omega)$ is also non-degenerate.
(3) V as a $\mathrm{Vir}(\omega)$-module is a direct sum of highest weight modules.

Let $V[n]$ be the sum of highest weight $\mathrm{Vir}(\omega)$-submodules of V with highest weight $n \in \frac{1}{2}\mathbb{Z}$. Then by (3) of Condition 2 one has

$$V = \oplus_{n \geq 0} V[n] \tag{17}$$

and we can define the projection map

$$\pi : V = \bigoplus_{n\geq 0} V[n] \longrightarrow V[0] = \mathrm{Vir}(\omega) = \langle \omega \rangle \tag{18}$$

which is a $\mathrm{Vir}(\omega)$-homomorphism.

4.1 Conformal Design and Casimir Vector

Let us recall the notion of the conformal design. Suppose U is a VOA satisfying (3) of Condition 2. Then we can define the projection $\pi : U \to \langle \omega \rangle$ as in (18).

Definition 1 ([11]). Let U be a VOA and M a U-module. An $L(0)$-homogeneous subspace X of M is called a *conformal t-design* based on U if $\mathrm{tr}_X\, o(a) = \mathrm{tr}_X\, o(\pi(a))$ holds for any $a \in \oplus_{0\leq n\leq t} U_n$.

The defining condition of conformal designs was initiated in Matsuo's paper [22] and it is related to the following condition.

Definition 2 ([22]). Let U be a VOA and G a subgroup of $\mathrm{Aut}(U)$. We say U is *of class $\mathscr{S}^n$* under G if $U_k^G \subset \langle \omega \rangle$ for $0 \leq k \leq n$. (We allow G to be $\mathrm{Aut}(U)$ itself.)

The above two conditions are in the following relation.

Lemma 3 ([11, 22]). *Let U be a VOA and M a G-stable U-module. Suppose further that we have a projective representation of G on M. If U is of class $\mathscr{S}^n$ under G then every $L(0)$-homogeneous subspace of M forms a conformal n-design.*

Let V be an SVOA and $g \in \mathrm{Aut}(V)$ satisfying Condition 2. The following is clear.

Lemma 4. *For $m > 0$ the components $V[m]$ and $V[0]$ in (17) are orthogonal with respect to the invariant bilinear form.*

Consider the extended Griess algebra $V_2^+ \oplus V_h$ of V. We set $d := \dim V_h$. Let $\{u_i\}_{1\leq i\leq d}$ be a basis of V_h and $\{u^i\}_{1\leq i\leq d}$ its dual basis. As in [22], we consider the *Casimir vector* of weight $m \in \mathbb{Z}$:

$$\kappa_m := \epsilon_h \sum_{i=1}^{d} u^i_{(2h-1-m)} u_i \in V_m, \tag{19}$$

where the signature ϵ_h is defined as in (11).

Lemma 5. *Let $a \in V$ be homogeneous. Then* $\mathrm{tr}_{V_h}\, o(a) = (-1)^{\mathrm{wt}(a)}(a \mid \kappa_{\mathrm{wt}(a)})$.

Proof. Without loss we may assume a is even, i.e., $a \in V^{0,+}$. We compute the trace as follows.

$$\begin{aligned}
\mathrm{tr}_{V_h}\,\mathrm{o}(a) &= \sum_{i=1}^{d}(\mathrm{o}(a)u^i|u_i) = \sum_{i=1}^{d}(a_{(\mathrm{wt}(a)-1)}u^i \mid u_i) \\
&= \sum_{i=1}^{d}\sum_{j=0}^{\infty}\frac{(-1)^{\mathrm{wt}(a)+j}}{j!}\left(L(-1)^j u^i_{(\mathrm{wt}(a)-1+j)}a \,\middle|\, u_i\right) \qquad \text{(by skew-symmetry)} \\
&= \sum_{i=1}^{d}\sum_{j=0}^{\infty}\frac{(-1)^{\mathrm{wt}(a)+j}}{j!}\epsilon_h\left(a \,\middle|\, u^i_{(2h-\mathrm{wt}(a)-1+j)}L(1)^j u_i\right) \text{(by invariance)} \\
&= \sum_{i=1}^{d}(-1)^{\mathrm{wt}(a)}\epsilon_h(a|u^i_{(2h-\mathrm{wt}(a)-1)}u_i) \qquad \text{(as } L(1)V_h = 0\text{)} \\
&= (-1)^{\mathrm{wt}(a)}(a \mid \kappa_{\mathrm{wt}(a)}).
\end{aligned}$$

Therefore, we obtain the desired equality. ∎

Proposition 5. *V_h is a conformal t-design based on $V^{0,+}$ if and only if $\kappa_m \in \langle\omega\rangle$ for $0 \le m \le t$.*

Proof. By Lemma 5, $\mathrm{tr}_{V_h}\,\mathrm{o}(a) = \mathrm{tr}_{V_h}\,\mathrm{o}(\pi(a))$ for any $a \in V_m^{0,+}$ if and only if $(a|\kappa_m) = (\pi(a)|\kappa_m)$. Since π is a projection, $\{a - \pi(a) \mid a \in V_m^{0,+}\} = V_m^{0,+} \cap \mathrm{Ker}\,\pi$. Then $(a - \pi(a)|\kappa_m) = 0$ if and only if $\kappa_m \in \pi(V) = V[0] = \langle\omega\rangle$ by Condition 2 and Lemma 4. Therefore the assertion holds. ∎

4.2 Derivation of Trace Formulae

We use Lemma 5 to derive the trace formulae. Recall the following associativity formula.

Lemma 6. *([19, Lemma 3.12]) Let $a, b \in V$, $v \in M$ and $p, q \in \mathbb{Z}$. Suppose $s \in \mathbb{Z}$ and $t \in \mathbb{N}$ satisfy $a_{(s+i)}v = b_{(q+t+i+1)}v = 0$ for all $i \ge 0$. Then for $p, q \in \mathbb{Z}$,*

$$a_{(p)}b_{(q)}v = \sum_{i=0}^{t}\sum_{j\ge 0}\binom{p-s}{i}\binom{s}{j}\left(a_{(p-s-i+j)}b\right)_{(q+s+i-j)}v. \tag{20}$$

Let $a, b \in V_2^+$ and $v \in V_h$. Since V_h is the top level of V^-, we can apply the lemma above with $p = q = 1$, $s = 2$ and $t = 0$ and obtain

$$\mathrm{o}(a)\mathrm{o}(b)v = a_{(1)}b_{(1)}v = \sum_{i=0}^{2}\binom{2}{i}\left(a_{(i-1)}b\right)_{(3-i)} = \mathrm{o}(a * b), \tag{21}$$

where $a * b$ is the product of Zhu algebra [31] defined as

$$a * b := \mathrm{Res}_z \frac{(1+z)^{\mathrm{wt}(a)}}{z} Y(a,z)b = \sum_{i \geq 0} \binom{\mathrm{wt}(a)}{i} a_{(i-1)}b. \tag{22}$$

Combining this with Lemma 5, we obtain

Lemma 7. *For $a^1, \dots, a^k \in V_2^+$ we have*

$$\begin{aligned} &\mathrm{tr}_{V_h} \mathrm{o}(a^1) \cdots \mathrm{o}(a^k) \\ &= \sum_{-1 \leq i_1, \cdots, i_{k-1} \leq 1} (-1)^{k+1-(i_1+\cdots+i_{k-1})} \left(a^1_{(i_1)} \cdots a^{k-1}_{(i_{k-1})} a^k \;\middle|\; \kappa_{k+1-(i_1+\cdots+i_{k-1})} \right). \end{aligned} \tag{23}$$

Proof. By (21) if $a \in V_2$ and $x \in V_n$ then $a * x = a_{(1)}x + 2a_{(0)}x + a_{(-1)}x \in V_n + V_{n+1} + V_{n+2}$. So by Lemma 5

$$\mathrm{tr}_{V_h} \mathrm{o}(a^1) \cdots \mathrm{o}(a^k) = \sum_{m=2}^{2k} (-1)^m \left(\mathrm{o}(a^1 * \cdots * a^k) \mid \kappa_m \right). \tag{24}$$

Expanding this we obtain the lemma. ∎

To describe the Casimir vector we need the following condition.

Condition 3. If V_h forms a conformal $2t$-design with $t \leq 5$ then the central charge of V is not a zero of the polynomial $D_{2t}(c)$ defined as follows.

$$\begin{aligned} &D_2(c) = c, \quad D_4(c) = c(5c+22), \quad D_6(c) = (2c-1)(7c+68)D_4(c), \\ &D_8(c) = (3c+46)(5c+3)D_6(c), \quad D_{10}(c) = (11c+232)D_8(c). \end{aligned} \tag{25}$$

The normalized polynomials $D_n(c)$ comes from the Shapovalov determinant of the Verma module $M(c,0)$. The following is well-known (cf. [14]).

Lemma 8. *If the central charge of V is not a zero of $D_n(c)$ in* (25) *then the degree m subspace of $\langle \omega \rangle$ with $m \leq n$ is isomorphic to that of $M(c,0)/M(c,1)$.*

We write $[n_1, \dots, n_k] \models m$ if $n_1 \geq \cdots \geq n_k \geq 2$ and $n_1 + \cdots + n_k = m$. If c is not a zero of $D_n(c)$ in (25) then the degree m subspace of $M(c,0)/M(c,1)$ with $m \leq n$ has a basis $\{L(-n_1) \cdots L(-n_k) \mid [n_1, \dots, n_k] \models m\}$. By definition $\kappa_0 = \epsilon_h \sum_{i=1}^d u^i_{(2h-1)} u_i = \sum_{i=1}^d (u^i | u_i) = d$ and $\kappa_1 = 0$, where $d = \dim V_h$. For $n > 0$ one can show

$$L(n)\kappa_m = (h(n-1) + m - n)\, \kappa_{m-n}. \tag{26}$$

Using this we can recursively compute $(L(-n_1) \cdots L(-n_k) \mid \kappa_m)$. As a result, the Casimir vector can be expressed as follows (cf. Proposition 2.5 of [22]).

Lemma 9. *Suppose the central charge of V is not a zero of $D_n(c)$ in (25) and the Casimir vector $\kappa_m \in \langle\omega\rangle$ with $m \leq n$. Then κ_m is uniquely written as*

$$\kappa_m = \frac{1}{D_{2\lfloor m/2 \rfloor}(c)} \sum_{[n_1,\ldots,n_k]\models m} A^{(m)}_{[n_1,\ldots,n_k]} L(-n_1)\cdots L(-n_k) \ ,$$

where $\lfloor m/2 \rfloor$ stands for the largest integer not exceeding $m/2$ and $A^{(m)}_{[n_1,\ldots,n_k]} \in \mathbb{Q}[c,d,h]$ are given in section "Coefficients in Generalized Casimir Vectors" in Appendix.

We present the main result of this paper.

Theorem 1. *Suppose V and $g \in \mathrm{Aut}(V)$ satisfy Conditions 2 and 3. Set $d = \dim V_h$.*

(1) If V_h forms a conformal 2-design based on $V^{0,+}$, then for any $a^0 \in V_2^+$,

$$\mathrm{tr}_{V_h}\, \mathrm{o}(a^0) = \frac{2hd}{c}(a^0|\omega).$$

(2) If V_h forms a conformal 4-design based on $V^{0,+}$, then for any $a^0, a^1 \in V_2^+$,

$$\mathrm{tr}_{V_h}\, \mathrm{o}(a^0)\mathrm{o}(a^1) = \frac{4hd(5h+1)}{c(5c+22)}(a^0|\omega)(a^1|\omega) + \frac{2hd(22h-c)}{c(5c+22)}(a^0|a^1).$$

(3) If V_h forms a conformal 6-design based on $V^{0,+}$, then for any $a^0, a^1, a^2 \in V_2^+$,

$$\begin{aligned}&\mathrm{tr}_{V_h}\, \mathrm{o}(a^0)\mathrm{o}(a^1)\mathrm{o}(a^2)\\ &= \frac{1}{D_6(c)}\Big(F_0^{(3)}(a^0|\omega)(a^1|\omega)(a^2|\omega) + F_1^{(3)}\,\mathrm{Sym}(a^0|\omega)(a^1|a^2) + F_2^{(3)}(a^0|a^1|a^2)\Big),\end{aligned}$$

where $(a^0|a^1|a^2) = (a^0a^1|a^2) = (a^0|a^1a^2)$ is a totally symmetric trilinear form, $\mathrm{Sym}(a^0|\omega)(a^1|a^2)$ is the sum of all $(a^{i_0}|\omega)(a^{i_1}|a^{i_2})$ which are mutually distinct, and $F_j^{(3)} \in \mathbb{Q}[c,d,h]$, $0 \leq j \leq 2$, are given in section "Coefficients in the Trace Formulae" in Appendix.

(4) If V_h forms a conformal 8-design based on $V^{0,+}$, then for any $a^0, a^1, a^2, a^3 \in V_2^+$,

$$\begin{aligned}&\mathrm{tr}_{V_h}\, \mathrm{o}(a^0)\mathrm{o}(a^1)\mathrm{o}(a^2)\mathrm{o}(a^3)\\ &= \frac{1}{D_8(c)}\Big(F_0^{(4)}(a^0|\omega)(a^1|\omega)(a^2|\omega)(a^3|\omega)\\ &\quad + F_1^{(4)}\,\mathrm{Sym}(a^0|\omega)(a^1|\omega)(a^2|a^3) + F_2^{(4)}\,\mathrm{Sym}(a^0|\omega)(a^1|a^2|a^3)\\ &\quad + F_3^{(4)}\,\mathrm{Sym}(a^0|a^1)(a^2|a^3) + F_4^{(4)}(a^0a^1|a^2a^3)\\ &\quad + F_5^{(4)}(a^0a^2|a^1a^3) + F_6^{(4)}(a^0a^3|a^1a^2)\Big),\end{aligned}$$

where Sym *denotes the sum over all possible permutations of* $(0,1,2,3)$ *for which we obtain mutually distinct terms, and* $F_j^{(4)} \in \mathbb{Q}[c,d,h]$, $0 \le j \le 6$, *are given in section "Coefficients in the Trace Formulae" in Appendix.*

(5) If V_h *forms a conformal 10-design based on* $V^{0,+}$, *then for any* $a^0, a^1, a^2, a^3, a^4 \in V_2^+$,

$$
\begin{aligned}
&\mathrm{tr}_{V_h}\, o(a^0)o(a^1)o(a^2)o(a^3)o(a^4)\\
&= \frac{1}{D_{10}(c)}\Big(F_0^{(5)}(a^0|\omega)(a^1|\omega)(a^2|\omega)(a^3|\omega)(a^4|\omega)\\
&\quad + F_1^{(5)}\,\mathrm{Sym}(a^0|\omega)(a^1|\omega)(a^2|\omega)(a^3|a^4)+F_2^{(5)}\mathrm{Sym}(a^0|\omega)(a^1|\omega)(a^2|a^3|a^4)\\
&\quad + F_3^{(5)}\mathrm{Sym}(a^0|\omega)(a^1|a^2)(a^3|a^4)\\
&\quad + F_4^{(5)}\big((a^0|\omega)(a^1a^2|a^3a^4) + (a^1|\omega)(a^0a^2|a^3a^4) + (a^2|\omega)(a^0a^1|a^3a^4)\\
&\qquad\qquad + (a^3|\omega)(a^0a^1|a^2a^4) + (a^4|\omega)(a^0a^1|a^2a^3)\big)\\
&\quad + F_5^{(5)}\big((a^0|\omega)(a^1a^3|a^2a^4) + (a^1|\omega)(a^0a^3|a^2a^4) + (a^2|\omega)(a^0a^3|a^1a^4)\\
&\qquad\qquad + (a^3|\omega)(a^0a^2|a^1a^4) + (a^4|\omega)(a^0a^2|a^1a^3)\big)\\
&\quad + F_6^{(5)}\big((a^0|\omega)(a^1a^4|a^2a^3) + (a^1|\omega)(a^0a^4|a^2a^3) + (a^2|\omega)(a^0a^4|a^1a^3)\\
&\qquad\qquad + (a^3|\omega)(a^0a^4|a^1a^2) + (a^4|\omega)(a^0a^3|a^1a^2)\big)\\
&\quad + F_7^{(5)}\,\mathrm{Sym}(a^0|a^1)(a^2|a^3|a^4) + F_8^{(5)}(a^0a^1a^2a^3a^4)\\
&\quad + \sum\nolimits^{*} F_{i_0i_1i_2i_3i_4}^{(5)}(a^{i_0}a^{i_1}|a^{i_2}|a^{i_3}a^{i_4})\Big)
\end{aligned}
$$

where Sym *denotes the sum over all possible permutations of* $(0,1,2,3,4)$ *for which we obtain mutually distinct terms,* $(a^0a^1a^2a^3a^4) = a^0_{(3)}a^1_{(2)}a^2_{(1)}a^3_{(0)}a^4$ *and the last summation* $\sum^*$ *is taken over all possible permutations* (i_0,i_1,i_2,i_3,i_4) *of* $(0,1,2,3,4)$ *such that* $(a^{i_0}a^{i_1}|a^{i_2}|a^{i_3}a^{i_4})$ *are mutually distinct. The coefficients* $F_\bullet^{(5)} \in \mathbb{Q}[c,d,h]$ *are given in section "Coefficients in the Trace Formulae" in Appendix.*

Proof. By Lemmas 7 and 9, it suffices to rewrite inner products

$$\left(a^0_{(i_0)}\cdots a^{k-1}_{(i_{k-1})}a^k \;\middle|\; L(-m_1)\cdots L(-m_l)\;\right),$$

with $-1 \le i_0,\cdots,i_{k-1} \le 1$, $k+1-(i_0+\cdots i_{k-1}) = m_1+\cdots+m_l$, in terms of the Griess algebra. By the invariance, this is equal to

$$\left(L(m_l)\cdots L(m_1)a^0_{(i_0)}\cdots a^{k-1}_{(i_{k-1})}a^k \;\middle|\;\right),$$

and by the commutation formula

$$[L(m),a_{(n)}] = (m-n+1)a_{(m+n)} + \delta_{m+n,1}\frac{m(m^2-1)}{6}(a|\omega),$$

we obtain a sum of $(a^{s_0}_{(j_0)}\cdots a^{s_{r-1}}_{(j_{r-1})}a^{s_r}\mid\;)$ with $r \le k$ and $-1 \le j_0,\cdots,j_{r-1} < 2k$. We will use the following relations to rewrite such a term further.

$$\begin{aligned}
(a_{(0)}b)_{(n)} &= [a_{(1)}, b_{(n-1)}] - (a_{(1)}b)_{(n-1)}, \\
\left[a_{(m)}\,,\, b_{(0)}\right] &= \left[a_{(1)}\,,\, b_{(m-1)}\right] + (m-1)(a_{(1)}b)_{(m-1)}, \\
a_{(-m-1)} &= \frac{1}{m!}L(-1)^m a, \quad a_{(m)}b_{(-m)} = m a_{(1)}b,
\end{aligned} \tag{27}$$

where $a, b \in V_2$, $m > 0$ and $n \in \mathbb{Z}$. For example, let us rewrite $(a^0_{(4)}a^1_{(1)}a^2_{(0)}a^3 \mid \ )$.

$$\begin{aligned}
\left(a^0_{(4)}a^1_{(1)}a^2_{(0)}a^3 \,\middle|\, \ \right) &= \left(a^2_{(0)}a^3 \,\middle|\, a^1_{(1)}a^0_{(-2)} \right) = \left(a^2_{(0)}a^3 \,\middle|\, \left[a^1_{(1)}\,,\, a^0_{(-2)}\right] \right) \\
&= \left(a^2_{(0)}a^3 \,\middle|\, \left(a^1_{(0)}a^0\right)_{(-1)} \right) + \left(a^2_{(0)}a^3 \,\middle|\, \left(a^1_{(1)}a^0\right)_{(-2)} \right) \\
&= \left(a^2_{(0)}a^3 \,\middle|\, a^1_{(0)}a^0\right) + \left(a^3 \,\middle|\, a^2_{(2)}\left(a^1_{(1)}a^0\right)_{(-2)} \right) \\
&= \left(a^0_{(3)}a^1_{(2)}a^2_{(0)}a^3 \,\middle|\, \ \right) + 2\left(a^3 \,\middle|\, a^2_{(1)}a^1_{(1)}a^0\right) \\
&= \left(a^0_{(3)}a^1_{(2)}a^2_{(0)}a^3 \,\middle|\, \ \right) + 2\left(a^2_{(1)}a^3 \,\middle|\, a^1_{(1)}a^0\right),
\end{aligned}$$

where $(a^0_{(3)}a^1_{(2)}a^2_{(0)}a^3| \ )$ can be simplified as

$$\begin{aligned}
\left(a^0_{(3)}a^1_{(2)}a^2_{(0)}a^3 \,\middle|\, \ \right) &= \left(a^1_{(2)}a^2_{(0)}a^3 \,\middle|\, a^0\right) = \left(\left[a^1_{(2)}\,,\, a^2_{(0)}\right]a^3 \,\middle|\, a^0\right) \\
&= \left(\left[a^1_{(1)}\,,\, a^2_{(1)}\right]a^3 \,\middle|\, a^0\right) + \left(\left(a^1_{(1)}a^2\right)_{(1)}a^3 \,\middle|\, a^0\right) \\
&= \left(a^2_{(1)}a^3 \,\middle|\, a^1_{(1)}a^0\right) - \left(a^1_{(1)}a^3 \,\middle|\, a^2_{(1)}a^0\right) + \left(a^1_{(1)}a^2 \,\middle|\, a^3_{(1)}a^0\right).
\end{aligned}$$

Therefore, we get

$$\left(a^0_{(4)}a^1_{(1)}a^2_{(0)}a^3 \,\middle|\, \ \right) = 3\left(a^0_{(1)}a^1 \,\middle|\, a^2_{(1)}a^3\right) - \left(a^0_{(1)}a^2 \,\middle|\, a^1_{(1)}a^3\right) + \left(a^0_{(1)}a^3 \,\middle|\, a^1_{(1)}a^2\right).$$

In this way, we can rewrite all terms and obtain the formulae of degrees 1 to 4. However, in the rewriting procedure of the trace of degree 5 we meet the expressions $a^{i_0}_{(3)}a^{i_1}_{(2)}a^{i_2}_{(1)}a^{i_3}_{(0)}a^{i_4} = (a^{i_0}a^{i_1}a^{i_2}a^{i_3}a^{i_4})$ which satisfy the following relations:

$$\begin{aligned}
&(a^0a^1a^2a^3a^4) + (a^1a^0a^2a^3a^4) \\
&= 3(a^0a^1|a^2|a^3a^4) - (a^0a^1|a^3|a^2a^4) + (a^0a^1|a^4|a^2a^3), \\
&(a^0a^1a^2a^3a^4) + (a^0a^2a^1a^3a^4) \\
&= (a^0a^1|a^2|a^3a^4) - (a^0a^1|a^3|a^2a^4) + (a^0a^1|a^4|a^2a^3) + (a^0a^2|a^1|a^3a^4) \\
&\quad - (a^0a^2|a^3|a^1a^4) + (a^0a^2|a^4|a^1a^3) - (a^0a^3|a^4|a^1a^2) + (a^0a^4|a^3|a^1a^2) \\
&\quad + (a^1a^2|a^0|a^3a^4), \\
&(a^0a^1a^2a^3a^4) + (a^0a^1a^3a^2a^4) \\
&= (a^0a^1|a^2|a^3a^4) + (a^0a^1|a^3|a^2a^4) + (a^0a^1|a^4|a^2a^3) - (a^0a^2|a^1|a^3a^4) \\
&\quad - (a^0a^3|a^1|a^2a^4) + (a^0a^4|a^1|a^2a^3) + (a^1a^2|a^0|a^3a^4) + (a^1a^3|a^0|a^2a^4) \\
&\quad - (a^1a^4|a^0|a^2a^3), \\
&(a^0a^1a^2a^3a^4) + (a^0a^1a^2a^4a^3) \\
&= 3(a^0a^1|a^2|a^3a^4) - (a^0a^2|a^1|a^3a^4) + (a^1a^2|a^0|a^3a^4).
\end{aligned} \tag{28}$$

Let R be the space of formal sums of $(a^{i_0}a^{i_1}|a^{i_2}|a^{i_3}a^{i_4})$ over $\mathbb{Z}$. Then it follows from (28) that

$$(a^{\sigma(0)}a^{\sigma(1)}a^{\sigma(2)}a^{\sigma(3)}a^{\sigma(4)}) \equiv \mathrm{sign}(\sigma)(a^0a^1a^2a^3a^4) \mod R$$

for $\sigma \in \mathrm{S}_5$. This shows the rewriting procedure is not unique, and in our rewriting procedure we have to include at least one term $(a^0a^1a^2a^3a^4)$ in the formula (cf. [22]). ∎

Remark 4. Our formulae are a variation of Matsuo-Norton trace formulae in [22] but there are some differences. The trace formula of degree 5 in (loc. cit) contains a totally anti-symmetric quinary form

$$\begin{aligned}&\sum_{\sigma\in \mathrm{S}_5} \mathrm{sgn}(\sigma)\big(a^{\sigma(0)}a^{\sigma(1)}a^{\sigma(2)}a^{\sigma(3)}a^{\sigma(4)}\big)\\ &= \sum_{\sigma\in \mathrm{S}_5} \mathrm{sgn}(\sigma)\left(a^{\sigma(0)}_{(3)}a^{\sigma(1)}_{(2)}a^{\sigma(2)}_{(1)}a^{\sigma(3)}_{(0)}a^{\sigma(4)}\ \middle|\ \right),\end{aligned}$$

which we do not have in ours. This is due to the non-uniqueness of the reduction procedure as explained in the proof. One can transform the formula to include this form using (28).

Remark 5. In the trace formula of degree n, if we put $a^i = \omega/h$ for one of $0 \le i < n$ then we obtain the trace formula of degree $n-1$. Even though we have derived the formulae for degree 4 and 5, the author does not know non-trivial examples of SVOAs which satisfy Conditions 2 and 3 and the odd top level forms a conformal 8- or 10-design. It is shown in [22] (see also [11] for related discussions) that if a VOA V satisfying Conditions 2 and 3 and is of class $\mathcal{S}^8$ (under $\mathrm{Aut}(V)$) and has a proper idempotent then $\dim V_2 = 196,884$ and $c = 24$, those of the moonshine VOA. By this fact, the author expected the non-existence of proper SVOAs (not VOAs) of class $\mathcal{S}^8$ and $\mathcal{S}^{10}$, but the reductions of the trace formula of degree 5 to degree 4 and of degree 4 to degree 3 are consistent and we cannot obtain any contradiction.

We will mainly use the formulae to compute traces of Virasoro vectors.

Corollary 1. *Suppose V and $g \in \mathrm{Aut}(V)$ satisfy Conditions 2 and 3. Let $e \in V_2^+$ be a Virasoro vector with central charge $c_e = 2(e|e)$. If V_h forms a conformal $2t$-design based on $V^{0,+}$, then $\mathrm{tr}_{V^h}\, \mathrm{o}(e)^t$ is given as follows.*

$$\begin{aligned}\mathrm{tr}_{V^h}\, \mathrm{o}(e) &= \frac{2hd}{c}(e|e) \quad \text{if } t=1,\\ \mathrm{tr}_{V^h}\, \mathrm{o}(e)^2 &= \frac{4hd(5h+1)}{c(5c+22)}(e|e)^2 + \frac{2hd(22h-c)}{c(5c+22)}(e|e) \quad \text{if } t=2,\\ \mathrm{tr}_{V^h}\, \mathrm{o}(e)^t &= D_{2t}(c)^{-1}\sum_{j=1}^{t} E_j^{(t)}(e|e)^j \quad \text{if } t=3,4,5,\end{aligned}$$

where $d = \dim V_h$ and $E_\bullet^{(t)} \in \mathbb{Q}[c,d,h]$ are given in section "Coefficients in the Trace Formulae" in Appendix.

5 Applications

In this section we show some applications of our formulae.

5.1 VOAs with $h = 1$

Let V be a VOA and $\theta \in \mathrm{Aut}(V)$ an involution satisfying Conditions 1 and 2. We assume the top weight h of V^- is 1 and denote $d = \dim V_1$. In this case the top level V_1 forms an abelian Lie algebra under 0-th product. For, $[V_1, V_1] = (V_1)_{(0)} V_1 \subset V_1^+ = 0$ as $V_1 \subset V^-$ and V^+ is of OZ-type. We have the following commutation relation for a and $b \in V_1$:

$$[a_{(m)}, b_{(n)}] = (a_{(0)}b)_{(m+n)} + m(a_{(1)}b)_{(m+n-1)} = -m(a|b)\delta_{m+n,0}. \tag{29}$$

Set $\mathfrak{h} := V_1$ and equip this with a symmetric bilinear form by $\langle a|b\rangle := -(a|b)$ for $a, b \in \mathfrak{h}$. Then this form is non-degenerate by Condition 2. Denote $\hat{\mathfrak{h}}$ the rank $d = \dim V_1$ Heisenberg algebra associated to $\mathfrak{h}$, the affinization of $\mathfrak{h}$. By (29) the sub VOA $\langle V_1\rangle$ generated by V_1 is isomorphic to a free bosonic VOA associated to $\hat{\mathfrak{h}}$ and the Casimir element κ_2 is twice the conformal vector of $\langle V_1\rangle$. Suppose V_1 forms a conformal 2-design based on V^+. Then κ_2 coincides with twice the conformal vector of V, and hence $\langle V_1\rangle$ is a full sub VOA of V. More precisely, we show the following.

Proposition 6. *Suppose a VOA V and its involution $\theta \in \mathrm{Aut}(V)$ satisfy Conditions 1and 2. If the top level of V^- has the top weight* 1 *and forms a conformal 2-design based on V^+ then $\langle V_1\rangle$ is isomorphic to the rank $d = \dim V_1$ free bosonic VOA and the restriction of θ on $\langle V_1\rangle$ is conjugate to a lift of the (-1)-map on $\mathfrak{h}$ in* $\mathrm{Aut}(\langle V_1\rangle)$. *Moreover, if $\langle V_1\rangle$ is a proper subalgebra of V then there is an even positive definite rootless lattice L of $\mathfrak{h}$ of rank less than or equal to d such that V is isomorphic to a tensor product of the lattice VOA V_L associated to L and a free bosonic VOA associated to the affinization of the orthogonal complement of $\mathbb{C}L$ in* $\mathfrak{h}$. *In this case the restriction of θ on V_L is conjugate to a lift of the (-1)-map on L in* $\mathrm{Aut}(V_L)$.

Proof. That $\langle V_1\rangle$ is isomorphic to a free bosonic VOA is already shown and θ is clearly a lift of the (-1)-map on it. Suppose $\langle V_1\rangle$ is a proper subalgebra. Then by [20] there is a even positive definite lattice L such that V is isomorphic to a tensor product of the lattice VOA V_L associated to L and a free bosonic VOA associated to the affinization of the orthogonal complement of $\mathbb{C}L$ in $\mathfrak{h}$. Since V_1 is abelian, L has no root. Let ρ be a lift of the (-1)-map on L to $\mathrm{Aut}(V_L)$. Since θ and ρ act by -1 on $(V_L)_1$, $\theta\rho$ is identity on it. Then $\theta\rho$ on V_L is a linear automorphism $\exp(a_{(0)})$ for some $a \in (V_L)_1$. Since $\theta \exp(\frac{1}{2}a_{(0)}) = \exp(\frac{1}{2}(\theta a)_{(0)})\theta = \exp(-\frac{1}{2}a_{(0)})\theta$, we have the conjugacy $\exp(-\frac{1}{2}a_{(0)})\theta \exp(\frac{1}{2}a_{(0)}) = \theta \exp(\frac{1}{2}a_{(0)}) \exp(\frac{1}{2}a_{(0)}) = \theta \exp(a_{(0)}) = \theta(\theta\rho) = \rho$. This completes the proof. ∎

If V is a free bosonic VOA then it is shown in [2] that V^+ is not of class $\mathscr{S}^4$. The case $V = \langle V_1 \rangle$ is not interesting and out of our focus. So we are reduced to the case when V is a lattice VOA V_L where L is an even positive definite rootless lattice and θ is a lift of the (-1)-map on L. (Clearly V_L and θ satisfy Conditions 1 and 2.) For such a lattice L the complete classification of simple $c = 1/2$ Virasoro vectors in V_L^+ is obtained in [27]. It is shown in [17,27] that V_L has a simple $c = 1/2$ Virasoro vector $e \in V_L^+$ if and only if there is a sublattice K isomorphic to $\sqrt{2}A_1$ or $\sqrt{2}E_8$ such that $e \in V_K^+ \subset V_L^+$. In particular, one can find a simple $c = 1/2$ Virasoro vector in V_L if (and only if) L has a norm 4 element.

Suppose we have a simple $c = 1/2$ Virasoro vector $e \in V_L$. Actually e is in V_L^+ in our case and one can find a sublattice K isomorphic to $\sqrt{2}A_1$ or $\sqrt{2}E_8$ such that $e \in V_K^+ \subset V_L^+$. Denote $X = (V_L)_1$. The zero-mode $\mathrm{o}(e)$ acts on X semisimply with possible eigenvalues 0, 1/2 and 1/16. We denote by d_λ the dimension of $\mathrm{o}(e)$-eigensubspace of X with eigenvalue λ. Then $d_0 + d_{1/2} + d_{1/16} = d$. By the trace formula in (1) of Theorem 1 we get

$$\mathrm{tr}_X\, \mathrm{o}(e) = 0 \cdot d_0 + \frac{1}{2} d_{1/2} + \frac{1}{16} d_{1/16} = \frac{1}{2}. \tag{30}$$

This has two possible solutions $(d_0, d_{1/2}, d_{1/16}) = (d-1, 1, 0)$ and $(d-8, 0, 8)$. Suppose further that X forms a conformal 4-design. Then by the trace formula in (2) of Theorem 1 we have

$$\mathrm{tr}_X\, \mathrm{o}(e)^2 = 0^2 \cdot d_0 + \frac{1}{2^2} d_{1/2} + \frac{1}{16^2} d_{1/16} = \frac{25-d}{2(5d+22)}. \tag{31}$$

The possible integer solution of (30) and (31) are $(d_0, d_{1/2}, d_{1/16}) = (3, 1, 0)$ and $(10, 0, 8)$. The case $(d_0, d_{1/2}, d_{1/16}) = (10, 0, 8)$ is impossible. For, if K is isomorphic to $\sqrt{2}A_1$ then $\mathrm{o}(e)$ does not have eigenvalue 1/16 on X, so K must be $\sqrt{2}E_8$. However, in this case we can also find another simple $c = 1/2$ Virasoro vector $e' \in V_K^+$ such that $\mathrm{o}(e')$ acts on V_K^- semisimply with eigenvalues only 0 and 1/2. (Take a norm 4 element $\alpha \in K$ and consider $V_{\mathbb{Z}\alpha}^+ \subset V_K^+$). But considering $\mathrm{tr}_X\, \mathrm{o}(e')^2$ we obtain a contradiction. Therefore, the possible solution is only $(d_0, d_{1/2}, d_{1/16}) = (3, 1, 0)$. We summarize the discussion here in the next theorem.

Theorem 2. *Let L be an even positive definite lattice without roots. If L contains a norm 4 vector and the weight 1 subspace of V_L^- forms a conformal 4-design then* $\mathrm{rank}\, L = \dim(V_L)_1 = 4$.

Remark 6. It is shown in [22] that if a $c = 4$ VOA of OZ-type satisfies: (i) it is of class $\mathscr{S}^4$, (ii) there is a simple $c = 1/2$ Virasoro vector such that its possible eigenvalues of the zero-mode on the Griess algebra are 0, 1/2 and 2, then its Griess algebra is 22-dimensional. The fixed point sub VOA $V_{\sqrt{2}D_4}^+$ would be an example with $c = 4$ and $\dim V_2 = 22$ and so $L = \sqrt{2}D_4$ would be an example of a lattice satisfying Theorem 2. (The author expects this is the unique example.) $V^+_{\sqrt{2}D_4}$ is isomorphic to the Hamming code VOA and its full automorphism group is $2^6{:}(\mathrm{GL}_3(2)\times \mathrm{S}_3)$ (cf. [21,26]). The top level of $V^-_{\sqrt{2}D_4}$ is not stable under this group [26].

Remark 7. Consider the case $L = \sqrt{2}E_8$. It is expected (cf. [22]) that $V^+_{\sqrt{2}E_8}$ is of class $\mathscr{S}^6$. It is shown in [7, 26] that $\mathrm{Aut}(V^+_{\sqrt{2}E_8}) = \mathrm{O}^+_{10}(2)$. The Griess algebra of $V^+_{\sqrt{2}E_8}$ is 156-dimensional which is a direct sum of a one-dimensional module and a 155-dimensional irreducible module over $\mathrm{O}^+_{10}(2)$ (cf. [1]). It is shown in [7, 17] that $V^+_{\sqrt{2}E_8}$ has totally 496 simple $c = 1/2$ Virasoro vectors which form an $\mathrm{O}^+_{10}(2)$-orbit in the Griess algebra. The 240 vectors of them are of A_1-type, and the remaining 256 vectors are of E_8-type (cf. [17]). All A_1-types and all E_8-types form mutually distinct $2^8{:}\mathrm{O}^+_8(2)$-orbits, where $2^8{:}\mathrm{O}^+_8(2)$ is the centralizer of θ in $\mathrm{Aut}(V_{\sqrt{2}E_8})$. The A_1-type vector has no eigenvalue 1/16 on X and corresponds to the solution $(d_0, d_{1/2}, d_{1/16}) = (7, 1, 0)$ of (30), whereas the E_8-type has only eigenvalue 1/16 on X and corresponds to the solution $(d_0, d_{1/2}, d_{1/16}) = (0, 0, 8)$ of (30). It is shown in [26] that under the conjugation of modules by automorphisms, the stabilizer of $V^-_{\sqrt{2}E_8}$ in $\mathrm{Aut}(V^+_{\sqrt{2}E_8}) = \mathrm{O}^+_{10}(2)$ is $2^8{:}\mathrm{O}^+_8(2)$. So X is $2^8{:}\mathrm{O}^+_8(2)$-stable but not $\mathrm{O}^+_{10}(2)$-stable.

5.2 The Baby-Monster SVOA: The Case $h = 3/2$

Here we consider the Baby-monster SVOA $V\!B^\natural = V\!B^{\natural,0} \oplus V\!B^{\natural,1}$ introduced by Höhn in [8] which affords an action of the Baby-monster sporadic finite simple group $\mathbb{B}$. Let $\theta = (-1)^{2L(0)} \in \mathrm{Aut}(V\!B^\natural)$ be the canonical $\mathbb{Z}_2$-symmetry. Then $V\!B^\natural$ and θ satisfy Conditions 1 and 2.

Remark 8. It is shown in [9,29] that $\mathrm{Aut}(V\!B^\natural) = \mathrm{Aut}(V\!B^{\natural,0})\times\langle\theta\rangle$, $\mathrm{Aut}(V\!B^{\natural,0}) \simeq \mathbb{B}$ and the even part $V\!B^{\natural,0}$ has three irreducible modules which are all $\mathbb{B}$-stable.

The top level of $V\!B^{\natural,1}$ is of dimension $d = 4,371$ and has the top weight $h = 3/2$. It is shown in Lemma 2.6 of [3] (see also [10, 11]) that $V\!B^{\natural,0}$ is of class $\mathscr{S}^6$. Therefore the top level of $V\!B^{\natural,1}$ forms a conformal 6-design based on $V\!B^{\natural,0}$. (Actually, it is shown in [10] and [11] that $V\!B^{\natural,0}$ is of class $\mathscr{S}^7$ and $V\!B^{\natural,1}_{3/2}$ is a conformal 7-design, respectively.)

Let t be a 2A-involution of $\mathbb{B}$. Then $C_{\mathbb{B}}(t) \simeq 2{\cdot}^2E_6(2){:}2$ (cf. [1]). The Griess algebra of $V\!B^{\natural,0}$ is of dimension 96,256 and we have the following decompositions as a module over $\mathbb{B}$ and $C_{\mathbb{B}}(t)$ (cf. [12]).

$$\begin{aligned} V\!B^{\natural,0}_2 &= \underline{\mathbf{1}} \oplus \underline{\mathbf{96255}} && \text{over } \mathbb{B}, \\ &= \underline{\mathbf{1}} \oplus \underline{\mathbf{1}} \oplus \underline{\mathbf{1938}} \oplus \underline{\mathbf{48620}} \oplus \underline{\mathbf{45696}} && \text{over } C_{\mathbb{B}}(t). \end{aligned} \tag{32}$$

The $\mathbb{B}$-invariant subalgebra of the Griess algebra is 1-dimensional spanned by the conformal vector ω of $VB^\natural$, but the $C_\mathbb{B}(t)$-invariant subalgebra of the Griess algebra forms a 2-dimensional commutative (and associative) subalgebra spanned by two mutually orthogonal Virasoro vectors. It is shown in [12] that central charges of these Virasoro vectors are $7/10$ and $114/5$. (The sum is the conformal vector of $VB^\natural$.) Let e be the shorter one, the simple $c = 7/10$ Virasoro vector fixed by $C_\mathbb{B}(t)$. For the odd part we have the following decompositions.

$$\begin{aligned} VB^{\natural,1}_{3/2} &= \underline{\mathbf{4371}} && \text{over } \mathbb{B}, \\ &= \underline{\mathbf{1}} + \underline{\mathbf{1938}} + \underline{\mathbf{2432}} && \text{over } C_\mathbb{B}(t). \end{aligned} \tag{33}$$

Since e is fixed by $C_\mathbb{B}(t)$, its zero-mode acts as scalars on $C_\mathbb{B}(t)$-irreducible components. As $c_{4,5} = 7/10$ belongs to the minimal series (16), $L(7/10, 0)$ has 6 irreducible modules $L(7/10, h)$ with $h = 0, 7/16, 3/80, 3/2, 3/5, 1/10$, and $\mathrm{o}(e)$ acts on each $C_\mathbb{B}(t)$-irreducible component in (33) by one of these values. Denote λ_1, λ_2 and λ_3 the eigenvalues of $\mathrm{o}(e)$ on $\underline{\mathbf{1}}$, $\underline{\mathbf{1938}}$ and $\underline{\mathbf{2432}}$ in (33), respectively. Applying Theorem 1 we will compute these eigenvalues. Set $X = VB^{\natural,1}_{3/2}$. Since $VB^{\natural,0}$ is of class $\mathscr{S}^6$ under $\mathbb{B}$ and $VB^{\natural,1}$ is $\mathbb{B}$-stable, by (1), (2), (3) of Theorem 1 we have

$$\mathrm{tr}_X\, \mathrm{o}(e) = \frac{1953}{10}, \quad \mathrm{tr}_X\, \mathrm{o}(e)^2 = \frac{2163}{100}, \quad \mathrm{tr}_X\, \mathrm{o}(e)^3 = \frac{5313}{1000}. \tag{34}$$

On the other hand, we have

$$\mathrm{tr}_X\, \mathrm{o}(e)^j = \lambda_1^j + 1938\lambda_2^j + 2432\lambda_3^j \tag{35}$$

for $j = 1, 2, 3$. Solving (34) and (35), we obtain a unique rational solution $\lambda_1 = 3/2$, $\lambda_2 = 1/10$ and $\lambda_3 = 0$ which are consistent with the representation theory of $L(7/10, 0)$.

$$\begin{array}{rcccc} VB^{\natural,1}_{3/2} & = & \underline{\mathbf{1}} + & \underline{\mathbf{1938}} + & \underline{\mathbf{2432}} \\ \mathrm{o}(e) & : & \frac{3}{2} & \frac{1}{10} & 0 \end{array} \tag{36}$$

Let x be a non-zero vector in the 1-dimensional $C_\mathbb{B}(t)$-invariant subspace of $VB^{\natural,1}_{3/2}$. Since both $\mathrm{o}(\omega)$ and $\mathrm{o}(e)$ act on x by 3/2, it follows $\mathrm{o}(\omega - e)x = 0$ and $x \in \mathrm{Ker}\,(\omega - e)_{(0)}$. This implies x is a square root of e in the extended Griess algebra $VB^{\natural,0}_2 \oplus VB^{\natural,1}_{3/2}$ and e is the conformal vector of the subalgebra generated by x. Thus as we discussed in Sect. 3.2 $\langle x\rangle$ is isomorphic to the $N = 1$ $c = 7/10$ Virasoro SVOA which is isomorphic to $L(7/10, 0) \oplus L(7/10, 3/2)$ as a $\langle e\rangle$-module.

Proposition 7. *Let t be a 2A-element of $\mathbb{B}$. Then $(VB^\natural)^{C_\mathbb{B}(t)}$ has a full sub SVOA isomorphic to $L(114/5, 0) \otimes (L(7/10, 0) \oplus L(7/10, 3/2))$.*

Let us recall the notion of Miyamoto involutions. A simple $c = 7/10$ Virasoro vector u of an SVOA V is called *of σ-type* on V if there is no irreducible $\langle u\rangle \simeq L(7/10,0)$-submodule of V isomorphic to either $L(7/10,7/16)$ or $L(7/10,3/80)$. If u is of σ-type on V, then define a linear automorphism σ_u of V acting on an irreducible $\langle u\rangle$-submodule M of V as follows.

$$\sigma_u|_M = \begin{cases} 1 & \text{if } M \simeq L(7/10,0),\ L(7/10,3/5), \\ -1 & \text{if } M \simeq L(7/10,1/10),\ L(7/10,3/2). \end{cases} \tag{37}$$

Then σ_u is well-defined and the fusion rules of $L(7/10,0)$-modules guarantees $\sigma_u \in \mathrm{Aut}(V)$ (cf. [23]). It is shown in [12] that the map $u \mapsto \sigma_u$ provides a one-to-one correspondence between the set of simple $c = 7/10$ Virasoro vectors of $VB^{\natural,0}$ of σ-type and the 2A-conjugacy class of $\mathbb{B} = \mathrm{Aut}(VB^{\natural,0})$. In this correspondence we have to consider only σ-type $c = 7/10$ Virasoro vectors, since we also have non σ-type ones in $VB^{\natural,0}$. We can reformulate this correspondence based on the SVOA $VB^{\natural}$. We say a simple $c = 7/10$ Virasoro vector u of $VB^{\natural,0}_2$ is *extendable* if it has a square root $v \in VB^{\natural,1}_{3/2}$ in the extended Griess algebra such that $\langle v\rangle \simeq L(7/10,0) \oplus L(7/10,3/2)$.

Suppose we have an extendable simple $c = 7/10$ Virasoro vector $u \in VB^{\natural,0}_2$ and its square root $v \in VB^{\natural,1}_{3/2}$. It is shown in [16] that the $\mathbb{Z}_2$-graded simple current extension $L(7/10,0) \oplus L(7/10,3/2)$ has two irreducible untwisted modules $L(7/10,0) \oplus L(7/10,3/2)$ and $L(7/10,1/10) \oplus L(7/10,3/5)$. Therefore, $\mathrm{o}(u)$ acts on $X = VB^{\natural,1}_{3/2}$ semisimply with possible eigenvalues 0, 1/10, 3/5 and 3/2. In particular, u is of σ-type. Let d_λ be the dimension of $\mathrm{o}(u)$-eigensubspace of $X = VB^{\natural,1}_{3/2}$ with eigenvalue λ. Then

$$\mathrm{tr}_X\, \mathrm{o}(u)^j = 0^j \cdot d_0 + \left(\frac{1}{10}\right)^j \cdot d_{1/10} + \left(\frac{3}{5}\right)^j \cdot d_{3/5} + \left(\frac{3}{2}\right)^j \cdot d_{3/2} \tag{38}$$

for $0 \le j \le 3$, where we understand $0^0 = 1$. By (34) one can solve this linear system and obtain $d_0 = 2,432$, $d_{1/10} = 1,938$, $d_{3/5} = 0$ and $d_{3/2} = 1$, recovering (36). (That $d_{3/5} = 0$ and $d_{3/2} = 1$ can be also shown by the representation theory of $L(7/10,0) \oplus L(7/10,3/2)$, see Remark 9 below.) The trace of σ_u on X is

$$\mathrm{tr}_X\, \sigma_u = 2,432 - 1 - 1,938 = 493. \tag{39}$$

By [1] we see that $-\sigma_u$ belongs to the 2A-conjugacy class of $\mathbb{B}$. Therefore $\sigma_u\theta \in \mathrm{Aut}(VB^{\natural})$ is a 2A-element of $\mathbb{B}$ by [9, 29]. Summarizing, we have the following reformulation of Theorem 5.13 of [12].

Theorem 3. *There is a one-to-one correspondence between the subalgebras of $VB^{\natural}$ isomorphic to the $N = 1$ $c = 7/10$ simple Virasoro SVOA $L(7/10,0) \oplus L(7/10,3/2)$ and the 2A-elements of the Baby-monster $\mathbb{B}$ given by the association $u \mapsto \sigma_u\theta$ where u*

is the conformal vector of the sub SVOA, σ_u *is defined as in* (37) *and* $\theta = (-1)^{2L(0)}$ *is the canonical* $\mathbb{Z}_2$*-symmetry of* $VB^\natural$.

Remark 9. Let $V = V^0 \oplus V^1$ be an SVOA such that V^1 has the top weight $3/2$. Suppose the top level $X = V^1_{3/2}$ forms a conformal 6-design based on V^0. It is shown in [11] that if $\dim X > 1$ then the central charge c of V is either 16 or $47/2$. Suppose further that there is a simple extendable $c = 7/10$ Virasoro vector e of V. Then e and its square root generate a sub SVOA W isomorphic to $L(^7\!/_{10}, 0) \oplus L(^7\!/_{10}, ^3\!/_2)$. By the representation theory of $L(^7\!/_{10}, 0) \oplus L(^7\!/_{10}, ^3\!/_2)$ (cf. [16]) V is a direct sum of irreducible W-submodules isomorphic to $L(^7\!/_{10}, 0) \oplus L(^7\!/_{10}, ^3\!/_2)$ or $L(^7\!/_{10}, ^1\!/_{10}) \oplus L(^7\!/_{10}, ^3\!/_5)$. Set $d := \dim X$ and let d_λ be the dimension of $o(e)$-eigensubspace of X with eigenvalue λ. Then $d = d_0 + d_{1/10} + d_{3/5} + d_{3/2}$. We know $d_{3/2} = 1$ as $L(^7\!/_{10}, ^3\!/_2)$ is a simple current $L(^7\!/_{10}, 0)$-module, and we also have $d_{3/5} = 0$ since X is the top level. Solving (38) in this case by Theorem 1 we obtain

$$\begin{aligned} d_{3/5} &= -\frac{7d(2c-47)(10c-7)(82c-37)}{80c(2c-1)(5c+22)(7c+68)}, \\ d_{3/2} &= \frac{d(800c^3 - 27588c^2 + 238596c - 112133)}{80c(2c-1)(5c+22)(7c+68)}. \end{aligned} \tag{40}$$

Combining (40) with $d_{3/5} = 0$ and $d_{3/2} = 1$ we get two possible solutions $(c, d) = (^7\!/_{10}, 1)$ and $(^{47}\!/_2, 4371)$. The case $V = W$ corresponds to the former, and $V = VB^\natural$ is an example corresponding to the latter case. The author expects that $VB^\natural$ is the unique example of class $\mathscr{S}^6$ corresponding to the latter case.

Acknowledgements The author wishes to thank Professor Atsushi Matsuo for stimulating discussions and for his Mathematica programs. He also thanks Professor Masahiko Miyamoto for valuable comments. Most of the results of this paper were obtained by using computer. The author used a computer algebra system Risa/Asir for Windows. This work was supported by JSPS Grant-in-Aid for Young Scientists (Start-up) No. 19840025 and (B) No. 21740011.

References

1. Conway, J.H., Curtis, R.T., Norton, S.P., Parker, R.A., Wilson, R.A.: ATLAS of Finite Groups. Clarendon Press, Oxford (1985)
2. Dong, C., Nagatomo, K.: Classification of irreducible modules for the vertex operator algebra $M(1)^+$. J. Algebra **216**, 384–404 (1999)
3. Dong, C. Li, H., Mason, G.: Some twisted sectors for the Moonshine module. In: Moonshine, the Monster, and Related Topics, South Hadley, 1994. Contemporary Mathematics, vol. 193, pp. 25–43. American Mathematical Society, Providence (1996)
4. Frenkel, I.B., Zhu, Y.: Vertex operator algebras associated to representation of affine and Virasoro algebras. Duke Math. J. **66**, 123–168 (1992)
5. Frenkel, I.B., Lepowsky, J. Meurman, A.: Vertex Operator Algebras and the Monster. Academic, New York (1988)

6. Frenkel, I., Huang, Y.-Z., Lepowsky, J.: On Axiomatic Approaches to Vertex Operator Algebras and Modules. Memoirs of the American Mathematical Society, vol. 104. American Mathematical Society, Providence (1993)
7. Griess, R.L.: The vertex operator algebra related to E_8 with automorphism group $O^+(10, 2)$, The Monster and Lie algebras, Ohio State Univ. Math. Res. Inst. Publ. **7**, 43–58 (1998)
8. Höhn, G.: Selbstduale Vertexoperatorsuperalgebren und das Babymonster. Ph.D. thesis, Bonn (1995), Bonner Mathematische Schriften, **286**, 1–85 (1996). arXiv:0706.0236
9. Höhn, G.: The group of symmetries of the shorter moonshine module. Abh. Math. Semin. Univ. Hambg. **80**(2), 275–283 (2010). arXiv:math/0210076
10. Höhn, G.: Generalized moonshine for the Baby Monster. Preprint, May 2003. http://www.math.ksu.edu/ gerald/papers/baby8.ps
11. Höhn, G.: Conformal designs based on vertex operator algebras. Adv. Math. **217**, 2301–2335 (2008)
12. Höhn, G., Lam, C.H., Yamauchi, H.: McKay's E_7 observation on the Babymonster. Int. Math. Res. Not. **2012**, 166–212 (2012). doi:10.1093/imrn/rnr009
13. Kac, V.G.: Vertex Algebras for Beginners. University Lecture Series, vol. 10. American Mathematical Society, Providence (1997)
14. Kac, V.G., Raina, A.K.: Bombay Lectures on Highest Weight Representations of Infinite Dimensional Lie algebras. World Scientific, Singapore (1987)
15. Lam, C.H.: Code vertex operator algebras under coordinate change. Commun. Algebra **27**, 4587–4605 (1999)
16. Lam, C.H., Lam, N., Yamauchi, H.: Extension of unitary Virasoro vertex operator algebra by a simple module. Int. Math. Res. Not. **11**, 577–611 (2003)
17. Lam, C.H., Sakuma, S., Yamauchi, H.: Ising vectors and automorphism groups of commutant subalgebras related to root systems. Math. Z. **255**(3), 597–626 (2007)
18. Li, H.: Symmetric invariant bilinear forms on vertex operator algebras. J. Pure Appl. Algebra **96**, 279–297 (1994)
19. Li, H.: Regular representation, Zhu's $A(V)$-theory, and induced modules. J. Algebra **238**, 159–193 (2001)
20. Li, H., Xu, X.: A characterization of vertex algebra associated to even lattices. J. Algebra **173**, 253–270 (1995)
21. Matsuo, A., Matsuo, M.: The automorphism group of the Hamming code vertex operator algebra. J. Algebra **228**, 204–226 (2000)
22. Matsuo, A.: Norton's trace formulae for the Griess algebra of a vertex operator algebra with larger symmetry. Commun. Math. Phys. **224**, 565–591 (2001)
23. Miyamoto, M.: Griess algebras and conformal vectors in vertex operator algebras. J. Algebra **179**, 528–548 (1996)
24. Primc, M.: Vertex algebras generalized by Lie algebras. J. Pure Appl. Algebra **135**, 253–293 (1999)
25. Scheithauer, N.: Vertex algebras Lie algebras, and superstrings. J. Algebra **200**, 363–403 (1998)
26. Shimakura, H.: The automorphism group of the vertex operator algebra V_L^+ for an even lattice without roots. J. Algebra **280**, 29–57 (2004)
27. Shimakura, H.: Classification of Ising vectors in the vertex operator algebra V_L^+. Pac. J. Math. **258**, 487–495 (2012)
28. Wang, W.: Rationality of Virasoro vertex operator algebras. Int. Math. Res. Not. **71**, 197–211 (1993)
29. Yamauchi, H.: 2A-orbifold construction and the baby-monster vertex operator superalgebra. J. Algebra **284**, 645–668 (2005)
30. Zamolodchikov, A.B.: Infinite additional symmetries in two dimensional conformal quantum field theory. Theor. Math. Phys. **65**, 1205 (1985)
31. Zhu, Y.: Modular invariance of characters of vertex operator algebras, J. Am. Math. Soc. **9**, 237–302 (1996)

Appendix

Coefficients in Generalized Casimir Vectors

$A^{(2)}_{[2]} = 2*h*d$, $A^{(3)}_{[3]} = h*d$, $A^{(4)}_{[4]} = 3*h*d*(c-2*h+4)$, $A^{(4)}_{[2,2]} = 2*h*(5*h+1)*d$, $A^{(5)}_{[5]} = 2*h*d*(c-2*h+4)$, $A^{(5)}_{[3,2]} = 2*h*(5*h+1)*d$, $A^{(6)}_{[6]} = 4*h*d*(5*c^3+(-15*h+65)*c^2+(-20*h^2-148*h+148)*c-26*h^2+98*h-92)$, $A^{(6)}_{[4,2]} = 2*h*d*((42*h+8)*c^2+(-84*h^2+349*h+65)*c-134*h^2-86*h-40)$, $A^{(6)}_{[3,3]} = (1/2)*h*d*((70*h+15)*c^2+(614*h+136)*c+248*h^2-464*h-64)$, $A^{(6)}_{[2,2,2]} = (4/3)*h*d*((70*h^2+42*h+8)*c+29*h^2-57*h-2)$, $A^{(7)}_{[7]} = 3*h*d*(5*c^3+(-15*h+65)*c^2+(-20*h^2-148*h+148)*c-26*h^2+98*h-92)$, $A^{(7)}_{[5,2]} = 2*h*d*((28*h+5)*c^2+(-56*h^2+243*h+41)*c-172*h^2-16*h-28)$, $A^{(7)}_{[4,3]} = 3*h*d*((14*h+3)*c^2+(-28*h^2+106*h+24)*c+38*h^2-70*h-12)$, $A^{(7)}_{[3,2,2]} = 2*h*d*((70*h^2+42*h+8)*c+29*h^2-57*h-2)$, $A^{(8)}_{[8]} = (1/2)*h*d*(350*c^5+(-1260*h+10080)*c^4+(-560*h^2-31735*h+85005)*c^3+(-5040*h^3-17240*h^2-192290*h+194494)*c^2+(-18520*h^3-43840*h^2+20928*h-8184)*c+4344*h^3-32496*h^2+76488*h-57744)$, $A^{(8)}_{[6,2]} = 2*h*d*((300*h+50)*c^4+(-900*h^2+7312*h+1176)*c^3+(-1200*h^3-18548*h^2+42969*h+6081)*c^2+(-4552*h^3-52960*h^2+32406*h-1466)*c-536*h^3-26880*h^2+5696*h-2808)$, $A^{(8)}_{[5,3]} = (1/2)*h*d*((840*h+175)*c^4+(-1680*h^2+19885*h+4188)*c^3+(-25392*h^2+107936*h+23184)*c^2+(-2016*h^3+1832*h^2+4060*h+968)*c+7792*h^3+1776*h^2-31312*h-6816)$, $A^{(8)}_{[4,4]} = (3/2)*h*d*((126*h+28)*c^4+(-504*h^2+2787*h+643)*c^3+(504*h^3-7156*h^2+13198*h+3338)*c^2+(3180*h^3+2372*h^2-2480*h+344)*c-2004*h^3+7248*h^2-6036*h-888)$, $A^{(8)}_{[4,2,2]} = 2*h*d*((630*h^2+366*h+68)*c^3+(-1260*h^3+9159*h^2+4793*h+958)*c^2+(-6942*h^3+11417*h^2-3187*h+210)*c+1114*h^3-654*h^2-3064*h-168)$, $A^{(8)}_{[3,3,2]} = h*d*((1050*h^2+645*h+125)*c^3+(16700*h^2+9170*h+1934)*c^2+(3720*h^3+15510*h^2-8662*h+716)*c-1016*h^3+6444*h^2-8692*h-264)$, $A^{(8)}_{[2,2,2,2]} = (2/3)*h*d*((1050*h^3+1260*h^2+606*h+108)*c^2+(3305*h^3-498*h^2-701*h+78)*c-251*h^3+918*h^2-829*h-6)$, $A^{(9)}_{[9]} = (2/3)*h*d*(210*c^5+(-756*h+6048)*c^4+(-756*h^2-19311*h+50949)*c^3+(-5544*h^3-19676*h^2-$

$120486*h+115622)*c^2+(-23508*h^3-30448*h^2+18372*h-5456)*c+4428*h^3-26232*h^2+52020*h-34536)$, $A^{(9)}_{[7,2]}=2*h*d*((225*h+35)*c^4+(-675*h^2+5565*h+826)*c^3+(-900*h^3-14615*h^2+33776*h+4267)*c^2+(-2910*h^3-49778*h^2+29666*h-1378)*c-2350*h^3-23916*h^2+6718*h-2196)$, $A^{(9)}_{[6,3]}=4*h*d*((75*h+15)*c^4+(-225*h^2+1747*h+350)*c^3+(-300*h^3-3933*h^2+9193*h+1814)*c^2+(-1642*h^3-3182*h^2+2740*h-88)*c+1814*h^3-2964*h^2-1022*h-612)$, $A^{(9)}_{[5,4]}=2*h*d*((126*h+28)*c^4+(-504*h^2+2787*h+643)*c^3+(504*h^3-7156*h^2+13198*h+3338)*c^2+(3180*h^3+2372*h^2-2480*h+344)*c-2004*h^3+7248*h^2-6036*h-888)$, $A^{(9)}_{[5,2,2]}=4*h*d*((210*h^2+117*h+21)*c^3+(-420*h^3+3208*h^2+1602*h+302)*c^2+(-3554*h^3+4166*h^2-784*h+40)*c+710*h^3-1032*h^2-746*h-60)$, $A^{(9)}_{[4,3,2]}=2*h*d*((630*h^2+381*h+73)*c^3+(-1260*h^3+8694*h^2+4780*h+1010)*c^2+(-3222*h^3+10336*h^2-4022*h+300)*c+98*h^3+1788*h^2-3890*h-156)$, $A^{(9)}_{[3,3,3]}=(1/6)*h*d*((1050*h^2+675*h+135)*c^3+(15770*h^2+9144*h+2038)*c^2+(11160*h^3+13348*h^2-10332*h+896)*c-3048*h^3+11328*h^2-10344*h-240)$, $A^{(9)}_{[3,2,2,2]}=(4/3)*h*d*((1050*h^3+1260*h^2+606*h+108)*c^2+(3305*h^3-498*h^2-701*h+78)*c-251*h^3+918*h^2-829*h-6)$, $A^{(10)}_{[10]}=(6/5)*h*d*(1050*c^6+(-4200*h+52290)*c^5+(-3150*h^2-195019*h+888199)*c^4+(-31500*h^3-160243*h^2-2900235*h+5888368)*c^3+(-33600*h^4-876400*h^3-2224448*h^2-13733560*h+11872408)*c^2+(-189616*h^4-3013900*h^3-3958988*h^2+2767600*h-800016)*c-29792*h^4+816800*h^3-3744448*h^2+6247744*h-3575424)$, $A^{(10)}_{[8,2]}=(1/5)*h*d*((19250*h+2800)*c^5+(-69300*h^2+881440*h+124040)*c^4+(-30800*h^3-2898185*h^2+12963179*h+1696856)*c^3+(-277200*h^4-1275240*h^3-35996682*h^2+64729982*h+6705400)*c^2+(-2026600*h^4-3142080*h^3-112130808*h^2+64543216*h-3235248)*c-335864*h^4-8601520*h^3-49036936*h^2+17453488*h-4052928)$, $A^{(10)}_{[7,3]}=(3/10)*h*d*((8250*h+1575)*c^5+(-24750*h^2+368615*h+70030)*c^4+(-33000*h^3-978660*h^2+5161264*h+966596)*c^3+(-814640*h^3-10273412*h^2+22761712*h+3992640)*c^2+(-103200*h^4-2524680*h^3-14390328*h^2+9889536*h-487008)*c+139456*h^4+5842880*h^3-11650816*h^2-447872*h-1475328)$, $A^{(10)}_{[6,4]}=(12/5)*h*d*((825*h+180)*c^5+(-4125*h^2+35248*h+7832)*c^4+(1650*h^3-149899*h^2+456809*h+104870)*c^3+(6600*h^4+60200*h^3-1328561*h^2+1698137*h+418354)*c^2+(67132*h^4+339830*h^3-202982*h^2-47264*h+18184)*c-33620*h^4-77560*h^3+609380*h^2-579320*h-120480)$, $A^{(10)}_{[6,2,2]}=(4/5)*h*d*((8250*h^2+4400*h+760)*c^4+(-24750*h^3+296210*h^2+151096*h+26364)*c^3+(-33000*h^4-801290*h^3+2704347*h^2+1191343*h+213790)*c^2+$

$(-232460*h^4-4589320*h^3+3842818*h^2-286646*h-11132)*c+28644*h^4+857640*h^3-1710804*h^2-212088*h-52032)$, $A^{(10)}_{[5,5]}=(1/5)*h*d*((4620*h+1050)*c^5+(-18480*h^2+198719*h+46201)*c^4+(18480*h^3-639607*h^2+2606009*h+632228)*c^3+(443400*h^3-5076574*h^2+9773262*h+2648692)*c^2+(41376*h^4+1074560*h^3+4481200*h^2-3129584*h+288608)*c+293648*h^4-2776160*h^3+7129072*h^2-5166496*h-672384)$, $A^{(10)}_{[5,3,2]}=(1/5)*h*d*((46200*h^2+27115*h+5075)*c^4+(-92400*h^3+1636765*h^2+922753*h+178072)*c^3+(-2371200*h^3+14798146*h^2+7158654*h+1515620)*c^2+(-110880*h^4-9063800*h^3+17179184*h^2-5440168*h+338704)*c-319088*h^4+2238560*h^3-652432*h^2-4969184*h-254976)$, $A^{(10)}_{[4,4,2]}=(3/5)*h*d*((6930*h^2+4180*h+800)*c^4+(-27720*h^3+226185*h^2+133283*h+26662)*c^3+(27720*h^4-631060*h^3+1836386*h^2+928514*h+211140)*c^2+(337140*h^4-1677780*h^3+2472564*h^2-812508*h+53544)*c+100532*h^4-226520*h^3+432268*h^2-745384*h-34656)$, $A^{(10)}_{[4,3,3]}=(3/10)*h*d*((11550*h^2+7425*h+1485)*c^4+(-23100*h^3+389020*h^2+243566*h+51294)*c^3+(-426920*h^3+3236462*h^2+1740898*h+429980)*c^2+(-81840*h^4+1025620*h^3+3455108*h^2-2154976*h+167248)*c-17296*h^4-522560*h^3+2125936*h^2-1996768*h-52992)$, $A^{(10)}_{[4,2,2,2]}=4*h*d*((2310*h^3+2706*h^2+1276*h+224)*c^3+(-4620*h^4+48797*h^3+50252*h^2+22925*h+4434)*c^2+(-47038*h^4+140169*h^3-6264*h^2-27525*h+2578)*c-4966*h^4+9340*h^3+15382*h^2-28252*h-288)$, $A^{(10)}_{[3,3,2,2]}=h*d*((11550*h^3+14025*h^2+6809*h+1222)*c^3+(274840*h^3+284503*h^2+133429*h+26346)*c^2+(40920*h^4+764986*h^3-116882*h^2-177916*h+18992)*c+8648*h^4-92288*h^3+251080*h^2-201520*h-1344)$, $A^{(10)}_{[2,2,2,2,2]}=(4/15)*h*d*((11550*h^4+23100*h^3+20130*h^2+8580*h+1440)*c^2+(76675*h^4+30590*h^3-25615*h^2-10898*h+1608)*c+3767*h^4-18410*h^3+29929*h^2-16342*h-24)$.

Coefficients in the Trace Formulae

$$\mathrm{Sym}(a^0|\omega)(a^1|a^2) = (a^0|\omega)(a^1|a^2) + (a^1|\omega)(a^0|a^2) + (a^2|\omega)(a^0|a^1),$$

$$\mathrm{Sym}(a^0|\omega)(a^1|\omega)(a^2|a^3) = (a^0|\omega)(a^1|\omega)(a^2|a^3) + (a^0|\omega)(a^2|\omega)(a^1|a^3) + (a^0|\omega)(a^3|\omega)(a^1|a^2) + (a^1|\omega)(a^2|\omega)(a^0|a^3) + (a^1|\omega)(a^3|\omega)(a^0|a^2) + (a^2|\omega)(a^3|\omega)(a^0|a^1),$$

$$\mathrm{Sym}(a^0|\omega)(a^1|a^2|a^3) = (a^0|\omega)(a^1|a^2|a^3) + (a^1|\omega)(a^0|a^2|a^3) + (a^2|\omega)(a^0|a^1|a^3) + (a^3|\omega)(a^0|a^1|a^2),$$

$$\mathrm{Sym}(a^0|a^1)(a^2|a^3) = (a^0|a^1)(a^2|a^3) + (a^0|a^2)(a^1|a^3) + (a^0|a^3)(a^1|a^2),$$

$$\mathrm{Sym}(a^0|\omega)(a^1|\omega)(a^2|\omega)(a^3|a^4) = (a^0|\omega)(a^1|\omega)(a^2|\omega)(a^3|a^4) + (a^0|\omega)(a^1|\omega)(a^3|\omega)(a^2|a^4) + (a^0|\omega)(a^1|\omega)(a^4|\omega)(a^2|a^3) + (a^0|\omega)(a^2|\omega)(a^3|\omega)(a^1|a^4) + (a^0|\omega)(a^2|\omega)(a^4|\omega)(a^1|a^3) + (a^0|\omega)(a^3|\omega)(a^4|\omega)(a^1|a^2) + (a^1|\omega)(a^2|\omega)(a^3|\omega)(a^0|a^4) + (a^1|\omega)(a^2|\omega)(a^4|\omega)(a^0|a^3) + (a^1|\omega)(a^3|\omega)(a^4|\omega)(a^0|a^2) + (a^2|\omega)(a^3|\omega)(a^4|\omega)(a^0|a^1),$$

$$\mathrm{Sym}(a^0|\omega)(a^1|\omega)(a^2|a^3|a^4) = (a^0|\omega)(a^1|\omega)(a^2|a^3|a^4) + (a^0|\omega)(a^2|\omega)(a^1|a^3|a^4) + (a^0|\omega)(a^3|\omega)(a^1|a^2|a^4) + (a^0|\omega)(a^4|\omega)(a^1|a^2|a^3) + (a^1|\omega)(a^2|\omega)(a^0|a^3|a^4) + (a^1|\omega)(a^3|\omega)(a^0|a^2|a^4) + (a^1|\omega)(a^4|\omega)(a^0|a^2|a^3) + (a^2|\omega)(a^3|\omega)(a^0|a^1|a^4) + (a^2|\omega)(a^4|\omega)(a^0|a^1|a^3) + (a^3|\omega)(a^4|\omega)(a^0|a^1|a^2),$$

$$\mathrm{Sym}(a^0|\omega)(a^1|a^2)(a^3|a^4) = (a^0|\omega)(a^1|a^2)(a^3|a^4) + (a^0|\omega)(a^1|a^3)(a^2|a^4) + (a^0|\omega)(a^1|a^4)(a^2|a^3) + (a^1|\omega)(a^0|a^2)(a^3|a^4) + (a^1|\omega)(a^0|a^3)(a^2|a^4) + (a^1|\omega)(a^0|a^4)(a^2|a^3) + (a^2|\omega)(a^0|a^1)(a^3|a^4) + (a^2|\omega)(a^0|a^3)(a^1|a^4) + (a^2|\omega)(a^0|a^4)(a^1|a^3) + (a^3|\omega)(a^0|a^1)(a^2|a^4) + (a^3|\omega)(a^0|a^2)(a^1|a^4) + (a^3|\omega)(a^0|a^4)(a^1|a^2) + (a^4|\omega)(a^0|a^1)(a^2|a^3) + (a^4|\omega)(a^0|a^2)(a^1|a^3) + (a^4|\omega)(a^0|a^3)(a^1|a^2),$$

$$\mathrm{Sym}(a^0|a^1)(a^2|a^3|a^4) = (a^0|a^1)(a^2|a^3|a^4) + (a^0|a^2)(a^1|a^3|a^4) + (a^0|a^3)(a^1|a^2|a^4) + (a^0|a^4)(a^1|a^2|a^3) + (a^1|a^2)(a^0|a^3|a^4) + (a^1|a^3)(a^0|a^2|a^4) + (a^1|a^4)(a^0|a^2|a^3) + (a^2|a^3)(a^0|a^1|a^4) + (a^2|a^4)(a^0|a^1|a^3) + (a^3|a^4)(a^0|a^1|a^2).$$

$F_0^{(3)} = 8*h*d*((70*h^2+42*h+8)*c+29*h^2-57*h-2)$, $F_1^{(3)} = -4*h*d*((14*h+4)*c^2+(-308*h^2-93*h-1)*c+170*h^2+34*h)$, $F_2^{(3)} = h*d*(4*c^3+(-222*h-1)*c^2+(3008*h^2+102*h)*c-1496*h^2)$, $F_0^{(4)} = 16*h*d*((1050*h^3+1260*h^2+606*h+108)*c^2+(3305*h^3-498*h^2-701*h+78)*c-251*h^3+918*h^2-829*h-6)$, $F_1^{(4)} = -8*h*d*((210*h^2+162*h+36)*c^3+(-4620*h^3-3227*h^2-861*h+26)*c^2+(-5614*h^3+2915*h^2-485*h-2)*c-1334*h^3+2622*h^2+92*h)$, $F_2^{(4)} = 2*h*d*(60*h*c^4+(-3330*h^2-523*h-487)*c^3+(45120*h^3+9648*h^2+13856*h-2336)*c^2+(36376*h^3-91186*h^2+43550*h-2232)*c-6760*h^3-47796*h^2+19756*h-696)$, $F_3^{(4)} = 4*h*d*((42*h+36)*c^4+(-1848*h^2-1279*h+513)*c^3+(20328*h^3+13052*h^2-14654*h+2334)*c^2+(-35836*h^3+98516*h^2-43320*h+2232)*c-16700*h^3+43104*h^2-19756*h+696)$, $F_4^{(4)} = (1/2)*h*d*((1128*h+199)*c^4+(-46392*h^2-3311*h-1768)*c^3+(497472*h^3-19488*h^2+73544*h-16440)*c^2+(351008*h^3-726256*h^2+326804*h-17160)*c-72848*h^3-344832*h^2+158048*h-5568)$, $F_5^{(4)} = (-1/2)*h*d*(60*c^5+(-2976*h+1023)*c^4+(44184*h^2-41669*h+2850)*c^3+(-164544*h^3+426432*h^2-65116*h-716)*c^2+(-22112*h^3+23984*h^2+13092*h-1528)*c+68816*h^3-150144*h^2+25024*h)$, $F_6^{(4)} = (1/2)*h*d*(60*c^5+(-2640*h+1311)*c^4+(29400*h^2-51901*h+6954)*c^3+(-1920*h^3+530848*h^2-182348*h+17956)*c^2+(-308800*h^3+812112*h^2-333468*h+16328)*c-64784*h^3+194688*h^2-133024*h+5568)$, $F_0^{(5)} = 32*h*d*((11550*h^4+23100*h^3+20130*h^2+8580*h+1440)*c^2+(76675*h^4+30590*h^3-25615*h^2-10898*h+1608)*c+3767*h^4-18410*h^3+29929*h^2-16342*h-24)$, $F_1^{(5)} = -16*h*d*((2310*h^3+3366*h^2+1848*h+360)*c^3+(-50820*h^4-64063*h^3-39624*h^2-9203*h+402)*c^2+(-190058*h^4+21757*h^3+50420*h^2-8593*h-6)*c+14558*h^4-53244*h^3+48082*h^2+348*h)$, $F_2^{(5)} = (4/5)*h*d*((3300*h^2+660*h-40)*c^4+(-183150*h^3-90835*h^2-94567*h-25578)*c^3+(2481600*h^4+1334700*h^3+2540131*h^2+285789*h-163830)*c^2+(7115560*h^4-13778670*h^3+2299334*h^2+2630452*h-245456)*c+858872*h^4+1045920*h^3-6623912*h^2+2211696*h-37056)$, $F_3^{(5)} = (8/5)*h*d*((2310*h^2+3300*h+920)*c^4+(-101640*h^3-123925*h^2+2681*h+13794)*c^3+(1118040*h^4+1178540*h^3-631298*h^2-179402*h+81900)*c^2+(-228580*h^4+4993420*h^3-750692*h^2-1313196*h+122728)*c+344284*h^4-2043720*h^3+3258596*h^2-1105848*h+18528)$, $F_4^{(5)} = (1/5)*h*d*(500*c^5+(62040*h^2+7735*h+25115)*c^4+(-2551560*h^3-564175*h^2-1452063*h+94428)*c^3+(27360960*h^4+2744160*h^3+30534534*h^2-6099454*h-348380)*c^2+(64210400*h^4-208744320*h^3+91532216*h^2+3799848*h-935504)*c-909872*h^4-64093920*h^3+8306672*h^2+7635744*h-148224)$,

$F_5^{(5)} = (-1/5) * h * d * ((3300 * h + 1500) * c^5 + (-163680 * h^2 - 585 * h + 39235) * c^4 + (2430120 * h^3 - 2017145 * h^2 - 1433609 * h + 240084) * c^3 + (-9049920 * h^4 + 31066560 * h^3 + 13487402 * h^2 - 7601082 * h + 456180) * c^2 + (-41190560 * h^4 - 16962080 * h^3 + 49902728 * h^2 - 9779816 * h - 107152) * c - 808336 * h^4 + 9987680 * h^3 - 17678384 * h^2 + 1778272 * h + 188928)$, $F_6^{(5)} = (1/5) * h * d * ((3300 * h + 1500) * c^5 + (-145200 * h^2 + 25815 * h + 46595) * c^4 + (1617000 * h^3 - 3008545 * h^2 - 1412161 * h + 350436) * c^3 + (-105600 * h^4 + 40494880 * h^3 + 8437018 * h^2 - 9036298 * h + 1111380) * c^2 + (-43019200 * h^4 + 22985280 * h^3 + 43897192 * h^2 - 20285384 * h + 874672) * c + 1945936 * h^4 - 6362080 * h^3 + 8390384 * h^2 - 7068512 * h + 337152)$, $F_7^{(5)} = (-2/5) * h * d * ((660 * h + 460) * c^5 + (-51150 * h^2 - 33647 * h + 6897) * c^4 + (1302180 * h^3 + 829156 * h^2 - 451426 * h + 40950) * c^3 + (-10919040 * h^4 - 7782640 * h^3 + 9315274 * h^2 - 2071698 * h + 61364) * c^2 + (21416272 * h^4 - 59346500 * h^3 + 27298188 * h^2 - 1866624 * h + 9264) * c + 9686000 * h^4 - 25000320 * h^3 + 11458480 * h^2 - 403680 * h)$, $F_8^{(5)} = (-1/2) * h * d * (100 * c^6 + (-8078 * h + 2861) * c^5 + (221174 * h^2 - 203081 * h + 19684) * c^4 + (-2214880 * h^3 + 4802538 * h^2 - 965274 * h + 52252) * c^3 + (4236288 * h^4 - 38346896 * h^3 + 13282628 * h^2 - 1695920 * h + 25584) * c^2 + (12825792 * h^4 - 32289856 * h^3 + 12276272 * h^2 - 758816 * h + 17536) * c - 155904 * h^4 - 1722368 * h^3 + 4176000 * h^2 - 215296 * h)$, $F_{01423}^{(5)} = (-1/10) * h * d * (500 * c^6 + (-33130 * h + 25625) * c^5 + (707230 * h^2 - 1434751 * h + 485426) * c^4 + (-4128560 * h^3 + 27827338 * h^2 - 20020070 * h + 4414912) * c^3 + (-15989760 * h^4 - 192127280 * h^3 + 258582588 * h^2 - 134080200 * h + 15339472) * c^2 + (136946816 * h^4 - 932999600 * h^3 + 1000619648 * h^2 - 291521120 * h + 11640256) * c + 25836352 * h^4 - 152808960 * h^3 + 252283328 * h^2 - 102259584 * h + 4475904)$, $F_{01324}^{(5)} = (1/10) * h * d * (500 * c^6 + (-35770 * h + 23785) * c^5 + (911830 * h^2 - 1300163 * h + 457838) * c^4 + (-9337280 * h^3 + 24510714 * h^2 - 18214366 * h + 4251112) * c^3 + (27686400 * h^4 - 160996720 * h^3 + 221321492 * h^2 - 125793408 * h + 15094016) * c^2 + (51281728 * h^4 - 695613600 * h^3 + 891426896 * h^2 - 284054624 * h + 11603200) * c - 12907648 * h^4 - 52807680 * h^3 + 206449408 * h^2 - 100644864 * h + 4475904)$, $F_{12034}^{(5)} = (1/10) * h * d * (100 * c^6 + (-3150 * h + 5575) * c^5 + (15650 * h^2 - 148721 * h + 119806) * c^4 + (-550800 * h^3 - 1490922 * h^2 - 4041146 * h + 949728) * c^3 + (14745600 * h^4 + 53833840 * h^3 + 29411876 * h^2 - 25511768 * h + 3284592) * c^2 + (-167754624 * h^4 + 73524400 * h^3 + 137639360 * h^2 - 61086944 * h + 2624448) * c + 3115968 * h^4 - 16839680 * h^3 + 31808832 * h^2 - 21690496 * h + 1007616)$, $F_{01234}^{(5)} = h * d * (50 * c^6 + (-5304 * h + 1238) * c^5 + (204604 * h^2 - 99615 * h + 13827) * c^4 + (-3383208 * h^3 + 2721160 * h^2 - 892294 * h + 59250) * c^3 + (20120832 * h^4 - 26868960 * h^3 + 17636364 * h^2 - 3522876 * h - 65568) * c^2 + (41237472 * h^4 - 107859720 * h^3 + 46229896 * h^2 - 423632 * h - 256192) * c - 772320 * h^4 - 27659904 * h^3 + 9574560 * h^2 + 1892928 * h - 31488)$, $F_{02134}^{(5)} = (-1/10) * h * d * (100 * c^6 + (150 * h + 7215) * c^5 + (-237790 * h^2 - 280007 * h + 130072) * c^4 + (5835360 * h^3 + 1550206 * h^2 -$

$5491014*h+812348)*c^3+(-38223360*h^4+30779440*h^3+63018188*h^2-27827312*h+1723824)*c^2+(-139274688*h^4-126848480*h^3+208934224*h^2-38241856*h-180160)*c+69888*h^4+19482880*h^3-41267328*h^2+3697664*h+692736)$, $F^{(5)}_{03214} = F^{(5)}_{04213} = (1/5)*h*d*((1650*h+820)*c^5+(-126720*h^2-65643*h+5133)*c^4+(3193080*h^3+1520564*h^2-724934*h-68690)*c^3+(-26484480*h^4-11527200*h^3+16803156*h^2-1157772*h-780384)*c^2+(14239968*h^4-100186440*h^3+35647432*h^2+11422544*h-1402304)*c-1523040*h^4+18161280*h^3-36538080*h^2+12694080*h-157440)$, $F^{(5)}_{02413} = F^{(5)}_{03412} = (-1/5)*h*d*((3630*h+5660)*c^5+(-199320*h^2-209673*h+193503)*c^4+(3472920*h^3+1907324*h^2-7596850*h+2076826)*c^3+(-18585600*h^4-196400*h^3+96084724*h^2-62800300*h+7605776)*c^2+(36408928*h^4-385775160*h^3+469619144*h^2-143863520*h+5776288)*c+13307936*h^4-72098560*h^3+115701664*h^2-50591552*h+2237952)$, $F^{(5)}_{02314} = F^{(5)}_{04312} = (1/5)*h*d*((2310*h+4740)*c^5+(-97020*h^2-142379*h+179709)*c^4+(868560*h^3+249012*h^2-6693998*h+1994926)*c^3+(3252480*h^4+15368880*h^3+77454176*h^2-58656904*h+7483048)*c^2+(-6423616*h^4-267082160*h^3+415022768*h^2-140130272*h+5757760)*c-6064064*h^4-22097920*h^3+92784704*h^2-49784192*h+2237952)$, $F^{(5)}_{04123} = F^{(5)}_{03124} = (-2/5)*h*d*(150*c^6+(-10060*h+5380)*c^5+(217020*h^2-323853*h+57123)*c^4+(-1309760*h^3+6390724*h^2-2579346*h+268402)*c^3+(-4260480*h^4-40238760*h^3+32357832*h^2-9076728*h+462936)*c^2+(-18786432*h^4-72074440*h^3+67578896*h^2-10508984*h-23120)*c-177408*h^4+2717760*h^3-5096832*h^2+655296*h+173184)$, $F^{(5)}_{14023} = F^{(5)}_{13024} = (1/5)*h*d*(300*c^6+(-21770*h+9940)*c^5+(560760*h^2-582063*h+109113)*c^4+(-5812600*h^3+11260884*h^2-4433758*h+605494)*c^3+(17963520*h^4-68950320*h^3+47912508*h^2-16995684*h+1706256)*c^2+(-51812832*h^4-43962440*h^3+99510360*h^2-32440512*h+1356064)*c+1168224*h^4-12725760*h^3+26344416*h^2-11383488*h+503808)$, $E^{(3)}_3 = 8*h*d*((70*h^2+42*h+8)*c+29*h^2-57*h-2)$, $E^{(3)}_2 = -12*h*d*((14*h+4)*c^2+(-308*h^2-93*h-1)*c+170*h^2+34*h)$, $E^{(3)}_1 = 2*h*d*(4*c^3+(-222*h-1)*c^2+(3008*h^2+102*h)*c-1496*h^2)$, $E^{(4)}_4 = 16*h*d*((1050*h^3+1260*h^2+606*h+108)*c^2+(3305*h^3-498*h^2-701*h+78)*c-251*h^3+918*h^2-829*h-6)$, $E^{(4)}_3 = -48*h*d*((210*h^2+162*h+36)*c^3+(-4620*h^3-3227*h^2-861*h+26)*c^2+(-5614*h^3+2915*h^2-485*h-2)*c-1334*h^3+2622*h^2+92*h)$, $E^{(4)}_2 = 4*h*d*((366*h+108)*c^4+(-18864*h^2-5929*h-409)*c^3+(241464*h^3+77748*h^2+11462*h-2342)*c^2+(37996*h^3-69196*h^2+44240*h-2232)*c-77140*h^3-61872*h^2+19756*h-696)$, $E^{(4)}_1 = 2*h*d*((1464*h+487)*c^4+(-61176*h^2-13543*h+2336)*c^3+(660096*h^3+84928*h^2-43688*h+2232)*c^2+(64320*h^3+61872*h^2-19756*h+696)*c-206448*h^3)$, $E^{(5)}_5 = 32*h*d*((11550*h^4+23100*h^3+20130*h^2+8580*h+1440)*c^2+(76675*h^4+30590*h^3-25615*h^2-10898*h+1608)*c+3767*h^4-18410*h^3+29929*h^2-16342*h-24)$,

$E_4^{(5)} = -160 * h * d * ((2310 * h^3 + 3366 * h^2 + 1848 * h + 360) * c^3 + (-50820 * h^4 - 64063 * h^3 - 39624 * h^2 - 9203 * h + 402) * c^2 + (-190058 * h^4 + 21757 * h^3 + 50420 * h^2 - 8593 * h - 6) * c + 14558 * h^4 - 53244 * h^3 + 48082 * h^2 + 348 * h)$, $E_3^{(5)} = 8 * h * d * ((13530 * h^2 + 11220 * h + 2680) * c^4 + (-671220 * h^3 - 553445 * h^2 - 181091 * h - 9774) * c^3 + (8317320 * h^4 + 6205020 * h^3 + 3186368 * h^2 + 33372 * h - 81960) * c^2 + (13545380 * h^4 - 12577080 * h^3 + 2346592 * h^2 + 1321316 * h - 122728) * c + 2750596 * h^4 - 4039320 * h^3 - 3472036 * h^2 + 1105848 * h - 18528)$, $E_2^{(5)} = -4 * h * d * ((1320 * h + 420) * c^5 + (-182820 * h^2 - 101429 * h - 18681) * c^4 + (5969040 * h^3 + 3213887 * h^2 + 527763 * h - 122880) * c^3 + (-58143360 * h^4 - 27737760 * h^3 - 6853602 * h^2 + 3391274 * h - 184092) * c^2 + (-19549216 * h^4 + 50103960 * h^3 - 30930304 * h^2 + 2972472 * h - 27792) * c + 17527600 * h^4 + 30443040 * h^3 - 11458480 * h^2 + 403680 * h)$, $E_1^{(5)} = -2 * h * d * (100 * c^6 + (1470 * h + 6495) * c^5 + (-501790 * h^2 - 424803 * h + 40956) * c^4 + (15693120 * h^3 + 8719374 * h^2 - 2073438 * h + 61364) * c^3 + (-141373440 * h^4 - 54143280 * h^3 + 27458268 * h^2 - 1866624 * h + 9264) * c^2 + (-12282432 * h^4 - 30443040 * h^3 + 11458480 * h^2 - 403680 * h) * c + 47895936 * h^4)$.

Mathieu Moonshine and Orbifold K3s

Matthias R. Gaberdiel and Roberto Volpato

Abstract The current status of 'Mathieu Moonshine', the idea that the Mathieu group $\mathbb{M}_{24}$ organises the elliptic genus of K3, is reviewed. While there is a consistent decomposition of all Fourier coefficients of the elliptic genus in terms of Mathieu $\mathbb{M}_{24}$ representations, a conceptual understanding of this phenomenon in terms of K3 sigma-models is still missing. In particular, it follows from the recent classification of the automorphism groups of arbitrary K3 sigma-models that (1) there is no single K3 sigma-model that has $\mathbb{M}_{24}$ as an automorphism group; and (2) there exist 'exceptional' K3 sigma-models whose automorphism group is not even a subgroup of $\mathbb{M}_{24}$. Here we show that all cyclic torus orbifolds are exceptional in this sense, and that almost all of the exceptional cases are realised as cyclic torus orbifolds. We also provide an explicit construction of a $\mathbb{Z}_5$ torus orbifold that realises one exceptional class of K3 sigma-models.

1 Introduction

In 2010, Eguchi et al. observed that the elliptic genus of K3 shows signs of an underlying Mathieu $\mathbb{M}_{24}$ group action [1]. In particular, they noted (see Sect. 2 below for more details) that the Fourier coefficients of the elliptic genus can be

M.R. Gaberdiel (✉)
Institut für Theoretische Physik, ETH Zurich, 8093 Zurich, Switzerland
e-mail: gaberdiel@itp.phys.ethz.ch

R. Volpato
Max-Planck-Institut für Gravitationsphysik, Am Mühlenberg 1, 14476 Golm, Germany
e-mail: roberto.volpato@aei.mpg.de

W. Kohnen and R. Weissauer (eds.), *Conformal Field Theory, Automorphic Forms and Related Topics*, Contributions in Mathematical and Computational Sciences 8,
DOI 10.1007/978-3-662-43831-2_5,

written as sums of dimensions of irreducible $\mathbb{M}_{24}$ representations.[1] This intriguing observation is very reminiscent of the famous realisation of McKay and Thompson who noted that the Fourier expansion coefficients of the J-function can be written in terms of dimensions of representations of the Monster group [2, 3]. This led to a development that is now usually referred to as 'Monstrous Moonshine', see [4] for a nice review. One important upshot of that analysis was that the J-function can be thought of as the partition function of a self-dual conformal field theory, the 'Monster conformal field theory' [5, 6], whose automorphism group is precisely the Monster group. The existence of this conformal field theory explains many aspects of Monstrous Moonshine although not all—in particular, the genus zero property is rather mysterious from this point of view (see [7] for recent progress on this issue).

In the Mathieu case, the situation is somewhat different compared to the early days of Monstrous Moonshine. It is by construction clear that the underlying conformal field theory *is* a K3 sigma-model (describing string propagation on the target space K3). However, this does not characterise the corresponding conformal field theory uniquely as there are many inequivalent such sigma-models—in fact, there is an 80-dimensional moduli space of such theories, all of which lead to the same elliptic genus. The natural analogue of the 'Monster conformal field theory' would therefore be a special K3 sigma-model whose automorphism group coincides with $\mathbb{M}_{24}$. Unfortunately, as we shall review here (see Sect. 3), such a sigma-model does not exist: we have classified the automorphism groups of all K3 sigma-models, and none of them contains $\mathbb{M}_{24}$ [8]. In fact, not even all automorphism groups are contained in $\mathbb{M}_{24}$: the exceptional cases are the possibilities (ii), (iii) and (iv) of the theorem in Sect. 3 (see [8]), as well as case (i) for nontrivial G'. Case (iii) was already shown in [8] to be realised by a specific Gepner model that is believed to be equivalent to a torus orbifold by $\mathbb{Z}_3$. Here we show that also cases (ii) and (iv) are realised by actual K3s—the argument in [8] for this relied on some assumption about the regularity of K3 sigma-models—and in both cases the relevant K3s are again torus orbifolds. More specifically, case (ii) is realised by an asymmetric $\mathbb{Z}_5$ orbifold of $\mathbb{T}^4$ (see Sect. 5),[2] while for case (iii) the relevant orbifold is by $\mathbb{Z}_3$ (see Sect. 6).

Cyclic torus orbifolds are rather special K3s since they always possess a quantum symmetry whose orbifold leads back to $\mathbb{T}^4$. Using this property of cyclic torus orbifolds, we show (see Sect. 4) that the group of automorphisms of K3s that are cyclic torus orbifolds is always exceptional; in particular, the quantum symmetry itself is never an element of $\mathbb{M}_{24}$. Although some 'exceptional' automorphism groups (contained in case (i) of the classification theorem) can also arise in K3

[1]Actually, they did not just look at the Fourier coefficients themselves, but at the decomposition of the elliptic genus with respect to the elliptic genera of irreducible $\mathcal{N} = 4$ superconformal representations. They then noted that these expansion coefficients (and hence in particular the usual Fourier coefficients) are sums of dimensions of irreducible $\mathbb{M}_{24}$ representations.

[2]Since the orbifold action is asymmetric, this evades various no-go-theorems (see e.g. [9]) that state that the possible orbifold groups are either $\mathbb{Z}_2$, $\mathbb{Z}_3$, $\mathbb{Z}_4$, or $\mathbb{Z}_6$.

models that are not cyclic torus orbifolds, our observation may go a certain way towards explaining why only $\mathbb{M}_{24}$ seems to appear in the elliptic genus of K3.

We should mention that Mathieu Moonshine can also be formulated in terms of a mock modular form that can be naturally associated to the elliptic genus of K3 [1, 10–12]; this point of view has recently led to an interesting class of generalisations [13]. Remarkably, it turns out that these mock modular forms can be expressed in terms of Rademacher sums [11]. The analogous property for the McKay-Thompson series has been shown to be equivalent to the 'genus zero property' in [7]. Rademacher sums and their applications to Monstrous and Mathieu Moonshine are described in detail in the contribution by Cheng and Duncan in this collection. There are also indications that, just as for Monstrous Moonshine, Mathieu Moonshine can possibly be understood in terms of an underlying Borcherds-Kac-Moody algebra [10, 14–17].

2 Mathieu Moonshine

Let us first review the basic idea of 'Mathieu Moonshine'. We consider a conformal field theory sigma-model with target space K3. This theory has $\mathcal{N} = (4,4)$ superconformal symmetry on the world-sheet. As a consequence, the space of states can be decomposed into representations of the $\mathcal{N} = 4$ superconformal algebra, both for the left- and the right-movers. (The left- and right-moving actions commute, and thus we can find a simultaneous decomposition.) The full space of states takes then the form

$$\mathcal{H} = \bigoplus_{i,j} N_{ij}\, \mathcal{H}_i \otimes \bar{\mathcal{H}}_j \,, \qquad (1)$$

where i and j label the different $\mathcal{N} = 4$ superconformal representations, and $N_{ij} \in \mathbb{N}_0$ denote the multiplicities with which these representations appear. The $\mathcal{N} = 4$ algebra contains, apart from the Virasoro algebra L_n at $c = 6$, four supercharge generators, as well as an affine $\hat{\mathfrak{su}}(2)_1$ subalgebra at level one; we denote the Cartan generator of the zero mode subalgebra $\mathfrak{su}(2)$ by J_0.

The full partition function of the conformal field theory is quite complicated, and is only explicitly known at special points in the moduli space. However, there exists some sort of partial index that is much better behaved. This is the so-called *elliptic genus* that is defined by

$$\phi_{\mathrm{K3}}(\tau,z) = \mathrm{Tr}_{\mathrm{RR}}\Big(q^{L_0-\frac{c}{24}}\, y^{J_0}\, (-1)^F\, \bar{q}^{\bar{L}_0-\frac{\bar{c}}{24}}\, (-1)^{\bar{F}}\Big) \equiv \phi_{0,1}(\tau,z)\,. \qquad (2)$$

Here the trace is only taken over the Ramond–Ramond part of the spectrum (1), and the right-moving $\mathcal{N} = 4$ modes are denoted by a bar. Furthermore, $q = \exp(2\pi i\tau)$ and $y = \exp(2\pi i z)$, F and $\bar{F}$ are the left- and right-moving fermion number operators, and the two central charges equal $c = \bar{c} = 6$. Note that the elliptic genus does not actually depend on $\bar{\tau}$, although $\bar{q} = \exp(-2\pi i \bar{\tau})$ does; the reason for this is that, with respect to the right-moving algebra, the elliptic genus is like a Witten index, and only the right-moving ground states contribute. To see this one notices that states that are not annihilated by a supercharge zero mode appear always as a boson–fermion pair; the contribution of such a pair to the elliptic genus however vanishes because the two states contribute with the opposite sign (as a consequence of the $(-1)^{\bar{F}}$ factor). Thus only the right-moving ground states, i.e. the states that are annihilated by all right-moving supercharge zero modes, contribute to the elliptic genus, and the commutation relations of the $\mathcal{N} = 4$ algebra then imply that they satisfy $(\bar{L}_0 - \frac{\bar{c}}{24})\phi_{\text{ground}} = 0$; thus it follows that the elliptic genus is independent of $\bar{\tau}$. Note that this argument does not apply to the left-moving contributions because of the y^{J_0} factor. (The supercharges are 'charged' with respect to the J_0 Cartan generator, and hence the two terms of a boson–fermion pair come with different powers of y. However, if we also set $y = 1$, the elliptic genus does indeed become a constant, independent of τ and $\bar{\tau}$.)

It follows from general string considerations that the elliptic genus defines a *weak Jacobi form of weight zero and index one* [18]. Recall that a weak Jacobi form of weight w and index m is a function [19]

$$\phi_{w,m} : \mathbb{H}_+ \times \mathbb{C} \to \mathbb{C} \,, \qquad (\tau, z) \mapsto \phi_{w,m}(\tau, z) \tag{3}$$

that satisfies

$$\phi_{w,m}\Big(\frac{a\tau + b}{c\tau + d}, \frac{z}{c\tau + d}\Big) = (c\tau + d)^w \, e^{2\pi i m \frac{cz^2}{c\tau + d}} \, \phi_{w,m}(\tau, z) \qquad \begin{pmatrix} a & b \\ c & d \end{pmatrix} \in \mathrm{SL}(2, \mathbb{Z}) \,, \tag{4}$$

$$\phi(\tau, z + \ell\tau + \ell') = e^{-2\pi i m(\ell^2\tau + 2\ell z)} \phi(\tau, z) \qquad \ell, \ell' \in \mathbb{Z} \,, \tag{5}$$

and has a Fourier expansion

$$\phi(\tau, z) = \sum_{n \geq 0,\, \ell \in \mathbb{Z}} c(n, \ell) q^n y^\ell \tag{6}$$

with $c(n, \ell) = (-1)^w c(n, -\ell)$. Weak Jacobi forms have been classified, and there is only one weak Jacobi form with $w = 0$ and $m = 1$. Up to normalisation ϕ_{K3} must therefore agree with this unique weak Jacobi form $\phi_{0,1}$, which can explicitly be written in terms of Jacobi theta functions as

$$\phi_{0,1}(\tau, z) = 8 \sum_{i=2,3,4} \frac{\vartheta_i(\tau, z)^2}{\vartheta_i(\tau, 0)^2} \,. \tag{7}$$

Note that the Fourier coefficients of ϕ_{K3} are integers; as a consequence they cannot change continuously as one moves around in the moduli space of K3 sigma-models, and thus $\phi_{\rm K3}$ must be actually independent of the specific K3 sigma-model that is being considered, i.e. independent of the point in the moduli space. Here we have used that the moduli space is connected. More concretely, it can be described as the double quotient

$$\mathscr{M}_{\rm K3} = {\rm O}(\Gamma^{4,20}) \setminus {\rm O}(4,20) \,/\, {\rm O}(4) \times {\rm O}(20)\ . \tag{8}$$

We can think of the Grassmannian on the right

$${\rm O}(4,20) \,/\, {\rm O}(4) \times {\rm O}(20) \tag{9}$$

as describing the choice of a positive-definite 4-dimensional subspace $\Pi \subset \mathbb{R}^{4,20}$, while the group on the left, ${\rm O}(\Gamma^{4,20})$, leads to discrete identifications among them. Here ${\rm O}(\Gamma^{4,20})$ is the group of isometries of a given fixed unimodular lattice $\Gamma^{4,20} \subset \mathbb{R}^{4,20}$. (In physics terms, the lattice $\Gamma^{4,20}$ can be thought of as the D-brane charge lattice of string theory on K3.)

Let us denote by $\mathscr{H}^{(0)} \subset \mathscr{H}_{\rm RR}$ the subspace of (1) that consists of those RR states for which the right-moving states are ground states. (Thus $\mathscr{H}^{(0)}$ consists of the states that contribute to the elliptic genus.) $\mathscr{H}^{(0)}$ carries an action of the left-moving $\mathscr{N} = 4$ superconformal algebra, and at any point in moduli space, its decomposition is of the form

$$\mathscr{H}^{(0)} = 20 \cdot \mathscr{H}_{h=\frac{1}{4},j=0} \,\oplus\, 2 \cdot \mathscr{H}_{h=\frac{1}{4},j=\frac{1}{2}} \,\oplus\, \bigoplus_{n=1}^{\infty} D_n\, \mathscr{H}_{h=\frac{1}{4}+n,j=\frac{1}{2}}\ , \tag{10}$$

where $\mathscr{H}_{h,j}$ denotes the irreducible $\mathscr{N} = 4$ representation whose Virasoro primary states have conformal dimension h and transform in the spin j representation of $\mathfrak{su}(2)$. The multiplicities D_n are *not* constant over the moduli space, but the above argument shows that

$$A_n = {\rm Tr}_{D_n}(-1)^{\bar F} \tag{11}$$

are (where D_n is now understood not just as a multiplicity, but as a representation of the right-moving $(-1)^{\bar F}$ operator that determines the sign with which these states contribute to the elliptic genus). In this language, the elliptic genus then takes the form

$$\phi_{\rm K3}(\tau,z) = 20 \cdot \chi_{h=\frac{1}{4},j=0}(\tau,z) - 2 \cdot \chi_{h=\frac{1}{4},j=\frac{1}{2}}(\tau,z) + \sum_{n=1}^{\infty} A_n \cdot \chi_{h=\frac{1}{4}+n,j=\frac{1}{2}}(\tau,z)\ , \tag{12}$$

where $\chi_{h,j}(\tau,z)$ is the 'elliptic' genus of the corresponding $\mathscr{N} = 4$ representation,

$$\chi_{h,j}(\tau,z) = {\rm Tr}_{\mathscr{H}_{h,j}}\left(q^{L_0-\frac{c}{24}}\, y^{J_0}(-1)^F\right), \tag{13}$$

and we have used that $(-1)^{\bar{F}}$ takes the eigenvalues $+1$ and -1 on the 20- and 2-dimensional multiplicity spaces of the first two terms in (10), respectively.

The key observation of Eguchi, Ooguri & Tachikawa (EOT) [1] was that the A_n are sums of dimensions of $\mathbb{M}_{24}$ representation, in striking analogy to the original Monstrous Moonshine conjecture of [3]; the first few terms are

$$A_1 = 90 = \mathbf{45} + \overline{\mathbf{45}} \tag{14}$$

$$A_2 = 462 = \mathbf{231} + \overline{\mathbf{231}} \tag{15}$$

$$A_3 = 1540 = \mathbf{770} + \overline{\mathbf{770}} , \tag{16}$$

where $\mathbf{N}$ denotes a representation of $\mathbb{M}_{24}$ of dimension N. Actually, they guessed correctly the first six coefficients; from A_7 onwards the guesses become much more ambiguous (since the dimensions of the $\mathbb{M}_{24}$ representations are not that large) and they actually misidentified the seventh coefficient in their original analysis. (We will come back to the question of why and how one can be certain about the 'correct' decomposition shortly, see Sect. 2.2.) The alert reader will also notice that the first two coefficients in (10), namely 20 and -2, are not directly $\mathbb{M}_{24}$ representations; the correct prescription is to introduce virtual representations and to write

$$20 = \mathbf{23} - 3 \cdot \mathbf{1} , \qquad -2 = -2 \cdot \mathbf{1} . \tag{17}$$

Recall that $\mathbb{M}_{24}$ is a sporadic finite simple group of order

$$|\mathbb{M}_{24}| = 2^{10} \cdot 3^3 \cdot 5 \cdot 7 \cdot 11 \cdot 23 = 244\,823\,040 . \tag{18}$$

It has 26 conjugacy classes (which are denoted by 1A, 2A, 3A, ..., 23A, 23B, where the number refers to the order of the corresponding group element)—see Eqs. (19) and (20) below for the full list—and therefore also 26 irreducible representations whose dimensions range from $N = 1$ to $N = 10{,}395$. The Mathieu group $\mathbb{M}_{24}$ can be defined as the subgroup of the permutation group S_{24} that leaves the extended Golay code invariant; equivalently, it is the quotient of the automorphism group of the $\mathfrak{su}(2)^{24}$ Niemeier lattice, divided by the Weyl group. Thought of as a subgroup of $\mathbb{M}_{24} \subset S_{24}$, it contains the subgroup $\mathbb{M}_{23}$ that is characterised by the condition that it leaves a given (fixed) element of $\{1, \ldots, 24\}$ invariant.

2.1 Classical Symmetries

The appearance of a Mathieu group in the elliptic genus of K3 is not totally surprising in view of the Mukai theorem [20, 21]. It states that any finite group of symplectic automorphisms of a K3 surface can be embedded into the Mathieu group $\mathbb{M}_{23}$. The symplectic automorphisms of a K3 surface define symmetries that act on the multiplicity spaces of the $\mathcal{N} = 4$ representations, and therefore explain part of the above findings. However, it is also clear from Mukai's argument that they do not even account for the full $\mathbb{M}_{23}$ group. Indeed, every symplectomorphism of K3 has at least five orbits on the set $\{1, \ldots, 24\}$, and thus not all elements of $\mathbb{M}_{23}$ can be realised as a symplectomorphism. More specifically, of the 26 conjugacy classes of $\mathbb{M}_{24}$, 16 have a representative in $\mathbb{M}_{23}$, namely

$$\text{repr. in } \mathbb{M}_{23}: \quad \begin{array}{l} \text{1A, 2A, 3A, 4B, 5A, 6A, 7A, 7B, 8A (geometric)} \\ \text{11A, 14A, 14B, 15A, 15B, 23A, 23B (non-geometric)}\,, \end{array} \tag{19}$$

where 'geometric' means that a representative can be (and in fact is) realised by a geometric symplectomorphism (i.e. that the representative has at least five orbits when acting on the set $\{1, \ldots, 24\}$), while 'non-geometric' means that this is not the case. The remaining conjugacy classes do *not* have a representative in $\mathbb{M}_{23}$, and are therefore not accounted for geometrically via the Mukai theorem

$$\text{no repr. in } \mathbb{M}_{23}: \quad \text{2B, 3B, 4A, 4C, 6B, 10A, 12A, 12B, 21A, 21B}\,. \tag{20}$$

The classical symmetries can therefore only explain the symmetries in the first line of (19). Thus an additional argument is needed in order to understand the origin of the other symmetries; we shall come back to this in Sect. 3.

2.2 *Evidence for Moonshine*

As was already alluded to above, in order to determine the 'correct' decomposition of the A_n multiplicity spaces in terms of $\mathbb{M}_{24}$ representations, we need to study more than just the usual elliptic genus. By analogy with Monstrous Moonshine, the natural objects to consider are the analogues of the McKay Thompson series [22]. These are obtained from the elliptic genus upon replacing

$$A_n = \dim R_n \;\to\; \mathrm{Tr}_{R_n}(g)\,, \tag{21}$$

where $g \in \mathbb{M}_{24}$, and R_n is the $\mathbb{M}_{24}$ representation whose dimension equals the coefficient A_n; the resulting functions are then [compare (12)]

$$\phi_g(\tau, z) = \mathrm{Tr}_{\mathbf{23}-3\cdot\mathbf{1}}(g)\, \chi_{h=\frac{1}{4}, j=0}(\tau, z) - 2\,\mathrm{Tr}_{\mathbf{1}}(g)\, \chi_{h=\frac{1}{4}, j=\frac{1}{2}}(\tau, z) + \sum_{n=1}^{\infty} \mathrm{Tr}_{R_n}(g)\, \chi_{h=\frac{1}{4}+n, j=\frac{1}{2}}(\tau, z)\,. \tag{22}$$

The motivation for this definition comes from the observation that if the underlying vector space $\mathcal{H}^{(0)}$, see Eq. (10), of states contributing to the elliptic genus were to carry an action of $\mathbb{M}_{24}$, $\phi_g(\tau,z)$ would equal the 'twining elliptic genus', i.e. the elliptic genus twined by the action of g

$$\phi_g(\tau,z) = \mathrm{Tr}_{\mathscr{H}^{(0)}}\Big(g\, q^{L_0-\frac{c}{24}}\, y^{J_0}\, (-1)^F\, \bar{q}^{\bar{L}_0-\frac{\bar{c}}{24}}\, (-1)^{\bar{F}}\Big)\,. \tag{23}$$

Obviously, a priori, it is not clear what the relevant R_n in (21) are. However, we have some partial information about them:

(i) For any explicit realisation of a symmetry of a K3 sigma-model, we can calculate (23) directly. (In particular, for some symmetries, the relevant twining genera had already been calculated in [23].)
(ii) The observation of EOT determines the first six coefficients explicitly.
(iii) The twining genera must have special modular properties.

Let us elaborate on (iii). Assuming that the functions $\phi_g(\tau,z)$ have indeed an interpretation as in (23), they correspond in the usual orbifold notation of string theory to the contribution

$$\phi_g(\tau,z) \;\longleftrightarrow\; e\,\underset{g}{\square} \tag{24}$$

where e is the identity element of the group. Under a modular transformation it is believed that these twining and twisted genera transform (up to a possible phase) as

$$h\,\underset{g}{\square} \quad \xrightarrow{\begin{pmatrix} a & b \\ c & d \end{pmatrix}} \quad h^d g^c\,\underset{g^a h^b}{\square} \tag{25}$$

The twining genera (24) are therefore invariant (possibly up to a phase) under the modular transformations with

$$\gcd(a, o(g)) = 1 \qquad \text{and} \qquad c = 0 \mod o(g), \tag{26}$$

where $o(g)$ is the order of the group element g and we used that for $\gcd(a,o(g)) = 1$, the group element g^a is in the same conjugacy class as g or g^{-1}. (Because of reality, the twining genus of g and g^{-1} should be the same.) Since $ad - bc = 1$, the second condition implies the first, and we thus conclude that $\phi_g(\tau,z)$ should be (up to a possible multiplier system) a weak Jacobi form of weight zero and index one under the subgroup of $\mathrm{SL}(2,\mathbb{Z})$

$$\Gamma_0(N) = \left\{ \begin{pmatrix} a & b \\ c & d \end{pmatrix} \in \mathrm{SL}(2,\mathbb{Z}) \;:\; c = 0 \bmod (N) \right\}\,, \tag{27}$$

Table 1 Value of h for the conjugacy classes in (28)

Class	2B	3B	4A	4C	6B	10A	12A	12B	21AB
h	2	3	2	4	6	2	2	12	3

where $N = o(g)$. This is a relatively strong condition, and knowing the first few terms (for a fixed multiplier system) determines the function uniquely. In order to use this constraint, however, it is important to know the multiplier system. An ansatz (that seems to work, see below) was made in [24]

$$\phi_g\Big(\frac{a\tau+b}{c\tau+d}, \frac{z}{c\tau+d}\Big) = e^{\frac{2\pi i cd}{Nh}}\, e^{\frac{2\pi i cz^2}{c\tau+d}}\, \phi_g(\tau,z)\,, \qquad \begin{pmatrix} a & b \\ c & d \end{pmatrix} \in \Gamma_0(N)\,, \tag{28}$$

where N is again the order of g and $h|\gcd(N,12)$. The multiplier system is trivial ($h = 1$) if and only if g contains a representative in $\mathbb{M}_{23} \subset \mathbb{M}_{24}$. For the other conjugacy classes, the values are tabulated in Table 1.

It was noted in [11] that h equals the length of the shortest cycle (when interpreted as a permutation in S_{24}, see Table 1 of [11]).

Using this ansatz, explicit expressions for all twining genera were determined in [24]; independently, the same twining genera were also found (using guesses based on the cycle shapes of the corresponding S_{24} representations) in [25]. (Earlier partial results had been obtained in [10] and [26].)

These explicit expressions for the twining genera then allow for a very non-trivial check of the EOT proposal. As is clear from their definition in (22), they determine the coefficients

$$\mathrm{Tr}_{R_n}(g) \qquad \text{for all } g \in \mathbb{M}_{24} \text{ and all } n \geq 1. \tag{29}$$

This information is therefore sufficient to *determine* the representations R_n, i.e. to calculate their decomposition into irreducible $\mathbb{M}_{24}$ representations, for all n. We have worked out the decompositions explicitly for the first 500 coefficients, and we have found that each R_n can be written as a direct sum of $\mathbb{M}_{24}$ representations with non-negative integer multiplicities [24]. (Subsequently [25] tested this property for the first 600 coefficients, and apparently Tachikawa has also checked it for the first 1,000 coefficients.) Terry Gannon has informed us that this information is sufficient to prove that the same will then happen for all n [27]. In some sense this then proves the EOT conjecture.

3 Symmetries of K3 Models

While the above considerations establish in some sense the EOT conjecture, they do not offer any insight into why the elliptic genus of K3 should exhibit an $\mathbb{M}_{24}$ symmetry. This is somewhat similar to the original situation in Monstrous

Moonshine, after Conway and Norton had found the various Hauptmodules by somewhat similar techniques. Obviously, in the case of Monstrous Moonshine, many of these observations were subsequently explained by the construction of the Monster CFT (that possesses the Monster group as its automorphism group) [5, 6]. So we should similarly ask for a microscopic explanation of these findings.

In some sense it is clear what the analogue of the Monster CFT in the current context should be: we know that the function in question *is* the elliptic genus of K3. However, there is one problem with this. As we mentioned before, there is not just one K3 sigma-model, but rather a whole moduli space (see Eq. (8)) of such CFTs. So the simplest explanation of the EOT observation would be if there is (at least) one special K3 sigma-model that has $\mathbb{M}_{24}$ as its automorphism group. Actually, the relevant symmetry group should commute with the action of the $\mathcal{N} = (4,4)$ superconformal symmetry (since it should act on the multiplicity spaces in $\mathcal{H}^{(0)}$, see Eq. (10)). Furthermore, since the two $\mathcal{N} = 4$ representations with $h = \frac{1}{4}$ and $j = \frac{1}{2}$ are singlets—recall that the coefficient -2 transforms as $-2 = -2 \cdot \mathbf{1}$, see (17)—the automorphism must act trivially on the four RR ground states that transform in the $(\mathbf{2}, \mathbf{2})$ representation of the $\mathfrak{su}(2)_L \times \mathfrak{su}(2)_R$ subalgebra of $\mathcal{N} = (4,4)$. Note that these four states generate the simultaneous half-unit spectral flows in the left- and the right-moving sector; the requirement that the symmetry leaves them invariant therefore means that spacetime supersymmetry is preserved.

Recall from (8) that the different K3 sigma-models are parametrised by the choice of a positive-definite 4-dimensional subspace $\Pi \subset \mathbb{R}^{4,20}$, modulo some discrete identifications. Let us denote by G_Π the group of symmetries of the sigma-model described by Π that commute with the action of $\mathcal{N} = (4,4)$ and preserve the RR ground states in the $(\mathbf{2}, \mathbf{2})$ (see above). It was argued in [8] that G_Π is precisely the subgroup of $\mathrm{O}(\Gamma^{4,20})$ consisting of those elements that leave Π pointwise fixed. The possible symmetry groups G_Π can then be classified following essentially the paradigm of the Mukai–Kondo argument for the symplectomorphisms of K3 surfaces [20, 21]. The outcome of the analysis can be summarised by the following theorem [8]:

Theorem. *Let G be the group of symmetries of a non-linear sigma-model on K3 preserving the $\mathcal{N} = (4,4)$ superconformal algebra as well as the spectral flow operators. One of the following possibilities holds:*

(i) $G = G'.G''$, *where G' is a subgroup of $\mathbb{Z}_2^{11}$, and G'' is a subgroup of $\mathbb{M}_{24}$ with at least four orbits when acting as a permutation on $\{1, \dots, 24\}$*
(ii) $G = 5^{1+2} : \mathbb{Z}_4$
(iii) $G = \mathbb{Z}_3^4 : A_6$
(iv) $G = 3^{1+4} : \mathbb{Z}_2.G''$, *where G'' is either trivial, $\mathbb{Z}_2$ or $\mathbb{Z}_2^2$.*

Here $G = N.Q$ means that N is a normal subgroup of G, and $G/N \cong Q$; when G is the semidirect product of N and Q, we denote it by $N : Q$. Furthermore, p^{1+2n} is an extra-special group of order p^{1+2n}, which is an extension of $\mathbb{Z}_p^{2n}$ by a central element of order p.

We will give a sketch of the proof below (see Sect. 3.1), but for the moment let us comment on the implications of this result. First of all, our initial expectation from above is not realised: none of these groups $G \equiv G_\Pi$ contains $\mathbb{M}_{24}$. In particular, the twining genera for the conjugacy classes 12B, 21A, 21B, 23A, 23B of $\mathbb{M}_{24}$ cannot be realised by any symmetry of a K3 sigma-model. Thus we cannot give a direct explanation of the EOT observation along these lines.

Given that the elliptic genus is constant over the moduli space, one may then hope that we can explain the origin of $\mathbb{M}_{24}$ by 'combining' symmetries from different points in the moduli space. As we have mentioned before, this is also similar to what happens for the geometric symplectomorphisms of K3: it follows from the Mukai theorem that the Mathieu group $\mathbb{M}_{23}$ is the smallest group that contains all symplectomorphisms, but there is no K3 surface where all of $\mathbb{M}_{23}$ is realised, and indeed, certain generators of $\mathbb{M}_{23}$ can never be symmetries, see (19). However, also this explanation of the EOT observation is somewhat problematic: as is clear from the above theorem, not all symmetry groups of K3 sigma-models are in fact subgroups of $\mathbb{M}_{24}$. In particular, none of the cases (ii), (iii) and (iv) (as well as case (i) with G' non-trivial) have this property, as can be easily seen by comparing the prime factor decompositions of their orders to (18). The smallest group that contains all groups of the theorem is the Conway group Co_1, but as far as we are aware, there is no evidence of any 'Conway Moonshine' in the elliptic genus of K3.

One might speculate that, generically, the group G must be a subgroup of $\mathbb{M}_{24}$, and that the models whose symmetry group is not contained in $\mathbb{M}_{24}$ are, in some sense, special or 'exceptional' points in the moduli space. In order to make this idea precise, it is useful to analyse the exceptional models in detail. In [8], some examples have been provided of case (i) with non-trivial G' (a torus orbifold $\mathbb{T}^4/\mathbb{Z}_2$ or the Gepner model 2^4, believed to be equivalent to a $\mathbb{T}^4/\mathbb{Z}_4$ orbifold), and of case (iii) (the Gepner model 1^6, which is believed to be equivalent to a $\mathbb{T}^4/\mathbb{Z}_3$ orbifold, see also [28]). For the cases (ii) and (iv), only an existence proof was given. In Sect. 5, we will improve the situation by constructing in detail an example of case (ii), realised as an asymmetric $\mathbb{Z}_5$-orbifold of a torus $\mathbb{T}^4$. Furthermore, in Sect. 6 we will briefly discuss the $\mathbb{Z}_3$-orbifold of a torus and the explicit realisation of its symmetry group, corresponding to cases (ii) and (iv) for any G''.

Notice that all the examples of exceptional models known so far are provided by torus orbifolds. In fact, we will show below (see Sect. 4) that all cyclic torus orbifolds have exceptional symmetry groups. Conversely, we will prove that the cases (ii)–(iv) of the theorem are always realised by (cyclic) torus orbifolds. On the other hand, as we shall also explain, some of the exceptional models in case (i) are not cyclic torus orbifolds.

3.1 Sketch of the Proof of the Theorem

In this subsection, we will describe the main steps in the proof of the above theorem; the details can be found in [8].

It was argued in [8] that the supersymmetry preserving automorphisms of the non-linear sigma-model characterised by Π generate the group $G \equiv G_\Pi$ that consists of those elements of $O(\Gamma^{4,20})$ that leave Π pointwise fixed. Let us denote by L^G the sublattice of G-invariant vectors of $L \equiv \Gamma^{4,20}$, and define L_G to be its orthogonal complement that carries a genuine action of G. Since $L^G \otimes \mathbb{R}$ contains the subspace Π, it follows that L^G has signature $(4, d)$ for some $d \geq 0$, so that L_G is a negative-definite lattice of rank $20-d$. In [8], it is proved that, for any consistent model, L_G can be embedded (up to a change of sign in its quadratic form) into the Leech lattice Λ, the unique 24-dimensional positive-definite even unimodular lattice containing no vectors of squared norm 2. Furthermore, the action of G on L_G can be extended to an action on the whole of Λ, such that the sublattice $\Lambda^G \subset \Lambda$ of vectors fixed by G is the orthogonal complement of L_G in Λ. This construction implies that G must be a subgroup of $\mathrm{Co}_0 \equiv \mathrm{Aut}(\Lambda)$ that fixes a sublattice Λ^G of rank $4 + d$. Conversely, it can be shown that any such subgroup of $\mathrm{Aut}(\Lambda)$ is the symmetry group of some K3 sigma-model.

This leaves us with characterising the possible subgroups of the finite group $\mathrm{Co}_0 \equiv \mathrm{Aut}(\Lambda)$ that stabilise a suitable sublattice; problems of this kind have been studied in the mathematical literature before. In particular, the stabilisers of sublattices of rank at least 4 are, generically, the subgroups of $\mathbb{Z}_2^{11} : \mathbb{M}_{24}$ described in case (i) of the theorem above. The three cases (ii), (iii), (iv) arise when the invariant sublattice Λ^G is contained in some $\mathscr{S}$-lattice $S \subset \Lambda$. An $\mathscr{S}$-lattice S is a primitive sublattice of Λ such that each vector of S is congruent modulo $2S$ to a vector of norm 0, 4 or 6. Up to isomorphisms, there are only three kind of $\mathscr{S}$-lattices of rank at least four; their properties are summarised in the following table:

Name	type	rk S	Stab(S)	Aut(S)
$(A_2 \oplus A_2)'(3)$	$2^9 3^6$	4	$\mathbb{Z}_3^4 : A_6$	$\mathbb{Z}_2 \times (S_3 \times S_3).\mathbb{Z}_2$
$A_4^*(5)$	$2^5 3^{10}$	4	$5^{1+2} : \mathbb{Z}_4$	$\mathbb{Z}_2 \times S_5$
$E_6^*(3)$	$2^{27} 3^{36}$	6	$3^{1+4} : \mathbb{Z}_2$	$\mathbb{Z}_2 \times W(E_6)$.

Here, S is characterised by the type $2^p 3^q$, which indicates that S contains p pairs of opposite vectors of norm 4 (type 2) and q pairs of opposite vectors of norm 6 (type 3). The group Stab(S) is the pointwise stabiliser of S in Co_0 and Aut(S) is the quotient of the setwise stabiliser of S modulo its pointwise stabiliser Stab(S). The group Aut(S) always contains a central $\mathbb{Z}_2$ subgroup, generated by the transformation that inverts the sign of all vectors of the Leech lattice. The lattice of type $2^{27}3^{36}$ is isomorphic to the weight lattice (the dual of the root lattice) of E_6 with quadratic form rescaled by 3 (i.e. the roots in $E_6^*(3)$ have squared norm 6), and $\mathrm{Aut}(S)/\mathbb{Z}_2$ is isomorphic to the Weyl group $W(E_6)$ of E_6. Similarly, the lattice of type $2^5 3^{10}$ is the weight lattice of A_4 rescaled by 5, and $\mathrm{Aut}(S)/\mathbb{Z}_2$ is isomorphic to the Weyl group $W(A_4) \cong S_5$ of A_4. Finally, the type $2^9 3^6$ is the three-rescaling of a lattice $(A_2 \oplus A_2)'$ obtained by adjoining to the root lattice $A_2 \oplus A_2$ an element $(e_1^*, e_2^*) \in A_2^* \oplus A_2^*$, with e_1^* and e_2^* of norm 2/3. The latter $\mathscr{S}$-lattice can also be described as the sublattice of vectors of $E_6^*(3)$ that are orthogonal to an $A_2(3)$

sublattice of $E_6^*(3)$. The group $\mathrm{Aut}(S)/\mathbb{Z}_2$ is the product $(S_3 \times S_3).\mathbb{Z}_2$ of the Weyl groups $W(A_2) = S_3$, and the $\mathbb{Z}_2$ symmetry that exchanges the two A_2 and maps e_1^* to e_2^*.

The cases (ii)–(iv) of the above theorem correspond to Λ^G being isomorphic to $A_4^*(5)$ (case ii), to $(A_2 \oplus A_2)'(3)$ (case iii) or to a sublattice of $E_6^*(3)$ different from $(A_2 \oplus A_2)'(3)$ (case iv). In the cases (ii) and (iii), G is isomorphic to $\mathrm{Stab}(S)$. In case (iv), $\mathrm{Stab}(S)$ is, generically, a normal subgroup of G, and $G'' \cong G/\mathrm{Stab}(S)$ is a subgroup of $\mathrm{Aut}(S) \cong \mathbb{Z}_2 \times W(E_6)$ that fixes a sublattice $\Lambda^G \subseteq E_6^*(3)$, with $\Lambda^G \neq (A_2 \oplus A_2)'(3)$, of rank at least 4. The only non-trivial subgroups of $\mathbb{Z}_2 \times W(E_6)$ with these properties are $G'' = \mathbb{Z}_2$, which corresponds to Λ^G being orthogonal to a single vector of norm 6 in $E_6^*(3)$ (a rescaled root), and $G'' = \mathbb{Z}_2^2$, which corresponds to Λ^G being orthogonal to two orthogonal vectors of norm 6.[3] If Λ^G is orthogonal to two vectors $v_1, v_2 \in E_6^*(3)$ of norm 6, with $v_1 \cdot v_2 = -3$, then $\Lambda^G \cong (A_2 \oplus A_2)'(3)$ and case (iii) applies.

4 Symmetry Groups of Torus Orbifolds

In this section we will prove that all K3 sigma-models that are realised as (possibly left-right asymmetric) orbifolds of $\mathbb{T}^4$ by a cyclic group have an 'exceptional' group of symmetries, i.e. their symmetries are not a subgroup of $\mathbb{M}_{24}$. Furthermore, these torus orbifolds account for most of the exceptional models (in particular, for all models in the cases (ii)–(iv) of the theorem). On the other hand, as we shall also explain, there are exceptional models in case (i) that are not cyclic torus orbifolds.

Our reasoning is somewhat reminiscent of the construction of [29, 30] in the context of Monstrous Moonshine. Any $\mathbb{Z}_n$-orbifold of a conformal field theory has an automorphism g of order n, called the *quantum symmetry*, which acts trivially on the untwisted sector and by multiplication by the phase $\exp(\frac{2\pi i k}{n})$ on the k-th twisted sector. Furthermore, the orbifold of the orbifold theory by the group generated by the quantum symmetry g, is equivalent to the original conformal field theory [31]. This observation is the key for characterising K3 models that can be realised as torus orbifolds:

A K3 model $\mathscr{C}$ is a $\mathbb{Z}_n$-orbifold of a torus model if and only if it has a symmetry g of order n such that $\mathscr{C}/\langle g\rangle$ is a consistent orbifold equivalent to a torus model.

In order to see this, suppose that $\mathscr{C}_{K3}$ is a K3 sigma-model that can be realised as a torus orbifold $\mathscr{C}_{K3} = \tilde{\mathscr{C}}_{\mathbb{T}^4}/\langle \tilde{g}\rangle$, where $\tilde{g}$ is a symmetry of order n of the torus model $\tilde{\mathscr{C}}_{\mathbb{T}^4}$. Then $\mathscr{C}_{K3}$ possesses a 'quantum symmetry' g of order n, such that the orbifold of $\mathscr{C}_{K3}$ by g describes again the original torus model, $\tilde{\mathscr{C}}_{\mathbb{T}^4} = \mathscr{C}_{K3}/\langle g\rangle$.

Conversely, suppose $\mathscr{C}_{K3}$ has a symmetry g of order n, such that the orbifold of $\mathscr{C}_{K3}$ by g is consistent, i.e. satisfies the level matching condition—this is the case

[3]The possibility $G'' = \mathbb{Z}_4$ that has been considered in [8] has to be excluded, since there are no elements of order 4 in $W(E_6)$ that preserve a four-dimensional sublattice of $E_6^*(3)$.

if and only if the twining genus ϕ_g has a trivial multiplier system—and leads to a torus model $\mathscr{C}_{K3}/\langle g\rangle = \tilde{\mathscr{C}}_{\mathbb{T}^4}$. Then $\mathscr{C}_{K3}$ itself is a torus orbifold since we can take the orbifold of $\tilde{\mathscr{C}}_{\mathbb{T}^4}$ by the quantum symmetry associated to g, and this will, by construction, lead back to $\mathscr{C}_{K3}$.

Thus we conclude that $\mathscr{C} \equiv \mathscr{C}_{K3}$ can be realised as a torus orbifold if and only if $\mathscr{C}$ contains a symmetry g such that (i) ϕ_g has a trivial multiplier system; and (ii) the orbifold of $\mathscr{C}$ by g leads to a torus model $\tilde{\mathscr{C}}_{\mathbb{T}^4}$. It is believed that the orbifold of $\mathscr{C}$ by any $\mathscr{N} = (4,4)$-preserving symmetry group, if consistent, will describe a sigma-model with target space either a torus $\mathbb{T}^4$ or a K3 manifold. The two cases can be distinguished by calculating the elliptic genus; in particular, if the target space is a torus, the elliptic genus vanishes. Actually, since the space of weak Jacobi forms of weight zero and index one is 1-dimensional, this condition is equivalent to the requirement that the elliptic genus $\tilde{\phi}(\tau,z)$ of $\tilde{\mathscr{C}} = \mathscr{C}/\langle g\rangle$ vanishes at $z=0$.

Next we recall that the elliptic genus of the orbifold by a group element g of order $n = o(g)$ is given by the usual orbifold formula

$$\tilde{\phi}(\tau,z) = \frac{1}{n}\sum_{i,j=1}^{n} \phi_{g^i,g^j}(\tau,z)\,, \tag{30}$$

where $\phi_{g^i,g^j}(\tau,z)$ is the twining genus for g^j in the g^i-twisted sector; this can be obtained by a modular transformation from the untwisted twining genus $\phi_{g^d}(\tau,z)$ with $d = \gcd(i,j,n)$. As we have explained above, it is enough to evaluate the elliptic genus for $z=0$. Then

$$\phi_{g^d}(\tau,z=0) = \mathrm{Tr}_{\mathbf{24}}(g^d)\,, \tag{31}$$

where $\mathrm{Tr}_{\mathbf{24}}(g^d)$ is the trace of g^d over the 24-dimensional space of RR ground states, and since (31) is constant (and hence modular invariant) we conclude that

$$\tilde{\phi}(\tau,0) = \frac{1}{n}\sum_{i,j=1}^{n} \mathrm{Tr}_{\mathbf{24}}(g^{\gcd(i,j,n)})\,. \tag{32}$$

According to the theorem in Sect. 3, all symmetry groups of K3 sigma-models are subgroups of Co_0 and, in fact, $\mathrm{Tr}_{\mathbf{24}}(g^d)$ coincides with the trace of $g^d \in \mathrm{Co}_0$ in the standard 24-dimensional representation of Co_0. Thus, the elliptic genus of the orbifold model $\tilde{\mathscr{C}} = \mathscr{C}/\langle g\rangle$ only depends on the conjugacy class of g in Co_0. The group Co_0 contains 167 conjugacy classes, but only 42 of them contain symmetries that are realised by some K3 sigma-model, i.e. elements that fix at least a four-dimensional subspace in the standard 24-dimensional representation of Co_0. If $\mathrm{Tr}_{\mathbf{24}}(g) \neq 0$ (this happens for 31 of the above 42 conjugacy classes), the twining genus $\phi_g(\tau,z)$ has necessarily a trivial multiplier system, and the orbifold $\mathscr{C}/\langle g\rangle$ is consistent. These classes are listed in the following table, together with the dimension of the space that is fixed by g, the trace over the 24-dimensional representation, and the elliptic genus $\tilde{\phi}(\tau, z = 0)$ of the orbifold model $\tilde{\mathscr{C}}$ (we underline the classes that restrict to $\mathbb{M}_{24}$ conjugacy classes):

Co_0-class	$\underline{1A}$	$\underline{2B}$	2C	$\underline{3B}$	3C	4B	$\underline{4E}$	4F	$\underline{5B}$	5C	6G	6H	6I	$\underline{6K}$	6L	6M	$\underline{7B}$
dim fix	24	16	8	12	6	8	10	6	8	4	6	6	6	8	4	4	6
$\mathrm{Tr}_{\mathbf{24}}(g)$	24	8	−8	6	−3	8	4	−4	4	−1	−4	4	5	2	−2	−1	3
$\tilde{\phi}(\tau, 0)$	24	24	0	24	0	24	24	0	24	0	0	24	24	24	0	0	24

Co_0-class	8D	$\underline{8G}$	8H	9C	10F	10G	10H	$\underline{11A}$	12I	12L	12N	12O	$\underline{14C}$	$\underline{15D}$
dim fix	4	6	4	4	4	4	4	4	4	4	4	4	4	4
$\mathrm{Tr}_{\mathbf{24}}(g)$	4	2	−2	3	−2	2	3	2	2	1	−2	2	1	1
$\tilde{\phi}(\tau, 0)$	24	24	0	24	0	24	24	24	24	24	0	24	24	24

Note that the elliptic genus of the orbifold theory $\tilde{\mathscr{C}}$ is always 0 or 24, corresponding to a torus or a K3 sigma-model, respectively. Out of curiosity, we have also computed the putative elliptic genus $\tilde{\phi}(\tau, 0)$ for the 11 classes of symmetries g with $\mathrm{Tr}_{24}(g) = 0$ for which we do not expect the orbifold to make sense—the corresponding twining genus ϕ_g will typically have a non-trivial multiplier system, and hence the orbifold will not satisfy level-matching. Indeed, for almost none of these cases is $\tilde{\phi}(\tau, 0)$ equal to 0 or 24, thus signaling an inconsistency of the orbifold model:

Co_0-class	$\underline{2D}$	$\underline{3D}$	4D	$\underline{4G}$	$\underline{4H}$	6O	$\underline{6P}$	8C	8I	$\underline{10J}$	$\underline{12P}$
dim fix	12	8	4	8	6	6	4	4	4	4	4
$\mathrm{Tr}_{\mathbf{24}}(g)$	0	0	0	0	0	0	0	0	0	0	0
$\tilde{\phi}(\tau, 0)$	12	8	0	12	6	12	4	12	6	12	12

The only exception is the class 4D, which might define a consistent orbifold (a torus model). It follows that a K3 model $\mathscr{C}$ is the $\mathbb{Z}_n$-orbifold of a torus model if and only if it contains a symmetry g in one of the classes

$$\begin{gathered}2C,\ 3C,\ 4F,\ 5C,\ 6G,\ 6L,\ 6M,\ 8H,\ 10F,\ 12N, \\ 4B,\ 4D,\ 6H,\ 6I,\ 8C,\ 8D,\ 9C,\ 10G,\ 10H,\ 12I,\ 12L,\ 12O\end{gathered} \tag{33}$$

of Co_0.[4] Here we have also included (in the second line) those classes of elements $g \in \mathrm{Co}_0$ for which $\mathscr{C}/\langle g^i \rangle$ is a torus model, for some power $i > 1$. Our main

[4]We should emphasise that for us the term 'orbifold' always refers to a conformal field theory (rather than a geometrical) construction. Although a non-linear sigma-model on a geometric orbifold $M/\mathbb{Z}_N$ always admits an interpretation as a CFT orbifold, the converse is not always true. In particular, there exist asymmetric orbifold constructions that do not have a direct geometric interpretation, see for example Sect. 5.

observation is now that none of the Co_0 classes in (33) restricts to a class in $\mathbb{M}_{24}$, i.e.

All K3 models that are realised as $\mathbb{Z}_n$-orbifolds of torus models are exceptional. In particular, the quantum symmetry is not an element of $\mathbb{M}_{24}$.

One might ask whether the converse is also true, i.e. whether all exceptional models are cyclic torus orbifolds. This is not quite the case: for example, the classification theorem of Sect. 3 predicts the existence of models with a symmetry group $G \cong GL_2(3)$ (the group of 2×2 invertible matrices on the field $\mathbb{F}_3$ with three elements). The group G contains no elements in the classes (33), so the model is not a cyclic torus orbifold; on the other hand, G contains elements in the class 8I of Co_0, which does not restrict to $\mathbb{M}_{24}$. A second counterexample is a family of models with a symmetry g in the class 6O of Co_0. A generic point of this family is not a cyclic torus model (although some special points are), since the full symmetry group is generated by g and contains no elements in (33). Both these counterexamples belong to case (i) of the general classification theorem. In fact, we can prove that

The symmetry group G of a K3 sigma-model $\mathscr{C}$ contains a subgroup $3^{1+4} : \mathbb{Z}_2$ [cases (iii) and (iv) of the theorem] if and only if $\mathscr{C}$ is a $\mathbb{Z}_3$-orbifold of a torus model. Furthermore, $G = 5^{1+2} : \mathbb{Z}_4$ [case (ii)] if and only if $\mathscr{C}$ is a $\mathbb{Z}_5$-orbifold of a torus model.

The proof goes as follows. All subgroups of Co_0 of the form $3^{1+4} : \mathbb{Z}_2$ (respectively, $5^{1+2} : \mathbb{Z}_4$) contain an element in the class 3C (resp., 5C), and therefore the corresponding models are $\mathbb{Z}_3$ (resp., $\mathbb{Z}_5$) torus orbifolds. Conversely, consider a $\mathbb{Z}_3$-orbifold of a torus model. Its symmetry group G contains the quantum symmetry g in class 3C of Co_0. (It must contain a symmetry generator of order three whose orbifold leads to a torus, and 3C is then the only possibility.) The sublattice $\Lambda^{\langle g\rangle} \subset \Lambda$ fixed by g is the $\mathscr{S}$-lattice $2^{27}3^{36}$ [32]. From the classification theorem, we know that G is the stabiliser of a sublattice $\Lambda^G \subset \Lambda$ of rank at least 4. Since $\Lambda^G \subseteq \Lambda^{\langle g\rangle}$, G contains as a subgroup the stabiliser of $\Lambda^{\langle g\rangle}$, namely $3^{1+4} : \mathbb{Z}_2$.

Analogously, a $\mathbb{Z}_5$ torus orbifold always has a symmetry in class 5C, whose fixed sublattice $\Lambda^{\langle g\rangle}$ is the $\mathscr{S}$-lattice $2^5 3^{10}$ [32]. Since $\Lambda^{\langle g\rangle}$ has rank 4 and is primitive, $\Lambda^G = \Lambda^{\langle g\rangle}$ and the symmetry group G must be the stabiliser $5^{1+2} : \mathbb{Z}_4$ of $\Lambda^{\langle g\rangle}$.

It was shown in [8] that the Gepner model $(1)^6$ corresponds to the case (ii) of the classification theorem. It thus follows from the above reasoning that it must indeed be equivalent to a $\mathbb{Z}_3$-orbifold of $\mathbb{T}^4$, see also [28]. (We shall also study these orbifolds more systematically in Sect. 6.) In the next section, we will provide an explicit construction of a $\mathbb{Z}_5$-orbifold of a torus model and show that its symmetry group is $5^{1+2} : \mathbb{Z}_4$, as predicted by the above analysis.

5 A K3 Model with Symmetry Group $5^{1+2} : \mathbb{Z}_4$

In this section we will construct a supersymmetric sigma-model on $\mathbb{T}^4$ with a symmetry g of order 5 commuting with an $\mathscr{N} = (4,4)$ superconformal algebra and acting asymmetrically on the left- and on the right-moving sector. The orbifold of this model by g will turn out to be a well-defined SCFT with $\mathcal{N} = (4,4)$ (in particular, the level matching condition is satisfied) that can be interpreted as a non-linear sigma-model on K3. We will argue that the group of symmetries of this model is $G = 5^{1+2}.\mathbb{Z}_4$, one of the exceptional groups considered in the general classification theorem.

5.1 The Torus Model

Let us consider a supersymmetric sigma-model on the torus $\mathbb{T}^4$. Geometrically, we can characterise the theory in terms of a metric and a Kalb–Ramond field, but it is actually more convenient to describe it simply as a conformal field theory that is generated by the following fields: four left-moving $u(1)$ currents $\partial X^a(z)$, $a = 1, \ldots, 4$, four free fermions $\psi^a(z)$, $a = 1, \ldots, 4$, their right-moving analogues $\bar\partial X^a(\bar z)$, $\tilde\psi^a(\bar z)$, as well as some winding-momentum fields $V_\lambda(z, \bar z)$ that are associated to vectors λ in an even unimodular lattice $\Gamma^{4,4}$ of signature $(4,4)$. The mode expansions of the left-moving fields are

$$\partial X^a(z) = \sum_{n\in\mathbb{Z}} \alpha_n z^{-n-1}\,, \qquad \psi^a = \sum_{n\in\mathbb{Z}+\nu} \psi_n z^{-n-\frac{1}{2}}\,, \tag{34}$$

where $\nu = 0, 1/2$ in the R- and NS-sector, respectively. Furthermore, we have the usual commutation relations

$$[\alpha^a_m, \alpha^b_n] = m\,\delta^{ab}\,\delta_{m,-n} \qquad \{\psi^a_m, \psi^b_n\} = \delta^{ab}\,\delta_{m,-n}\,. \tag{35}$$

Analogous statements also hold for the right-moving modes $\tilde\alpha_n$ and $\tilde\psi_n$. The vectors $\lambda \equiv (\lambda_L, \lambda_R) \in \Gamma^{4,4}$ describe the eigenvalues of the corresponding state with respect to the left- and right-moving zero modes α^a_0 and $\tilde\alpha^a_0$, respectively. In these conventions the inner product on $\Gamma^{4,4}$ is given as

$$(\lambda, \lambda') = \lambda_L \cdot \lambda'_L - \lambda_R \cdot \lambda'_R\,. \tag{36}$$

Continuous and Discrete Symmetries

Any torus model contains an $\hat{\mathfrak{su}}(2)_1 \oplus \hat{\mathfrak{su}}(2)_1 \oplus \hat{\mathfrak{u}}(1)^4$ current algebra, both on the left and on the right. Here, the $\hat{\mathfrak{u}}(1)^4$ currents are the ∂X^a, $a = 1, \ldots, 4$, while $\hat{\mathfrak{su}}(2)_1 \oplus \hat{\mathfrak{su}}(2)_1 = \hat{\mathfrak{so}}(4)_1$ is generated by the fermionic bilinears

$$a^3 := \bar\psi^{(1)}\psi^{(1)} + \bar\psi^{(2)}\psi^{(2)} \qquad a^+ := \bar\psi^{(1)}\bar\psi^{(2)} \qquad a^- := -\psi^{(1)}\psi^{(2)}\,, \tag{37}$$

$$\hat a^3 := \bar\psi^{(1)}\psi^{(1)} - \bar\psi^{(2)}\psi^{(2)} \qquad \hat a^+ := \bar\psi^{(1)}\psi^{(2)} \qquad \hat a^- := -\psi^{(1)}\bar\psi^{(2)}\,, \tag{38}$$

where

$$\psi^{(1)} = \tfrac{1}{\sqrt{2}}(\psi^1 + i\psi^2) \qquad\qquad \psi^{(2)} = \tfrac{1}{\sqrt{2}}(\psi^3 + i\psi^4) \tag{39}$$

$$\bar\psi^{(1)} = \tfrac{1}{\sqrt{2}}(\psi^1 - i\psi^2) \qquad\qquad \bar\psi^{(2)} = \tfrac{1}{\sqrt{2}}(\psi^3 - i\psi^4)\,. \tag{40}$$

At special points in the moduli space, where the $\Gamma^{4,4}$ lattice contains vectors of the form $(\lambda_L, 0)$ with $\lambda_L^2 = 2$, the bosonic $\mathfrak{u}(1)^4$ algebra is enhanced to some non-abelian algebra $\mathfrak{g}$ of rank 4. There are generically 16 (left-moving) supercharges; they form four $(\mathbf{2},\mathbf{2})$ representations of the $\mathfrak{su}(2)\oplus\mathfrak{su}(2)$ zero mode algebra from (37) and (38). Altogether, the chiral algebra at generic points is a *large* $\mathcal{N}=4$ superconformal algebra.

We want to construct a model with a symmetry g of order 5, acting non-trivially on the fermionic fields, and commuting with the *small* $\mathcal{N}=4$ subalgebras both on the left and on the right. A small $\mathcal{N}=4$ algebra contains an $\hat{\mathfrak{su}}(2)_1$ current algebra and four supercharges in two doublets of $\mathfrak{su}(2)$. The symmetry g acts by an $O(4,\mathbb{R})$ rotation on the left-moving fermions ψ^a, preserving the anti-commutation relations (35). Without loss of generality, we may assume that $\psi^{(1)}$ and $\bar\psi^{(1)}$ are eigenvectors of g with eigenvalues ζ and ζ^{-1}, where ζ is a primitive fifth root of unity

$$\zeta^5 = 1\,, \tag{41}$$

and that the $\hat{\mathfrak{su}}(2)_1$ algebra preserved by g is (37). This implies that g acts on all the fermionic fields by

$$\psi^{(1)} \mapsto \zeta\,\psi^{(1)}\,,\quad \bar\psi^{(1)} \mapsto \zeta^{-1}\bar\psi^{(1)}\,,\quad \psi^{(2)} \mapsto \zeta^{-1}\psi^{(2)}\,,\quad \bar\psi^{(2)} \mapsto \zeta\,\bar\psi^{(2)}\,. \tag{42}$$

Note that the action of g on the fermionic fields can be described by $e^{\frac{2\pi i k}{5}\hat a^3_0}$ for some $k = 1,\ldots,4$, where $\hat a^3$ is the current in the algebra (38). The four g-invariant supercharges can then be taken to be

$$\begin{aligned}&\sqrt{2}\sum_{i=1}^{2} J^{(i)}\bar\psi^{(i)}\,,\quad \sqrt{2}\sum_{i=1}^{2}\bar J^{(i)}\psi^{(i)}\,,\\ &\sqrt{2}(\bar J^{(1)}\bar\psi^{(2)} - \bar J^{(2)}\bar\psi^{(1)})\,,\quad \sqrt{2}(J^{(1)}\psi^{(2)} - J^{(2)}\psi^{(1)})\,,\end{aligned} \tag{43}$$

where $J^{(1)}, \bar J^{(1)}, J^{(2)}, \bar J^{(2)}$ are suitable (complex) linear combinations of the left-moving currents ∂X^a, $a = 1,\ldots,4$. In order to preserve the four supercharges, g must act with the same eigenvalues on the bosonic currents

$$J^{(1)} \mapsto \zeta\,J^{(1)}\,,\quad \bar J^{(1)} \mapsto \zeta^{-1}\bar J^{(1)}\,,\quad J^{(2)} \mapsto \zeta^{-1}J^{(2)}\,,\quad \bar J^{(2)} \mapsto \zeta\,\bar J^{(2)}\,. \tag{44}$$

A similar reasoning applies to the right-moving algebra with respect to an eigenvalue $\tilde\zeta$, with $\tilde\zeta^5 = 1$. For the symmetries with a geometric interpretation, the action on the left- and right-moving bosonic currents is induced by an $\mathrm{O}(4,\mathbb{R})$-transformation on the scalar fields X^a, $a, = 1, \ldots, 4$, representing the coordinates on the torus; then ζ and $\tilde{\zeta}$ are necessarily equal. In our treatment, we want to allow for the more general case where $\zeta \neq \tilde{\zeta}$.

The action of g on J^a and $\tilde{J}^a$ induces an $\mathrm{O}(4,4,\mathbb{R})$-transformation on the lattice $\Gamma^{4,4}$. The transformation g is a symmetry of the model if and only if it induces an automorphism on $\Gamma^{4,4}$. In particular, it must act by an invertible integral matrix on any lattice basis. The requirement that the trace of this matrix (and of any power of it) must be integral, leads to the condition that

$$2(\zeta^i + \zeta^{-i} + \tilde{\zeta}^i + \tilde{\zeta}^{-i}) \in \mathbb{Z}\,, \tag{45}$$

for all $i \in \mathbb{Z}$. For g of order 5, this condition is satisfied by

$$\zeta = e^{\frac{2\pi i}{5}} \qquad \text{and} \qquad \tilde{\zeta} = e^{\frac{4\pi i}{5}}\,, \tag{46}$$

and this solution is essentially unique (up to taking powers of it or exchanging the definition of ζ and ζ^{-1}). Equation (46) shows that a supersymmetry preserving symmetry of order 5 is necessarily left-right asymmetric, and hence does not have a geometric interpretation.

It is now clear how to construct a torus model with the symmetries (42) and (44). First of all, we need an automorphism g of $\Gamma^{4,4}$ of order five. Such an automorphism must have eigenvalues $\zeta, \zeta^2, \zeta^3, \zeta^4$, each corresponding to two independent eigenvectors $v^{(1)}_{\zeta^i}, v^{(2)}_{\zeta^i}$, $i = 1, \ldots, 4$, in $\Gamma^{4,4} \otimes \mathbb{C}$. Given the discussion above, see in particular (46), we now require that the vectors

$$v^{(1)}_{\zeta^1}\,, \quad v^{(2)}_{\zeta^1}\,, \quad v^{(1)}_{\zeta^4}\,, \quad v^{(2)}_{\zeta^4} \tag{47}$$

span a positive-definite subspace of $\Gamma^{4,4} \otimes \mathbb{C}$ (i.e. correspond to the left-movers), while the vectors

$$v^{(1)}_{\zeta^2}\,, \quad v^{(2)}_{\zeta^2}\,, \quad v^{(1)}_{\zeta^3}\,, \quad v^{(2)}_{\zeta^3} \tag{48}$$

span a negative-definite subspace of $\Gamma^{4,4} \otimes \mathbb{C}$ (i.e. correspond to the right-movers).

An automorphism g with the properties above can be explicitly constructed as follows. Let us consider the real vector space with basis vectors $x_1, \ldots, x_4$, and $y_1, \ldots, y_4$, and define a linear map g of order 5 by

$$g(x_i) = x_{i+1}\,, \qquad g(y_i) = y_{i+1}\,, \qquad\qquad i = 1, \ldots, 3\,, \tag{49}$$

and

$$g(x_4) = -(x_1 + x_2 + x_3 + x_4)\,, \qquad g(y_4) = -(y_1 + y_2 + y_3 + y_4)\,. \tag{50}$$

A g-invariant bilinear form on the space is uniquely determined by the conditions

$$(x_i, x_i) = 0 = (y_i, y_i) , \qquad i = 1, \ldots, 4 \tag{51}$$

and

$$\begin{aligned} (x_1, x_2) = 1 , \quad (x_1, x_3) = (x_1, x_4) = -1 , \\ (y_1, y_2) = 1 , \quad (y_1, y_3) = (y_1, y_4) = -1 , \end{aligned} \tag{52}$$

as well as

$$(x_1, y_1) = 1 , \quad \text{and} \quad (x_i, y_1) = 0 , \ (i = 2, 3, 4) . \tag{53}$$

The lattice spanned by these basis vectors is an indefinite even unimodular lattice of rank 8 and thus necessarily isomorphic to $\Gamma^{4,4}$. The g-eigenvectors can be easily constructed in terms of the basis vectors and one can verify that the eigenspaces have the correct signature.

This torus model has an additional $\mathbb{Z}_4$ symmetry group that preserves the superconformal algebra and normalises the group generated by g. The generator h of this group acts by

$$h(x_i) := g^{1-i}(x_1 + x_4 + 2y_1 + y_2 + y_3 + y_4) , \tag{54}$$

$$h(y_i) := g^{1-i}(-2x_1 - x_2 - x_3 - x_4 - y_1 - y_3 - y_4) , \quad i = 1, \ldots, 4 , \tag{55}$$

on the lattice vectors. The g-eigenvectors $v^{(a)}_{\zeta^i}$, $a = 1, 2, i = 1, \ldots, 4$ can be defined as

$$v^{(1)}_{\zeta^i} := \sum_{j=0}^{4} \zeta^{-ij} g^j (x_1 + h(x_1)) , \qquad v^{(2)}_{\zeta^i} := \sum_{j=0}^{4} \zeta^{-ij} g^j (x_1 - h(x_1)) , \tag{56}$$

so that

$$h(v^{(1)}_{\zeta^i}) = -v^{(2)}_{\zeta^{-i}} , \qquad h(v^{(2)}_{\zeta^i}) = v^{(1)}_{\zeta^{-i}} . \tag{57}$$

Correspondingly, the action of h on the left-moving fields is

$$\psi^{(1)} \mapsto -\psi^{(2)} , \quad \psi^{(2)} \mapsto \psi^{(1)} , \quad \bar{\psi}^{(1)} \mapsto -\bar{\psi}^{(2)} , \quad \bar{\psi}^{(2)} \mapsto \bar{\psi}^{(1)} , \tag{58}$$

$$J^{(1)} \mapsto -J^{(2)} , \quad J^{(2)} \mapsto J^{(1)} , \quad \bar{J}^{(1)} \mapsto -\bar{J}^{(2)} , \quad \bar{J}^{(2)} \mapsto \bar{J}^{(1)} , \tag{59}$$

and the action on the right-moving fields is analogous. It is immediate to verify that the generators of the superconformal algebra are invariant under this transformation.

5.2 *The Orbifold Theory*

Next we want to consider the orbifold of this torus theory by the group $\mathbb{Z}_5$ that is generated by g.

The Elliptic Genus

The elliptic genus of the orbifold theory can be computed by summing over the $\mathrm{SL}(2,\mathbb{Z})$ images of the untwisted sector contribution, which in turn is given by

$$\phi^U(\tau,z) = \frac{1}{5}\sum_{k=0}^{4}\phi_{1,g^k}(\tau,z)\,, \tag{60}$$

where

$$\phi_{1,g^k}(\tau,z) = \mathrm{Tr}_{\mathrm{RR}}(g^k\, q^{L_0-\frac{c}{24}}\bar{q}^{\bar{L}_0-\frac{\bar{c}}{24}}y^{2J_0}(-1)^{F+\tilde{F}})\,. \tag{61}$$

The $k=0$ contribution, i.e. the elliptic genus of the original torus theory, is zero. Each g^k-contribution, for $k=1,\ldots,4$, is the product of a factor coming from the ground states, one from the oscillators and one from the momenta

$$\phi_{1,g^k}(\tau,z) = \phi^{\mathrm{gd}}_{1,g^k}(\tau,z)\,\phi^{\mathrm{osc}}_{1,g^k}(\tau,z)\,\phi^{\mathrm{mom}}_{1,g^k}(\tau,z)\,. \tag{62}$$

These contributions are, respectively,

$$\phi^{\mathrm{gd}}_{1,g^k}(\tau,z) = y^{-1}(1-\zeta^k y)(1-\zeta^{-k}y)(1-\zeta^{2k})(1-\zeta^{-2k}) = 2y^{-1}+2y+1\,, \tag{63}$$

$$\phi^{\mathrm{osc}}_{1,g^k}(\tau,z) = \prod_{n=1}^{\infty}\frac{(1-\zeta^k yq^n)(1-\zeta^{-k}yq^n)(1-\zeta^k y^{-1}q^n)(1-\zeta^{-k}y^{-1}q^n)}{(1-\zeta^k q^n)^2(1-\zeta^{-k}q^n)^2}\,, \tag{64}$$

and

$$\phi^{\mathrm{mom}}_{1,g^k}(\tau,z) = 1\,, \tag{65}$$

where the last result follows because the only g-invariant state of the form (k_L,k_R) is the vacuum $(0,0)$. Thus we have

$$\phi_{1,g^k}(\tau,z) = 5\,\frac{\vartheta_1(\tau,z+\frac{k}{5})\,\vartheta_1(\tau,z-\frac{k}{5})}{\vartheta_1(\tau,\frac{k}{5})\,\vartheta_1(\tau,-\frac{k}{5})}\,, \tag{66}$$

where

$$\vartheta_1(\tau,z) = -iq^{1/8}y^{-\frac{1}{2}}(y-1)\prod_{n=1}^{\infty}(1-q^n)(1-yq^n)(1-y^{-1}q^n)\,, \tag{67}$$

is the first Jacobi theta function. Modular transformations of $\phi_{1,g^k}(\tau,z)$ reproduce the twining genera in the twisted sector

$$\phi_{g^l,g^k}(\tau,z) = \mathrm{Tr}_{\mathscr{H}^{(l)}}\big(g^k\, q^{L_0-\frac{c}{24}}\,\bar{q}^{\tilde{L}_0-\frac{\tilde{c}}{24}}\,y^{2J_0}(-1)^{F+\tilde{F}}\big)\,, \tag{68}$$

and using the modular properties of the theta function we obtain

$$\phi_{g^l,g^k}(\tau,z) = 5\,\frac{\vartheta_1(\tau,z+\frac{k}{5}+\frac{l\tau}{5})\,\vartheta_1(\tau,z-\frac{k}{5}-\frac{l\tau}{5})}{\vartheta_1(\tau,\frac{k}{5}+\frac{l\tau}{5})\,\vartheta_1(\tau,-\frac{k}{5}-\frac{l\tau}{5})}\,, \tag{69}$$

for $k,l\in\mathbb{Z}/5\mathbb{Z}$, $(k,l)\neq(0,0)$. The elliptic genus of the full orbifold theory is then

$$\phi_{\rm orb}(\tau,z) = \frac{1}{5}\sum_{k,l\in\mathbb{Z}/5\mathbb{Z}}\phi_{g^l,g^k}(\tau,z) = \sum_{\substack{k,l\in\mathbb{Z}/5\mathbb{Z}\\(k,l)\neq(0,0)}}\frac{\vartheta_1(\tau,z+\frac{k}{5}+\frac{l\tau}{5})\,\vartheta_1(\tau,z-\frac{k}{5}-\frac{l\tau}{5})}{\vartheta_1(\tau,\frac{k}{5}+\frac{l\tau}{5})\,\vartheta_1(\tau,-\frac{k}{5}-\frac{l\tau}{5})}\,. \tag{70}$$

Since $\phi_{g^k,g^l}(\tau,0)=5$ for all $(k,l)\neq(0,0)$, we have

$$\phi_{\rm orb}(\tau,0) = \frac{1}{5}\sum_{\substack{k,l\in\mathbb{Z}/5\mathbb{Z}\\(k,l)\neq(0,0)}} 5 = 24\,, \tag{71}$$

which shows that the orbifold theory is a non-linear sigma-model on K3. In particular, the untwisted sector has four RR ground states, while each of the four twisted sectors contains five RR ground states. For the following it will be important to understand the structure of the various twisted sectors in detail.

The Twisted Sectors

In the g^k-twisted sector, let us consider a basis of g-eigenvectors for the currents and fermionic fields. For a given eigenvalue ζ^i, $i\in\mathbb{Z}/5\mathbb{Z}$, of g^k, the corresponding currents $J^{i,a}$ and fermionic fields $\psi^{i,b}$ (where a,b labels distinct eigenvectors with the same eigenvalue) have a mode expansion

$$J^{i,a}(z) = \sum_{n\in\frac{i}{5}+\mathbb{Z}}\alpha_n^{i,a}z^{-n-1}\,,\qquad \psi^{i,a}(z) = \sum_{r\in\frac{i}{5}+\nu+\mathbb{Z}}\psi_r^{i,a}z^{-r-1/2}\,, \tag{72}$$

where $\nu = 1/2$ in the NS- and $\nu = 0$ in the R-sector. The ground states of the g^k-twisted sector are characterised by the conditions

$$\alpha_n^{i,a}|m,k\rangle = \tilde{\alpha}_n^{i,a}|m,k\rangle = 0\,, \qquad \forall\, n > 0,\ i,\ a\,, \tag{73}$$

$$\psi_r^{i,b}|m,k\rangle = \tilde{\psi}_r^{i,b}|m,k\rangle = 0\,, \qquad \forall\, r > 0,\ i,\ b\,, \tag{74}$$

where $|m,k\rangle$ denotes the mth ground state in the g^k-twisted sector. Note that since none of the currents are g-invariant, there are no current zero modes in the g^k-twisted sector, and similarly for the fermions. For a given k, the states $|m,k\rangle$ have then all the same conformal dimension, which can be calculated using the commutation relation $[L_{-1}, L_1] = 2L_0$ or read off from the leading term of the modular transform of the twisted character (68). In the g^k-twisted NS-NS-sector the ground states have conformal dimension

$$\text{NS-NS } g^k\text{-twisted:} \qquad h = \frac{k}{5} \qquad \text{and} \qquad \tilde{h} = \frac{2k}{5}\,, \tag{75}$$

while in the RR-sector we have instead

$$\text{R-R } g^k\text{-twisted:} \qquad h = \tilde{h} = \frac{1}{4}\,. \tag{76}$$

In particular, level matching is satisfied, and thus the asymmetric orbifold is consistent [33]. The full g^k-twisted sector is then obtained by acting with the negative modes of the currents and the fermionic fields on the ground states $|m,k\rangle$.

Let us have a closer look at the ground states of the g^k twisted sector; for concreteness we shall restrict ourselves to the case $k = 1$, but the modifications for general k are minor (see below). The vertex operators $V_\lambda(z,\bar{z})$ associated to $\lambda \in \Gamma^{4,4}$, act on the ground states $|m,1\rangle$ by

$$\lim_{z\to 0} V_\lambda(z,\bar{z})|m,1\rangle = e_\lambda|m,1\rangle\,, \tag{77}$$

where e_λ are operators commuting with all current and fermionic oscillators and satisfying

$$e_\lambda\, e_\mu = \epsilon(\lambda,\mu)\, e_{\lambda+\mu}\,, \tag{78}$$

for some fifth root of unity $\epsilon(\lambda,\mu)$. The vertex operators V_λ and V_μ must be local relative to one another, and this is the case provided that (see the appendix)

$$\frac{\epsilon(\lambda,\mu)}{\epsilon(\mu,\lambda)} = C(\lambda,\mu)\,, \tag{79}$$

where

$$C(\lambda,\mu)=\prod_{i=1}^{4}(\zeta^i)^{(g^i(\lambda),\mu)}=\zeta^{(P_g(\lambda),\mu)}\qquad\text{with}\qquad P_g(\lambda)=\sum_{i=1}^{4} i g^i(\lambda)\,. \tag{80}$$

The factor $C(\lambda,\mu)$ has the properties

$$C(\lambda,\mu_1+\mu_2)=C(\lambda,\mu_1)\,C(\lambda,\mu_2)\,,\qquad C(\lambda_1+\lambda_2,\mu)=C(\lambda_1,\mu)\,C(\lambda_2,\mu)\,, \tag{81}$$

$$C(\lambda,\mu)=C(\mu,\lambda)^{-1}\,, \tag{82}$$

$$C(\lambda,\mu)=C(g(\lambda),g(\mu))\,. \tag{83}$$

Because of (81), $C(\lambda,0)=C(0,\lambda)=1$ for all $\lambda\in\Gamma^{4,4}$, and we can set

$$e_0=1\,, \tag{84}$$

so that $\epsilon(0,\lambda)=1=\epsilon(\lambda,0)$. More generally, for the vectors λ in the sublattice

$$R:=\{\lambda\in\Gamma^{4,4}\mid C(\lambda,\mu)=1\,,\ \text{for all }\mu\in\Gamma^{4,4}\}\subset\Gamma^{4,4}\,, \tag{85}$$

we have $C(\lambda+\mu_1,\mu_2)=C(\mu_1,\mu_2)$, for all $\mu_1,\mu_2\in\Gamma^{4,4}$, so that we can set

$$e_{\mu+\lambda}=e_\mu\,,\qquad\forall\lambda\in R\,,\quad\mu\in\Gamma^{4,4}\,. \tag{86}$$

Thus, we only need to describe the operators corresponding to representatives of the group $\Gamma^{4,4}/R$. The vectors $\lambda\in R$ are characterised by

$$(P_g(\lambda),\mu)\equiv 0\mod 5\,,\qquad\text{for all }\mu\in\Gamma^{4,4}\,, \tag{87}$$

and since $\Gamma^{4,4}$ is self-dual this condition is equivalent to

$$P_g(\lambda)\in 5\,\Gamma^{4,4}\,. \tag{88}$$

Since $\Gamma^{4,4}$ has no g-invariant subspace, we have the identity

$$1+g+g^2+g^3+g^4=0 \tag{89}$$

that implies [see (80)]

$$P_g\circ(1-g)=(1-g)\circ P_g=-5\cdot\mathbf{1}\,. \tag{90}$$

Thus, $\lambda\in R$ if and only if

$$P_g(\lambda)=P_g\circ(1-g)(\tilde\lambda)\,, \tag{91}$$

for some $\tilde{\lambda} \in \Gamma^{4,4}$, and since P_g has trivial kernel [see (90)], we finally obtain

$$R = (1-g)\,\Gamma^{4,4}\,. \tag{92}$$

Since also $(1-g)$ has trivial kernel, R has rank 8 and $\Gamma^{4,4}/R$ is a finite group. Furthermore,

$$|\Gamma^{4,4}/R| = \det(1-g) = 25\,, \tag{93}$$

and, since $5\,\Gamma^{4,4} \subset R$, the group $\Gamma^{4,4}/R$ has exponent 5. The only possibility is

$$\Gamma^{4,4}/R \cong \mathbb{Z}_5 \times \mathbb{Z}_5\,. \tag{94}$$

Let $x, y \in \Gamma^{4,4}$ be representatives for the generators of $\Gamma^{4,4}/R$. By (82), we know that $C(x,x) = C(y,y) = 1$, so that $C(x,y) \neq 1$ (otherwise C would be trivial over the whole $\Gamma^{4,4}$), and we can choose x, y such that

$$C(x,y) = \zeta\,. \tag{95}$$

Thus, the ground states form a representation of the algebra of operators generated by e_x, e_y, satisfying

$$e_x^5 = 1 = e_y^5\,, \qquad e_x e_y = \zeta e_y e_x\,. \tag{96}$$

The group generated by e_x and e_y is the extra-special group 5^{1+2}, and all its non-abelian irreducible representations[5] are five dimensional.

In particular, for the representation on the g-twisted ground states, we can choose a basis of e_x-eigenvectors

$$|m;1\rangle \quad \text{with} \quad e_x|m;1\rangle = \zeta^m|m;1\rangle\,, \qquad m \in \mathbb{Z}/5\mathbb{Z}\,, \tag{97}$$

and define the action of the operators e_y by

$$e_y|m;1\rangle = |m+1;1\rangle\,. \tag{98}$$

For any vector $\lambda \in \Gamma^{4,4}$, there are unique $a, b \in \mathbb{Z}/5\mathbb{Z}$ such that $\lambda = ax + by + (1-g)(\mu)$ for some $\mu \in \Gamma^{4,4}$ and we define[6]

$$e_\lambda := e_x^a e_y^b\,. \tag{99}$$

[5] We call a representation non-abelian if the central element does not act trivially.

[6] The ordering of e_x and e_y in this definition is arbitrary; however, any other choice corresponds to a redefinition $\tilde{e}_\lambda = c(\lambda)e_\lambda$, for some fifth root of unity $c(\lambda)$, that does not affect the commutation relations $\tilde{e}_\lambda\tilde{e}_\mu = C(\lambda,\mu)\tilde{e}_\mu\tilde{e}_\lambda$.

Since $g(\lambda) = ax + by + (1-g)(g(\mu) - ax - by)$, by (86) we have

$$e_{g(\lambda)} = e_\lambda \,, \tag{100}$$

so that, with respect to the natural action $g(e_\lambda) := e_{g(\lambda)}$, the algebra is g-invariant. This is compatible with the fact that all ground states have the same left and right conformal weights h and $\tilde{h}$, so that the action of $g = e^{2\pi i(h-\tilde{h})}$ is proportional to the identity.

The construction of the g^k-twisted sector, for $k = 2,3,4$, is completely analogous to the g^1-twisted case, the only difference being that the root ζ in the definition of $C(\lambda,\mu)$ should be replaced by ζ^k. Thus, one can define operators $e_x^{(k)}, e_y^{(k)}$ on the g^k-twisted sector, for each $k = 1,\ldots,4$, satisfying

$$(e_x^{(k)})^5 = 1 = (e_y^{(k)})^5 \,, \qquad e_x^{(k)} e_y^{(k)} = \zeta^k e_y^{(k)} e_x^{(k)} \,. \tag{101}$$

The action of these operators on the analogous basis $|m;k\rangle$ with $m \in \mathbb{Z}/5\mathbb{Z}$ is then

$$e_x^{(k)}|m;k\rangle = \zeta^m |m;k\rangle \,, \quad e_y^{(k)}|m;k\rangle = |m+k;k\rangle \,. \tag{102}$$

Spectrum and Symmetries

The spectrum of the actual orbifold theory is finally obtained from the above twisted sectors by projecting onto the g-invariant states; technically, this is equivalent to restricting to the states for which the difference of the left- and right- conformal dimensions is integer, $h - \tilde{h} \in \mathbb{Z}$. In particular, the RR ground states (102) in each (twisted) sector have $h = \tilde{h} = 1/4$, so that they all survive the projection. Thus, the orbifold theory has four RR ground states in the untwisted sector (the spectral flow generators), forming a $(\mathbf{2},\mathbf{2})$ representation of $\mathfrak{su}(2)_L \oplus \mathfrak{su}(2)_R$, and five RR ground states in each twisted sector, which are singlets of $\mathfrak{su}(2)_L \oplus \mathfrak{su}(2)_R$. In total there are therefore 24 RR ground states, as expected for a non-linear sigma-model on K3. (Obviously, we are here just reproducing what we already saw in (71).)

Next we want to define symmetry operators acting on the orbifold theory. First we can construct operators e_λ associated to $\lambda \in \Gamma^{4,4}$, that will form the extra special group 5^{1+2}. They are defined to act by $e_\lambda^{(k)}$ on the g^k-twisted sector. The action of the untwisted sector preserves the subspaces $\mathscr{H}_m^U$, $m \in \mathbb{Z}/5\mathbb{Z}$, of states with momentum of the form $\lambda = nx + my + (1-g)(\mu)$, for some $n \in \mathbb{Z}$ and $\mu \in \Gamma^{4,4}$. Let us denote by $T_{m;k}$ a generic vertex operator associated with a g^k-twisted state, $k = 1,\ldots,4$, with e_x-eigenvalue ζ^m, $m \in \mathbb{Z}/5\mathbb{Z}$, and by $T_{m;0}$ a vertex operator associated with a state in $\mathscr{H}_m^U$. Consistency of the OPE implies the fusion rules

$$T_{m;k} \times T_{m';k'} \to T_{m+m';k+k'} \,. \tag{103}$$

These rules are preserved by the maps

$$T_{m;k} \mapsto e_\lambda \, T_{m;k} \, e_\lambda^{-1} \,, \qquad \lambda \in \Gamma^{4,4} \,, \tag{104}$$

which therefore define symmetries of the orbifold theory. As we have explained above, these symmetries form the extra-special group 5^{1+2}.

Finally, the symmetries (54), (55), (58) and (59) of the original torus theory induce a $\mathbb{Z}_4$-group of symmetries of the orbifold. Since $h^{-1}gh = g^{-1}$, the space of g-invariant states of the original torus theory is stabilised by h, so that h restricts to a well-defined transformation on the untwisted sector of the orbifold. Furthermore, h maps the g^k- to the g^{5-k}-twisted sector. Equations (54) and (55) can be written as

$$h(x_1) = 2x_1 + (1-g)(-x_1 - x_2 - x_3 + y_1 + y_2 + y_3 + y_4) \,, \tag{105}$$

$$h(y_1) = 2y_1 + (1-g)(-x_1 - x_2 - x_3 - x_4 - 2y_1 - y_2 - y_3 - y_4) \,. \tag{106}$$

It follows that the action of h on the operators $e_\lambda^{(k)}$, $k = 1, \ldots, 4$ must be

$$h e_x^{(k)} h^{-1} = e_{2x}^{(5-k)} \,, \qquad h e_y^{(k)} h^{-1} = e_{2y}^{(5-k)} \,, \tag{107}$$

and it is easy to verify that this transformation is compatible with (101). Correspondingly, the action on the twisted sector ground states is

$$h|m;k\rangle = |3m; 5-k\rangle \,, \tag{108}$$

and it is consistent with (103).

Thus the full symmetry group is the semi-direct product

$$G = 5^{1+2} : \mathbb{Z}_4 \,, \tag{109}$$

where the generator $h \in \mathbb{Z}_4$ maps the central element $\zeta \in 5^{1+2}$ to ζ^{-1}.

All of these symmetries act trivially on the superconformal algebra and on the spectral flow generators, and therefore define symmetries in the sense of the general classification theorem. Indeed, G agrees precisely with the group in case (ii) of the theorem. Thus our orbifold theory realises this possibility.

6 Models with Symmetry Group Containing $3^{1+4} : \mathbb{Z}_2$

Most of the torus orbifold construction described in the previous section generalises to symmetries g of order different than 5. In particular, one can show explicitly that orbifolds of $\mathbb{T}^4$ models by a symmetry g of order 3 contain a group of symmetries $3^{1+4} : \mathbb{Z}_2$, so that they belong to one of the cases (iii) and (iv) of the theorem, as expected from the discussion in Sect. 4.

We take the action of the symmetry g on the left-moving currents and fermionic fields to be of the form (42) and (44), where ζ is a now a third root of unity; analogous transformations hold for the right-moving fields with respect to a third root of unity $\tilde{\zeta}$. In this case, Eq. (45) can be satisfied by

$$\zeta = \tilde{\zeta} = e^{\frac{2\pi i}{3}} , \tag{110}$$

so that the action is left-right symmetric and g admits an interpretation as a geometric $\mathrm{O}(4,\mathbb{R})$-rotation of order 3 of the torus $\mathbb{T}^4$. For example, the torus $\mathbb{R}^4/(A_2 \oplus A_2)$, where A_2 is the root lattice of the $su(3)$ Lie algebra, with vanishing Kalb–Ramond field, admits such an automorphism.

The orbifold by g contains six RR ground states in the untwisted sector. In the kth twisted sector, $k = 1, 2$, the ground states form a representation of an algebra of operators $e^{(k)}_\lambda$, $\lambda \in \Gamma^{4,4}$, satisfying the commutation relations

$$e^{(k)}_\lambda e^{(k)}_\mu = C(\lambda,\mu)^k e^{(k)}_\mu e^{(k)}_\lambda , \tag{111}$$

where

$$C(\lambda,\mu) = \zeta^{(P_g(\lambda),\mu)} , \qquad P_g = g + 2g^2 . \tag{112}$$

As discussed in Sect. 5.2, we can set

$$e^{(k)}_{\lambda+\mu} = e^{(k)}_\mu , \qquad \forall \lambda \in R, \mu \in \Gamma^{4,4} , \tag{113}$$

where

$$R = (1-g)\Gamma^{4,4} . \tag{114}$$

(Note that $\Gamma^{4,4}$ contains no g-invariant vectors). The main difference with the analysis of Sect. 5.2 is that, in this case,

$$\Gamma^{4,4}/R \cong \mathbb{Z}_3^4 . \tag{115}$$

In particular, we can find vectors $x_1, x_2, y_1, y_2 \in \Gamma^{4,4}$ such that

$$C(x_i, y_j) = \zeta^{\delta_{ij}} , \qquad C(x_i, x_j) = C(y_i, y_j) = 1 . \tag{116}$$

The corresponding operators obey the relations

$$e^{(k)}_{x_i} e^{(k)}_{y_j} = \zeta^{k\delta_{ij}} e^{(k)}_{y_j} e^{(k)}_{x_i} , \quad e^{(k)}_{x_i} e^{(k)}_{x_j} = e^{(k)}_{x_j} e^{(k)}_{x_i} , \quad e^{(k)}_{y_i} e^{(k)}_{y_j} = e^{(k)}_{y_j} e^{(k)}_{y_i} , \tag{117}$$

as well as

$$(e^{(k)}_{x_i})^3 = 1 = (e^{(k)}_{y_i})^3 . \tag{118}$$

These operators generate the extra-special group 3^{1+4} of exponent 3, and the kth-twisted ground states form a representation of this group. We can choose a basis $|m_1, m_2; k\rangle$, with $m_1, m_2 \in \mathbb{Z}/3\mathbb{Z}$, of simultaneous eigenvectors of $e_{x_1}^{(k)}$ and $e_{x_2}^{(k)}$, so that

$$e_{x_i}^{(k)}|m_1, m_2; k\rangle = \zeta^{m_i}|m_1, m_2; k\rangle \,, \quad e_{y_i}^{(k)}|m_1, m_2; k\rangle = |m_1 + k\delta_{1i}, m_2 + k\delta_{2i}; k\rangle \,. \tag{119}$$

The resulting orbifold model has nine RR ground states in each twisted sector, for a total of $6 + 9 + 9 = 24$ RR ground states, as expected for a K3 model. As in Sect. 5.2, the group 3^{1+4} generated by $e_\lambda^{(k)}$ extends to a group of symmetries of the whole orbifold model. Furthermore, the $\mathbb{Z}_2$-symmetry that flips the signs of the coordinates in the original torus theory induces a symmetry h of the orbifold theory, which acts on the twisted sectors by

$$h\,|m_1, m_2; k\rangle = |-m_1, -m_2; k\rangle \,. \tag{120}$$

We conclude that the group G of symmetries of any torus orbifold $\mathbb{T}^4/\mathbb{Z}_3$ contains a subgroup $3^{1+4} : \mathbb{Z}_2$, and is therefore included in the cases (iii) or (iv) of the classification theorem. This obviously ties in nicely with the general discussion of Sect. 4.

7 Conclusions

In this paper we have reviewed the current status of the EOT conjecture concerning a possible $\mathbb{M}_{24}$ symmetry appearing in the elliptic genus of K3. We have explained that, in some sense, the EOT conjecture has already been proven since twining genera, satisfying the appropriate modular and integrality properties, have been constructed for all conjugacy classes of $\mathbb{M}_{24}$. However, the analogue of the Monster conformal field theory that would 'explain' the underlying symmetry has not yet been found. In fact, no single K3 sigma-model will be able to achieve this since none of them possesses an automorphism group that contains $\mathbb{M}_{24}$.

Actually, the situation is yet further complicated by the fact that there are K3 sigma-models whose automorphism group is *not even a subgroup of* $\mathbb{M}_{24}$; on the other hand, the elliptic genus of K3 does not show any signs of exhibiting 'Moonshine' with respect to any larger group. As we have explained in this paper, most of the exceptional automorphism groups (i.e. automorphism groups that are not subgroups of $\mathbb{M}_{24}$) appear for K3s that are torus orbifolds. In fact, all cyclic torus orbifolds are necessarily exceptional in this sense, and (cyclic) torus orbifolds account for all incarnations of the cases (ii)–(iv) of the classification theorem of [8] (see Sect. 4). We have checked these predictions by explicitly constructing an asymmetric $\mathbb{Z}_5$ orbifold that realises case (ii) of the theorem (see Sect. 5), and a family of $\mathbb{Z}_3$ orbifolds realising cases (iii) and (iv) of the theorem (see Sect. 6). Incidentally, these constructions also demonstrate that the exceptional cases (ii)–(iv) actually appear in the K3 moduli space—in the analysis of [8] this conclusion relied on some assumption about the regularity of K3 sigma-models.

The main open problem that remains to be understood is why precisely $\mathbb{M}_{24}$ is 'visible' in the elliptic genus of K3, rather than any smaller (or indeed bigger) group. Recently, we have constructed (some of) the twisted twining elliptic genera of K3 [34] (see also [35]), i.e. the analogues of Simon Norton's generalised Moonshine functions [36]. We hope that they will help to shed further light on this question.

8 Commutation Relations in the Twisted Sector

The vertex operators $V_\lambda(z,\bar{z})$ in the g-twisted sector can be defined in terms of formal exponentials of current oscillators

$$E_\lambda^\pm(z,\bar{z}) := \exp\Big(\sum_{\substack{r\in\frac{1}{5}\mathbb{Z}\\ \pm r>0}} (\lambda_L\cdot\alpha)_r^{(r)}\frac{z^{-r}}{r}\Big)\exp\Big(\sum_{\substack{r\in\frac{1}{5}\mathbb{Z}\\ \pm r>0}} (\lambda_R\cdot\tilde{\alpha})_r^{(r)}\frac{\bar{z}^{-r}}{r}\Big)\,, \tag{121}$$

where $(\lambda_L\cdot\alpha)_r^{(r)}$ and $(\lambda_R\cdot\tilde{\alpha})_r^{(r)}$ are the r-modes of the currents

$$(\lambda_L\cdot\partial X)^{(r)} := \frac{1}{5}\sum_{i=0}^{4}\zeta^{5ir}\lambda_L\cdot g^i(\partial X) = \frac{1}{5}\sum_{i=0}^{4}\zeta^{5ir}g^{-i}(\lambda_L)\cdot\partial X\,, \tag{122}$$

$$(\lambda_R\cdot\bar{\partial} X)^{(r)} := \frac{1}{5}\sum_{i=0}^{4}\bar{\zeta}^{5ir}\lambda_R\cdot g^i(\bar{\partial} X) = \frac{1}{5}\sum_{i=0}^{4}\bar{\zeta}^{5ir}g^{-i}(\lambda_R)\cdot\bar{\partial} X\,. \tag{123}$$

Then we can define

$$V_\lambda(z,\bar{z}) := E_\lambda^-(z,\bar{z})\,E_\lambda^+(z,\bar{z})\,e_\lambda\,, \tag{124}$$

where the operators e_λ commute with all current oscillators and satisfy

$$e_\lambda\,e_\mu = \epsilon(\lambda,\mu)\,e_{\lambda+\mu}\,, \tag{125}$$

for some fifth root of unity $\epsilon(\lambda,\mu)$. The commutator factor

$$C(\lambda,\mu) := \frac{\epsilon(\lambda,\mu)}{\epsilon(\mu,\lambda)}\,, \tag{126}$$

can be determined by imposing the locality condition

$$V_\lambda(z_1,\bar{z}_1)\,V_\mu(z_2,\bar{z}_2) = V_\mu(z_2,\bar{z}_2)\,V_\lambda(z_1,\bar{z}_1)\,. \tag{127}$$

To do so, we note that the commutation relations between the operators $E_\lambda^\pm$ can be computed, as in [37], using the Campbell–Baker–Hausdorff formula

$$E_\lambda^+(z_1,\bar{z}_1)E_\mu^-(z_2,\bar{z}_2) = E_\mu^-(z_2,\bar{z}_2)E_\lambda^+(z_1,\bar{z}_1)$$
$$\prod_{i=0}^{4}[(1-\zeta^{-i}(\tfrac{z_1}{z_2})^{\frac{1}{5}})^{g^i(\lambda)_L\cdot\mu_L}(1-\bar{\zeta}^{-i}(\tfrac{\bar{z}_1}{\bar{z}_2})^{\frac{1}{5}})^{g^i(\lambda)_R\cdot\mu_R}] . \quad (128)$$

Using (128) and $e_\lambda e_\mu = C(\lambda,\mu)e_\mu e_\lambda$, the locality condition is then equivalent to

$$C(\lambda,\mu)\prod_{i=0}^{4}\frac{(1-\zeta^{-i}(\frac{z_1}{z_2})^{\frac{1}{5}})^{g^i(\lambda)_L\cdot\mu_L}(1-\bar{\zeta}^{-i}\frac{\bar{z}_1^{1/5}}{\bar{z}_2^{1/5}})^{g^i(\lambda)_R\cdot\mu_R}}{(1-\zeta^{i}(\frac{z_2}{z_1})^{\frac{1}{5}})^{g^i(\lambda)_L\cdot\mu_L}(1-\bar{\zeta}^{i}(\frac{\bar{z}_2}{\bar{z}_1})^{\frac{1}{5}})^{g^i(\lambda)_R\cdot\mu_R}} = 1 , \quad (129)$$

that is

$$C(\lambda,\mu)\Big(-\frac{z_1^{1/5}}{z_2^{1/5}}\Big)^{\sum_i g^i(\lambda)_L\cdot\mu_L}\Big(-\frac{\bar{z}_1^{1/5}}{\bar{z}_2^{1/5}}\Big)^{\sum_i g^i(\lambda)_R\cdot\mu_R}\prod_{i=0}^{4}[(\zeta^{-i})^{g^i(\lambda)_L\cdot\mu_L}(\bar{\zeta}^{-i})^{g^i(\lambda)_R\cdot\mu_R}] = 1 . \quad (130)$$

Since $\Gamma^{4,4}$ has no g-invariant vector, we have the identities

$$\sum_{i=0}^{4} g^i(\lambda)_L = 0 = \sum_{i=0}^{4} g^i(\lambda)_R , \quad (131)$$

and hence finally obtain

$$C(\lambda,\mu) = \prod_{i=0}^{4}(\zeta^i)^{g^i(\lambda_L)\cdot\mu_L - g^i(\lambda_R)\cdot\mu_R} . \quad (132)$$

Acknowledgements We thank Terry Gannon, Gerald Hoehn, and Michael Tuite for useful conversations. The research of MRG is supported in parts by the Swiss National Science Foundation.

References

1. Eguchi, T., Ooguri, H., Tachikawa, Y.: Notes on the K3 surface and the Mathieu group M_{24}. Exp. Math. **20**, 91–96 (2011). arXiv:1004.0956 [hep-th]
2. Thompson, J.G.: Some numerology between the Fischer-Griess Monster and the elliptic modular function. Bull. Lond. Math. Soc. **11**, 352–353 (1979)
3. Conway, J.H., Norton, S.: Monstrous moonshine. Bull. Lond. Math. Soc. **11**, 308–339 (1979)

4. Gannon, T.: Moonshine Beyond the Monster: The Bridge Connecting Algebra, Modular Forms and Physics. Cambridge University Press, Cambridge (2006)
5. Borcherds, R.: Vertex algebras, Kac-Moody algebras and the Monster. Proc. Natl. Acad. Sci. USA **83**, 3068–3071 (1986)
6. Frenkel, I., Lepowski, J., Meurman, A.: Vertex Operator Algebras and the Monster. Academic, Boston (1986)
7. Duncan, J.F., Frenkel, I.B.: Rademacher sums, Moonshine and gravity. Commun. Number Theory Phys. **5**(4), 1–128 (2011). arXiv:0907.4529 [math.RT]
8. Gaberdiel, M.R., Hohenegger, S., Volpato, R.: Symmetries of K3 sigma models. Commun. Number Theory Phys. **6**, 1–50 (2012). arXiv:1106.4315 [hep-th]
9. Walton, M.A.: Heterotic string on the simplest Calabi-Yau manifold and its orbifold limits. Phys. Rev. D **37**, 377–390 (1988)
10. Cheng, M.C.N.: K3 Surfaces, N=4 dyons, and the Mathieu group M_{24}. Commun. Number Theory Phys. **4**, 623–658 (2010). arXiv:1005.5415 [hep-th]
11. Cheng, M.C.N., Duncan, J.F.R.: On Rademacher sums, the largest Mathieu group, and the holographic modularity of moonshine. Commun. Number Theory Phys. **6**, 697–758 (2012). arXiv:1110.3859 [math.RT]
12. Cheng, M.C.N., Duncan, J.F.R.: The largest Mathieu group and (mock) automorphic forms. (2012) arXiv:1201.4140 [math.RT]
13. Cheng, M.C.N., Duncan, J.F.R., Harvey, J.A.: Umbral moonshine. (2012) arXiv:1204.2779 [math.RT]
14. Govindarajan, S.: BKM Lie superalgebras from counting twisted CHL dyons. J. High Energy Phys. **1105**, 089 (2011). arXiv:1006.3472 [hep-th]
15. Govindarajan, S.: Brewing moonshine for Mathieu. (2010) arXiv:1012.5732 [math.NT]
16. Govindarajan, S.: Unravelling Mathieu Moonshine. Nucl. Phys. B **864**, 823–839 (2012). arXiv:1106.5715 [hep-th]
17. Hohenegger, S., Stieberger, S.: BPS saturated string amplitudes: K3 elliptic genus and Igusa cusp form. Nucl. Phys. B **856**, 413–448 (2012). arXiv:1108.0323 [hep-th]
18. Kawai, T., Yamada, Y., Yang, S.-K.: Elliptic genera and $\mathcal{N} = 2$ superconformal field theory. Nucl. Phys. B **414**, 191–212 (1994). arXiv:hep-th/9306096
19. Eichler, M., Zagier, D.: The Theory of Jacobi Forms. Birkhäuser, Boston (1985)
20. Mukai, S.: Finite groups of automorphisms of K3 surfaces and the Mathieu group. Invent. Math. **94**, 183–221 (1988)
21. Kondo, S.: Niemeier lattices, Mathieu groups and finite groups of symplectic automorphisms of K3 surfaces. Duke Math. J. **92**, 593–603 (1998). Appendix by S. Mukai
22. Thompson, J.G.: Finite groups and modular functions. Bull. Lond. Math. Soc. **11**, 347–351 (1979)
23. David, J.R., Jatkar, D.P., Sen, A.: Product representation of dyon partition function in CHL models. J. High Energy Phys. **0606**, 064 (2006). arXiv:hep-th/0602254
24. Gaberdiel, M.R., Hohenegger, S., Volpato, R.: Mathieu moonshine in the elliptic genus of K3. J. High Energy Phys. **1010**, 062 (2010). arXiv:1008.3778 [hep-th]
25. Eguchi, T., Hikami, K.: Note on twisted elliptic genus of K3 surface. Phys. Lett. B **694**, 446–455 (2011). arXiv:1008.4924 [hep-th]
26. Gaberdiel, M.R., Hohenegger, S., Volpato, R.: Mathieu twining characters for K3. J. High Energy Phys. **1009**, 058 (2010). arXiv:1006.0221 [hep-th]
27. Gannon, T.J.: Much ado about Mathieu. (2012) arXiv:1211.5531 [math.RT]
28. Fluder, M.: Symmetries of non-linear sigma models on K3. Master thesis, ETH Zürich (2011)
29. Tuite, M.P.: Monstrous moonshine and the uniqueness of the moonshine module. (1992) arXiv:hep-th/9211069
30. Tuite, M.P.: On the relationship between monstrous moonshine and the uniqueness of the moonshine module. Commun. Math. Phys. **166**, 495–532 (1995). arXiv:hep-th/9305057
31. Ginsparg, P.H.: Applied conformal field theory. (1988) arXiv:hep-th/9108028
32. Curtis, R.T.: On subgroups of $\cdot O$. II: local structure. J. Algebra **63**, 413–434 (1980)

33. Narain, K.S., Sarmadi, M.H., Vafa, C.: Asymmetric orbifolds. Nucl. Phys. B **288**, 551–577 (1987)
34. Gaberdiel, M.R., Persson, D., Ronellenfitsch, H., Volpato, R.: Generalised Mathieu moonshine. Commun. Number Theory Phys. **7**, 145–223 (2013). arXiv:1211.7074 [hep-th]
35. Gaberdiel, M.R., Persson, D., Volpato, R.: Generalised Moonshine and holomorphic orbifolds. (2013) arXiv:1302.5425 [hep-th]
36. Norton, S.P.: Generalised moonshine. Proc. Symp. Pure Math. **47**, 208–209 (1987)
37. Lepowsky, J.: Calculus of twisted vertex operators. Proc. Natl. Acad. Sci. USA **82**, 8295–8299 (1985)

Rademacher Sums and Rademacher Series

Miranda C.N. Cheng and John F.R. Duncan

Abstract We exposit the construction of Rademacher sums in arbitrary weights and describe their relationship to mock modular forms. We introduce the notion of Rademacher series and describe several applications, including the determination of coefficients of Rademacher sums and a very general form of Zagier duality. We then review the application of Rademacher sums and series to moonshine both monstrous and umbral and highlight several open problems. We conclude with a discussion of the interpretation of Rademacher sums in physics.

1 Introduction

Modular forms are fundamental objects in number theory which have many applications in geometry, combinatorics, string theory, and other branches of mathematics and physics. One may wonder "what are the natural ways are to obtain a modular form?" In general we can construct a symmetric function from a non-symmetric one by summing its images under the desired group of symmetries, although if infinite symmetry is required convergence may be a problem. A refinement of this idea, pioneered by Poincaré (cf. Sect. 2.1), is to build in the required symmetry by summing over the images of a function f that is already invariant under a (large enough) subgroup of the full group of symmetries. Then we may restrict the summation to representatives of cosets of the subgroup fixing f and still expect to obtain a fully symmetric function.

M.C.N. Cheng (✉)
CNRS and Université Paris 7, UMR CNRS 7586, Paris, France
e-mail: chengm@math.jussieu.fr

J.F.R. Duncan
Department of Mathematics, Case Western Reserve University, Cleveland, OH 44106, USA
e-mail: john.duncan@case.edu

W. Kohnen and R. Weissauer (eds.), *Conformal Field Theory, Automorphic Forms and Related Topics*, Contributions in Mathematical and Computational Sciences 8,
DOI 10.1007/978-3-662-43831-2_6,

For instance, to obtain a modular form of even integral weight $w = 2k$ we may, following Poincaré (cf. (14)), take $f(\tau) = \mathrm{e}(m\tau)$ where m is an integer, τ is a parameter on the upper-half plane $\mathbb{H}$, and here and everywhere else in the article we employ the notation

$$\mathrm{e}(x) = e^{2\pi \mathrm{i} x}. \tag{1}$$

Then the subgroup of $\Gamma = SL_2(\mathbb{Z})$ leaving f invariant is just the subgroup of upper-triangular matrices, which we denote Γ_∞ since its elements are precisely those that fix the infinite cusp of Γ (cf. Sect. 2.1). Thus we are led to consider the sum

$$\sum_{\left(\begin{smallmatrix} a & b \\ c & d \end{smallmatrix}\right) \in \Gamma_\infty \backslash \Gamma} \mathrm{e}\left(m \frac{a\tau + b}{c\tau + d}\right) \frac{1}{(c\tau + d)^w}, \tag{2}$$

for $w = 2k$, taken over a set of representatives for the right cosets of Γ_∞ in Γ. When $k > 1$ this sum is absolutely convergent, locally uniformly for $\tau \in \mathbb{H}$, and thus defines a holomorphic function on $\mathbb{H}$ which is invariant for the weight $w = 2k$ action of Γ by construction. If $m \geq 0$ then it remains bounded as $\Im(\tau) \to \infty$ and is thus a modular form of weight $2k$ for $\Gamma = SL_2(\mathbb{Z})$. This result was obtained by Poincaré in [58]. (See [45] for a historical discussion.)

For many choices of w and m, however (e.g. for $w \leq 2$), the infinite sum in (2) is not absolutely convergent (and not even conditionally convergent if $w < 1$). Nontheless, we may ask if there is some way to regularise (2) in the case that $w \leq 2$. One solution to this problem, for the case that $w = 0$, was established by Rademacher in [61]. Let $J(\tau)$ denote the *elliptic modular invariant* normalised to have vanishing constant term, so that $J(\tau)$ is the unique holomorphic function on $\mathbb{H}$ satisfying $J\left(\frac{a\tau+b}{c\tau+d}\right) = J(\tau)$ whenever $\left(\begin{smallmatrix} a & b \\ c & d \end{smallmatrix}\right) \in SL_2(\mathbb{Z})$ and also $J(\tau) = q^{-1} + O(q)$ as $\Im(\tau) \to \infty$ for $q = \mathrm{e}(\tau)$.

$$J(\tau) = q^{-1} + 196884q + 21493760q^2 + 864299970q^3 + \cdots \tag{3}$$

In [61] Rademacher established the validity of the expression

$$J(\tau) + 12 = \mathrm{e}(-\tau) + \lim_{K \to \infty} \sum_{\substack{\left(\begin{smallmatrix} a & b \\ c & d \end{smallmatrix}\right) \in \Gamma_\infty \backslash \Gamma \\ 0 < c < K \\ -K^2 < d < K^2}} \mathrm{e}\left(-\frac{a\tau + b}{c\tau + d}\right) - \mathrm{e}\left(-\frac{a}{c}\right) \tag{4}$$

for $J(\tau)$ as a conditionally convergent sum, where $\Gamma = SL_2(\mathbb{Z})$, and one can recognise the right hand side of (4) as a modification of the $w = 0$ case of (2) with $m = -1$. This result has been generalised to other groups Γ, and ultimately to negative (and some positive) weights, in various works, including [11,25,37–40,54]. (We refer to Sect. 2 for more details.)

These regularised Poincaré series, which we refer to as Rademacher sums, have several important applications. Perhaps the most obvious of these is the construction of modular forms. We will see in Sect. 2 that modular invariance sometimes but not always survives the regularisation procedure (to be described in general in Sect. 2.2). More generally, a convergent Rademacher sum (cf. (32)) defines a mock modular form (cf. Sect. 2.3); a generalisation of the notion of modular form in which the usual weight w action of a discrete group Γ is twisted by a modular form of weight $2-w$ (cf. (39)).

Another application is to the computation of coefficients of modular forms. We will see in Sect. 2—by way of an example, cf. (25)—that the Rademacher sum construction leads quite naturally to series expressions for its Fourier coefficients. This in turn leads to the notion of *Rademacher series*; a construction which we introduce in Sect. 3. To a given discrete group, multiplier system and weight, the Rademacher series construction attaches, in the convergent cases, a two-dimensional grid of values. Some of these values appear as coefficients of Rademacher sums, but this typically accounts for just half of the values in the grid; the remaining values admit other interesting interpretations. For example, certain Rademacher series encode the Fourier coefficients of Eichler integrals of modular (and mock modular) forms, as we will show in Sect. 3.2. The Rademacher series construction also serves to highlight a very general version of Zagier duality for Rademacher sums, whereby the set of coefficients of two families of mock modular forms in dual weights are shown to coincide, up to sign (cf. Sect. 3.2).

Moreover, as we will discuss in great length in Sect. 4, Rademacher sums play a crucial role in the study of moonshine. We treat the monstrous case in Sect. 4.1, the case of Mathieu moonshine in Sect. 4.2, and the recently discovered umbral moonshine in Sect. 4.3. We will also highlight some important open problems in this section.

Finally, an important application to physics was first pointed out in [24]. It was argued there that some Rademacher sums admit a natural physical interpretation in terms of quantum gravity via the so-called AdS/CFT correspondence. This interpretation has led to various work relating Rademacher sums to physical theories, and in particular to the article [25] which applied the Rademacher sum construction to monstrous moonshine. One of the main results of [25] is the reformulation of the genus zero property of monstrous moonshine in terms of Rademacher sums. The importance of this development has been reinforced recently by further applications [11–13]. The applications of Rademacher sums in physics will be discussed in Sect. 5.

2 Rademacher Sums

2.1 Preliminaries

The group $SL_2(\mathbb{R})$ acts naturally on the upper-half plane $\mathbb{H}$ by orientation preserving isometries according to the rule

$$\begin{pmatrix} a & b \\ c & d \end{pmatrix} \tau = \frac{a\tau + b}{c\tau + d}. \tag{5}$$

For $\gamma \in SL_2(\mathbb{R})$ define $\mathrm{j}(\gamma, \tau)$ to be the derivative (with respect to τ) of this action, so that

$$\mathrm{j}(\gamma, \tau) = (c\tau + d)^{-2} \tag{6}$$

when (c, d) is the lower row of γ. Let Γ be a subgroup of $SL_2(\mathbb{R})$ that contains $\pm I$ and is commensurable with the modular group $SL_2(\mathbb{Z})$ and write Γ_∞ for the subgroup of Γ consisting of upper-triangular matrices. Then Γ_∞ is a subgroup of Γ isomorphic to $\mathbb{Z} \times \mathbb{Z}/2$ and is precisely the set of $\gamma \in \Gamma$ for which the limit of $\gamma\tau$ as $\Im(\tau) \to \infty$ fails to be finite. (We write $\Im(\tau)$ for the imaginary part of τ.) We set

$$T = \begin{pmatrix} 1 & 1 \\ 0 & 1 \end{pmatrix} \tag{7}$$

so that $T\tau = \tau + 1$ for $\tau \in \mathbb{H}$, and we write T^h for $\left(\begin{smallmatrix} 1 & h \\ 0 & 1 \end{smallmatrix}\right)$. Then there is a unique $h > 0$ such that $\Gamma_\infty = \langle T^h, -I \rangle$ and we call this h the *width* of Γ at infinity. Evidently $\mathrm{j}(\gamma, \tau) = 1$ for $\gamma \in \Gamma_\infty$.

The groups we encounter in applications typically contain and normalise the *Hecke congruence group* $\Gamma_0(n)$ for some n.

$$\Gamma_0(n) = \left\{ \begin{pmatrix} a & b \\ c & d \end{pmatrix} \in SL_2(\mathbb{Z}) \mid c \equiv 0 \pmod{n} \right\} \tag{8}$$

Observe that $\Gamma_0(n)$ has width 1 at infinity. A description of the normaliser $N(\Gamma_0(n))$ of $\Gamma_0(n)$ is given in [14, Sect. 3], and from this one can see that the width of $N(\Gamma_0(n))$ at infinity is $1/h$ where h is the largest divisor of 24 for which h^2 divides n.

For $w \in \mathbb{R}$ say that a function $\psi : \Gamma \to \mathbb{C}$ is a *multiplier system* for Γ with weight w if

$$\psi(\gamma_1)\psi(\gamma_2)\,\mathrm{j}(\gamma_1, \gamma_2\tau)^{w/2}\,\mathrm{j}(\gamma_2, \tau)^{w/2} = \psi(\gamma_1\gamma_2)\,\mathrm{j}(\gamma_1\gamma_2, \tau)^{w/2} \tag{9}$$

for all $\gamma_1, \gamma_2 \in \Gamma$ where here and everywhere else in this paper we choose the principal branch of the logarithm (cf. (100)) in order to define the exponential $x \mapsto x^s$ in case s is not an integer.

Note that a multiplier system of weight w is also a multiplier system of weight $w + 2k$ for any integer k since $\mathrm{j}(\gamma_1, \gamma_2\tau)\,\mathrm{j}(\gamma_2, \tau) = \mathrm{j}(\gamma_1\gamma_2, \tau)$ for any $\gamma_1, \gamma_2 \in SL_2(\mathbb{R})$. Given a multiplier system ψ for Γ with weight w we may define the (ψ, w)-*action* of Γ on the space $\mathscr{O}(\mathbb{H})$ of holomorphic functions on the upper-half plane by setting

$$(f|_{\psi,w}\gamma)(\tau) = f(\gamma\tau)\psi(\gamma)\,\mathrm{j}(\gamma, \tau)^{w/2} \tag{10}$$

for $f \in \mathcal{O}(\mathbb{H})$ and $\gamma \in \Gamma$. We then say that $f \in \mathcal{O}(\mathbb{H})$ is an *unrestricted modular form* with multiplier ψ and weight w for Γ in the case that f is invariant for this action; i.e. $f|_{\psi,w}\gamma = f$ for all $\gamma \in \Gamma$. Since $(-\gamma)\tau = \gamma\tau$ and $\mathrm{j}(-I,\tau)^{w/2} = \mathrm{e}(-w/2)$ the multiplier ψ must satisfy the *consistency condition*

$$\psi(-I) = \mathrm{e}\left(\frac{w}{2}\right) \tag{11}$$

in order that the corresponding space(s) of unrestricted modular forms be non-vanishing. (Recall that $\mathrm{e}(x)$ is used as a shorthand for $e^{2\pi \mathrm{i} x}$ throughout the article.)

Since Γ is assumed to be commensurable with $SL_2(\mathbb{Z})$ its natural action on the boundary $\hat{\mathbb{R}} = \mathbb{R} \cup \{\mathbf{i}\infty\}$ of $\mathbb{H}$ restricts to $\hat{\mathbb{Q}} = \mathbb{Q} \cup \{\mathbf{i}\infty\}$. The orbits of Γ on $\hat{\mathbb{Q}}$ are called the *cusps* of Γ. The quotient space

$$X_\Gamma = \Gamma\backslash\mathbb{H} \cup \hat{\mathbb{Q}} \tag{12}$$

is naturally a compact Riemann surface (cf. e.g. [66, Sect. 1.5]). We adopt the common practice of saying that Γ has *genus zero* in case X_Γ is a genus zero surface.

We assume throughout that if Γ does not act transitively on $\hat{\mathbb{Q}}$—i.e. if Γ has more than one cusp—then it is contained in a group $\tilde{\Gamma} < SL_2(\mathbb{R})$ that is commensurable with $SL_2(\mathbb{Z})$ and does act transitively on $\hat{\mathbb{Q}}$, and we assume that the multiplier ψ for Γ is of the form $\psi = \rho\tilde{\psi}$ where $\rho : \Gamma \to \mathbb{C}^\times$ is a morphism of groups and $\tilde{\psi}$ is a multiplier for $\tilde{\Gamma}$. With this understanding we say that an unrestricted modular form f for Γ with multiplier ψ and weight w is a *weak modular form* in case f has at most exponential growth at the cusps of Γ; i.e. in case there exists $C > 0$ such that $(f|_{\tilde{\psi},w}\sigma)(\tau) = O(e^{C\Im(\tau)})$ as $\Im(\tau) \to \infty$ for any $\sigma \in \tilde{\Gamma}$. We say that f is a *modular form* if $(f|_{\tilde{\psi},w}\sigma)(\tau)$ remains bounded as $\Im(\tau) \to \infty$ for any $\sigma \in \tilde{\Gamma}$, and we say f is a *cusp form* if $(f|_{\tilde{\psi},w}\sigma)(\tau) \to 0$ as $\Im(\tau) \to \infty$ for any $\sigma \in \tilde{\Gamma}$.

If Γ has width h at infinity then any multiplier ψ for Γ restricts to a character on $\langle T^h \rangle < \Gamma_\infty$ and so we have

$$\psi(T^h) = e(\alpha) \tag{13}$$

for some $\alpha \in \mathbb{R}$, uniquely determined subject to $0 \leq \alpha < 1$. Then $q^\mu = \mathrm{e}(\mu\tau)$ is a Γ_∞-invariant function for the (ψ, w)-action so long as $h\mu + \alpha \in \mathbb{Z}$, and we may attempt to construct a Γ-invariant function—a modular form with multiplier ψ and weight w for Γ—by summing the images of q^μ over a set of coset representatives for Γ_∞ in Γ.

$$\begin{aligned} P^{[\mu]}_{\Gamma,\psi,w}(\tau) &= \sum_{\gamma \in \Gamma_\infty\backslash\Gamma} q^\mu|_{\psi,w}\gamma \\ &= \sum_{\gamma \in \Gamma_\infty\backslash\Gamma} \mathrm{e}(\mu\gamma\tau)\psi(\gamma)\,\mathrm{j}(\gamma,\tau)^{w/2} \end{aligned} \tag{14}$$

This is the method that was pioneered by Poincaré in [58]. If $w > 2$ then this sum (14) converges absolutely, locally uniformly in τ, so that $P^{[\mu]}_{\Gamma,\psi,w}$ is a well-defined holomorphic function on $\mathbb{H}$, invariant under the (ψ, w)-action of Γ by construction. Although it is not immediately obvious, $P^{[\mu]}_{\Gamma,\psi,w}$ is a weak modular form in general, a modular form in case $\mu \geq 0$ and a cusp form when $\mu > 0$. Poincaré considered the special case of this construction where $\Gamma = SL_2(\mathbb{Z})$, the multiplier ψ is trivial and the weight w is an even integer not less than 4 in [58]. The more general expression (14) was introduced by Petersson in [55], and following him—Petersson called $P^{[\mu]}_{\Gamma,\psi,w}$ a "kind of Poincaré series"—we call $P^{[\mu]}_{\Gamma,\psi,w}$ the *Poincaré series* of weight w and index μ attached to the group Γ and the multiplier ψ.

For example, in the case that Γ is the modular group $SL_2(\mathbb{Z})$ the constant multiplier $\psi \equiv 1$ is a multiplier of weight $w = 2k$ on Γ for any integer k. Taking $\mu = 0$ and $k > 1$ we obtain the function

$$\begin{aligned} P^{[0]}_{\Gamma,1,2k}(\tau) &= \sum_{\gamma \in \Gamma_\infty \backslash \Gamma} \mathrm{j}(\gamma, \tau)^k \\ &= 1 + \sum_{\substack{c,d \in \mathbb{Z} \\ c>0 \\ (c,d)=1}} (c\tau + d)^{-2k} \end{aligned} \tag{15}$$

which is the *Eisenstein series* of weight $2k$, often denoted E_{2k}, with Fourier expansion

$$P^{[0]}_{\Gamma,1,2k}(\tau) = 1 - \frac{4k}{B_{2k}} \sum_{n>0} \sigma_{2k-1}(n) q^n \tag{16}$$

where $\sigma_p(n)$ denotes the sum of the p-th powers of the divisors of n and B_m denotes the m-th Bernoulli number (cf. (95)). One of the main results of [55]—and a principal application of the Poincaré series construction—is that, when $w > 2$, the $P^{[\mu]}_{\Gamma,\psi,w}$ for varying $\mu > 0$ linearly span the space of cusp forms with multiplier ψ and weight w for Γ.

2.2 Regularisation

We may ask if there is a natural way to regularise the simple summation of (14) in the generally divergent case when $w \leq 2$; the following method, inspired by work of Rademacher, is just such a procedure.

First consider the case that $w = 2$. Then the sum in (14) is generally not absolutely convergent, but can be ordered in such a way that the result is conditionally convergent and locally uniformly so in τ, thus yielding a holomorphic function on $\mathbb{H}$. The ordering is obtained as follows. Observe that left multiplication of a matrix $\gamma \in \Gamma$ by $\pm T^h$ has no effect on the lower row of γ other than to change its sign in the case of $-T^h$. So the non-trivial right-cosets of $\Gamma_\infty = \langle T^h, -I\rangle$ in Γ are indexed by pairs (c, d) such that $c > 0$ and (c, d) is the lower row of some element of Γ. For $K > 0$ we define Γ_{K,K^2} to be the set of elements of Γ having lower rows (c, d) satisfying $|c| < K$ and $|d| < K^2$.

$$\Gamma_{K,K^2} = \left\{ \begin{pmatrix} a & b \\ c & d \end{pmatrix} \in \Gamma \mid |c| < K,\ |d| < K^2 \right\} \tag{17}$$

Observe that Γ_{K,K^2} is a union of cosets of Γ_∞ for any K. Now for ψ a multiplier of weight 2 we define the *index* μ *Rademacher sum* $R^{[\mu]}_{\Gamma,\psi,2}$ formally by setting

$$\begin{aligned} R^{[\mu]}_{\Gamma,\psi,2}(\tau) &= \lim_{K\to\infty} \sum_{\gamma\in\Gamma_\infty\backslash\Gamma_{K,K^2}} q^\mu|_{\psi,2}\gamma \\ &= \lim_{K\to\infty} \sum_{\gamma\in\Gamma_\infty\backslash\Gamma_{K,K^2}} \mathrm{e}(\mu\gamma\tau)\psi(\gamma)\,\mathrm{j}(\gamma,\tau), \end{aligned} \tag{18}$$

and we may regard $R^{[\mu]}_{\Gamma,\psi,2}(\tau)$ as a holomorphic function on $\mathbb{H}$ in case the limit in (18) converges locally uniformly in τ.

As an example we take $\Gamma = SL_2(\mathbb{Z})$ and $\psi \equiv 1$ and $\mu = 0$ in analogy with (16). Then we obtain the expression

$$\begin{aligned} R^{[0]}_{\Gamma,1,2}(\tau) &= \lim_{K\to\infty} \sum_{\gamma\in\Gamma_\infty\backslash\Gamma_{K,K^2}} \mathrm{j}(\gamma,\tau) \\ &= 1 + \lim_{K\to\infty} \sum_{\substack{0<c<K \\ -K^2<d<K^2 \\ (c,d)=1}} (c\tau+d)^{-2}. \end{aligned} \tag{19}$$

We will show now that this expression converges. For fixed $K > 0$ let $R(K)$ denote the sum in (19) so that $R^{[0]}_{\Gamma,1,0} = 1 + \lim_{K\to\infty} R(K)$. Then we have

$$\begin{aligned} R(K) &= \sum_{0<c<K} c^{-2} \sum_{\substack{|d|<K^2 \\ (c,d)=1}} (\tau + d/c)^{-2} \\ &= \sum_{0<c<K} c^{-2} \sum_{\substack{0\leq d<c \\ (c,d)=1}} \left(\sum_{|n|<K^2/c} (\tau + d/c + n)^{-2} + O(c/K^2) \right) \end{aligned} \tag{20}$$

where the term $O(c/K^2)$ accounts for the difference between summing over n such that $|d + nc| < K^2$ and summing over n such that $|n| < K^2/c$, and the implied constant holds locally uniformly in τ. The difference between the sum over n in the second line of (20) and its limit $\sum_{n\in\mathbb{Z}}(\tau+d/c+n)^{-2}$ as $K\to\infty$ is also $O(c/K^2)$, locally uniformly for $\tau\in\mathbb{H}$, so we obtain

$$R(K)=\sum_{0<c<K}(-4\pi^2)c^{-2}\sum_{\substack{0\leq d<c\\(c,d)=1}}\left(\sum_{n>0}n\,\mathrm{e}(nd/c)\,\mathrm{e}(n\tau)+O(c/K^2)\right) \tag{21}$$

after an application of the Lipschitz summation formula (108) with $s=2$, $\alpha=0$. We may now estimate $\sum_{0\leq d<c}O(c/K^2)=O(c^2/K^2)$ and $\sum_{0<c<K}c^{-2}$ $O(c^2/K^2)=O(1/K)$ and so obtain

$$\lim_{K\to\infty}R(K)=\lim_{K\to\infty}\sum_{0<c<K}(-4\pi^2)c^{-2}\sum_{\substack{0\leq d<c\\(c,d)=1}}\sum_{n>0}n\,\mathrm{e}(nd/c)\,\mathrm{e}(n\tau). \tag{22}$$

Now let $R'(K)$ denote the summation over c in (22). Then $R'(K)$ is an absolutely convergent sum for fixed $K>0$ (locally uniformly so for $\tau\in\mathbb{H}$) and so we may reorder the terms and write

$$R'(K)=(-4\pi^2)\sum_{n>0}n\,\mathrm{e}(n\tau)\sum_{0<c<K}c^{-2}\sum_{\substack{0\leq d<c\\(c,d)=1}}\mathrm{e}\left(n\frac{d}{c}\right). \tag{23}$$

The summation over d in (23) is the sum of the n-th powers of the primitive c-th roots of unity, which is to say, it is a *Ramanujan sum*. The associated Dirichlet series (for fixed n and varying c) converges absolutely for $\Re(s)>1$ and admits the explicit formula

$$\sum_{c>0}\sum_{\substack{0\leq d<c\\(c,d)=1}}\mathrm{e}\left(n\frac{d}{c}\right)c^{-s}=n^{1-s}\frac{\sigma_{s-1}(n)}{\zeta(s)} \tag{24}$$

in this region (cf. [67, Sect. IX.1]), where $\zeta(s)$ is the Riemann zeta function. Taking $s=2$ in (24) we conclude that $\lim_{K\to\infty}R'(K)=\sum_{n>0}(-4\pi^2)\zeta(2)^{-1}\sigma_1(n)q^n$, and in particular, (19) converges, locally uniformly for $\tau\in\mathbb{H}$. Applying the identity $\zeta(2)=\pi^2/6$ we obtain the Fourier expansion

$$R^{[0]}_{\Gamma,1,2}(\tau)=1-24\sum_{n>0}\sigma_1(n)q^n \tag{25}$$

and recognise $R^{[0]}_{\Gamma,1,2}$ as the *quasi-modular* Eisenstein series, often denoted E_2. (Another common normalisation is $G_2=2\zeta(2)E_2$, cf. [2, Sect. 3.10].)

The argument just given may be readily generalised. For example, let Γ be an arbitrary group commensurable with $SL_2(\mathbb{Z})$ that contains $-I$ and suppose for simplicity that Γ has width one at infinity. Applying a method directly similar to the above we obtain the identity $R^{[0]}_{\Gamma,1,2} = 1 + \lim_{K\to\infty} R'(K)$ where now

$$R'(K) = \sum_{n>0}(-4\pi^2)n\,\mathrm{e}(n\tau) \sum_{\gamma\in\Gamma_\infty\backslash\Gamma_K^\times/\Gamma_\infty} \mathrm{e}\left(n\frac{d}{c}\right)c^{-2}. \tag{26}$$

In (26) we write $\Gamma_K^\times$ for the set of elements $\gamma = \left(\begin{smallmatrix} a & b \\ c & d \end{smallmatrix}\right) \in \Gamma$ satisfying $0 < |c| < K$

$$\Gamma_K^\times = \left\{ \begin{pmatrix} a & b \\ c & d \end{pmatrix} \in \Gamma \mid 0 < |c| < K \right\}, \tag{27}$$

the summation is over a (complete and irredundant) set of representatives for the double cosets of Γ_∞ in $\Gamma_K^\times$, and in each summand in the right most summation of (26) the values c and d are chosen so that (c,d) is the lower row of the representative γ. Then the convergence of $R^{[0]}_{\Gamma,1,2}$, locally uniform for $\tau \in \mathbb{H}$, follows in case the Dirichlet series

$$Z_{0,n}(s) = \lim_{K\to\infty} \sum_{\gamma\in\Gamma_\infty\backslash\Gamma_K^\times/\Gamma_\infty} \mathrm{e}\left(n\frac{d}{c}\right)c^{-2s} \tag{28}$$

converges at $s = 1$. This series $Z_{0,n}(s)$ is a special case of a more general construction—the *Kloosterman zeta function*—due to Selberg [65] that we will discuss further in Sect. 3 (cf. (52)). It is argued in [65] that (28) converges absolutely for $\Re(s) > 1$; we refer to [25] for a verification of the convergence of (28) at $s = 1$ in the case that Γ is commensurable with $SL_2(\mathbb{Z})$ and contains $-I$. Applying this result we obtain the convergence of $R^{[0]}_{\Gamma,1,2}$ for such groups Γ.

Specifying the order of summation as in (18) we may, for suitable choices of Γ and ψ, obtain conditionally convergent sums

$$R^{[\mu]}_{\Gamma,\psi,w}(\tau) = \lim_{K\to\infty} \sum_{\gamma\in\Gamma_\infty\backslash\Gamma_{K,K^2}} \mathrm{e}(\mu\gamma\tau)\psi(\gamma)\,\mathrm{j}(\gamma,\tau)^{w/2}, \tag{29}$$

converging locally uniformly for $\tau \in \mathbb{H}$, with weights in the range $w \geq 1$. However, the technical difficulties can be expected to increase as w tends to 1 for generally the convergence of (29) requires the convergence of a Kloosterman zeta function similar to (28) at $s = w/2$, which is close to the critical line $\Re(s) = 1/2$ in case w is close to 1. The convergence of some Rademacher sums with $w = 3/2$ is established in [11].

Theorem 2.1 ([11]). *Let $\Gamma = \Gamma_0(n)$ for n a positive integer, let h be a divisor of n that also divides 24 and set $\psi = \rho_{n|h}\epsilon^{-3}$ where ϵ and $\rho_{n|h}$ are defined by (104) and (87). Then the Rademacher sum $R^{[1/8]}_{\Gamma,\bar{\psi},3/2}$ converges, locally uniformly for $\tau \in \mathbb{H}$.*

In order to regularise the Poincaré series (14) for weights strictly less than 1 we require to modify the terms in the sum as well as the order in which they are taken. In general, and supposing for now that $\alpha \neq 0$ (cf. (13)), we define the *Rademacher sum* $R^{[\mu]}_{\Gamma,\psi,w}$, for μ such that $h\mu + \alpha \in \mathbb{Z}$, by setting

$$R^{[\mu]}_{\Gamma,\psi,w}(\tau) = \lim_{K\to\infty} \sum_{\gamma\in\Gamma_\infty\backslash\Gamma_{K,K^2}} \mathrm{e}(\mu\gamma\tau)\, \mathrm{r}^{[\mu]}_w(\gamma,\tau)\psi(\gamma)\,\mathrm{j}(\gamma,\tau)^{w/2} \tag{30}$$

where $\mathrm{r}^{[\mu]}_w(\gamma,\tau)$ is defined to be 1 in case $w \geq 1$ or γ is upper-triangular, and is given otherwise, in terms of the complete and lower incomplete Gamma functions (cf. (96)–(99)), by setting

$$\mathrm{r}^{[\mu]}_w(\gamma,\tau) = \frac{1}{\Gamma(1-w)}\gamma(1-w, 2\pi\mathbf{i}\mu(\gamma\tau - \gamma\infty)). \tag{31}$$

In (31) we write $\gamma\infty$ for the limit of $\gamma\tau$ as $\tau \to \mathbf{i}\infty$, so $\gamma\infty$ is none other than a/c in case $\gamma = \left(\begin{smallmatrix} a & b \\ c & d \end{smallmatrix}\right)$ for $c \neq 0$ and is undefined when $\gamma \in \Gamma_\infty$. We trust the reader will not be confused by the two different uses of the symbol γ in (31). Note that since we employ the principal branch of the logarithm (100) everywhere in this article, and, in particular, in the definition (99) of the lower incomplete Gamma function, we should restrict μ to be a non-positive real number when constructing Rademacher sums $R^{[\mu]}_{\Gamma,\psi,w}$ with $w < 1$, for if μ is positive then $\tau \mapsto 2\pi\mathbf{i}\mu(\gamma\tau - \gamma\infty)$ covers the left-half plane and $\mathrm{r}^{[\mu]}_w(\gamma,\tau)$ can fail to be continuous with respect to τ.

In the case that $w < 1$ and $\alpha = 0$ we need a constant term correction to the specification (30) so that the a complete definition is given by

$$R^{[\mu]}_{\Gamma,\psi,w}(\tau) = \delta_{\alpha,0}\frac{1}{2}c_{\Gamma,\psi,w}(\mu,0) + \lim_{K\to\infty} \sum_{\gamma\in\Gamma_\infty\backslash\Gamma_{K,K^2}} \mathrm{e}(\mu\gamma\tau)\, \mathrm{r}^{[\mu]}_w(\gamma,\tau)\psi(\gamma)\,\mathrm{j}(\gamma,\tau)^{w/2} \tag{32}$$

where $c_{\Gamma,\psi,w}(\mu,0)$ is zero in case $w \geq 1$ and is given otherwise by

$$c_{\Gamma,\psi,w}(\mu,0) = \frac{1}{h}\,\mathrm{e}\left(-\frac{w}{4}\right)\frac{(2\pi)^{2-w}(-\mu)^{1-w}}{\Gamma(2-w)} \lim_{K\to\infty} \sum_{\gamma\in\Gamma_\infty\backslash\Gamma^\times_K/\Gamma_\infty} \frac{\mathrm{e}(\mu\gamma\infty)}{c(\gamma)^{2-w}}\psi(\gamma) \tag{33}$$

where h is again the width of Γ, the lower-left-hand entry of a matrix $\gamma \in SL_2(\mathbb{R})$ is denoted $c(\gamma)$, and $\Gamma^\times_K$ is as in (27). As in (26) the summation in (33) is to be taken over a (complete and irredundant) set of representatives for the double cosets of Γ_∞ in $\Gamma^\times_K$, chosen so that $c(\gamma) > 0$. The condition $\alpha = 0$ is necessary in order that the sum in (33) not depend on the choice of representatives. As we will see in due course, the constant term correction in (32) is included so as to improve the modularity of the resulting function $R^{[\mu]}_{\Gamma,\psi,w}$.

As a concrete example of a Rademacher sum with weight less than 1 we may consider the case that $\Gamma = SL_2(\mathbb{Z})$ is again the modular group, $\psi \equiv 1$ and $w = 0$. Then $\gamma(1,x) = 1 - e^{-x}$ according to (99) so that when $\mu = -1$ the general term in the Rademacher sum (32) becomes, for γ non-upper-triangular,

$$\mathrm{e}(\mu\gamma\tau)\, \mathrm{r}_w^{[\mu]}(\gamma,\tau)\psi(\gamma)\,\mathrm{j}(\gamma,\tau)^{w/2} = \mathrm{e}(-\gamma\tau) - \mathrm{e}(-\gamma\infty), \tag{34}$$

and we obtain

$$R^{[-1]}_{\Gamma,1,0}(\tau) = \mathrm{e}(-\tau) + \frac{1}{2} c_{\Gamma,1,0}(-1,0) + \lim_{K\to\infty} \sum_{\gamma\in\Gamma_\infty\backslash\Gamma^\times_{K,K^2}} \mathrm{e}(-\gamma\tau) - \mathrm{e}(-\gamma\infty) \tag{35}$$

where the superscript $\times$ in the summation indicates a restriction to non-trivial cosets of Γ_∞. The right-hand side of (35) is in fact (but for the constant correction term) the original Rademacher sum, introduced by Rademacher in [61]. Rademacher's main result in [61] is that the sum

$$\mathrm{e}(-\tau) + \lim_{K\to\infty} \sum_{\gamma\in\Gamma_\infty\backslash\Gamma^\times_{K,K^2}} \mathrm{e}(-\gamma\tau) - \mathrm{e}(-\gamma\infty) \tag{36}$$

converges to a holomorphic function on $\mathbb{H}$ that is invariant for the ($\psi \equiv 1$, $w = 0$) action of the modular group and has constant term 12 in its Fourier expansion. To calculate $c_{\Gamma,1,0}(-1,0)$ we observe that the non-trivial double cosets of Γ_∞ in $\Gamma = SL_2(\mathbb{Z})$ are represented irredundantly by the matrices $\left(\begin{smallmatrix} a & b \\ c & d \end{smallmatrix}\right)$ with $c > 0$ and d (necessarily coprime to c) satisfying $0 \le d < c$. So we have

$$c_{\Gamma,1,0}(-1,0) = 4\pi^2 \sum_{c>0} \sum_{\substack{0\le d<c \\ (c,d)=1}} \mathrm{e}\left(-\frac{a}{c}\right)\frac{1}{c^2} \tag{37}$$

where in each term in the summation a is chosen so that ad is congruent to 1 modulo c. Now each summation over d is the sum of the primitive c-th roots of unity for some c, and so the summation over c in (37) coincides with the special case of (24) in which $n = 1$ and $s = 2$. So we have $c_{\Gamma,1,0}(-1,0) = 4\pi^2\zeta(2)^{-1} = 24$ and thus we conclude that

$$R^{[-1]}_{\Gamma,1,0}(\tau) = J(\tau) + 24 \tag{38}$$

where J denotes the elliptic modular invariant (cf. (3)). We refer to [43] for a nice review of Rademacher's treatment of (35).

Generalisations of Rademacher's construction (35) have been developed by various authors, including Knopp, who attached weight 0 Rademacher sums to various groups $\Gamma < SL_2(\mathbb{R})$ in [37, 38, 40], and Niebur, who established a very general convergence result for Rademacher sums of arbitrary negative weight in [54].

Theorem 2.2 ([54]). *Let Γ be a discrete subgroup of $SL_2(\mathbb{R})$ having exactly one cusp. Let ψ be a multiplier for Γ and let w be a compatible weight. If $w < 0$ then the Rademacher sum $R^{[\mu]}_{\Gamma,\psi,w}$ converges for any $\mu < 0$ such that $h\mu + \alpha \in \mathbb{Z}$.*

We remark that the method of [54] used to demonstrate convergence certainly applies to groups having more than one cusp.

It will develop in Sect. 3 that the convergence of Rademacher sums is generally more delicate for weights in the range $0 \le w \le 2$ than for $|w-1| > 1$. In [25] it is shown that the weight 0 Rademacher sum $R^{[\mu]}_{\Gamma,1,0}$ converges for any negative integer μ, for any group $\Gamma < SL_2(\mathbb{R})$ that is commensurable with $SL_2(\mathbb{Z})$ and contains $-I$, and certain Rademacher sums of weight $1/2$ (of relevance to Mathieu moonshine, cf. Sect. 4.2) are shown to converge in [11].

Theorem 2.3 ([25]). *Let Γ be a subgroup of $SL_2(\mathbb{R})$ that is commensurable with $SL_2(\mathbb{Z})$ and contains $-I$. Then the Rademacher sum $R^{[\mu]}_{\Gamma,1,0}$ converges, locally uniformly for $\tau \in \mathbb{H}$, for any negative integer μ.*

Theorem 2.4 ([11]). *Let $\Gamma = \Gamma_0(n)$ for n a positive integer, let h be a divisor of n that also divides 24 and set $\psi = \rho_{n|h}\epsilon^{-3}$ where $\rho_{n|h}$ is defined by (87). Then the Rademacher sum $R^{[-1/8]}_{\Gamma,\psi,1/2}$ converges, locally uniformly for $\tau \in \mathbb{H}$.*

2.3 Mock Modularity

The reader will have noticed from the examples presented so far that Γ-invariance sometimes, but not always, survives the Rademacher regularisation procedure; the Rademacher sum $R^{[0]}_{\Gamma,1,2} = E_2$ is not invariant when $\Gamma = SL_2(\mathbb{Z})$—the Eisenstein series E_2 is only quasi-modular (cf. (41))—whilst the original Rademacher sum $R^{[-1]}_{\Gamma,1,0} = J + 24$ is invariant. In a word, the Γ-invariance (with respect to the (ψ, w)-action) of a (convergent) sum $R^{[\mu]}_{\Gamma,\psi,w}$ depends upon the geometry of the group Γ. For example, supposing that Γ is a subgroup of $SL_2(\mathbb{R})$ containing $-I$ and commensurable with $SL_2(\mathbb{Z})$, the Rademacher sum $R^{[-1]}_{\Gamma,1,0}$ fails to be Γ-invariant exactly when Γ does not define a genus zero quotient of $\mathbb{H}$ (i.e. when the genus of X_Γ is not zero, cf. (12)) and in this case there is a function $\omega : \Gamma \to \mathbb{C}$ such that $R^{[-1]}_{\Gamma,1,0}(\gamma\tau) + \omega(\gamma) = R^{[-1]}_{\Gamma,1,0}(\tau)$ for each $\gamma \in \Gamma$ (cf. [25, Thm. 3.4.4]). The sensitivity to the genus of Γ in this example is a consequence of the choices $\psi \equiv 1$ and $w = 0$, as we shall see presently. For other choices of ψ and w the modularity or otherwise of $R^{[\mu]}_{\Gamma,\psi,w}$ will be determined by some other geometric feature of Γ.

In general the Rademacher regularisation defines a *weak mock modular form* which is a function on $\mathbb{H}$ that is invariant for a certain twist of the usual Γ-action, where the twisting is determined by a(n honest) modular form with the dual weight and inverse multiplier. More precisely, suppose that ψ is a multiplier system for Γ with weight w and g is a modular form for Γ with the inverse multiplier system

$\bar{\psi} : \gamma \mapsto \overline{\psi(\gamma)}$ and *dual* weight $2 - w$. Then we can use g to twist the (ψ, w)-action of Γ on $\mathscr{O}(\mathbb{H})$ by setting

$$\left(f|_{\psi,w,g}\gamma\right)(\tau) = f(\gamma\tau)\psi(\gamma)\mathrm{j}(\gamma,\tau)^{w/2} + (2\pi\mathbf{i})^{1-w}\int_{-\gamma^{-1}\infty}^{\mathbf{i}\infty}(z+\tau)^{-w}\overline{g(-\bar{z})}\mathrm{d}z. \tag{39}$$

A weak mock modular form for Γ with multiplier ψ, weight w, and *shadow* g is a holomorphic function f on $\mathbb{H}$ that is invariant for the (ψ, w, g)-action of Γ defined in (39) and which has at most exponential growth at the cusps of Γ (i.e. there exists $C > 0$ such that $(f|_{\tilde{\psi},w}\sigma) = O(e^{C\Im(\tau)})$ for all $\sigma \in \tilde{\Gamma}$ as $\Im(\tau) \to \infty$ where $\tilde{\Gamma}$ and $\tilde{\psi}$ are as in Sect. 2.1). A weak mock modular form is called a *mock modular form* in case it is bounded at every cusp. From this point of view a (weak) modular form is a (weak) mock modular form with vanishing shadow. The notion of mock modular form developed from Zwegers' ground breaking work [72] on Ramanujan's mock theta functions. It is very closely related to the notion of *automorphic integral* which was introduced by Niebur to describe the Rademacher sums of negative weight he constructed in [54]: an automorphic integral of weight w in the sense of Niebur is a weak mock modular form whose shadow is a cusp form.

Given that convergent Rademacher sums are (weak) mock modular forms we may ask for an explicit description of the corresponding shadow functions. In fact, the Rademacher machinery itself provides such a description (cf. e.g. [25, Sect. 3.4], [11, Sect. 7]). Indeed, we can expect that the Rademacher sum $R^{[\mu]}_{\Gamma,\psi,w}$, supposing it converges, is a mock modular form whose shadow $S^{[\mu]}_{\Gamma,\psi,w}$ is also given by a Rademacher sum; namely,

$$S^{[\mu]}_{\Gamma,\psi,w} = \frac{(-\mu)^{1-w}}{\Gamma(1-w)}R^{[-\mu]}_{\Gamma,\bar{\psi},2-w}. \tag{40}$$

Niebur established the identity (40) for arbitrary negative weights and a large class of groups.

Theorem 2.5 ([54]). *Let Γ be a discrete subgroup of $SL_2(\mathbb{R})$ having exactly one cusp. Let ψ be a multiplier for Γ and let w be a compatible weight. If $w < 0$ and $\mu < 0$ is such that $h\mu + \alpha \in \mathbb{Z}$ then the Rademacher sum $R^{[\mu]}_{\Gamma,\psi,w}$ is a weak mock modular form for Γ with shadow given by (40).*

Again, we remark that the method of [54] used to demonstrate mock modularity certainly applies to groups having more than one cusp. The case that $\psi \equiv 1$ and $w = 0$ in (40) was considered in [25] and results for $w = 1/2$ were established in [11].

Theorem 2.6 ([25]). *Let Γ be a subgroup of $SL_2(\mathbb{R})$ that is commensurable with $SL_2(\mathbb{Z})$ and contains $-I$. Then for μ a negative integer the Rademacher sum $R^{[\mu]}_{\Gamma,1,0}$ is a weak mock modular form with shadow $S^{[\mu]}_{\Gamma,1,0}$ given by (40).*

Theorem 2.7 ([11]). *Let $\Gamma = \Gamma_0(n)$ for n a positive integer, let h be a divisor of n that also divides 24 and set $\psi = \rho_{n|h}\epsilon^{-3}$ where $\rho_{n|h}$ is defined by (87). Then the Rademacher sum $R^{[-1/8]}_{\Gamma,\psi,1/2}$ is a weak mock modular form with shadow $S^{[-1/8]}_{\Gamma,\psi,1/2}$ given by (40).*

We can see using Theorem 2.6 why $R^{[-1]}_{\Gamma,1,0}$ has to be Γ-invariant in case Γ has genus zero, for the shadow $S^{[-1]}_{\Gamma,1,0} = R^{[1]}_{\Gamma,1,2}$ is a modular form of weight 2 with trivial multiplier, and in fact a cusp form since it is obtained by summing images of $q = \mathrm{e}(\tau)$ under the weight 2 action of Γ. (We refer the reader to [25] and [11] for more on the behavior of Rademacher sums at arbitrary cusps.) The cusp forms of weight 2 with trivial multiplier for Γ are in correspondence with holomorphic 1-forms on the Riemann surface X_Γ (cf. (12)) and the dimension of the space of holomorphic 1-forms on a Riemann surface is equal to its genus. So if Γ has genus zero then X_Γ has no non-zero 1-forms and we must have $g = S^{[-1]}_{\Gamma\,1,0} = 0$ in (39).

As a second example consider the sum $R^{[0]}_{\Gamma,1,2}$ which we found in Sect. 2.2 to be the Eisenstein series E_2 when $\Gamma = SL_2(\mathbb{Z})$. To compute the right-hand side of (40) when $\mu = 0$ and $w = 2$ we consider a one-parameter family of multipliers $\psi_\delta = \epsilon^\delta$, with corresponding weights $w_\delta = 2 + \delta/2$, where $\epsilon : \Gamma \to \mathbb{C}$ is the multiplier system of the Dedekind eta function (cf. (102)–(103)). Substituting $\delta/24$ for μ and $w_\delta = 2 + \delta/2$ for w in (40) we obtain $-12R^{[0]}_{\Gamma,1,0}$ in the limit as $\delta \to 0$. Recalling the definition of $\mathrm{r}^{[\mu]}_w(\gamma,\tau)$ and using (31) and (97) we see that $\mathrm{r}^{[0]}_0(\gamma,\tau) = 0$ unless γ belongs to Γ_∞ in which case $\mathrm{r}^{[0]}_0(\gamma,\tau) = 1$, so we arrive at the suggestion that the shadow of $R^{[0]}_{\Gamma,1,2}$ should be given by $S^{[0]}_{\Gamma,1,2} = -12R^{[0]}_{\Gamma,1,0} \equiv -12$; that is, $R^{[0]}_{\Gamma,1,2}$ is a mock modular form with constant shadow -12. Taking $g \equiv -12$ in (39), and writing $R(\tau)$ for $R^{[0]}_{\Gamma,1,2}(\tau)$ to ease notation, we find that

$$\begin{aligned} R(\tau) = (R|_{1,2,1}\gamma)(\tau) &= R(\gamma\tau)\,\mathrm{j}(\gamma,\tau) + \frac{6\mathbf{i}}{\pi}\int_{-\gamma^{-1}\infty}^{\mathbf{i}\infty}(z+\tau)^{-2}\mathrm{d}z \\ &= R(\gamma\tau)\,\mathrm{j}(\gamma,\tau) + \frac{6\mathbf{i}}{\pi}\frac{1}{(\tau-\gamma^{-1}\infty)} \end{aligned} \tag{41}$$

for $\gamma \in \Gamma$, which is in agreement with the known quasi-modularity of E_2 (cf. [2, p. 69]).

Before concluding this section we remark on an alternative approach to studying the mock modular forms we have obtained above using Rademacher sums. An equivalent and more common definition of the notion of mock modular form, more closely related to Zwegers' original treatment in [72], is to say that a holomorphic function $f : \mathbb{H} \to \mathbb{C}$ is a weak mock modular form for the group Γ with multiplier ψ, weight w, and shadow g if the *completion* of f, denoted $\hat{f}$ and defined as

$$\hat{f}(\tau) = f(\tau) - (2\pi\mathbf{i})^{1-w}\int_{-\bar{\tau}}^{\mathbf{i}\infty}(z+\tau)^{-w}\overline{g(-\bar{z})}\mathrm{d}z, \tag{42}$$

is invariant for the usual (untwisted) (ψ, w)-action of Γ (cf. (10)) on real-analytic functions on $\mathbb{H}$. From (42) one can check that $\hat{f}$ is annihilated by the differential operator

$$\frac{\partial}{\partial \tau}(\mathrm{Im}\tau)^w \frac{\partial}{\partial \bar{\tau}} \tag{43}$$

and hence is a *harmonic weak Maaß form* of weight w, which is to say, $\hat{f}$ is a (non-holomorphic) modular form for Γ with at most exponential growth at the cusps which is also an eigenfunction for the weight w Laplace operator with eigenvalue $\frac{w}{2}(1-\frac{w}{2})$. (We refer to [71, Sect. 5] for an exposition of this.) For a suitably defined Poincaré series (adapted to the construction of Maaß forms) the function $R^{[\mu]}_{\Gamma,\psi,w}$ may then be recovered as its *holomorphic part*. We refer to [4] for the pioneering example of this approach; further examples appear in [5–7]. The harmonic weak Maaß form whose holomorphic part is $R^{[-1/8]}_{\Gamma,\epsilon^{-3},1/2}$ was investigated in [26] in the cases that $\Gamma = SL_2(\mathbb{Z})$ and $\Gamma = \Gamma_0(2)$.

3 Rademacher Series

The Rademacher sums of the previous section are indexed by cosets of Γ_∞ in Γ. In this section we consider a construction—also inspired by work of Rademacher, among others, and hinted at in the definition of the constant correction term in (32)—of series indexed by double coset spaces $\Gamma_\infty\backslash\Gamma^\times/\Gamma_\infty$. It will develop that these series—we call them *Rademacher series*— recover the Fourier coefficients of the Rademacher sums of the previous section, but also admit other applications, such as recovering Fourier coefficients of false theta series (cf. (67)), and Eichler integrals of (mock) modular forms more generally (cf. (63)). In addition, the Rademacher series construction serves to illuminate a form of *Zagier duality* for Rademacher sums: the coincidence (up to a root of unity depending only on w) of the Fourier coefficients attached to the *dual families*

$$\left\{R^{[\mu]}_{\Gamma,\psi,w} \mid h\mu+\alpha\in\mathbb{Z},\ \mu<0\right\},\quad \left\{R^{[\nu]}_{\Gamma,\bar{\psi},2-w} \mid h\nu-\alpha\in\mathbb{Z},\ \nu<0\right\}, \tag{44}$$

(cf. (13)) of Rademacher sums.

We now detail the Rademacher series construction. Suppose as before that $\Gamma < SL_2(\mathbb{R})$ contains $-I$ and is commensurable with $SL_2(\mathbb{Z})$. Recall that $h>0$ is chosen so that $\Gamma_\infty = \langle T^h, -I\rangle$ (cf. (7)). Given a multiplier system ψ of weight w for such a group Γ, and given also $\mu,\nu\in\frac{1}{h}(\mathbb{Z}-\alpha)$ where $\psi(T^h)=\mathrm{e}(\alpha)$ (cf. (13)), we define the *Rademacher series* $c_{\Gamma,\psi,w}(\mu,\nu)$ by setting

$$c_{\Gamma,\psi,w}(\mu,\nu) = \frac{1}{h}\lim_{K\to\infty}\sum_{\gamma\in\Gamma_\infty\backslash\Gamma^\times_K/\Gamma_\infty} K_{\gamma,\psi}(\mu,\nu)B_{\gamma,w}(\mu,\nu) \tag{45}$$

where $\Gamma_K^\times$ is defined as in (27) and $K_{\gamma,\psi}$ and $B_{\gamma,w}$ are given by

$$K_{\gamma,\psi}(\mu,\nu) = \mathrm{e}\left(\mu\frac{a}{c}\right)\mathrm{e}\left(\nu\frac{d}{c}\right)\psi(\gamma), \tag{46}$$

$$B_{\gamma,w}(\mu,\nu) = \begin{cases} \mathrm{e}\left(-\frac{w}{4}\right)\sum_{k\geq 0}\left(\frac{2\pi}{c}\right)^{2k+w}\frac{(-\mu)^k}{k!}\frac{\nu^{k+w-1}}{\Gamma(k+w)}, & w\geq 1,\\ \mathrm{e}\left(-\frac{w}{4}\right)\sum_{k\geq 0}\left(\frac{2\pi}{c}\right)^{2k+2-w}\frac{(-\mu)^{k+1-w}}{\Gamma(k+2-w)}\frac{\nu^k}{k!}, & w\leq 1,\end{cases} \tag{47}$$

in case $\gamma = \left(\begin{smallmatrix} a & b\\ c & d\end{smallmatrix}\right)$ and $c>0$. Observe that the restriction $\mu,\nu\in\frac{1}{h}(\mathbb{Z}-\alpha)$ is necessary in order that the map $\gamma\mapsto K_{\gamma,\psi}(\mu,\nu)B_{\gamma,\psi}(\mu,\nu)$ descend to the double coset space $\Gamma_\infty\backslash\Gamma_K^\times/\Gamma_\infty$; assuming convergence we may regard $c_{\Gamma,\psi,w}$ as a function on the *grid*

$$\frac{1}{h}\mathbb{Z}\times\frac{1}{h}\mathbb{Z}-\left(\frac{\alpha}{h},\frac{\alpha}{h}\right)\subset\mathbb{R}^2. \tag{48}$$

Note that the convergence of the expression (45) defining $c_{\Gamma,\psi,w}(\mu,\nu)$ is not obvious when w lies in the range $0\leq w\leq 2$ but is relatively easy to show for $w<0$ and $2<w$. For example, if $\Gamma = SL_2(\mathbb{Z})$ and $w\geq 1$ then we have the simple estimate

$$\begin{aligned} |c_{\Gamma,\psi,w}(\mu,\nu)| &\leq \sum_{\gamma\in\Gamma_\infty\backslash\Gamma^\times/\Gamma_\infty}|K_{\gamma,\psi}(\mu,\nu)||B_{\gamma,w}(\mu,\nu)|\\ &\leq \sum_{c>0}c\sum_{k\geq 0}\left(\frac{2\pi}{c}\right)^{2k+w}\frac{|\mu|^k|\nu|^{k+w-1}}{k!\Gamma(k+w)}\end{aligned} \tag{49}$$

where both c and k are restricted to be integers and the factor c appearing between the two summations serves as a crude upper bound for the number of double cosets in $\Gamma_\infty\backslash\Gamma/\Gamma_\infty$ with representatives having lower-left entry equal to c. Consider the result of interchanging the two summations in the right-hand side of (49). If $w>2$ then we obtain

$$\begin{aligned} &\sum_{k\geq 0}(2\pi)^{2k+w}\frac{|\mu|^k|\nu|^{k+w-1}}{k!\Gamma(k+w)}\sum_{c>0}c^{1-2k-w}\\ &\leq \sum_{k\geq 0}(2\pi)^{2k+w}\frac{|\mu|^k|\nu|^{k+w-1}}{k!\Gamma(k+w)}\frac{1}{w-2}\\ &= \frac{2\pi}{w-2}|\mu|^{(1-w)/2}|\nu|^{(w-1)/2}I_{w-1}(4\pi|\mu\nu|^{1/2})\end{aligned} \tag{50}$$

where $I_\alpha(z)$ denotes the *modified Bessel function of the first kind* and we have used its series expression (101) in the second line of (50). In particular, the left-hand side of (50) is absolutely convergent for $w > 2$. This verifies the coincidence of the left-hand side of (50) with the right-hand side of (49) and thus we obtain the absolute convergence of the Rademacher series $c_{\Gamma,\psi,w}(\mu,\nu)$ for $w > 2$. The case that $w < 0$ is similar, and for a more general group Γ, being a union of finitely many cosets of a finite-index subgroup of $SL_2(\mathbb{Z})$, the necessary adjustments to the above argument are not unduly complicated. We refer to [25] for the case that ψ is trivial and w is an even integer. (See also Theorem 3.1 below.) We refer to [54] for a treatment of the case that $w < 0$.

The question of convergence is more subtle in the cases that $0 \leq w \leq 2$. To establish convergence for weights in this region one has to replace the c appearing between the two summations in (49) with a more careful estimate for the *Kloosterman sum*

$$S_{\Gamma,\psi}(\mu,\nu,c) = \sum_{\substack{\gamma\in\Gamma_\infty\backslash\Gamma/\Gamma_\infty \\ c(\gamma)=c}} K_{\gamma,\psi}(\mu,\nu). \tag{51}$$

In (51) we again write $c(\gamma)$ for the lower-left entry of γ. A beautiful approach to analysing Kloosterman sums was pioneered by Selberg in [65]. Selberg introduced the *Kloosterman zeta function*

$$Z_{\mu,\nu}(s) = \sum_{\gamma\in\Gamma_\infty\backslash\Gamma^\times/\Gamma_\infty} K_{\gamma,\psi}(\mu,\nu)c(\gamma)^{-2s} = \sum_{c>0} S_{\Gamma,\psi}(\mu,\nu,c)c^{-2s} \tag{52}$$

and demonstrated that it admits an analytic continuation that is holomorphic in the half-plane $\Re(s) > 1/2$ but for finitely many poles on the real line segment $1/2 < s < 1$. Further, these poles are determined by the vanishing or otherwise of particular Fourier coefficients of particular cusp forms for Γ. Using this together with the growth estimates for $Z_{\mu,\nu}(s)$ due to Goldfeld–Sarnak [35] (see also [21]) one may, for suitable choices of μ and ν, obtain the convergence of the series defining $c_{\Gamma,\psi,w}(\mu,\nu)$. Such an approach was first implemented by Knopp in [41,42]. It was applied in [25] so as to establish the convergence of $c_{\Gamma,1,w}(\mu,\nu)$ in weights $w = 0$ and $w = 2$ for arbitrary Γ commensurable with $SL_2(\mathbb{Z})$ and arbitrary $\mu,\nu \in \mathbb{Z}$, and it was applied in [11] to demonstrate the convergence of $c_{\Gamma,\psi,1/2}(\mu,\nu)$ for $\mu = -1/8$ and $\nu > 0$ when $\Gamma = \Gamma_0(n)$ for some n, and ψ is one of the multipliers relevant for Mathieu moonshine (cf. Sect. 4.2).

Theorem 3.1 ([25]). *Let Γ be a subgroup of $SL_2(\mathbb{R})$ that is commensurable with $SL_2(\mathbb{Z})$ and contains $-I$. Then the Rademacher series $c_{\Gamma,1,0}(\mu,\nu)$ and $c_{\Gamma,1,2}(\mu,\nu)$ converge for all $\mu,\nu \in \mathbb{Z}$.*

Theorem 3.2 ([11]). *Let $\Gamma = \Gamma_0(n)$ for n a positive integer, let h be a divisor of n that also divides 24 and set $\psi = \rho_{n|h}\epsilon^{-3}$ where $\rho_{n|h}$ is defined by (87). Then the Rademacher series $c_{\Gamma,\psi,1/2}(-1/8,\nu)$ converges for all $\nu \in \mathbb{Z} - 1/8$ such that $\nu > 0$, and the Rademacher series $c_{\Gamma,\bar{\psi},3/2}(1/8,\nu')$ converges for all $\nu' \in \mathbb{Z} + 1/8$ such that $\nu' > 0$.*

At this point we may recognise the expression (33), defining the constant term correction to Rademacher sums with $w = 0$, as a specialisation of the Rademacher series construction (45). In particular, we can confirm that $c_{\Gamma,\psi,w}(\mu, 0) = 0$ when $w \geq 1$ and $c_{\Gamma,\psi,w}(\mu, 0)$ should not be defined unless $\alpha = 0$. Note also that $B_{\gamma,w}$ can be expressed conveniently in terms of Bessel functions (cf. Sect. 6) in case $xy \neq 0$. For example, if $x < 0$ or $y > 0$ then we have

$$B_{\gamma,w}(\mu, \nu) = \mathrm{e}\left(-\frac{w}{4}\right)(-\mu)^{(1-w)/2}\nu^{(w-1)/2}\frac{2\pi}{c}I_{|w-1|}\left(\frac{4\pi}{c}(-\mu\nu)^{1/2}\right) \tag{53}$$

for any weight $w \in \mathbb{R}$. (In the case that $y < 0 < x$ the right-hand side of (53) should be multiplied by $e^{\pi \mathrm{i}|w-1|}$.)

In the remainder of this section we consider some applications of the Rademacher series.

3.1 *Coefficients of Rademacher Sums*

Expressions like that defined by (45)–(47) first appeared in the aforementioned work [58] of Poincaré where he considered the case that $\Gamma = SL_2(\mathbb{Z})$, the multiplier ψ is trivial, w is an even integer greater than 2, and μ is a non-negative integer. Poincaré obtained an expression equivalent to $c_{\Gamma,\psi,w}(m, n) + \delta_{m,n}$ for the Fourier coefficient of q^n in $P^{[m]}_{\Gamma,\psi,w}(\tau)$, for m and n non-negative integers. The series of [58] were generalised by Petersson in [55], where he obtained the analogous expression

$$P^{[\mu]}_{\Gamma,\psi,w} = q^{\mu} + \sum_{\substack{h\nu+\alpha\in\mathbb{Z}\\ \nu\geq 0}} c_{\Gamma,\psi,w}(\mu, \nu)q^{\nu} \tag{54}$$

when Γ is the principal congruence group $\Gamma(N)$ (the kernel of the map $SL_2(\mathbb{Z}) \to SL_2(\mathbb{Z}/N\mathbb{Z})$) for some N. Thus we see many instances in which the Rademacher series recover the Fourier coefficients of a Poincaré series.

The formula (54) was established for more general subgroups $\Gamma < SL_2(\mathbb{Z})$ and for weights $w \geq 2$ in [56,57], and on the strength of this, together with his result that an arbitrary modular form may be written as a linear combination of Poincaré series, Petersson essentially solved the problem of finding convergent series expressions for the Fourier coefficients of modular forms with weight $w \geq 2$. Using the fact that the derivative of the elliptic modular invariant $J(\tau)$ is a weak modular form of weight 2, and thus a function whose coefficients can be written in terms of the $c_{\Gamma,\psi,w}$ according to his results, Petersson was able to derive series expressions for the coefficients of the function $J(\tau)$ itself, by integration. To see such expressions consider the values $c_{\Gamma,\psi,w}(\mu, \nu)$ for $\Gamma = SL_2(\mathbb{Z})$, $\psi \equiv 1$ and $w = 0$. Then $h = 1$, $\alpha = 0$ and $(\mu, \nu) \in \mathbb{Z} \times \mathbb{Z}$ (cf. (48)). Observing that the non-trivial double cosets of Γ_∞ in $\Gamma = SL_2(\mathbb{Z})$ are represented, irredundantly, by the matrices $\left(\begin{smallmatrix} a & b \\ c & d \end{smallmatrix}\right)$ with $c > 0$ and d coprime to c satisfying $0 \leq d < c$ we find that

$$c_{\Gamma,1,0}(-1,n) = \sum_{\substack{c>0 \\ 0\leq d<c \\ (c,d)=1}} \mathrm{e}\left(\frac{a+nd}{c}\right) n^{-1/2} \frac{2\pi}{c} I_1\left(\frac{4\pi}{c} n^{1/2}\right) \tag{55}$$

in agreement with Petersson's formula [56, p. 202] for the n-th coefficient of $J(\tau)$, so that, according to Rademacher's identity $R_{\Gamma,1,0}^{[-1]} = J + 24$ (cf. (38)), we have

$$R_{\Gamma,1,0}^{[-1]}(\tau) = q^{-1} + \sum_{n\geq 0} c_{\Gamma,1,0}(-1,n) q^n. \tag{56}$$

In particular, the Rademacher series $c_{\Gamma,1,0}$ recover the Fourier coefficients of the Rademacher sum $R_{\Gamma,1,0}^{[-1]}$.

In independent work Rademacher solved the problem of providing an exact formula for the partition function [59] and this furnishes another instructive example, for if $p(n)$ denotes the number of partitions of the positive integer n then we have

$$\frac{1}{\eta(\tau)} = q^{-1/24} + \sum_{n>0} p(n) q^{n-1/24} \tag{57}$$

where η denotes the Dedekind eta function (cf. (102)). So it suffices to compute expressions for the Fourier coefficients of the (weak) modular form $1/\eta$ of weight $-1/2$. Let $\Gamma = SL_2(\mathbb{Z})$ and let $\epsilon : \Gamma \to \mathbb{C}$ denote the multiplier system of η (cf. (103)). Then $\bar{\epsilon} = \epsilon^{-1}$ is a multiplier system in weight $w = -1/2$ for Γ with $h = 1$ and $\alpha = 1/24$ and so we may consider the values $c_{\Gamma,\bar{\epsilon},-1/2}(-1/24, n - 1/24)$ for n a positive integer. Comparing with the explicit formula (104) for ϵ we find that

$$\begin{aligned} &c_{\Gamma,\bar{\epsilon},-1/2}\left(-\frac{1}{24}, n - \frac{1}{24}\right) \\ &= \sum_{\substack{c>0 \\ 0\leq d<c \\ (c,d)=1}} \mathrm{e}\left(n\frac{d}{c} - \frac{s(d,c)}{4}\right) (24n-1)^{-3/4} \frac{2\pi}{c} I_{3/2}\left(\frac{\pi}{6c}(24n-1)^{1/2}\right) \end{aligned} \tag{58}$$

which is in agreement with the formula for $p(n)$ derived in [59]. (The right-hand side of (58) is more immediately recognised in the subsequent work [62] which gives a general description of coefficients of modular forms of negative weight for the modular group in terms of the $c_{\Gamma,\psi,w}$ defined above and revisits the case of $1/\eta(\tau)$ as a specific example on p. 455.)

Rademacher went on to determine an analogue of (58) for the coefficients of J in [60]. Using a completely different method to that of [56] he independently rediscovered the formula (55). Rademacher's motivation for the subsequent work [61], and the introduction of the original Rademacher sum $R^{[-1]}_{\Gamma,1,0}$ (cf. (35)), was to derive the modular invariance of the function $q^{-1} + \sum_{n>0} c_{\Gamma,1,0}(-1,n)q^n$, and thereby establish its coincidence with J directly, using just the expression (55) for $c_{\Gamma,1,0}(-1,n)$.

We have seen now several examples in which the series $c_{\Gamma,\psi,w}$ serve to recover coefficients of a modular form, and a Rademacher sum in particular. In general we can expect the direct relationship

$$R^{[\mu]}_{\Gamma,\psi,w}(\tau) = q^\mu + \sum_{\substack{h\nu+\alpha\in\mathbb{Z}\\ \nu\geq 0}} c_{\Gamma,\psi,w}(\mu,\nu)q^\nu \tag{59}$$

between Rademacher sums and Rademacher series, assuming that $R^{[\mu]}_{\Gamma,\psi,w}$ and all the $c_{\Gamma,\psi,w}(\mu,\nu)$ with $\nu \geq 0$ are convergent. To see how this relationship can be derived we may begin by replacing $\mathrm{e}(\mu\gamma\tau)$ with $\mathrm{e}(\mu\gamma\infty)\,\mathrm{e}(\mu(\gamma\tau - \gamma\infty))$ in (32) and rewriting $\gamma\tau - \gamma\infty$ as $-c^{-1}(c\tau + d)^{-1}$ in case (c,d) is the lower row of γ. Then we may proceed in a way similar to that employed in the discussion leading to (25), applying the Lipschitz summation formula (108) (and typically also its non-absolutely convergent version, Lemma 6.1) together with the fact that

$$\psi(\gamma T^h)e(\mu\gamma T^h\infty)\,\mathrm{j}(\gamma T^h,\tau)^{w/2} = \psi(\gamma)e(\mu\gamma\infty)\,\mathrm{j}(\gamma,\tau+h)^{w/2} \tag{60}$$

for $h\mu + \alpha \in \mathbb{Z}$, and this brings us quickly to the required expression for $R^{[\mu]}_{\Gamma,\psi,w}$ as a sum of sums over the double coset space $\Gamma_\infty\backslash\Gamma^\times/\Gamma_\infty$. We refer to [25] and [11] for detailed implementations of this approach, including careful consideration of convergence.

Since the Rademacher sum $R^{[\mu]}_{\Gamma,\psi,w}$ is precisely the Poincaré series $P^{[\mu]}_{\Gamma,\psi,w}$ when $w > 2$ we have (59) for $w > 2$ according to the aforementioned work of Petersson. Niebur established (59) for arbitrary weights $w < 0$ in [54] (and thus we have that $1/\eta$ is also a Rademacher sum—namely, $1/\eta = R^{[-1/24]}_{\Gamma,\bar{\epsilon},-1/2}$—according to the Rademacher's formula for $p(n)$ and the identity (58)). We have illustrated above that the convergence of the Rademacher series $c_{\Gamma,\psi,w}$ is more subtle in case $0 \leq w \leq 2$. As we have mentioned, Petersson and Rademacher independently gave the first instance of (59) for $w = 0$; other examples were established by Knopp in [37, 38, 40]. The general case that Γ is commensurable with $SL_2(\mathbb{Z})$ and contains $-I$, the multiplier ψ is trivial and $w = 0$ was proven in [25], and examples with $w = 1/2$ and $w = 3/2$ were established in [11].

Theorem 3.3 ([25]). *Let Γ be a subgroup of $SL_2(\mathbb{R})$ that is commensurable with $SL_2(\mathbb{Z})$ and contains $-I$. Then the Fourier expansion of the Rademacher sum $R^{[\mu]}_{\Gamma,1,0}$ is given by (59).*

Theorem 3.4 ([11]). *Let $\Gamma = \Gamma_0(n)$ for n a positive integer, let h be a divisor of n that also divides 24 and set $\psi = \rho_{n|h}\epsilon^{-3}$ where $\rho_{n|h}$ is defined by (87). Then the Fourier expansions of the Rademacher sums $R^{[-1/8]}_{\Gamma,\psi,1/2}$ and $R^{[1/8]}_{\Gamma,\bar{\psi},3/2}$ are given by (59).*

Results closely related to (59) for weights in the range $0 < w < 2$ have been established by Knopp [41,42], Pribitkin [19,20], and Bringmann–Ono [4,7].

3.2 Dualities

The Bessel function expression (53) emphasises a symmetry in $B_{\gamma,w}$ under the exchange of a weight w with it's dual weight $2 - w$; namely, $-\mathrm{e}(-w/2) B_{\gamma,2-w}(-\nu,-\mu) = B_{\gamma,w}(\mu,\nu)$. Replacing γ with $-\gamma^{-1}$ in (46)–(47) we observe that $\mathrm{e}(w/2)K_{-\gamma^{-1},\bar{\psi}}(-\nu,-\mu) = K_{\gamma,w}(\mu,\nu)$ and $B_{-\gamma^{-1},w} = B_{\gamma,w}$, and thus we obtain the *Zagier duality* identity

$$c_{\Gamma,\bar{\psi},2-w}(-\nu,-\mu) = c_{\Gamma,\psi,w}(\mu,\nu) \tag{61}$$

in case $\mu,\nu \in \frac{1}{h}(\mathbb{Z}-\alpha)$ (cf. (48)). This may be regarded as a generalisation of the coincidence, up to a minus sign, of coefficients in certain families of modular forms in dual weights that was observed by Zagier in [70]. Much of the interest in Zagier duality derives from its power to give novel interpretations to coefficients of modular forms, such as in terms of traces of singular moduli in the original example [70]; for other generalisations and applications we refer to [5,9,29,36,64].

The duality (61) demonstrates that dual Rademacher series—attached to mutually inverse multiplier systems in dual weights—coincide up to transposition and negation of their arguments. In other words, the vertical lines in the grid of values $(\mu,\nu) \mapsto c_{\Gamma,\psi,w}(\mu,\nu)$ are, up to sign, the horizontal lines in the corresponding grid $(\mu',\nu') \mapsto c_{\Gamma,\bar{\psi},2-w}(\mu',\nu')$ for the dual Rademacher series. Consequently, when considering Fourier coefficients of Rademacher sums with a given weight and multiplier system one is simultaneously considering the Fourier coefficients of Rademacher sums in the dual weight. As an application of this we see that the Rademacher series $c_{\Gamma,\psi,w}$ encode not only the Fourier expansions of the $R^{[\mu]}_{\Gamma,\psi,w}$ but also the Fourier expansions of their shadows $S^{[\mu]}_{\Gamma,\psi,w}$. For by applying (59) to the formula (40), which relates the shadow $S^{[\mu]}_{\Gamma,\psi,w}$ of $R^{[\mu]}_{\Gamma,\psi,w}$ to the dual Rademacher sum $R^{[-\mu]}_{\Gamma,\bar{\psi},2-w}$, we obtain

$$S^{[\mu]}_{\Gamma,\psi,w}(\tau) = \frac{(-\mu)^{1-w}}{\Gamma(1-w)}\left(q^{-\mu}\sum_{\substack{h\nu-\alpha\in\mathbb{Z}\\ \nu\geq 0}} c_{\Gamma,\psi,w}(-\nu,\mu)q^{\nu}\right). \tag{62}$$

The *Eichler integral* of a cusp form $f(\tau) = \sum_{\nu>0} c(\nu)q^{\nu}$ with weight w for some group Γ is the function $\tilde{f}(\tau)$ defined by the q-series

$$\tilde{f}(\tau) = \sum_{\nu>0} \nu^{1-w}c(\nu)q^{\nu}. \tag{63}$$

Let us consider the effect of transposing μ with ν and replacing γ with $-\gamma^{-1}$ in (46)–(47). We obtain $\mathrm{e}(w/2)\overline{K_{-\gamma^{-1},\psi}(\nu,\mu)} = K_{\gamma,\psi}(\mu,\nu)$ and $\mathrm{e}(-w/2)\overline{B_{-\gamma^{-1},w}(\nu,\mu)\mu^{1-w}} = B_{\gamma,w}(\mu,\nu)\nu^{1-w}$ for $w \geq 1$, and this, together with an application of (61), leads us to the *Eichler duality* identity

$$-\overline{c_{\Gamma,\bar{\psi},2-w}(-\mu,-\nu)\mu^{1-w}} = c_{\Gamma,\psi,w}(\mu,\nu)\nu^{1-w}, \tag{64}$$

valid for $w \geq 1$. (A similar but slightly different expression obtains when $w < 1$.) The relation (64) demonstrates another application of the Rademacher series construction: the Eichler integral of the Rademcher sum $R^{[\mu]}_{\Gamma,\psi,w}$, assuming $\alpha \neq 0$ or $c_{\Gamma,\psi,w}(\mu,0) = 0$, is computed, up to conjugation and a scalar factor, by the Rademacher series attached to the inverse multiplier system in the dual weight.

$$\tilde{R}^{[\mu]}_{\Gamma,\psi,w}(\tau) = -\overline{\mu^{1-w}} \sum_{\nu>0} \overline{c_{\Gamma,\bar{\psi},2-w}(-\mu,-\nu)} q^{\nu} \tag{65}$$

As an example consider the case that $\Gamma = SL_2(\mathbb{Z})$ is the modular group, $\psi = \epsilon^3$ and $w = 3/2$. Then $R^{[1/8]}_{\Gamma,\epsilon^3,3/2}$ is, up to a scalar factor, the shadow of the weak mock modular form $R^{[-1/8]}_{\Gamma,\epsilon^{-3},1/2}$. It is shown in [11] that $R^{[1/8]}_{\Gamma,\epsilon^3,3/2} = -12\eta^3$ and we have $\eta(\tau)^3 = \sum_{n\geq 0}(-1)^n(2n+1)q^{(2n+1)^2/8}$ according to an identity due to Euler. Thus we find that

$$\tilde{R}^{[1/8]}_{\Gamma,\epsilon^3,3/2}(\tau) = -24\sqrt{2}\sum_{n\geq 0}(-1)^n q^{(2n+1)^2/8}, \tag{66}$$

and applying (65) to this we obtain the beautiful formula

$$c_{\Gamma,\epsilon^{-3},1/2}(-1/8,-n-1/8) = \begin{cases} 12(-1)^m & \text{if } n = \binom{m}{2} \text{ for some } m > 0, \\ 0 & \text{else,} \end{cases} \tag{67}$$

when $n \geq 0$. Compare this to the fact that the values $c_{\Gamma,\epsilon^{-3},1/2}(-1/8,-n-1/8)$ for $n < 0$ are the coefficients of the weak mock modular form $R^{[-1/8]}_{\Gamma,\epsilon^{-3},1/2}$ according to Theorem 3.4. (This weak mock modular form will play a special rôle in Sect. 4.2.) The function $\sum_{n\geq 0}(-1)^n q^{(2n+1)^2/8}$, appearing here as (a rescaling of) the Eichler integral of η^3, is one of the *false theta series* studied by Rogers in [63] (cf. [1]).

4 Moonshine

Some of the most fascinating and powerful applications of Rademacher sums have appeared in moonshine. To describe them we shall start with a short discussion of the relevant modular objects. The study of monstrous moonshine was initiated with the realisation (cf. [68,69]) that the Fourier coefficients of the elliptic modular invariant J (cf. (4)) encode positive integer combinations of dimensions of irreducible representations of the monster group $\mathbb{M}$. More generally, monstrous moonshine attaches a holomorphic function $T_g = q^{-1} + \sum_{n>0} c_g(n)q^n$ on the upper-half plane to each element g in the Monster group $\mathbb{M}$. This association is such that the Fourier coefficients of the *McKay–Thompson series* T_g furnish characters $g \mapsto c_g(n)$ of non-trivial representations of $\mathbb{M}$ (thus the function T_g depends only on the conjugacy class of g), and such that the T_g all have the following *genus zero property*:

If Γ_g is the invariance group of T_g then the natural map $T_g : \Gamma_g\backslash\mathbb{H} \to \mathbb{C}$ extends to an isomorphism of Riemann surfaces $X_{\Gamma_g} \to \hat{\mathbb{C}}$.

Here $\hat{\mathbb{C}}$ denotes the Riemann sphere and X_Γ is the Riemann surface $\Gamma\backslash\mathbb{H} \cup \hat{\mathbb{Q}}$ (cf. (12)). We are using the weight 0 action of $SL_2(\mathbb{R})$ with trivial multiplier, $(f|_{1,0}\gamma)(\tau) = f(\gamma\tau)$ (cf. (10)), to define the invariance.

Conway–Norton introduced the term moonshine in [14] and detailed many interesting features and properties of the—at that time conjectural—correspondence $g \mapsto T_g$. An explicit monster module conjecturally realizing the T_g of [14] as graded-traces was constructed by Frenkel–Lepowsky–Meurman in [30–32], and a proof of the Conway–Norton *moonshine conjectures*—that these graded traces do determine functions T_g with the genus zero property formulated above—was given by Borcherds in [3]. All that notwithstanding, a clear conceptual explanation for the genus zero property of monstrous moonshine is yet to be established. A step towards this goal was made in [25] by employing the Rademacher sum machinery, as we shall see presently in Sect. 4.1. In particular, we will show that the genus zero property is actually equivalent to fact that T_g coincides (up to a constant) with the relevant Rademacher sum (cf. (79)).

In [28] a remarkable observation was made relating the elliptic genus of a $K3$ surface to the largest Mathieu group M_{24} via a decomposition of the former into a linear combination of characters of irreducible representations of the small $N = 4$ superconformal algebra. The elliptic genus is a topological invariant and for any $K3$ surface it is given by the weak Jacobi form

$$Z_{K3}(\tau,z) = 8\left(\left(\frac{\theta_2(\tau,z)}{\theta_2(\tau,0)}\right)^2 + \left(\frac{\theta_3(\tau,z)}{\theta_3(\tau,0)}\right)^2 + \left(\frac{\theta_4(\tau,z)}{\theta_4(\tau,0)}\right)^2\right) \tag{68}$$

of weight 0 and index 1. The θ_i are Jacobi theta functions (cf. (106)). When decomposed into $N = 4$ characters we obtain

$$\begin{aligned} Z_{K3}(\tau,z) &= 20\,\mathrm{ch}^{(2)}_{\frac{1}{4},0} - 2\,\mathrm{ch}^{(2)}_{\frac{1}{4},\frac{1}{2}} + \sum_{n\geq 0} t_n\,\mathrm{ch}^{(2)}_{\frac{1}{4}+n,\frac{1}{2}} \\ &= \frac{\theta_1(\tau,z)^2}{\eta(\tau)^3}\left(24\,\mu(\tau,z) + q^{-1/8}\Big(-2 + \sum_{n=1}^{\infty} t_n q^n\Big)\right) \end{aligned} \tag{69}$$

for some $t_n \in \mathbb{Z}$ where $\theta_1(\tau,z)$ and $\mu(\tau,z)$ are defined in (106)–(107). In the above equation we write

$$\mathrm{ch}_{h,j}^{(\ell)}(\tau,z) = \mathrm{tr}_{V_{h,j}^{(\ell)}}\left((-1)^{J_0^3} y^{J_0^3} q^{L_0 - c/24}\right) \tag{70}$$

for the Ramond sector character of the unitary highest weight representation $V_{h,j}^{(\ell)}$ of the small $N = 4$ superconformal algebra with central charge $c = 6(\ell - 1)$. By inspection, the first five t_n are given by $t_1 = 90$, $t_2 = 462$, $t_3 = 1{,}540$, $t_4 = 4{,}554$, and $t_5 = 11{,}592$. The surprising observation of [28] is that each of these t_n is twice the dimension of an irreducible representation of M_{24}. See the contribution by Gaberdiel and Volpato in this proceeding for a discussion of this Mathieu observation from the viewpoint of conformal field theories.

One is thus compelled to conjecture that every t_n may be interpreted as the dimension of an M_{24}-module $K_{n-1/8}$. If we define $H(\tau)$ by requiring

$$Z_{K3}(\tau,z)\eta(\tau)^3 = \theta_1(\tau,z)^2(a\mu(\tau,z) + H(\tau)) \tag{71}$$

then $a = 24$ and

$$H(\tau) = q^{-\frac{1}{8}}\left(-2 + \sum_{n=1}^{\infty} t_n q^n\right) \tag{72}$$

is a slight modification of the generating function of the t_n. The inclusion of the term -2 and the factor $q^{-1/8}$ has the effect of improving the modularity: $H(\tau)$ is a weak mock modular form (cf. Sect. 2.3) for $SL_2(\mathbb{Z})$ with multiplier ϵ^{-3} (cf. (104)), weight $1/2$, and shadow $-\frac{12}{\sqrt{2\pi}}\eta^3$ (cf. (102)). If the t_n really do encode the dimensions of M_{24}-modules $K_{n-1/8}$ then we can expect to obtain interesting functions $H_g(\tau)$—*McKay–Thompson series* for M_{24}—by replacing t_n with $\mathrm{tr}_{K_{n-1/8}}(g)$ in (72). In other words, we should also consider

$$H_g(\tau) = -2q^{-1/8} + \sum_{n=1}^{\infty} \mathrm{tr}_{K_{n-1/8}}(g) q^{n-1/8}. \tag{73}$$

Strictly speaking, to determine H_g requires knowledge of the M_{24}-module $K = \bigoplus_{n=1}^{\infty} K_{n-1/8}$ whose existence remains conjectural, but one can attempt to formulate conjectural expressions for H_g by identifying a suitably distinguishing modular property that they should satisfy. If the property is well-chosen then it will be strong enough for us to determine concrete expressions for the H_g, and compatibility between the low order terms amongst the Fourier coefficients of H_g with the character table of M_{24} will serve as evidence for both the validity of H_g and the existence of the module K. Exactly this was done in a series of papers, starting with [10], and the independent work [34], and concluding with [33] and [27]. Despite this progress no construction of the conjectured M_{24}-module K is yet known. To find such a construction is probably the most important open problem in Mathieu moonshine at the present time. Similar remarks also apply to the more general umbral moonshine that we will describe shortly.

The strong evidence for the conjecture that $H(\tau)$ encodes the graded dimension of an M_{24}-module invites us to consider the M_{24} analogue of the Conway–Norton moonshine conjectures—this will justify the use of the term moonshine in the M_{24} setting—except that it is not immediately obvious what the analogue should be. Whilst the McKay–Thompson series H_g is a mock modular form of weight $1/2$ on some $\Gamma_g < SL_2(\mathbb{Z})$ for every g in M_{24} [27], it is not the case that Γ_g is a genus zero group for every g, and even if it were, there is no obvious sense in which a weak mock modular form of weight $1/2$ can induce an isomorphism $X_\Gamma \to \hat{\mathbb{C}}$, and thus no obvious analogue of the genus zero property formulated above. A solution to this problem—the formulation of the moonshine conjecture for M_{24}—was recently found in [11]. As we shall explain in Sect. 4.2, the correct analogue of the genus zero property is that the McKay–Thompson series H_g should coincide with a certain Rademacher sum attached to its invariance group Γ_g.

It is striking that, despite the very different modular properties the two sets of McKay–Thompson series H_g and T_g display they can be constructed in completely analogous ways in terms of Rademacher sums. We are hence led to believe that Rademacher sums are an integral element of the moonshine phenomenon. And such a belief has in fact been instrumental in the discovery of *umbral moonshine* [13], whereby a finite group $G^{(\ell)}$ and a family of vector-valued mock modular forms $H_g^{(\ell)}$ for $g \in G^{(\ell)}$ is specified for each ℓ in $\Lambda = \{2, 3, 4, 5, 7, 13\}$—the set of positive integers ℓ such that $\ell - 1$ divides 12—and these groups $G^{(\ell)}$ and vector-valued mock modular forms $H_g^{(\ell)}$ are conjectured to be related in a way that we shall describe presently.

Following [13] we say that a weak Jacobi form $\phi(\tau, z)$ of weight 0 and index $\ell - 1$ is *extremal* if it admits a decomposition

$$\phi = a_{\frac{\ell-1}{4},0}\mathrm{ch}^{(\ell)}_{\frac{\ell-1}{4},0} + a_{\frac{\ell-1}{4},\frac{1}{2}}\mathrm{ch}^{(\ell)}_{\frac{\ell-1}{4},\frac{1}{2}} + \sum_{0<r<\ell} \sum_{\substack{n\in\mathbb{Z} \\ r^2-4\ell n<0}} a_{\frac{\ell-1}{4}+n,\frac{r}{2}}\mathrm{ch}^{(\ell)}_{\frac{\ell-1}{4}+n,\frac{r}{2}} \tag{74}$$

for some $a_{h,j} \in \mathbb{C}$ where the $\mathrm{ch}^{(\ell)}_{h,j}$ are as in (70). In [13] it was shown that an extremal Jacobi form is unique (up to scalar multiplication) if it exists. Moreover, it was speculated that there are no extremal Jacobi forms of index $\ell - 1$ unless $\ell - 1$ divides 12, and this was shown to be true for indexes in the range $1 \le \ell - 1 \le 24$. As was discussed in detail in [13], the above decomposition of an extremal Jacobi form $\phi^{(\ell)}$ of index $\ell - 1$ leads naturally to a vector-valued mock modular form $H^{(\ell)}$ with $\ell - 1$ components $H_r^{(\ell)}$, $r \in \{1, \ldots, \ell - 1\}$. Equivalently, the components of the vector-valued mock modular form $H^{(\ell)} = (H_r^{(\ell)})$ are the coefficients of the theta-decomposition of the pole-free part (cf. [18]) of a meromorphic Jacobi form of weight 1 and index ℓ with a simple pole at $z = 0$ that is closely related to $\phi^{(\ell)}$.

Table 1 The groups of umbral moonshine

ℓ	2	3	4	5	7	13
$G^{(\ell)}$	M_{24}	$2.M_{12}$	$2.AGL_3(2)$	$GL_2(5)/2$	$SL_2(3)$	$\mathbb{Z}/4\mathbb{Z}$

In [13] it was observed that the mock modular form $H^{(\ell)}$ obtained in this way has a close relation to a certain finite group $G^{(\ell)}$ (specified in Table 1) and it was conjectured that for ℓ such that $\ell - 1$ divides 12 there exists a naturally defined $\mathbb{Z} \times \mathbb{Q}$-graded $G^{(\ell)}$-module

$$K^{(\ell)} = \bigoplus_{\substack{r \in \mathbb{Z} \\ 0<r<\ell}} K_r^{(\ell)} = \bigoplus_{\substack{r,k \in \mathbb{Z} \\ 0<r<\ell}} K^{(\ell)}_{r,k-r^2/4\ell} \tag{75}$$

such that the graded dimension of $K^{(\ell)}$ is related to the vector-valued mock modular form $H^{(\ell)}$ via

$$H_r^{(\ell)}(\tau) = -2\delta_{r,1} q^{-1/4\ell} + \sum_{\substack{k \in \mathbb{Z} \\ r^2-4k\ell<0}} \dim K^{(\ell)}_{r,k-r^2/4\ell} q^{k-r^2/4\ell}. \tag{76}$$

Moreover, as in monstrous and Mathieu moonshine we expect to encounter interesting functions if we replace $\dim K^{(\ell)}_{r,k-r^2/4\ell}$ with $\operatorname{tr}_{K^{(\ell)}_{r,k-r^2/4\ell}}(g)$ in (76) for $g \in G^{(\ell)}$. Consider the *umbral McKay–Thompson series* $H_g^{(\ell)} = (H_{g,r}^{(\ell)})$ for $g \in G^{(\ell)}$ and $\ell \in \{2, 3, 4, 5, 7, 13\}$ defined, modulo a definition of $K^{(\ell)}$, by setting

$$H_{g,r}^{(\ell)}(\tau) = -2\delta_{r,1} q^{-1/4\ell} + \sum_{\substack{k \in \mathbb{Z} \\ r^2-4k\ell<0}} \operatorname{tr}_{K^{(\ell)}_{r,k-r^2/4\ell}}(g) q^{k-r^2/4\ell}. \tag{77}$$

It was conjectured in [13] that the $G^{(\ell)}$ module $K^{(\ell)}$ has the property that all the $H_g^{(\ell)}$ defined above transform as vector-valued mock modular forms with specified (vector-valued) shadows. We refer to [13] for various explicit expressions for $H_g^{(\ell)}$. The fact that all the McKay–Thompson series are mock modular forms and thus come attached with shadows is the origin of the term *umbral moonshine*. Notice that $G^{(2)} = M_{24}$. When $\ell = 2$ the umbral moonshine conjecture stated above recovers the Mathieu moonshine conjecture relating $H(\tau)$ and M_{24}. The Rademacher sums of relevance for umbral moonshine will be discussed in Sect. 4.3.

This series of examples clearly demonstrates the importance of Rademacher sums in understanding connections between finite groups and (mock) modular forms, and yet it seems likely that the examples presented here are not exhaustive. A complete understanding of the relationships between finite groups and mock modular forms arising from Rademacher sums would be highly desirable.

4.1 Monstrous Moonshine

Consider the Rademacher sums $R^{[\mu]}_{\Gamma,1,0}$ attached to groups $\Gamma < SL_2(\mathbb{R})$ equipped with the trivial multiplier $\psi \equiv 1$ in weight 0, and let us specialise momentarily to the index $\mu = -1$. As was shown in Sect. 2, the formula (32) for $R^{[-1]}_{\Gamma,1,0}$ reduces in this case to

$$R^{[-1]}_{\Gamma,1,0}(\tau) = \mathrm{e}(-\tau) + \frac{1}{2}c_{\Gamma_g,1,0}(-1,0) + \lim_{K\to\infty} \sum_{\Gamma_\infty\backslash\Gamma^{\times}_{<K}} \mathrm{e}(-\gamma\tau) - \mathrm{e}(-\gamma\infty). \quad (78)$$

As was also discussed in Sect. 2, it was shown in [25] that the expression (78) defining $R^{[-1]}_{\Gamma,1,0}(\tau)$ converges locally uniformly in τ for Γ commensurable with $SL_2(\mathbb{Z})$ and containing $-I$, thus yielding a holomorphic function on $\mathbb{H}$. Moreover, there is a function $\omega : \Gamma \to \mathbb{C}$ such that $R^{[-1]}_{\Gamma,1,0}(\gamma\tau) + \omega(\gamma) = R^{[-1]}_{\Gamma,1,0}(\tau)$ for all $\tau \in \mathbb{H}$, and the function ω is identically zero whenever Γ defines a genus zero quotient of the upper-half plane. This last fact suggests a connection between Rademacher sums and the genus zero property of monstrous moonshine: the groups Γ_g are all of this specific type (commensurable with the modular group, containing $-I$ and having genus zero) so that $R^{[-1]}_{\Gamma_g,1,0}$ converges and is Γ_g-invariant for every $g \in \mathbb{M}$. Furthermore, for $\Gamma = \Gamma_g$ the Rademacher sum $R^{[-1]}_{\Gamma,1,0}$ induces an isomorphism $X_\Gamma \to \hat{\mathbb{C}}$ (cf. (12), [25]).

In fact, the connection between Rademacher sums and monstrous moonshine is even stronger. Given any group element g of the monster, the function T_g may be characterised as the unique Γ_g-invariant holomorphic function on $\mathbb{H}$ with Fourier expansion of the form $T_g(\tau) = q^{-1} + O(q)$ and no poles at any non-infinite cusps of Γ_g. In particular, the Fourier expansion (at the infinite cusp) has vanishing constant term. It follows then that the Rademacher construction with $\mu = -1$ recovers the T_g exactly, up to their constant terms, so that we have

$$T_g(\tau) = R^{[-1]}_{\Gamma_g,1,0}(\tau) - c_{\Gamma_g,1,0}(-1,0) \quad (79)$$

for each $g \in \mathbb{M}$ according to (59). Hence we see that the Rademacher sum furnishes a uniform group-theoretic construction of the monstrous McKay–Thompson series, a fact that is equivalent to the genus zero property of monstrous moonshine which is yet to be fully explained. This leads to the expectation that a suitable physical interpretation of the Rademacher sum construction should be an integral part of a conceptual understanding of the genus zero property, and perhaps moonshine itself. We refer to Sect. 5 for more on the rôle of Rademacher sums in physics, and to [25, Sect. 7] for a speculative discussion of the rôle that physics may play in explicating monstrous moonshine.

Given the power of Rademacher sums, one might wonder if it is possible to use them to characterise the groups Γ_g relevant for monstrous moonshine. At first glance this seems to be unlikely for there are many more genus zero groups[1] commensurable with $SL_2(\mathbb{Z})$ than there are functions T_g. Nevertheless, a natural answer to the characterisation question is found in [25, Sect. 6], following earlier work [15] by Conway–McKay–Sebbar. Following [15] we employ the natural notion of groups of $n|h$-*type*, whose definition is carefully discussed in [25, Sect. 6] and will be suppressed here. Assuming the notion of $n|h$-type, the characterisation of [25] reads as follows. A group $\Gamma < SL_2(\mathbb{R})$ that is of $n|h$-type and is such that $\Gamma/\Gamma_0(nh)$ has exponent 2 coincides with Γ_g for some $g \in \mathbb{M}$ if and only if

- the Rademacher sum $R^{[-1]}_{\Gamma,1,0}$ is Γ-invariant, and
- the expansion of $R^{[-1]}_{\Gamma,1,0}$ at any cusp of Γ is $\Gamma_0(nh)$-invariant.

We regard the simplicity of this formulation as further evidence that Rademacher sums have an important rôle to play in elucidating the nature of moonshine. (The condition that $\Gamma/\Gamma_0(N)$ be a group of exponent 2 can also be formulated in terms of Rademacher sums. We refer the reader to [25, Sect. 6] for more details.)

Finally we discuss Zagier duality for the monstrous Rademacher sums. So far we have only considered the Rademacher sums $R^{[\mu]}_{\Gamma,1,0}$ for $\mu = -1$ but the families

$$\left\{R^{[\mu]}_{\Gamma,1,0} \mid \mu \in \mathbb{Z},\ \mu < 0\right\} \tag{80}$$

for Γ a monstrous group are also relevant for moonshine. Set $T^{[\mu]}_{\Gamma} = R^{[\mu]}_{\Gamma,1,0} - c_{\Gamma,1,0}(\mu,0)$ so that $T^{[-1]}_{\Gamma} = T_g$ when $\Gamma = \Gamma_g$. In [25, Sects. 5, 7] it is argued (with detail in the case of $\Gamma = SL_2(\mathbb{Z})$) that the exponential of the generating function $\sum_{m>0} T^{[-m]}_{\Gamma}(\tau)p^m$ furnishes the graded dimension of a certain generalised Kac–Moody algebra attached to g by Carnahan in [8] when $\Gamma = \Gamma_g$ for $g \in \mathbb{M}$. According to the Zagier duality (61) specialised to $w = 0$ the Fourier coefficients of the family $\{T^{[-m]}_{\Gamma} \mid m \in \mathbb{Z},\ m > 0\}$ coincide, up to a minus sign, with those of the dual family

$$\left\{R^{[\nu]}_{\Gamma,1,2} \mid \nu \in \mathbb{Z},\ \nu < 0\right\}. \tag{81}$$

It is interesting to observe that the subtraction of the constant terms from the $R^{[\mu]}_{\Gamma,1,0}$, which is necessary in order to obtain the functions $T^{[\mu]}_{\Gamma}$ that are of direct relevance

[1]Norton, in unpublished work (cf. [16]), has found 616 groups Γ such that $\Gamma_\infty = \langle T, -I\rangle$, the congruence group $\Gamma_0(N)$ is contained in Γ for some N, and the coefficients of the corresponding Rademacher sum $R^{[-1]}_{\Gamma,1,0}$ are rational, and Cummins has shown [16] that 6,486 genus zero groups are obtained by dropping the condition of rationality. On the other hand, there are 194 conjugacy classes in the monster, but the two classes of order 27 are related by inversion and thus determine the same McKay–Thompson series. There are no other coincidences amongst the T_g but there are some linear relations, and curiously, the space of functions spanned linearly by the T_g for $g \in \mathbb{M}$ is 163 dimensional.

to moonshine, has a natural reinterpretation under Zagier duality: it corresponds to the omission of the Rademacher sum $R^{[0]}_{\Gamma,1,2}$—an Eisenstein series that fails to be modular, as was observed in Sect. 2—from the family (81).

As a final remark, we observe that the coefficients $c_{\Gamma,1,0}$ and $c_{\Gamma,1,2}$ are related in another way as one can see by inspecting (45)–(47); namely, $-mc_{\Gamma,1,2}(-m,n) = nc_{\Gamma,1,0}(-m,n)$ for m and n positive integers, so the Rademacher sums $R^{[-m]}_{\Gamma,1,2}$ dual to the functions $T^{[-m]}_{\Gamma} = R^{[-m]}_{\Gamma,1,0} - c_{\Gamma,1,0}(-m,0)$ of relevance to monstrous moonshine are just their normalised derivatives,

$$R^{[-m]}_{\Gamma,1,2} = -\frac{1}{m} q \frac{\mathrm{d}}{\mathrm{d}q} T^{[-m]}_{\Gamma}. \tag{82}$$

4.2 Mathieu Moonshine

Consider the Rademacher sums $R^{[\mu]}_{\Gamma,\psi,w}$ with $\Gamma = SL_2(\mathbb{Z})$, $\psi = \epsilon^{-3}$ and $w = 1/2$. We have $\alpha = 1/8$ when $\psi = \epsilon^{-3}$ so the smallest non-positive possibility for the index is $\mu = -1/8$. Substituting into (30) we find that $R^{[-1/8]}_{\Gamma,\epsilon^{-3},1/2}(\tau)$ is given by

$$\begin{aligned} &\lim_{K\to\infty} \sum_{\substack{0<c<K\\ -K^2<d<K^2\\ (c,d)=1}} \mathrm{e}\left(\frac{1}{8c(c\tau+d)} + \frac{d}{8c} - \frac{3s(d,c)}{2}\right) \\ &\times \frac{-\sqrt{\mathbf{i}}}{\sqrt{\pi}(c\tau+d)^w} \gamma\left(\frac{1}{2}, \frac{-\pi\mathbf{i}}{4c(c\tau+d)}\right) \end{aligned} \tag{83}$$

where $s(d,c)$ is as in (104). In deriving (83) we have used the identities $\Gamma(1/2) = \sqrt{\pi}$ and $\gamma\tau - \gamma\infty = c^{-1}(c\tau+d)^{-1}$, the latter being valid in case (c,d) is the lower row of $\gamma \in SL_2(\mathbb{R})$. For the Rademacher series $c_{\Gamma,\epsilon^{-3},1/2}$ we have

$$\begin{aligned} &c_{\Gamma,\epsilon^{-3},\frac{1}{2}}\left(-\frac{1}{8}, n-\frac{1}{8}\right) \\ &= -2\pi \sum_{\substack{c>0\\ 0\leq d<c\\ (c,d)=1}} \mathrm{e}\left(n\frac{d}{c} - \frac{3s(d,c)}{2}\right) \frac{1}{c(8n-1)^{\frac{1}{4}}} I_{\frac{1}{2}}\left(\frac{\pi}{2c}(8n-1)^{\frac{1}{2}}\right) \end{aligned} \tag{84}$$

according to (45)–(53) when n is a positive integer. As discussed in Sects. 2 and 3, If the expressions (83) and (84) are convergent then the latter furnishes the Fourier expansion of the former,

$$R^{[-1/8]}_{\Gamma,\epsilon^{-3},1/2}(\tau) = q^{-1/8} + \sum_{n>0} c_{\Gamma,\epsilon^{-3},1/2}(-1/8, n-1/8)\, q^{n-1/8}. \tag{85}$$

On the other hand, the right-hand side of (84) appeared (up to a scalar factor) earlier in [26] as a proposal for an explicit formula for t_n. This suggests that the function $H(\tau)$ may be a scalar multiple of the Rademacher sum $R^{[-1/8]}_{\Gamma,\epsilon^{-3},1/2}(\tau)$. In fact, more is true, for in [11] it is shown that for each $g \in M_{24}$ there is a character ρ_g on $\Gamma_0(n_g)$, for n_g the order of g, such that the Rademacher sum $R^{[-1/8]}_{\Gamma_0(n_g),\rho_g\epsilon^{-3},1/2}$ converges, locally uniformly for $\tau \in \mathbb{H}$, and is related to the McKay–Thompson series $H_g(\tau)$ by

$$H_g(\tau) = -2R^{[-1/8]}_{\Gamma_0(n_g),\rho_g\epsilon^{-3},1/2}(\tau). \tag{86}$$

The Rademacher series $c_{\Gamma_0(n_g),\rho_g\epsilon^{-3},1/2}$ are also shown to converge in [11], and we recover (85) upon taking g to be the identity. In a word then, Rademacher sums furnish a uniform construction of the (candidate) H_g determined earlier in [10, 27, 33, 34], which constitutes further evidence in support of their validity. The character ρ_g may be specified easily: if $n = n_g$ is the order of g and $h = h_g$ is the minimal length among cycles in the cycle shape of g (regarded as a permutation in the unique non-trivial permutation action on 24 points) then $\rho_g = \rho_{n|h}$ where

$$\rho_{n|h}(\gamma) = \mathrm{e}\left(-\frac{cd}{nh}\right) \tag{87}$$

when (c, d) is the lower row of $\gamma \in \Gamma_0(n)$. The fact that (87) defines a morphism of groups $\Gamma_0(n_g) \to \mathbb{C}^\times$ relies upon the result that if h is a divisor of 24 then $x^2 \equiv 1 \pmod{h}$ whenever x is coprime to h together with the fact that all the h_g for $g \in M_{24}$ are divisors of 24. We refer to [11, 12] for more detailed discussions on the multiplier $\rho_{n|h}$, as well as all the other material in this section.

As briefly mentioned before, beyond furnishing a uniform construction of the H_g the result (86) demonstrates the correct analogue of the genus zero property that is relevant to this *Mathieu moonshine* relating representations of M_{24} to $K3$ surfaces. The rest of this subsection will be devoted to the explanation of this fact. Recall that there is in this case no obvious analogue of the genus zero property which holds for the monstrous McKay–Thompson series T_g since some of the groups $\Gamma_0(n_g)$ arising in Mathieu moonshine do not define genus zero quotients of $\mathbb{H}$ (viz., $n_g \in \{11, 14, 15, 23\}$). On the other hand, from the discussion of Sect. 4.1 we see that the genus zero property of the T_g is equivalent to the fact that they are modular functions recovered from Rademacher sums as in (79). Therefore, the identity (86) proven in [11]—the property of H_g to be uniformly expressible as a Rademacher sum—serves as the natural analogue of the genus zero property that is relevant for Mathieu moonshine (modulo a proof that the H_g really are the McKay–Thompson series attached to a suitably defined M_{24}-module $K = \bigoplus_{n>0} K_{n-1/8}$).

In more detail, we note that the identity (86) implies that the Rademacher sums $R^{[-1/8]}_{\Gamma,\psi,1/2}$ with $\Gamma = \Gamma_0(n_g)$ and $\psi = \rho_{n_g|h_g}\epsilon^{-3}$ have the special property that they are mock modular forms whose shadows lie in the one-dimensional space spanned by the cusp form η^3. This must be the case because every proposed McKay–Thompson series in Mathieu moonshine has shadow proportional to η^3, a fact that is equivalent to their relation to weak Jacobi forms generalising (71). Indeed, from (40) we see that the shadow of the mock modular form $H_g(\tau)$ is a weight 3/2 modular form given by

$$-2S^{[-1/8]}_{\Gamma_0(n_g),\rho_g\epsilon^{-3},1/2} = -\frac{1}{\sqrt{2\pi}}R^{[1/8]}_{\Gamma_0(n_g),\rho_g^{-1}\epsilon^3,3/2}. \tag{88}$$

Moreover, it is proven in [11] that

$$-2S^{[-1/8]}_{\Gamma_0(n_g),\rho_g\epsilon^{-3},1/2} = -\frac{\chi_g}{2}\frac{1}{\sqrt{2\pi}}\eta^3 \tag{89}$$

where χ_g denotes the number of fixed points of g (in the unique non-trivial permutation representation of M_{24} on 24 points).

As is observed in [11], it is not typical behavior of the Rademacher sum $R^{[-1/8]}_{\Gamma,\psi,1/2}$ to have shadow lying in this particular one-dimensional space. For $n = 9$, for example—note that 9 is not the order of an element in M_{24}—the shadow of the Rademacher sum $R^{[-1/8]}_{\Gamma_0(n),\epsilon^{-3},1/2}$ is not proportional to η^3, at least according to experimental evidence. It is natural then to ask if there is a characterisation of the modular groups and the multipliers of the McKay–Thompson series H_g expressible in terms of Rademacher sums, in analogy with that of [25] (derived following [15]) for the monstrous case as discussed in Sect. 4.1. In such a characterisation the pairs $(\Gamma_0(n), \rho_{n|h})$ for h a divisor of n dividing 24 would replace the groups of $n|h$-type, and the condition

- the Rademacher sum $-2R^{[-1/8]}_{\Gamma_0(n),\rho_{n|h}\epsilon^{-3},1/2}$ has shadow proportional to η^3

would replace the Γ-invariance condition in Sect. 4.1. So far we do not know of any examples that do not arise as H_g for some $g \in M_{24}$. It would be very interesting to determine whether or not the above conditions are sufficient to characterise the McKay–Thompson series of Mathieu moonshine.

4.3 Umbral Moonshine

In Sect. 2,3 we have described a regularisation procedure attaching Rademacher sums $R^{[\mu]}_{\Gamma,\psi,w}$ to a group $\Gamma < SL_2(\mathbb{R})$, a multiplier ψ for Γ, a compatible weight w and a compatible index μ. This procedure can be generalised to the vector-valued case with a higher-dimensional ψ and μ. To be precise, we suppose that $\psi = (\psi_{ij})$ is a matrix-valued multiplier system, satisfying (9) as before, for some weight w, and we suppose also that $\psi_{ij}(T^h) = \delta_{ij}\,\mathrm{e}(\alpha_i)$ for some $0 < \alpha_i < 1$ where h is such that $\Gamma_\infty = \langle T^h, -I\rangle$. Then to a vector-valued index $\mu = [\mu_i]$ such that $h\mu_i + \alpha_i \in \mathbb{Z}$

for all i (and $\mu_i < 0$ in case $w < 1$) we attach the *(row) vector-valued Rademacher sum*

$$R^{[\mu]}_{\Gamma,\psi,w}(\tau) = \lim_{K\to\infty} \sum_{\Gamma_\infty\backslash\Gamma_{K,K^2}} \mathrm{e}(\mu\gamma\tau)\,\mathrm{r}^{[\mu]}_w(\gamma,\tau)\psi(\gamma)\,\mathrm{j}(\gamma,\tau)^{w/2} \tag{90}$$

where $\mathrm{e}(\mu\gamma\tau)$ now denotes the (row) vector-valued function whose i-th component is $\mathrm{e}(\mu_i\gamma\tau)$ and $\mathrm{r}^{[\mu]}_w(\gamma,\tau)$ denotes the diagonal matrix-valued function whose (i,i)-th entry is $\mathrm{r}^{[\mu_i]}_w(\gamma,\tau)$ (cf. (31)). For the sake of simplicity we exclude the case that some $\alpha_i = 0$ in (90). In such a case one can expect constant term corrections analogous to (32).

In order to apply the above construction to the vector-valued mock modular forms relevant for umbral moonshine we have to specify the appropriate (matrix-valued) multiplier. Recall that the vector-valued mock modular forms $H^{(\ell)}$ are obtained from the decomposition of extremal Jacobi forms into $N = 4$ characters. As is explained in detail in [13], the relation to the weak Jacobi form immediately implies that the mock modular form $H^{(\ell)}$ has shadow (proportional to) $S^{(\ell)} = (S^{(\ell)}_r)$, whose components are the *unary theta series*

$$S^{(\ell)}_r(\tau) = \sum_{k\in\mathbb{Z}}(2\ell k + r)q^{\frac{(2\ell k+r)^2}{4\ell}}, \tag{91}$$

while the extremality condition implies that $H^{(\ell)}$ has a single polar (non-vanishing as $\tau \to i\infty$) term $-2q^{-\frac{1}{4\ell}}$ in its first component $H^{(\ell)}_1$ (cf. (76)). Notice that in the case that $\ell = 2$ we have $S^{(2)} = (S^{(2)}_1) = (\eta^3)$ by an identity due to Euler, and this is in part a reflection of the fact that $S^{(\ell)}$ is a (vector-valued) cusp form of weight $3/2$ for $SL_2(\mathbb{Z})$ for all $\ell \geq 2$.

Let $\sigma^{(\ell)} = (\sigma^{(\ell)}_{ij})$ be the multiplier system for $S^{(\ell)}$. Then from the above discussion, we would like to consider the $(\ell-1)$-vector-valued Rademacher sum $R^{[\mu]}_{\Gamma,\psi^{(\ell)},1/2}$ where $\Gamma = SL_2(\mathbb{Z})$, we take $\psi^{(\ell)}$ to be the inverse of $\sigma^{(\ell)}$, and where we set $\mu = \mu^{(\ell)} = (-\frac{1}{4\ell}, 0, \ldots, 0)$. As was uncovered in [13], the Rademacher sum $R^{[\mu]}_{\Gamma,\psi^{(\ell)},1/2}$ (denoted $R^{(\ell)}$ in [13]) has special properties when $\ell - 1$ is a divisor of 12. First, in these cases $R^{[\mu]}_{\Gamma,\psi^{(\ell)},1/2}$ turns out to be a vector-valued mock modular form with shadow proportional to the vector-valued cusp form $S^{(\ell)}$ defined in (91). This means that, for a suitably chosen constant $C^{(\ell)}$, the vector-valued function $R^{[\mu]}_{\Gamma,\psi^{(\ell)},1/2}$ is invariant for the (ψ, w, G)-action of $\Gamma = SL_2(\mathbb{Z})$ on $(\ell-1)$-vector-valued holomorphic functions $F(\tau) = (F_1(\tau), \ldots, F_{\ell-1}(\tau))$ defined, in direct analogy with (39), by setting

$$\left(F|_{\psi,w,G}\gamma\right)(\tau) = F(\gamma\tau)\psi(\gamma)\,\mathrm{j}(\gamma,\tau)^{w/2} + (2\pi\mathbf{i})^{1-w}\int_{-\gamma^{-1}\infty}^{i\infty}(z+\tau)^{-w}\overline{G(-\bar{z})}\mathrm{d}z, \tag{92}$$

when $\psi = \psi^{(\ell)}$, $w = 1/2$, and $G(\tau) = C^{(\ell)} S^{(\ell)}(\tau) = C^{(\ell)}(S_1^{(\ell)}(\tau), \ldots, S_{\ell-1}^{(\ell)}(\tau))$. Second, it appears to have a close relation to the group $G^{(\ell)}$ as described in (77).

As the reader might have noticed, in case $\ell = 2$ the function $R^{[\mu]}_{\Gamma,\psi^{(\ell)},1/2}$ has a single component which by definition coincides with $R^{[-1/8]}_{\Gamma,\epsilon^{-3},1/2}$. Thus $-2R^{[\mu]}_{\Gamma,\psi^{(\ell)},1/2}$ recovers the mock modular form $H(\tau)$ of importance in Mathieu moonshine (and discussed in Sect. 4.2) in case $\ell = 2$.

Recall that in the case of monstrous moonshine the genus zero property—that each T_g should induce an isomorphism $X_\Gamma \to \hat{\mathbb{C}}$ (cf. (12)) for some group $\Gamma < SL_2(\mathbb{R})$—was the primary tool for predicting the McKay–Thompson series, and we have seen in Sect. 4.1 that this is equivalent to the property that T_g coincide (up to an additive constant, cf. (79)) with the Rademacher sum $R^{[-1]}_{\Gamma,1,0}$ for some Γ. In the case of Mathieu moonshine we have seen that each H_g may recovered as $-2R^{[-1/8]}_{\Gamma,\rho_g\epsilon^{-3},1/2}$ for a suitable character ρ_g, and this is evidently a powerful analogue of the genus zero property of monstrous moonshine. Analogously, in the case of umbral moonshine it is conjectured [13] that each umbral McKay–Thompson series $H_g^{(\ell)}$ is recovered from a vector-valued Rademacher sum according to

$$H_g^{(\ell)} = -2R^{[\mu]}_{\Gamma_0(n_g),\psi^{(\ell)}\rho_g^{(\ell)},1/2} \tag{93}$$

where $\mu = \mu^{(\ell)}$ and $\psi^{(\ell)}$ are as before, $\rho_g^{(\ell)}$ is a suitably defined (matrix-valued) function on $\Gamma_0(n_g)$ and n_g is a suitably chosen integer. (We refer to [13, Sect. 4.8] for more details on $\rho_g^{(\ell)}$ and n_g.) The conjectural identity (93) was the primary tool used in determining the concrete expressions for the $H_g^{(\ell)}$ that were furnished in [13].

5 Physical Applications

In the previous sections we have described the Rademacher summing procedure that produces a (mock) modular form by computing a certain regularised sum over the representatives of the cosets $\Gamma_\infty\backslash\Gamma$, where $\Gamma < SL_2(\mathbb{R})$ is the modular group and Γ_∞ is its subgroup fixing the infinite cusp. These (mock) modular forms are often closely related to the partition function or the twisted partition function of certain two-dimensional conformal field theories in physics. Hence, one might wonder if the associated Rademacher sum also has a physical meaning. The answer to this question is positive and in fact constituted an important part of the motivation to explore the relation between moonshine and Rademacher sums [11, 12, 25].

A compelling physical interpretation of the Rademacher sum is provided by the so-called AdS/CFT correspondence [47] (also referred to as the gauge/gravity duality or the holographic duality in more general contexts), which asserts, among many other things, that the partition function of a given two dimensional CFT "with an AdS dual" equals the partition function of another physical theory in three

Euclidean dimensions with gravitational interaction and with asymptotically anti de Sitter (AdS) boundary condition. The correspondence, when applicable, provides both deep intuitive insights and powerful computational tools for the study of the theory. From the fact that the only smooth three-manifold with asymptotically AdS torus boundary condition is a solid torus, it follows that the saddle points of such a partition function are labeled by the different possible ways to "fill in the torus;" that is, the different choices of primitive cycle on the boundary torus which may become contractible in a solid torus that fills it [48]. These different saddle points are therefore labeled by the coset space $\Gamma_\infty\backslash\Gamma$, where $\Gamma = SL_2(\mathbb{Z})$ [24]. From a bulk, gravitational point of view, the group $SL_2(\mathbb{Z})$ has an interpretation as the group of large diffeomorphisms, and Γ_∞ is the subgroup that leaves the contractible cycle invariant and therefore can be described by a mere change of coordinates. Such considerations underlie the previous use of Rademacher sums in the physics literature [22–24,46,50,51,53]. See also [52] for a refinement of this interpretation using localisation techniques.

In the presence of a discrete symmetry of the conformal field theory, apart from the partition function one can also compute the twisted (or equivariant) partition function. In more details, recall that the partition function computes the dimension of the Hilbert space graded by the basic charges (the energy, for instance) of the theory. In the presence of a discrete symmetry whose action on the Hilbert space commutes with the operators associated with the basic conserved charges, more refined information can be gained by studying the twisted partition function (a trace over the Hilbert space with a group element inserted) which computes the graded group characters of the Hilbert space. In the Lagrangian formulation of quantum field theories this twisting corresponds to a modification of the boundary condition. For a two dimensional CFT with an AdS gravity dual, this translates into a corresponding modification of the boundary condition in the gravitational path integral by an insertion of a group element g, which changes the set of allowed saddle points. as a result, the allowed large diffeomorphisms is now given by a discrete group $\Gamma_g \subset SL_2(\mathbb{R})$, generally different from $SL_2(\mathbb{Z})$.

Note that when $\Gamma \not\subset SL_2(\mathbb{Z})$, in particular when $\Gamma = \Gamma_0(n|h) + S$ where S in a non-trivial subgroup of the group of exact divisors of n/h (see [25] for details), the above interpretation suggests that certain orbifold geometries should be included in the path integral as well as smooth geometries. We do not have a precise understanding from the gravity viewpoint as for when these extra contributions should be included. Some interpretation in terms of a $\mathbb{Z}/n\mathbb{Z}$-generalisation of the spin structure ($n = 2$) have been put forward in [25]. See also [49] for a related discussion. We hope further developments will shed light on this question in the future.

We have explained above how the sum over $\Gamma_\infty\backslash\Gamma$ for $\Gamma = SL_2(\mathbb{Z})$ can be thought of as a sum over the smooth, asymptotically AdS_3 geometries. Moreover, recent progress in the exact computation of path integrals in quantum gravity in AdS backgrounds suggests that the precise form of the regulator itself is also natural from the gravitational viewpoint. Recall the use of the Lipschitz summation formula (108) in reducing the Rademacher sum (19) to a sum (26) of sums over (representatives of the non-trivial) double cosets of Γ_∞ in Γ. This procedure can be applied quite generally and verifies the relationship (59) between Fourier coefficients of Rademacher sums and the Rademacher series. In practice then, instead of a sum over a pair of co-prime integers (c,d) we can write a Rademacher sum as a generating function of sums over a single integer c. This readily renders the following form for the Fourier coefficient $c_\Gamma(\Delta)$ of the term q^Δ in the Rademacher sum. It is the infinite sum

$$c_\Gamma(\Delta) = \sum_{c=1}^{\infty} c_\Gamma(\Delta; c), \tag{94}$$

where $c_\Gamma(\Delta; c)$ takes the form of a product of a modified Bessel function with argument $\pi\sqrt{\Delta}/c$ and a Kloosterman sum (cf. (45)).

In [17] an example has been provided where the gravity path integral is argued to localise on configurations giving precisely the contribution of the above form to the gravity partition function. First, the sum over c has the interpretation as a sum over gravitational instantons obtained from orbifolding the configuration corresponding to $c = 1$ by a symmetry group $\mathcal{G} \cong \mathbb{Z}/c\mathbb{Z}$. Second, the Bessel function arises naturally as the result of the finite-dimensional integral obtained from localising the infinite-dimensional path integral on the given instanton configuration. This result is argued to be independent of the details of the orbifold and depends only on the order c of the symmetry. Finally, the Kloosterman sum and the extra numerical factor is speculated to arise from summing over different possibilities of order c orbifold group $\mathcal{G} \cong \mathbb{Z}/c\mathbb{Z}$. It would be very interesting to see if further developments in localising the gravity path integral will lead to a more complete understanding of quantum gravity utilising Rademacher sums.

6 Appendix: Special Functions

The *Bernoulli numbers* B_m may be defined by the following Taylor expansion.

$$\frac{t}{e^t - 1} = \sum_{m\geq 0} B_m \frac{t^m}{m!} \tag{95}$$

The *Gamma function* $\Gamma(s)$ and *lower incomplete Gamma function* $\gamma(s,x)$ are defined by the integrals

$$\Gamma(s) = \int_0^\infty t^{s-1} e^{-t}\,\mathrm{d}t \tag{96}$$

$$\gamma(s,x) = \int_0^x t^{s-1} e^{-t}\,\mathrm{d}t \tag{97}$$

for s real and positive. The expression (97) is well defined for positive real x but this situation can be improved, for integration by parts yields the recurrence relation

$$\gamma(s,x) = (s-1)\gamma(s-1,x) - x^{s-1}e^{-x}, \tag{98}$$

and this in turn leads to a power series expansion

$$\gamma(s,x) = \frac{\Gamma(s)}{e^x}\sum_{n\geq 0}\frac{x^{n+s}}{\Gamma(n+s+1)} \tag{99}$$

which converges absolutely and locally uniformly for x in $\mathbb{C}$.

For the exponential x^s we employ the principal branch of the logarithm, so that

$$x^s = |x|^s e^{\mathrm{i}\theta s} \quad \text{whenever} \quad x = |x|e^{\mathrm{i}\theta},\ -\pi < \theta \leq \pi. \tag{100}$$

The *modified Bessel function of the first kind* is denoted $I_\alpha(x)$ and may be defined by the power series expression

$$I_\alpha(z) = \sum_{n\geq 0}\frac{1}{\Gamma(m+\alpha+1)m!}\left(\frac{z}{2}\right)^{2m+\alpha} \tag{101}$$

which converges absolutely and locally uniformly in z so long as z avoids the negative reals (cf. (100)). We consider only non-negative real values of α in this article.

The *Dedekind eta function*, denoted $\eta(\tau)$, is a holomorphic function on the upper half-plane defined by the infinite product

$$\eta(\tau) = q^{1/24}\prod_{n>0}(1-q^n) \tag{102}$$

where $q = \mathrm{e}(\tau) = e^{2\pi \mathrm{i}\tau}$. It is a modular form of weight $1/2$ for the modular group $SL_2(\mathbb{Z})$ with multiplier $\epsilon : SL_2(\mathbb{Z}) \to \mathbb{C}$ so that

$$\eta(\gamma\tau)\epsilon(\gamma)\,\mathrm{j}(\gamma,\tau)^{1/4} = \eta(\tau) \tag{103}$$

for all $\gamma = \left(\begin{smallmatrix} a & b \\ c & d \end{smallmatrix}\right) \in SL_2(\mathbb{Z})$, where $\mathrm{j}(\gamma,\tau) = (c\tau+d)^{-2}$. The *multiplier system* ϵ may be described explicitly as

$$\epsilon\begin{pmatrix} a & b \\ c & d \end{pmatrix} = \begin{cases} \mathrm{e}(-b/24), & c = 0,\ d = 1 \\ \mathrm{e}(-(a+d)/24c + s(d,c)/2 + 1/8), & c > 0 \end{cases} \tag{104}$$

where $s(d,c) = \sum_{m=1}^{c-1}(d/c)((md/c))$ and $((x))$ is 0 for $x \in \mathbb{Z}$ and $x - \lfloor x \rfloor - 1/2$ otherwise. We can deduce the values $\epsilon(a,b,c,d)$ for $c < 0$, or for $c = 0$

and $d = -1$, by observing that $\epsilon(-\gamma) = \epsilon(\gamma)\,\mathrm{e}(1/4)$ for $\gamma \in SL_2(\mathbb{Z})$. Observe that

$$\epsilon(T^m\gamma) = \epsilon(\gamma T^m) = \mathrm{e}(-m/24)\epsilon(\gamma) \tag{105}$$

for $m \in \mathbb{Z}$.

Setting $q = \mathrm{e}(\tau)$ and $y = \mathrm{e}(z)$ we use the following conventions for the four standard *Jacobi theta functions*.

$$\begin{aligned}
\theta_1(\tau,z) &= -iq^{1/8}y^{1/2}\prod_{n=1}^{\infty}(1-q^n)(1-yq^n)(1-y^{-1}q^{n-1})\\
\theta_2(\tau,z) &= q^{1/8}y^{1/2}\prod_{n=1}^{\infty}(1-q^n)(1+yq^n)(1+y^{-1}q^{n-1})\\
\theta_3(\tau,z) &= \prod_{n=1}^{\infty}(1-q^n)(1+y\,q^{n-1/2})(1+y^{-1}q^{n-1/2})\\
\theta_4(\tau,z) &= \prod_{n=1}^{\infty}(1-q^n)(1-y\,q^{n-1/2})(1-y^{-1}q^{n-1/2})
\end{aligned} \tag{106}$$

We write $\mu(\tau,z)$ for the *Appell-Lerch sum* defined by setting

$$\mu(\tau,z) = \frac{-iy^{1/2}}{\theta_1(\tau,z)}\sum_{\ell=-\infty}^{\infty}\frac{(-1)^{\ell}y^n q^{\ell(\ell+1)/2}}{1-yq^{\ell}}. \tag{107}$$

The *Lipschitz summation formula* is the identity

$$\frac{(-2\pi i)^s}{\Gamma(s)}\sum_{k=1}^{\infty}(k-\alpha)^{s-1}\,\mathrm{e}((k-\alpha)\tau) = \sum_{\ell\in\mathbb{Z}}\mathrm{e}(\alpha\ell)(\tau+\ell)^{-s}, \tag{108}$$

valid for $\Re(s) > 1$ and $0 \le \alpha < 1$, where $\mathrm{e}(x) = e^{2\pi \mathrm{i}x}$. A nice proof of this using Poisson summation appears in [44]. Observe that both sides of (108) converge absolutely and uniformly in τ on compact subsets of $\mathbb{H}$. For applications to Rademacher sums of weight less than 1 one requires an extension of (108) to $s = 1$. Absolute convergence on the right hand side breaks down at this point but we may consider the following useful analogue. The reader may consult [11, Sect. C], for example, for a proof of (109), and may see [25, Sect. 3.3] for a proof of (110).

Lemma 6.1. *For* $0 < \alpha < 1$ *we have*

$$\sum_{k=1}^{\infty}\mathrm{e}((k-\alpha)\tau) = \sum_{-K<\ell<K}\mathrm{e}(\alpha\ell)(-2\pi i)^{-1}(\tau+\ell)^{-1} + E_K(\tau) \tag{109}$$

where $E_K(\tau) = \mathcal{O}(1/K^2)$*, locally uniformly for* $\tau \in \mathbb{H}$*. For* $\alpha = 0$ *we have*

$$\frac{1}{2} + \sum_{k>0} \mathrm{e}(k\tau) = \lim_{K\to\infty} \sum_{-K<\ell<K} (-2\pi \mathbf{i})^{-1}(\tau + \ell)^{-1}. \tag{110}$$

References

1. Andrews, G.E.: Ramunujan's "lost" notebook. III. The Rogers-Ramanujan continued fraction. Adv. Math. **41**(2), 186–208 (1981)
2. Apostol, T.M.: Modular Functions and Dirichlet Series in Number Theory. Graduate Texts in Mathematics, vol. 41, 2nd edn. Springer, New York (1990)
3. Borcherds, R.E.: Monstrous moonshine and monstrous Lie superalgebras. Invent. Math. **109**(2), 405–444 (1992)
4. Bringmann, K., Ono, K.: The $f(q)$ mock theta function conjecture and partition ranks. Invent. Math. **165**(2), 243–266 (2006)
5. Bringmann, K., Ono, K.: Arithmetic properties of coefficients of half-integral weight Maass-Poincaré series. Math. Ann. **337**(3), 591–612 (2007)
6. Bringmann, K., Ono, K.: Dyson's ranks and Maass forms. Ann. Math. (2) **171**(1), 419–449 (2010)
7. Bringmann, K., Ono, K.: Coefficients of harmonic maass forms. In: Alladi, K., Garvan, F. (eds.) Partitions, q-Series, and Modular Forms. Developments in Mathematics, vol. 23, pp. 23–38. Springer, New York (2012)
8. Carnahan, S.: Monstrous Lie algebras and generalized moonshine. Ph.D. thesis, University of California, Berkeley (2007)
9. Cho, B., Choie, Y.: Zagier duality for harmonic weak Maass forms of integral weight. Proc. Am. Math. Soc. **139**(3), 787–797 (2011)
10. Cheng, M.C.N.: $K3$ Surfaces, $N = 4$ Dyons, and the Mathieu Group M_{24} (May 2010). http://inspirehep.net/record/856699/export/hx and http://inspirehep.net/record/856699/export/hlxe
11. Cheng, M.C.N., Duncan, J.F.R.: On Rademacher Sums, the Largest Mathieu Group, and the Holographic Modularity of Moonshine (October 2011). http://inspirehep.net/record/940401/export/hx and http://inspirehep.net/record/940401/export/hlxu
12. Cheng, M.C.N., Duncan, J.F.R.: Mathieu 24 and (Mock) automorphic forms. In: Contribution to AMS Proceedings of Symposia in Pure Mathematics (String-Math 2011). arXiv:1201.4140 [math.RT]
13. Cheng, M.C.N., Duncan, J.F.R., Harvey, J.: Umbral moonshine. arXiv:1204.2779 [math.RT] (submitted CNTP); Cheng, M.C.N., Duncan, J.F.R., Harvey, J.: Umbral moonshine and the Niemeier lattices. Res. Math. Sci. (Springer) 1, 3 (2014). arXiv:1307.5793 [math.RT]. http://www.resmathsci.com/content/1/1/3
14. Conway, J.H., Norton, S.P.: Monstrous Moonshine. Bull. Lond. Math. Soc. **11**, 308–339 (1979)
15. Conway, J., McKay, J., Sebbar, A.: On the discrete groups of Moonshine. Proc. Am. Math. Soc. **132**, 2233–2240 (2004)
16. Cummins, C.J.: Congruence subgroups of groups commensurable with $\mathrm{PSL}(2,\mathbb{Z})$ of genus 0 and 1. Exp. Math. **13**(3), 361–382 (2004)
17. Dabholkar, A., Gomes, J., Murthy, S.: Localization and exact holography. J High Energy Phys. 1304, 062 (2011). doi:10.1007/JHEP04(2013)062
18. Dabholkar, A., Murthy, S., Zagier, D.: Quantum Black holes, wall-crossing, and mock modular forms (2012). arXiV:1208.4074
19. de Azevedo Pribitkin, W.: The Fourier coefficients of modular forms and Niebur modular integrals having small positive weight. I. Acta Arith. **91**(4), 291–309 (1999)

20. de Azevedo Pribitkin, W.: The Fourier coefficients of modular forms and Niebur modular integrals having small positive weight. II. Acta Arith. **93**(4), 343–358 (2000)
21. de Azevedo Pribitkin, W.: A generalization of the Goldfeld-Sarnak estimate on Selberg's Kloosterman zeta-function. Forum Math. **12**(4), 449–459 (2000)
22. de Boer, J., Cheng, M.C.N., Dijkgraaf, R., Manschot, J., Verlinde, E.: A farey tail for attractor black holes. J. High Energy Phys. **11**, 024 (2006)
23. Denef, F., Moore, G.W.: Split states, entropy enigmas, holes and halos. J. High Energy Phys. **1111**, 129, 149 pp. (2011) [21 figures]
24. Dijkgraaf, R., Maldacena, J.M., Moore, G.W., Verlinde, E.P.: A Black hole Farey tail (2000). hep-th/0005003
25. Duncan, J.F.R., Frenkel, I.B.: Rademacher sums, moonshine and gravity. Commun. Number Theory Phys. **5**(4), 1–128 (2011)
26. Eguchi, T., Hikami, K.: Superconformal algebras and mock theta functions 2. Rademacher expansion for K3 surface. Commun. Number Theory Phys. **3**, 531–554 (2009)
27. Eguchi, T., Hikami, K.: Note on twisted elliptic genus of K3 surface. Phys. Lett. **B694**, 446–455 (2011)
28. Eguchi, T., Ooguri, H., Tachikawa, Y.: Notes on the K3 Surface and the Mathieu group M_{24}. Exp. Math. **20**, 91–96 (2011)
29. Folsom, A., Ono, K.: Duality involving the mock theta function $f(q)$. J. Lond. Math. Soc. (2) **77**(2), 320–334 (2008)
30. Frenkel, I.B., Lepowsky, J., Meurman, A.: A natural representation of the Fischer-Griess Monster with the modular function J as character. Proc. Natl. Acad. Sci. USA **81**(10, Phys. Sci.), 3256–3260 (1984)
31. Frenkel, I.B., Lepowsky, J., Meurman, A.: A moonshine module for the Monster. In: Vertex Operators in Mathematics and Physics (Berkeley, Calif., 1983). Mathematical Sciences Research Institute Publications, vol. 3, pp. 231–273. Springer, New York (1985)
32. Frenkel, I., Lepowsky, J., Meurman, A.: Vertex Operator Algebras and the Monster. Pure and Applied Mathematics, vol. 134. Academic, Boston (1988)
33. Gaberdiel, M.R., Hohenegger, S., Volpato, R.: Mathieu Moonshine in the elliptic genus of K3. J. High Energy Phys. **1010**, 062 (2010)
34. Gaberdiel, M.R., Hohenegger, S., Volpato, R.: Mathieu twining characters for K3. J. High Energy Phys. **1009**, 058, 19 pp. (2010)
35. Goldfeld, D., Sarnak, P.: Sums of Kloosterman sums. Invent. Math. **71**(2), 243–250 (1983)
36. Guerzhoy, P.: On weak harmonic Maass-modular grids of even integral weights. Math. Res. Lett. **16**(1), 59–65 (2009)
37. Knopp, M.I.: Construction of a class of modular functions and forms. Pacific J. Math. **11**, 275–293 (1961)
38. Knopp, M.I.: Construction of a class of modular functions and forms. II. Pacific J. Math. **11**, 661–678 (1961)
39. Knopp, M.I.: Construction of automorphic forms on H-groups and supplementary Fourier series. Trans. Am. Math. Soc. **103**, 168–188 (1962)
40. Knopp, M.I.: On abelian integrals of the second kind and modular functions. Am. J. Math. **84**, 615–628 (1962)
41. Knopp, M.I.: On the Fourier coefficients of small positive powers of $\theta(\tau)$. Invent. Math. **85**(1), 165–183 (1986)
42. Knopp, M.I.: On the Fourier coefficients of cusp forms having small positive weight. In: Theta Functions—Bowdoin 1987, Part 2 (Brunswick, ME, 1987). Proceedings of Symposia in Pure Mathematics, vol. 49, pp. 111–127. American Mathematical Society, Providence (1989)
43. Knopp, M.I.: Rademacher on $J(\tau)$, Poincaré series of nonpositive weights and the Eichler cohomology. Not. Am. Math. Soc. **37**(4), 385–393 (1990)
44. Knopp, M., Robins, S.: Easy proofs of Riemann's functional equation for $\zeta(s)$ and of Lipschitz summation. Proc. Am. Math. Soc. **129**(7), 1915–1922 (2001)
45. Kowalski, E.: Poincaré and analytic number theory. In: The Scientific Legacy of Poincaré. History of Mathematics, vol. 36, pp. 73–85. American Mathematical Society, Providence (2010)

46. Kraus, P., Larsen, F.: Partition functions and elliptic genera from supergravity. J. High Energy Phys. **0701**, 002 (2007)
47. Maldacena, J.M.: The large N limit of superconformal field theories and supergravity. Adv. Theor. Math. Phys. **2**, 231–252 (1998)
48. Maldacena, J., Strominger, A.: Ads3 black holes and a stringy exclusion principle. J. High Energy Phys. **9812**, 005 (1998)
49. Maloney, A., Witten, E.: Quantum gravity partition functions in three dimensions. J. High Energy Phys. **2**, 029, 58 (2010)
50. Manschot, J., Moore, G.W.: A modern fareytail. Commun. Number Theory Phys. **4**, 103–159 (2010)
51. Moore, G.W.: Les Houches Lectures on Strings and Arithmetic (2004). hep-th/0401049
52. Murthy, S., Nawata, S.: Which AdS3 configurations contribute to the SCFT2 elliptic genus? (2011). arXiV:1112.4844
53. Murthy, S., Pioline, B.: A Farey tale for N=4 dyons. J. High Energy Phys. **0909**, 022 (2009)
54. Niebur, D.: Construction of automorphic forms and integrals. Trans. Am. Math. Soc. **191**, 373–385 (1974)
55. Petersson, H.: Theorie der automorphen Formen beliebiger reeller Dimension und ihre Darstellung durch eine neue Art Poincaréscher Reihen. Math. Ann. **103**(1), 369–436 (1930)
56. Petersson, H.: Über die Entwicklungskoeffizienten der automorphen Formen. Acta Math. **58**(1), 169–215 (1932)
57. Petersson, H.: Über die Entwicklungskoeffizienten einer allgemeinen Klasse automorpher Formen. Math. Ann. **108**(1), 370–377 (1933)
58. Poincaré, H.: Fonctions modulaires et fonctions fuchsiennes. Ann. Fac. Sci. Toulouse Sci. Math. Sci. Phys. (3) **3**, 125–149 (1911)
59. Rademacher, H.: On the partition function $p(n)$. Proc. Lond. Math. Soc. (2) **43**, 241–254 (1937)
60. Rademacher, H.: The Fourier coefficients of the modular invariant J(τ). Am. J. Math. **60**(2), 501–512 (1938)
61. Rademacher, H.: The Fourier series and the functional equation of the absolute modular invariant J(τ). Am. J. Math. **61**(1), 237–248 (1939)
62. Rademacher, H., Zuckerman, H.S.: On the Fourier coefficients of certain modular forms of positive dimension. Ann. Math. (2) **39**(2), 433–462 (1938)
63. Rogers, L.J.: On two theorems of combinatory analysis and some allied identities. Proc. Lond. Math. Soc. **16**, 315–316 (1917)
64. Rouse, J.: Zagier duality for the exponents of Borcherds products for Hilbert modular forms. J. Lond. Math. Soc. (2) **73**(2), 339–354 (2006)
65. Selberg, A.: On the estimation of Fourier coefficients of modular forms. In: Proceedings of Symposia in Pure Mathematics, vol. VIII, pp. 1–15. American Mathematical Society, Providence (1965)
66. Shimura, G.: Introduction to the Arithmetic Theory of Automorphic Functions. Publications of the Mathematical Society of Japan, vol. 11. Iwanami Shoten, Publishers, Tokyo (1971). Kanô Memorial Lectures, No. 1
67. Sivaramakrishnan, R.: Classical Theory of Arithmetic Functions. Monographs and Textbooks in Pure and Applied Mathematics, vol. 126. Marcel Dekker, New York (1989)
68. Thompson, J.G.: Finite groups and modular functions. Bull. Lond. Math. Soc. **11**(3), 347–351 (1979)
69. Thompson, J.G.: Some numerology between the Fischer-Griess Monster and the elliptic modular function. Bull. Lond. Math. Soc. **11**(3), 352–353 (1979)
70. Zagier, D.: Traces of singular moduli. In: Motives, Polylogarithms and Hodge Theory, Part I (Irvine, CA, 1998). International Press Lecture Series, vol. 3, pp. 211–244. International Press, Somerville (2002)
71. Zagier, D.: Ramanujan's mock theta functions and their applications (after Zwegers and Ono-Bringmann), Séminaire Bourbaki. Vol. 2007/2008. Astérisque **326** (2009), Exp. No. 986, vii–viii, 143–164 (2010)
72. Zwegers, S.: Mock Theta Functions. Ph.D. thesis, Utrecht University (2002)

Free Bosonic Vertex Operator Algebras on Genus Two Riemann Surfaces II

Geoffrey Mason and Michael P. Tuite

Abstract We study n-point correlation functions for a vertex operator algebra V on a Riemann surface of genus 2 obtained by attaching a handle to a torus. We obtain closed formulas for the genus two partition function for free bosonic theories and lattice vertex operator algebras V_L and describe their holomorphic and modular properties. We also compute the genus two Heisenberg vector n-point function and the Virasoro vector one point function. Comparing with the companion paper, when a pair of tori are sewn together, we show that the partition functions are not compatible in the neighborhood of a two-tori degeneration point. The *normalized* partition functions of a lattice theory V_L *are* compatible, each being identified with the genus two Siegel theta function of L.

1 Introduction

In previous work [17–20, 34] we developed the general theory of n-point functions for a Vertex Operator Algebra (VOA) on a compact Riemann surface $\mathcal{S}$ obtained by sewing together two surfaces of lower genus, and applied this theory to obtain detailed results in the case that $\mathcal{S}$ is obtained by sewing a pair of complex tori—the so-called ϵ-formalism discussed in the companion paper[1] [20]. In the

[1]Reference [20] together with the present paper constitute a much expanded version of [21].

G. Mason
Department of Mathematics, University of California Santa Cruz, Santa Cruz, CA 95064, USA
e-mail: gem@ucsc.edu

M.P. Tuite (✉)
School of Mathematics, Statistics and Applied Mathematics, National University of Ireland Galway, University Road, Galway, Ireland
e-mail: michael.tuite@nuigalway.ie

W. Kohnen and R. Weissauer (eds.), *Conformal Field Theory, Automorphic Forms and Related Topics*, Contributions in Mathematical and Computational Sciences 8, DOI 10.1007/978-3-662-43831-2_7, © Springer-Verlag Berlin Heidelberg 2014

present paper we consider in detail the situation when $\mathcal{S}$ results from self-sewing a complex torus, i.e., attaching a handle, which we refer to as the ρ-formalism. We describe the nature of the resulting n-point functions, paying particular attention to the 0-point function, i.e., the genus 2 *partition function*, in the ρ-formalism. We find the explicit form of the partition function for the Heisenberg free bosonic string and for lattice vertex operator algebras, and show that these functions are holomorphic on the parameter domain defined by the sewing. We study the generating function for genus two Heisenberg n-point functions and show that the Virasoro vector 1-point function satisfies a genus two Ward identity. Many of these results are analogous to those found in the ϵ-formalism discussed in [20] but with significant technical differences. Finally, we compare the results in the two formalisms, and show that the partition functions (and hence all n-point functions) are *incompatible*. We introduce *normalized* partition functions, and in the case of V_L show that they *are* compatible; in both formalisms the normalized partition function is the genus two Siegel theta function $\theta_L^{(2)}$.

We now discuss the contents of the paper in more detail. Our approach to genus two correlation functions in both formalisms is to define them in terms of genus one data coming from a VOA V. In Sect. 2 we review the ρ-formalism introduced in [18]. There, we constructed a genus two surface by self-sewing a torus, and obtained explicit expressions for the genus two normalized 2-form of the second kind $\omega^{(2)}$, a basis of normalized holomorphic 1-forms ν_1, ν_2, and the period matrix Ω, in terms of genus one data. In particular, we constructed a holomorphic map

$$\begin{aligned} F^{\rho} : \mathcal{D}^{\rho} &\longrightarrow \mathbb{H}_2 \\ (\tau, w, \rho) &\longmapsto \Omega(\tau, w, \rho) \end{aligned} \tag{1}$$

Here, and below, $\mathbb{H}_g$ $(g \geq 1)$ is the genus g *Siegel upper half-space*, and $\mathcal{D}^{\rho} \subseteq \mathbb{H}_1 \times \mathbb{C}^2$ is the domain defined in terms of data (τ, w, ρ) needed to self-sew a torus of modulus τ. Sewing produces a surface $\mathcal{S} = \mathcal{S}(\tau, w, \rho)$ of genus 2, and the map F^{ρ} assigns to $\mathcal{S}$ its period matrix. We also introduce some diagrammatic techniques which provide a convenient way of describing $\omega^{(2)}, \nu_1, \nu_2$ and Ω in the ρ-formalism.

Section 3 consists of a brief review of relevant background material on VOA theory, with particular attention paid to the Li-Zamolodchikov or LiZ metric. In Sect. 4, motivated by ideas in conformal field theory [6, 29, 31, 32], we introduce n-point functions (at genus one and two) in the ρ-formalism for a general VOA with nondegenerate LiZ metric. In particular, the genus two *partition function* $Z_V^{(2)} : \mathcal{D}^{\rho} \rightarrow \mathbb{C}$ is formally defined as

$$Z_V^{(2)}(\tau, w, \rho) = \sum_{n \geq 0} \rho^n \sum_{u \in V_{[n]}} Z_V^{(1)}(\overline{u}, u, w, \tau), \tag{2}$$

where the inner sum is taken over any basis for a homogeneous space $V_{[n]}$ of weight $wt[n]$, $Z_V^{(1)}(\overline{u}, u, w, \tau)$ is a genus one 2-point function and $\overline{u}$ is the LiZ metric dual of u. In Sect. 4.1 we consider an example of self-sewing a sphere (Theorem 6), while in

Sect. 4.2 we show (Theorem 7) that a particular degeneration of the genus 2 partition function of a VOA V can be described in terms of genus 1 data. Of particular interest here is the interesting relationship between the quasiprimary decomposition of V and the Catalan series.

In Sects. 5 and 6 we consider in detail the case of the Heisenberg free bosonic theory M^l corresponding to l free bosons, and lattice VOAs V_L associated with a positive-definite even lattice L. Although (2) is a priori a formal power series in ρ, w and $q = e^{2\pi i\tau}$, we will see that for these two theories it is a holomorphic function on $\mathcal{D}^\rho$. We expect that this result holds in much wider generality. Although our calculations in these two sections generally parallel those for the ϵ-formalism [20], the ρ-formalism is far from being a simple translation. Several issues require additional attention, so that the ρ-formalism is rather more complicated than its ϵ-counterpart. This arises in part from the fact that F^ρ involves a logarithmic term that is absent in the ϵ-formalism. The moment matrices employed are also more unwieldy.

We establish (Theorem 8) a fundamental formula describing $Z_M^{(2)}(\tau, w, \rho)$ as a quotient of the genus one partition function for M by a certain infinite determinant. This determinant was already introduced in [18], and its holomorphy and nonvanishing in D^ρ (loc. cit.) implies the holomorphy of $Z_M^{(2)}$. We also obtain a product formula for the infinite determinant (Theorem 9), and establish the automorphic properties of $Z_{M^2}^{(2)}$ with respect to the action of a group $\Gamma_1 \cong \mathrm{SL}(2, \mathbb{Z})$ (Theorem 11) that naturally acts on D^ρ. In particular, we find that $Z_{M^{24}}^{(2)}$ is a form of weight -12 with respect to the action of Γ_1. These are the analogs in the ρ-formalism of results obtained in Sect. 6 of [20] for the genus two partition function of M in the ϵ-formalism.

We also calculate some genus two n-point functions for the rank one Heisenberg VOA M, specifically the n-point function for the weight 1 Heisenberg vector and the 1-point function for the Virasoro vector $\tilde{\omega}$. We show that, up to an overall factor of the genus two partition function, the formal differential forms associated with these n-point functions are described in terms of the global symmetric 2-form $\omega^{(2)}$ [33] and the genus two projective connection [11] respectively. Once again, these results are analogous to results obtained in [20] in the ϵ-formalism.

In Sect. 6.1 we establish (Theorem 14) a basic formula for the genus two partition function for lattice theories in the ρ-formalism. The result is

$$Z_{V_L}^{(2)}(\tau, w, \rho) = Z_{M^l}^{(2)}(\tau, w, \rho)\theta_L^{(2)}(\Omega), \tag{3}$$

where $\theta_L^{(2)}(\Omega)$ is the genus two Siegel theta function attached to L [7] and $\Omega = F^\rho(\tau, w, \rho)$; indeed, (3) is an identity of formal power series. The holomorphy and automorphic properties of $Z_{V_L,\rho}^{(2)}$ follow from (3) and those of $Z_{M^l}^{(2)}$ and $\Theta_L^{(2)}$. Heisenberg n-point functions and a genus two Ward identity involving the Virasoro 1-point function are also discussed.

Section 7 is devoted to a *comparison* of genus two n-point functions, and especially partition functions, in the ϵ- and ρ-formalisms. There are strong formal similarities between $Z^{(2)}_{M^l,\epsilon}(\tau_1,\tau_2,\epsilon)$ and $Z^{(2)}_{M^l,\rho}(\tau,w,\rho)$ so it is natural to ask if they are equal in some sense.[2] In the very special case that V is holomorphic (i.e., it has a *unique* irreducible module), one knows (e.g., [33]) that the genus 2 conformal block is one-dimensional, in which case an identification of the two partition functions might seem inevitable. On the other hand, the partition functions are defined on quite different domains, so there is no question of them being literally equal. Indeed, we argue in Sect. 7 that $Z^{(2)}_{M^l,\epsilon}(\tau_1,\tau_2,\epsilon)$ and $Z^{(2)}_{M^l,\rho}(\tau,w,\rho)$ are *incompatible*, i.e., there is *no* sensible way in which they can be identified.

We therefore introduce *normalized* partition functions, defined as

$$\hat{Z}^{(2)}_{V,\rho}(\tau,w,\rho) := \frac{Z^{(2)}_{V,\rho}(\tau,w,\rho)}{Z^{(2)}_{M^l,\rho}(\tau,w,\rho)}, \quad \hat{Z}^{(2)}_{V,\epsilon}(\tau_1,\tau_2,\epsilon) := \frac{Z^{(2)}_{V,\epsilon}(\tau_1,\tau_2,\epsilon)}{Z^{(2)}_{M^l,\epsilon}(\tau_1,\tau_2,\epsilon)},$$

associated to a VOA V of central charge l. For M^l, the normalized partition functions are equal to 1. The relation between the normalized partition functions for lattice theories V_L ($\mathrm{rk}\, L = l$) in the two formalisms can be displayed in the diagram

$$\begin{array}{ccccc} D^{\epsilon} & \xrightarrow{F^{\epsilon}} & \mathbb{H}_2 & \xleftarrow{F^{\rho}} & D^{\rho} \\ & \hat{Z}^{(2)}_{V,\epsilon} \searrow & \downarrow \theta^{(2)}_L & \swarrow \hat{Z}^{(2)}_{V,\rho} & \\ & & \mathbb{C} & & \end{array} \tag{4}$$

That this is a *commuting* diagram combines formula (3) in the ρ-formalism, and Theorem 14 of [20] for the analogous result in the ϵ-formalism. Thus, the *normalized* partition functions for V_L are *independent of the sewing scheme*. They can be identified, via the sewing maps $F^{\bullet}$, with *a genus two Siegel modular form of weight* $l/2$, the Siegel theta function. It is therefore the normalized partition function(s) which can be identified with an element of the conformal block, and with each other. It would obviously be useful to have available a result that provides an a priori guarantee of this fact. A partial confirmation of this fact is described in [12] where it is shown that the normalized partition functions for any VOA V agree in the degeneration limit where one torus is pinched down to a Riemann sphere. Section 8 contains a brief further discussion of these issues in the light of related ideas in string theory and algebraic geometry.

[2] Here we include an additional subscript of either ϵ or ρ to distinguish between the two formalisms.

2 Genus Two Riemann Surface from Self-sewing a Torus

In this section we review some relevant results of [18] based on a general sewing formalism due to Yamada [36]. In particular, we review the construction of a genus two Riemann surface formed by self-sewing a twice-punctured torus. We refer to this sewing scheme as the ρ-formalism. We discuss the explicit form of various genus two structures such as the period matrix Ω. We also review the convergence and holomorphy of an infinite determinant that naturally arises later on. An alternative genus two surface formed by sewing together two tori, which we refer to as the ϵ-formalism, is utilised in the companion paper [20].

2.1 *Some Elliptic Function Theory*

We begin with the definition of various modular and elliptic functions [17, 18]. We define

$$\begin{aligned} P_2(\tau, z) &= \wp(\tau, z) + E_2(\tau) \\ &= \frac{1}{z^2} + \sum_{k=2}^{\infty}(k-1)E_k(\tau)z^{k-2}, \end{aligned} \tag{5}$$

where $\tau \in \mathbb{H}_1$, the complex upper half-plane and where $\wp(\tau, z)$ is the Weierstrass function (with periods $2\pi \mathrm{i}$ and $2\pi \mathrm{i}\tau$) and $E_k(\tau) = 0$ for k odd, and for k even is the Eisenstein series. Here and below, we take $q = \exp(2\pi \mathrm{i}\tau)$. We define $P_0(\tau, z)$, up to a choice of the logarithmic branch, and $P_1(\tau, z)$ by

$$P_0(\tau, z) = -\log(z) + \sum_{k\geq 2}\frac{1}{k}E_k(\tau)z^k, \tag{6}$$

$$P_1(\tau, z) = \frac{1}{z} - \sum_{k\geq 2}E_k(\tau)z^{k-1}. \tag{7}$$

P_0 is related to the elliptic prime form $K(\tau, z)$, by [27]

$$K(\tau, z) = \exp\left(-P_0(\tau, z)\right). \tag{8}$$

Define elliptic functions $P_k(\tau, z)$ for $k \geq 3$

$$P_k(\tau, z) = \frac{(-1)^{k-1}}{(k-1)!}\frac{d^{k-1}}{dz^{k-1}}P_1(\tau, z). \tag{9}$$

Define for $k, l \geq 1$

$$C(k,l) = C(k,l,\tau) = (-1)^{k+1}\frac{(k+l-1)!}{(k-1)!(l-1)!}E_{k+l}(\tau), \tag{10}$$

$$D(k,l,z) = D(k,l,\tau,z) = (-1)^{k+1}\frac{(k+l-1)!}{(k-1)!(l-1)!}P_{k+l}(\tau,z). \tag{11}$$

2.2 *The ρ-Formalism for Self-sewing a Torus*

Consider a compact Riemann surface $\mathcal{S}$ of genus 2 with standard homology basis a_1, a_2, b_1, b_2. Let

$$\omega(x,y) = \left(\frac{1}{(x-y)^2} + \text{regular terms}\right) dxdy \tag{12}$$

be the normalized differential of the second kind [4, 36] for local coordinates x, y with normalization $\oint_{a_i} \omega(x,\cdot) = 0$ for $i = 1, 2$. Then

$$\nu_i(x) = \oint_{b_i} \omega(x,\cdot), \tag{13}$$

for $i = 1, 2$ is a basis of holomorphic 1-forms with normalization $\oint_{a_i} \nu_j = 2\pi \mathrm{i}\delta_{ij}$. The genus 2 period matrix $\Omega \in \mathbb{H}_2$ is defined by

$$\Omega_{ij} = \frac{1}{2\pi \mathrm{i}}\oint_{b_i} \nu_j. \tag{14}$$

We now review a general method due to Yamada [36], and discussed at length in [18], for calculating $\omega(x,y)$, $\nu_i(x)$ and Ω_{ij} on the Riemann surface formed by sewing a handle to an oriented torus $\mathcal{S} = \mathbb{C}/\Lambda$ with lattice $\Lambda = 2\pi\mathrm{i}(\mathbb{Z}\tau \oplus \mathbb{Z})$ and $\tau \in \mathbb{H}_1$. Consider discs centered at $z = 0$ and $z = w$ with local coordinates $z_1 = z$ and $z_2 = z - w$, and positive radius $r_a < \frac{1}{2}D(q)$ with $1 \leq a \leq 2$. Here, we have introduced the minimal lattice distance

$$D(q) = \min_{(m,n)\neq(0,0)} 2\pi|m + n\tau| > 0. \tag{15}$$

Note that r_1, r_2 must be sufficiently small to ensure that the discs do not intersect on $\mathcal{S}$. Introduce a complex parameter ρ where $|\rho| \leq r_1 r_2$ and excise the discs $\{z_a, |z_a| \leq |\rho|/r_{\overline{a}}\}$ to obtain a twice-punctured torus (illustrated in Fig. 1)

$$\hat{\mathcal{S}} = \mathcal{S}\backslash\{z_a, |z_a| \leq |\rho|/r_{\overline{a}}\} \ (1 \leq a \leq 2).$$

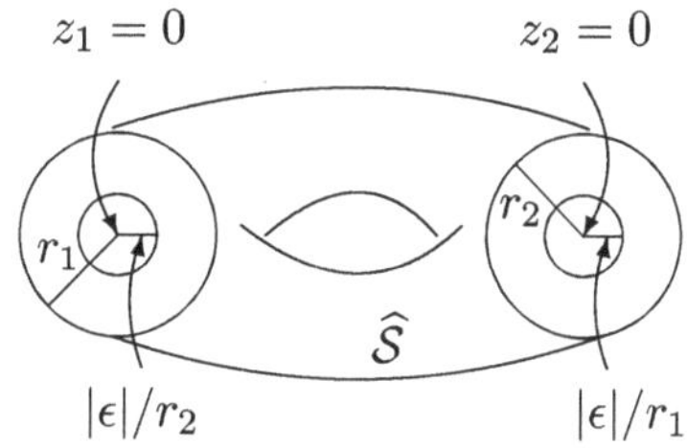

Fig. 1 Self-sewing a torus

Here, and below, we use the convention

$$\overline{1} = 2, \qquad \overline{2} = 1. \tag{16}$$

Define annular regions $\mathcal{A}_a = \{z_a, |\rho| r_{\overline{a}}^{-1} \leq |z_a| \leq r_a\} \in \hat{\mathcal{S}}$ $(1 \leq a \leq 2)$, and identify $\mathcal{A}_1$ with $\mathcal{A}_2$ as a single region via the sewing relation

$$z_1 z_2 = \rho. \tag{17}$$

The resulting genus two Riemann surface (excluding the degeneration point $\rho = 0$) is parameterized by the domain

$$\mathcal{D}^\rho = \left\{(\tau, w, \rho) \in \mathbb{H}_1 \times \mathbb{C} \times \mathbb{C} : \ |w - \lambda| > 2|\rho|^{1/2} > 0 \text{ for all } \lambda \in \Lambda\right\}, \tag{18}$$

where the first inequality follows from the requirement that the annuli do not intersect. The Riemann surface inherits the genus one homology basis a_1, b_1. The cycle a_2 is defined to be the anti-clockwise contour surrounding the puncture at w, and b_2 is a path between identified points $z_1 = z_0$ to $z_2 = \rho/z_0$ for some $z_0 \in \mathcal{A}_1$.

ω, ν_i and Ω are expressed as a functions of $(\tau, w, \rho) \in \mathcal{D}^\rho$ in terms of an infinite matrix of 2×2 blocks $R(\tau, w, \rho) = (R(k, l, \tau, w, \rho))$ $(k, l \geq 1)$ where [18]

$$R(k, l, \tau, w, \rho) = -\frac{\rho^{(k+l)/2}}{\sqrt{kl}} \begin{pmatrix} D(k, l, \tau, w) & C(k, l, \tau) \\ C(k, l, \tau) & D(l, k, \tau, w) \end{pmatrix}, \tag{19}$$

for C, D of (10) and (11). $I - R$ and $\det(I - R)$ play a central rôle in our discussion, where I denotes the doubly-indexed identity matrix and $\det(I - R)$ is defined by

$$\log \det(I - R) = \operatorname{Tr} \log(I - R) = -\sum_{n \geq 1} \frac{1}{n} \operatorname{Tr} R^n. \tag{20}$$

In particular (op. cit., Proposition 6 and Theorem 7)

Theorem 1. *We have*

(a)

$$(I-R)^{-1}=\sum_{n\geq 0}R^{n} \tag{21}$$

is convergent in $\mathcal{D}^{\rho}$.

(b) $\det(I-R)$ *is nonvanishing and holomorphic in* $\mathcal{D}^{\rho}$. □

We define a set of 1-forms on $\hat{\mathcal{S}}$ given by

$$\begin{aligned} a_1(k,x) &= a_1(k,x,\tau,\rho)=\sqrt{k}\rho^{k/2}P_{k+1}(\tau,x)dx,\\ a_2(k,x) &= a_2(k,x,\tau,\rho)=a_1(k,x-w), \end{aligned} \tag{22}$$

indexed by integers $k\geq 1$. We also define the infinite row vector $a(x)=(a_a(k,x))$ and infinite column vector $\overline{a}(x)^T=(a_{\overline{a}}(k,x))^T$ for $k\geq 1$ and block index $1\leq a\leq 2$. We find (op. cit., Lemma 11, Proposition 6 and Theorem 9):

Theorem 2.

$$\omega(x,y)=P_2(\tau,x-y)dxdy-a(x)(I-R)^{-1}\overline{a}(y)^T. \quad \square \tag{23}$$

Applying (13) results in (op. cit., Lemma 12 and Theorem 9)

Theorem 3.

$$\begin{aligned} \nu_1(x) &= dx-\rho^{1/2}\sigma\left((a(x)(I-R)^{-1})(1)\right)\\ \nu_2(x) &= (P_1(\tau,x-w)-P_1(\tau,x))\,dx-a(x)(I-R)^{-1}\overline{d}^{\,T}. \end{aligned} \tag{24}$$

$d=(d_a(k))$ *is a doubly-indexed infinite row vector*[3]

$$\begin{aligned} d_1(k) &= -\frac{\rho^{k/2}}{\sqrt{k}}\left(P_k(\tau,w)-E_k(\tau)\right),\\ d_2(k) &= (-1)^k\frac{\rho^{k/2}}{\sqrt{k}}\left(P_k(\tau,w)-E_k(\tau)\right), \end{aligned} \tag{25}$$

with $\overline{d}_a=d_{\overline{a}}$. (1) *refers to the* $(k)=(1)$ *entry of a row vector and* $\sigma(M)$ *denotes the sum over the finite block indices for a given* 1×2 *block matrix* M. □

Ω is determined (op. cit., Proposition 11) by (14) as follows:

[3]Note that d is denoted by β in [18].

Theorem 4. *There is a holomorphic map*

$$\begin{aligned} F^\rho : \mathcal{D}^\rho &\to \mathbb{H}_2, \\ (\tau, w, \rho) &\mapsto \Omega(\tau, w, \rho), \end{aligned} \tag{26}$$

where $\Omega = \Omega(\tau, w, \rho)$ *is given by*

$$2\pi i\Omega_{11} = 2\pi i\tau - \rho\sigma\left((I - R)^{-1}(1,1)\right), \tag{27}$$

$$2\pi i\Omega_{12} = w - \rho^{1/2}\sigma\left(d(I - R)^{-1}(1)\right), \tag{28}$$

$$2\pi i\Omega_{22} = \log\left(-\frac{\rho}{K(\tau,w)^2}\right) - d(I - R)^{-1}\overline{d}^T. \tag{29}$$

K is the elliptic prime form (8), $(1,1)$ and (1) refer to the $(k,l) = (1,1)$, respectively, $(k) = (1)$ entries of an infinite matrix and row vector respectively. $\sigma(M)$ denotes the sum over the finite block indices for a given 2×2 or 1×2 block matrix M. □

$\mathcal{D}^\rho$ admits an action of the Jacobi group $J = \mathrm{SL}(2,\mathbb{Z}) \ltimes \mathbb{Z}^2$ as follows:

$$(a,b).(\tau, w, \rho) = (\tau, w + 2\pi i a\tau + 2\pi i b, \rho) \quad ((a,b) \in \mathbb{Z}^2), \tag{30}$$

$$\gamma_1.(\tau, w, \rho) = (\frac{a_1\tau + b_1}{c_1\tau + d_1}, \frac{w}{c_1\tau + d_1}, \frac{\rho}{(c_1\tau + d_1)^2}) \quad (\gamma_1 \in \Gamma_1), \tag{31}$$

with $\Gamma_1 = \left\{\begin{pmatrix} a_1 & b_1 \\ c_1 & d_1 \end{pmatrix}\right\} = \mathrm{SL}(2,\mathbb{Z})$. Due to the branch structure of the logarithmic term in (29), F^ρ is not equivariant with respect to J. (See Sect. 6.3 of [18] for details.)

There is a natural injection $\Gamma_1 \to \mathrm{Sp}(4,\mathbb{Z})$ defined by

$$\begin{pmatrix} a_1 & b_1 \\ c_1 & d_1 \end{pmatrix} \mapsto \begin{pmatrix} a_1 & 0 & b_1 & 0 \\ 0 & 1 & 0 & 0 \\ c_1 & 0 & d_1 & 0 \\ 0 & 0 & 0 & 1 \end{pmatrix}, \tag{32}$$

through which Γ_1 acts on $\mathbb{H}_2$ by the standard action

$$\gamma.\Omega = (A\Omega + B)(C\Omega + D)^{-1}, \quad \left(\gamma = \begin{pmatrix} A & B \\ C & D \end{pmatrix} \in \mathrm{Sp}(4,\mathbb{Z})\right). \tag{33}$$

We then have (op. cit., Theorem 11, Corollary 2)

Theorem 5. *F^ρ is equivariant with respect to the action of Γ_1, i.e. there is a commutative diagram for $\gamma_1 \in \Gamma_1$,*

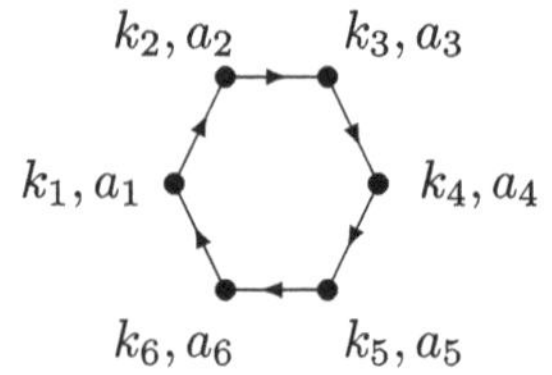

Fig. 2 Doubly-indexed cycle

$$\begin{array}{ccc} \mathcal{D}^{\rho} & \stackrel{F^{\rho}}{\rightarrow} & \mathbb{H}_2 \\ \gamma_1 \downarrow & & \downarrow \gamma_1 \\ \mathcal{D}^{\rho} & \stackrel{F^{\rho}}{\rightarrow} & \mathbb{H}_2 \end{array} \qquad \square$$

2.3 Graphical Expansions

We present a graphical approach to describing the expressions for $\omega, \nu_i, \Omega_{ij}$ reviewed above. These also play an important rôle in the analysis of genus two partition functions for the Heisenberg vertex operator algebra. A similar approach is described in [20] suitable for the ϵ-sewing scheme. Here we introduce *doubly-indexed* cycles construed as (clockwise) oriented, labelled polygons L with n nodes for some integer $n \geq 1$, nodes being labelled by a pair of integers k, a where $k \geq 1$ and $a \in \{1, 2\}$. Thus, a typical doubly-indexed cycle looks as in Fig. 2.

We define a weight function[4] ζ with values in the ring of elliptic functions and quasi-modular forms $\mathbb{C}[P_2(\tau, w), P_3(\tau, w), E_2(\tau), E_4(\tau), E_6(\tau)]$ as follows: if L is a doubly-indexed cycle then L has edges E labelled as $\overset{k,a}{\bullet} \rightarrow \overset{l,b}{\bullet}$, and we set

$$\zeta(E) = R_{ab}(k, l, \tau, w, \rho), \tag{34}$$

with $R_{ab}(k, l)$ as in (19) and

$$\zeta(L) = \prod \zeta(E),$$

where the product is taken over all edges of L.

We also introduce *doubly-indexed necklaces* $\mathcal{N} = \{N\}$. These are connected graphs with $n \geq 2$ nodes, $(n - 2)$ of which have valency 2 and two of which have valency 1 together with an orientation, say from left to right, on the edges. In this case, each vertex carries two integer labels k, a with $k \geq 1$ and $a \in \{1, 2\}$. We define the degenerate necklace N_0 to be a single node with no edges, and set $\zeta(N_0) = 1$.

[4]Denoted by ω in Sect. 6.2 of [18].

We define necklaces with distinguished end nodes labelled $k, a; l, b$ as follows:

$$\underset{k,a}{\bullet} \longrightarrow \underset{k_1,a_1}{\bullet} \cdots \underset{k_2,a_2}{\bullet} \longrightarrow \underset{l,b}{\bullet} \qquad \text{(type } k,a;l,b)$$

and set[5]

$$\mathcal{N}(k,a;l,b) = \{\text{isomorphism classes of necklaces of type } k,a;l,b\}. \tag{35}$$

We define

$$\begin{aligned}
\zeta(1;1) &= \sum_{a_1,a_2=1,2} \sum_{N \in \mathcal{N}(1,a_1;1,a_2)} \zeta(N),\\
\zeta(d;1) &= \sum_{a_1,a_2=1,2} \sum_{k\geq 1} d_{a_1}(k) \sum_{N \in \mathcal{N}(k,a_1;1,a_2)} \zeta(N),\\
\zeta(d;\overline{d}) &= \sum_{a_1,a_2=1,2} \sum_{k,l\geq 1} d_{a_1}(k)\overline{d}_{a_2}(l) \sum_{N \in \mathcal{N}(k,a_1;l,a_2)} \zeta(N).
\end{aligned} \tag{36}$$

Then we find

Proposition 1 ([18], Proposition 12). *The period matrix is given by*

$$\begin{aligned}
2\pi i\Omega_{11} &= 2\pi i\tau - \rho\zeta(1;1),\\
2\pi i\Omega_{12} &= w - \rho^{1/2}\zeta(d;1),\\
2\pi i\Omega_{22} &= \log\left(-\frac{\rho}{K(\tau,w)^2}\right) - \zeta(d;\overline{d}). \quad \square
\end{aligned}$$

We can similarly obtain necklace graphical expansions for the bilinear form $\omega(x,y)$ and the holomorphic one forms $\nu_i(x)$. We introduce further distinguished valence one nodes labelled by $x \in \hat{\mathcal{S}}$, the punctured torus. The set of edges $\{E\}$ is augmented by edges with weights defined by:

$$\begin{aligned}
\zeta(\overset{x}{\bullet} \longrightarrow \overset{y}{\bullet}) &= P_2(\tau, x-y)dxdy,\\
\zeta(\overset{x}{\bullet} \longrightarrow \overset{k,a}{\bullet}) &= a_a(k,x),\\
\zeta(\overset{k,a}{\bullet} \longrightarrow \overset{y}{\bullet}) &= -a_{\overline{a}}(k,y),
\end{aligned} \tag{37}$$

for 1-forms (22).

We also consider doubly-indexed necklaces where one or both end points are x, y-labeled nodes. We thus define for $x, y \in \hat{\mathcal{S}}$ two isomorphism classes of oriented

[5] Two graphs are isomorphic if they have the same labelled vertices and directed edges.

doubly-indexed necklaces denoted by $\mathcal{N}(x; y)$, and $\mathcal{N}(x; k, a)$ with the following respective typical configurations

$$\{\overset{x}{\bullet} \longrightarrow \overset{k_1,a_1}{\bullet} \ldots \overset{k_2,a_2}{\bullet} \longrightarrow \overset{y}{\bullet}\}, \tag{38}$$

$$\{\overset{x}{\bullet} \longrightarrow \overset{k_1,a_1}{\bullet} \ldots \overset{k_2,a_2}{\bullet} \longrightarrow \overset{k,a}{\bullet}\}. \tag{39}$$

Furthermore, we define the weights

$$\begin{aligned}
\zeta(x; y) &= \sum_{N \in \mathcal{N}(x;y)} \zeta(N), \\
\zeta(x; 1) &= \sum_{a=1,2} \sum_{N \in \mathcal{N}(x;1,a)} \zeta(N), \\
\zeta(x; \overline{d}) &= \sum_{a=1,2} \sum_{k \geq 1} \sum_{N \in \mathcal{N}(x;k,a)} \zeta(N)\overline{d}_a(k).
\end{aligned} \tag{40}$$

Comparing to (23) and (24) we find the following graphical expansions for the bilinear form $\omega(x, y)$ and the holomorphic one forms $\nu_i(x)$

Proposition 2. *For* $x, y \in \hat{\mathcal{S}}$

$$\omega(x, y) = \zeta(x; y), \tag{41}$$

$$\nu_1(x) = dx - \rho^{1/2}\zeta(x; 1), \tag{42}$$

$$\nu_2(x) = (P_1(\tau, x - w) - P_1(\tau, x))dx - \zeta(x; \overline{d}). \tag{43}$$

3 Vertex Operator Algebras and the Li-Zamolodchikov Metric

3.1 Vertex Operator Algebras

We review some relevant aspects of vertex operator algebras [8, 9, 13, 15, 22, 23]. A vertex operator algebra (VOA) is a quadruple $(V, Y, \mathbf{1}, \omega)$ consisting of a $\mathbb{Z}$-graded complex vector space $V = \bigoplus_{n \in \mathbb{Z}} V_n$, a linear map $Y : V \to (\mathrm{End}\, V)[[z, z^{-1}]]$, for formal parameter z, and a pair of distinguished vectors (states), the vacuum $\mathbf{1} \in V_0$, and the conformal vector $\omega \in V_2$. For each state $v \in V$ the image under the Y map is the vertex operator

$$Y(v, z) = \sum_{n \in \mathbb{Z}} v(n) z^{-n-1}, \tag{44}$$

with modes $v(n) \in \mathrm{End}\, V$ where $\mathrm{Res}_{z=0} z^{-1} Y(v,z)\mathbf{1} = v(-1)\mathbf{1} = v$. Vertex operators satisfy the Jacobi identity or equivalently, operator locality or Borcherds's identity for the modes (loc. cit.).

The vertex operator for the conformal vector ω is defined as

$$Y(\omega, z) = \sum_{n\in\mathbb{Z}} L(n) z^{-n-2}.$$

The modes $L(n)$ satisfy the Virasoro algebra of central charge c:

$$[L(m), L(n)] = (m-n)L(m+n) + (m^3 - m)\frac{c}{12}\delta_{m,-n}.$$

We define the homogeneous space of weight k to be $V_k = \{v \in V | L(0)v = kv\}$ where we write $wt(v) = k$ for v in V_k. Then as an operator on V we have

$$v(n) : V_m \to V_{m+k-n-1}.$$

In particular, the *zero mode* $o(v) = v(wt(v) - 1)$ is a linear operator on V_m. A non-zero vector v is said to be *quasi-primary* if $L(1)v = 0$ and *primary* if additionally $L(2)v = 0$.

The subalgebra $\{L(-1), L(0), L(1)\}$ generates a natural action on vertex operators associated with $\mathrm{SL}(2, \mathbb{C})$ Möbius transformations [2, 3, 9, 13]. In particular, we note the inversion $z \mapsto 1/z$, for which

$$Y(v, z) \mapsto Y^{\dagger}(v, z) = Y\left(e^{zL(1)}\left(-\frac{1}{z^2}\right)^{L(0)} v, \frac{1}{z}\right). \tag{45}$$

$Y^{\dagger}(v, z)$ is the *adjoint* vertex operator [9].

We consider in particular the Heisenberg free boson VOA and lattice VOAs. Consider an l-dimensional complex vector space (i.e., abelian Lie algebra) $\mathfrak{H}$ equipped with a non-degenerate, symmetric, bilinear form $(\ ,\ )$ and a distinguished orthonormal basis $a_1, a_2, \ldots a_l$. The corresponding affine Lie algebra is the Heisenberg Lie algebra $\hat{\mathfrak{H}} = \mathfrak{H} \otimes \mathbb{C}[t, t^{-1}] \oplus \mathbb{C}k$ with brackets $[k, \hat{\mathfrak{H}}] = 0$ and

$$[a_i \otimes t^m, a_j \otimes t^n] = m\delta_{i,j}\delta_{m,-n}k. \tag{46}$$

Corresponding to an element λ in the dual space $\mathfrak{H}^*$ we consider the Fock space defined by the induced (Verma) module

$$M^{(\lambda)} = U(\hat{\mathfrak{H}}) \otimes_{U(\mathfrak{H}\otimes\mathbb{C}[t]\oplus\mathbb{C}k)} \mathbb{C},$$

where $\mathbb{C}$ is the one-dimensional space annihilated by $\mathfrak{H} \otimes t\mathbb{C}[t]$ and on which k acts as the identity and $\mathfrak{H} \otimes t^0$ via the character λ; U denotes the universal enveloping algebra. There is a canonical identification of linear spaces

$$M^{(\lambda)} = S(\mathfrak{H} \otimes t^{-1}\mathbb{C}[t^{-1}]),$$

where S denotes the (graded) symmetric algebra. The Heisenberg free boson VOA M^l corresponds to the case $\lambda = 0$. The Fock states

$$v = a_1(-1)^{e_1}.a_1(-2)^{e_2} \ldots .a_1(-n)^{e_n} \ldots .a_l(-1)^{f_1}.a_l(-2)^{f_2} \ldots a_l(-p)^{f_p}.\mathbf{1}, \quad (47)$$

for non-negative integers $e_i, \ldots, f_j$ form a basis of M^l. The vacuum $\mathbf{1}$ is canonically identified with the identity of $M_0^l = \mathbb{C}$, while the weight 1 subspace M_1^l may be naturally identified with $\mathfrak{H}$. M^l is a simple VOA of central charge l.

Next we consider the case of a lattice vertex operator algebra V_L associated to a positive-definite even lattice L (cf. [2, 8]). Thus L is a free abelian group of rank l equipped with a positive definite, integral bilinear form $(\ ,\ ) : L \otimes L \to \mathbb{Z}$ such that (α, α) is even for $\alpha \in L$. Let $\mathfrak{H}$ be the space $\mathbb{C} \otimes_{\mathbb{Z}} L$ equipped with the $\mathbb{C}$-linear extension of $(\ ,\ )$ to $\mathfrak{H} \otimes \mathfrak{H}$ and let M^l be the corresponding Heisenberg VOA. The Fock space of the lattice theory may be described by the linear space

$$V_L = M^l \otimes \mathbb{C}[L] = \sum_{\alpha \in L} M^l \otimes \mathrm{e}^{\alpha}, \quad (48)$$

where $\mathbb{C}[L]$ denotes the group algebra of L with canonical basis e^{α}, $\alpha \in L$. M^l may be identified with the subspace $M^l \otimes \mathrm{e}^0$ of V_L, in which case M^l is a subVOA of V_L and the rightmost equation of (48) then displays the decomposition of V_L into irreducible M^l-modules. V_L is a simple VOA of central charge l. Each $\mathbf{1} \otimes \mathrm{e}^{\alpha} \in V_L$ is a primary state of weight $\frac{1}{2}(\alpha, \alpha)$ with vertex operator (loc. cit.)

$$\begin{aligned} Y(\mathbf{1} \otimes \mathrm{e}^{\alpha}, z) &= Y_-(\alpha, z) Y_+(\alpha, z) \mathrm{e}^{\alpha} z^{\alpha}, \\ Y_{\pm}(\alpha, z) &= \exp\left(\mp \sum_{n>0} \frac{\alpha(\pm n)}{n} z^{\mp n} \right). \end{aligned} \quad (49)$$

The operators $\mathrm{e}^{\alpha} \in \mathbb{C}[L]$ obey

$$\mathrm{e}^{\alpha}\mathrm{e}^{\beta} = \epsilon(\alpha, \beta)\mathrm{e}^{\alpha+\beta} \quad (50)$$

for a bilinear 2-cocycle $\epsilon(\alpha, \beta)$ satisfying $\epsilon(\alpha, \beta)\epsilon(\beta, \alpha) = (-1)^{(\alpha,\beta)}$.

3.2 The Li-Zamolodchikov Metric

A bilinear form $\langle\ ,\ \rangle : V \times V \longrightarrow \mathbb{C}$ is called *invariant* in case the following identity holds for all $a, b, c \in V$ [9]:

$$\langle Y(a, z)b, c \rangle = \langle b, Y^{\dagger}(a, z)c \rangle, \quad (51)$$

with $Y^{\dagger}(a,z)$ the adjoint operator (45). If $V_0 = \mathbb{C}\mathbf{1}$ and V is self-dual (i.e. V is isomorphic to the contragradient module V' as a V-module) then V has a unique non-zero invariant bilinear form up to scalar [16]. Note that $\langle\ ,\rangle$ is necessarily symmetric by a theorem of [9]. Furthermore, if V is simple then such a form is necessarily non-degenerate. All of the VOAs that occur in this paper satisfy these conditions, so that normalizing $\langle \mathbf{1}, \mathbf{1}\rangle = 1$ implies that $\langle\ ,\rangle$ is unique. We refer to such a bilinear form as the *Li-Zamolodchikov metric* on V, or LiZ-metric for short [20]. We also note that the LiZ-metric is multiplicative over tensor products in the sense that LiZ metric of the tensor product $V_1 \otimes V_2$ of a pair of simple VOAs satisfying the above conditions is by uniqueness, the tensor product of the LiZ metrics on V_1 and V_2.

For a quasi-primary vector a of weight $wt(a)$, the component form of (51) becomes

$$\langle a(n)b, c\rangle = (-1)^{wt(a)} \langle b, a(2wt(a) - n - 2)c\rangle . \tag{52}$$

In particular, for the conformal vector ω we obtain

$$\langle L(n)b, c\rangle = \langle b, L(-n)c\rangle. \tag{53}$$

Taking $n = 0$, it follows that the homogeneous spaces V_n and V_m are orthogonal if $n \neq m$.

Consider the rank one Heisenberg VOA $M = M^1$ generated by a weight one state a with $(a,a) = 1$. Then $\langle a, a\rangle = -\langle \mathbf{1}, a(1)a(-1)\mathbf{1}\rangle = -1$. Using (46), it is straightforward to verify that the Fock basis (47) is orthogonal with respect to the LiZ-metric and

$$\langle v, v\rangle = \prod_{1\le i\le n} (-i)^{e_i} e_i!. \tag{54}$$

This result generalizes in an obvious way to the rank l free boson VOA M^l because the LiZ metric is multiplicative over tensor products.

We consider next the lattice vertex operator algebra V_L for a positive-definite even lattice L. We take as our Fock basis the states $\{v \otimes \mathrm{e}^{\alpha}\}$ where v is as in (47) and α ranges over the elements of L.

Lemma 1. *If $u, v \in M^l$ and $\alpha, \beta \in L$, then*

$$\begin{aligned}\left\langle u \otimes \mathrm{e}^{\alpha}, v \otimes \mathrm{e}^{\beta}\right\rangle &= \langle u, v\rangle \langle \mathbf{1} \otimes \mathrm{e}^{\alpha}, \mathbf{1} \otimes \mathrm{e}^{\beta}\rangle \\ &= (-1)^{\frac{1}{2}(\alpha,\alpha)}\epsilon(\alpha, -\alpha)\langle u, v\rangle \delta_{\alpha,-\beta}.\end{aligned}$$

Proof. It follows by successive applications of (52) that the first equality in the lemma is true, and that it is therefore enough to prove it in the case that $u = v = \mathbf{1}$. We identify the primary vector $\mathbf{1} \otimes \mathrm{e}^{\alpha}$ with e^{α} in the following. Then $\left\langle \mathrm{e}^{\alpha}, \mathrm{e}^{\beta}\right\rangle = \left\langle \mathrm{e}^{\alpha}(-1)\mathbf{1}, \mathrm{e}^{\beta}\right\rangle$ is given by

$$
\begin{aligned}
&(-1)^{\frac{1}{2}(\alpha,\alpha)}\left\langle \mathbf{1}, \mathrm{e}^{\alpha}((\alpha,\alpha)-1)\mathrm{e}^{\beta}\right\rangle \\
&= (-1)^{\frac{1}{2}(\alpha,\alpha)}\,\mathrm{Res}_{z=0} z^{(\alpha,\alpha)-1}\left\langle \mathbf{1}, Y(\mathrm{e}^{\alpha},z)\mathrm{e}^{\beta}\right\rangle \\
&= (-1)^{\frac{1}{2}(\alpha,\alpha)}\epsilon(\alpha,\beta)\,\mathrm{Res}_{z=0} z^{(\alpha,\beta)+(\alpha,\alpha)-1}\left\langle \mathbf{1}, Y_{-}(\alpha,z).\mathrm{e}^{\alpha+\beta}\right\rangle .
\end{aligned}
$$

Unless $\alpha+\beta=0$, all states to the left inside the bracket $\langle\ ,\ \rangle$ on the previous line have positive weight, hence are orthogonal to $\mathbf{1}$. So $\langle \mathrm{e}^{\alpha}, \mathrm{e}^{\beta}\rangle = 0$ if $\alpha+\beta \neq 0$. In the contrary case, the exponential operator acting on the vacuum yields just the vacuum itself among weight zero states, and we get $\langle \mathrm{e}^{\alpha}, \mathrm{e}^{-\alpha}\rangle = (-1)^{\frac{1}{2}(\alpha,\alpha)}\epsilon(\alpha,-\alpha)$ in this case. □

Corollary 1. *We may choose the cocycle so that* $\epsilon(\alpha,-\alpha) = (-1)^{\frac{1}{2}(\alpha,\alpha)}$ *(cf. (132) in Appendix). In this case, we have*

$$\left\langle u\otimes \mathrm{e}^{\alpha}, v\otimes \mathrm{e}^{\beta}\right\rangle = \langle u,v\rangle \delta_{\alpha,-\beta}. \tag{55}$$

4 Partition and n-Point Functions for Vertex Operator Algebras on a Genus Two Riemann Surface

In this section we consider the partition and n-point functions for a VOA on Riemann surface of genus one or two, formed by attaching a handle to a surface of lower genus. We assume that V has a non-degenerate LiZ metric $\langle\ ,\ \rangle$. Then for any V basis $\{u^{(a)}\}$, we may define the *dual basis* $\{\overline{u}^{(a)}\}$ with respect to the LiZ metric where

$$\left\langle u^{(a)}, \overline{u}^{(b)}\right\rangle = \delta_{ab}. \tag{56}$$

4.1 Genus One

It is instructive to first consider an alternative approach to defining the genus one partition function. In order to define n-point correlation functions on a torus, Zhu introduced [37] a second VOA $(V, Y[\ ,\ ], \mathbf{1}, \tilde{\omega})$ isomorphic to $(V, Y(\ ,\ ), \mathbf{1}, \omega)$ with vertex operators

$$Y[v,z] = \sum_{n\in\mathbb{Z}} v[n]z^{-n-1} = Y\left(q_z^{L(0)}v, q_z - 1\right), \tag{57}$$

and conformal vector $\tilde{\omega} = \omega - \frac{c}{24}\mathbf{1}$. Let

$$Y[\tilde{\omega},z] = \sum_{n\in\mathbb{Z}} L[n]z^{-n-2}, \tag{58}$$

and write $wt[v] = k$ if $L[0]v = kv$, $V_{[k]} = \{v \in V | wt[v] = k\}$. Similarly, we define the square bracket LiZ metric $\langle\ ,\ \rangle_{\mathrm{sq}}$ which is invariant with respect to the square bracket adjoint.

The (genus one) 1-point function is now defined as

$$Z_V^{(1)}(v, \tau) = \mathrm{Tr}_V \left(o(v) q^{L(0)-c/24}\right). \tag{59}$$

An n-point function can be expressed in terms of 1-point functions [17, Lemma 3.1] as follows:

$$\begin{aligned} &Z_V^{(1)}(v_1, z_1; \ldots v_n, z_n; \tau) \\ &= Z_V^{(1)}(Y[v_1, z_1] \ldots Y[v_{n-1}, z_{n-1}] Y[v_n, z_n]\mathbf{1}, \tau) \qquad (60) \\ &= Z_V^{(1)}(Y[v_1, z_{1n}] \ldots Y[v_{n-1}, z_{n-1n}] v_n, \tau), \qquad (61) \end{aligned}$$

where $z_{in} = z_i - z_n$ $(1 \le i \le n-1)$. In particular, $Z_V^{(1)}(v_1, z_1; v_2, z_2; \tau)$ depends only one z_{12}, and we denote this 2-point function by

$$\begin{aligned} Z_V^{(1)}(v_1, v_2, z_{12}, \tau) &= Z_V^{(1)}(v_1, z_1; v_2, z_2; \tau) \\ &= \mathrm{Tr}_V \left(o(Y[v_1, z_{12}] v_2) q^{L(0)}\right). \end{aligned} \tag{62}$$

Now consider a torus obtained by self-sewing a Riemann sphere with punctures located at the origin and an arbitrary point w on the complex plane (cf. [18, Sect. 5.2.2]). Choose local coordinates z_1 in the neighborhood of the origin and $z_2 = z - w$ for z in the neighborhood of w. For a complex sewing parameter ρ, identify the annuli $|\rho| r_{\bar{a}}^{-1} \le |z_a| \le r_a$ for $1 \le a \le 2$ and $|\rho| \le r_1 r_2$ via the sewing relation

$$z_1 z_2 = \rho. \tag{63}$$

Define

$$\chi = -\frac{\rho}{w^2}. \tag{64}$$

Then the annuli do not intersect provided $|\chi| < \frac{1}{4}$, and the torus modular parameter is

$$q = f(\chi), \tag{65}$$

where $f(\chi)$ is the Catalan series

$$f(\chi) = \frac{1 - \sqrt{1 - 4\chi}}{2\chi} - 1 = \sum_{n \ge 1} \frac{1}{n} \binom{2n}{n+1} \chi^n. \tag{66}$$

$f = f(\chi)$ satisfies $f = \chi(1+f)^2$ and the following identity, which can be proved by induction on m:

Lemma 2. *$f(\chi)$ satisfies*

$$f(\chi)^m = \sum_{n\geq m} \frac{m}{n}\binom{2n}{n+m}\chi^n \quad (m \geq 1). \quad \square$$

We now define the genus one partition function in the ρ-sewing scheme (63) by

$$Z^{(1)}_{V,\rho}(\rho, w) = \sum_{n\geq 0}\rho^n \sum_{u\in V_n} \mathrm{Res}_{z_2=0} z_2^{-1} \mathrm{Res}_{z_1=0} z_1^{-1} \left\langle \mathbf{1}, Y(u, w+z_2)Y(\overline{u}, z_1)\mathbf{1}\right\rangle, \tag{67}$$

where the inner sum is taken over any basis for V_n. This partition function is directly related to the standard one $Z^{(1)}_V(q) = \mathrm{Tr}_V\left(q^{L(0)-c/24}\right)$ as follows:

Theorem 6. *In the sewing scheme (63), we have*

$$Z^{(1)}_{V,\rho}(\rho, w) = q^{c/24} Z^{(1)}_V(q), \tag{68}$$

where $q = f(\chi)$ is given by (65).

Proof. The summand in (67) for $u \in V_n$ is

$$\begin{aligned}\langle \mathbf{1}, Y(u, w)\overline{u}\rangle &= \left\langle Y^{\dagger}(u, w)\mathbf{1}, \overline{u}\right\rangle \\ &= (-w^{-2})^n \left\langle Y(e^{wL(1)}u, w^{-1})\mathbf{1}, \overline{u}\right\rangle \\ &= (-w^{-2})^n \left\langle e^{w^{-1}L(-1)} e^{wL(1)}u, \overline{u}\right\rangle,\end{aligned}$$

where we have used (45) and $Y(v,z)\mathbf{1} = \exp(zL(-1))v$ (e.g [13,22,23]). Hence we find that

$$\begin{aligned}Z^{(1)}_{V,\rho}(\rho, w) &= \sum_{n\geq 0}\left(-\frac{\rho}{w^2}\right)^n \sum_{u\in V_n}\left\langle e^{w^{-1}L(-1)} e^{wL(1)}u, \overline{u}\right\rangle \\ &= \sum_{n\geq 0}\chi^n\, \mathrm{Tr}_{V_n}\left(e^{w^{-1}L(-1)} e^{wL(1)}\right).\end{aligned}$$

Expanding the exponentials yields

$$Z^{(1)}_{V,\rho}(\rho, w) = \mathrm{Tr}_V\left(\sum_{r\geq 0}\frac{L(-1)^r L(1)^r}{(r!)^2}\chi^{L(0)}\right), \tag{69}$$

an expression which depends only on χ.

In order to compute (69) we consider the quasi-primary decomposition of V. Let $Q_m = \{v \in V_m | L(1)v = 0\}$ denote the space of quasiprimary states of weight $m \geq 1$. Then $\dim Q_m = p_m - p_{m-1}$ with $p_m = \dim V_m$. Consider the decomposition of V into $L(-1)$-descendents of quasi-primaries

$$V_n = \bigoplus_{m=1}^{n} L(-1)^{n-m} Q_m. \tag{70}$$

Lemma 3. *Let $v \in Q_m$ for $m \geq 1$. For an integer $n \geq m$,*

$$\sum_{r \geq 0} \frac{L(-1)^r L(1)^r}{(r!)^2} L(-1)^{n-m} v = \binom{2n-1}{n-m} L(-1)^{n-m} v.$$

Proof. First use induction on $t \geq 0$ to show that

$$L(1)L(-1)^t v = t(2m+t-1)L(-1)^{t-1} v.$$

Then by induction in r it follows that

$$\frac{L(-1)^r L(1)^r}{(r!)^2} L(-1)^{n-m} v = \binom{n-m}{r}\binom{n+m-1}{r} L(-1)^{n-m} v.$$

Hence

$$\sum_{r \geq 0} \frac{L(-1)^r L(1)^r}{(r!)^2} L(-1)^{n-m} v = \sum_{r \geq 0}^{n-m} \binom{n-m}{r}\binom{n+m-1}{r} L(-1)^{n-m} v,$$
$$= \binom{2n-1}{n-m} L(-1)^{n-m} v,$$

where the last equality follows from a comparison of the coefficient of x^{n-m} in the identity $(1+x)^{n-m}(1+x)^{n+m-1} = (1+x)^{2n-1}$. □

Lemma 3 and (70) imply that for $n \geq 1$,

$$\mathrm{Tr}_{V_n}\left(\sum_{r \geq 0} \frac{L(-1)^r L(1)^r}{(r!)^2}\right) = \sum_{m=1}^{n} \mathrm{Tr}_{Q_m}\left(\sum_{r \geq 0} \frac{L(-1)^r L(1)^r}{(r!)^2} L(-1)^{n-m}\right)$$
$$= \sum_{m=1}^{n} (p_m - p_{m-1}) \binom{2n-1}{n-m}.$$

The coefficient of p_m is

$$\binom{2n-1}{n-m} - \binom{2n-1}{n-m-1} = \frac{m}{n}\binom{2n}{m+n},$$

and hence

$$\mathrm{Tr}_{V_n}\left(\sum_{r\geq 0}\frac{L(-1)^r L(1)^r}{(r!)^2}\right) = \sum_{m=1}^{n}\frac{m}{n}\binom{2n}{m+n}p_m.$$

Using Lemma 2, we find that

$$\begin{aligned}
Z^{(1)}_{V,\rho}(\rho,w) &= 1+\sum_{n\geq 1}\chi^n\sum_{m=1}^{n}\frac{m}{n}\binom{2n}{m+n}p_m,\\
&= 1+\sum_{m\geq 1}p_m\sum_{n\geq m}\frac{m}{n}\binom{2n}{m+n}\chi^n\\
&= 1+\sum_{m\geq 1}p_m(f(\chi))^m\\
&= \mathrm{Tr}_V\left(f(\chi)^{L(0)}\right),
\end{aligned}$$

and Theorem 6 follows. □

4.2 Genus Two

We now turn to the case of genus two. Following Sect. 2.2, we employ the ρ-sewing scheme to self-sew a torus $\mathcal{S}$ with modular parameter τ via the sewing relation (17). For $x_1,\ldots,x_n \in \mathcal{S}$ with $|x_i| \geq |\epsilon|/r_2$ and $|x_i - w| \geq |\epsilon|/r_1$, we define the genus two n-point function in the ρ-formalism by

$$\begin{aligned}
&Z^{(2)}_V(v_1,x_1;\ldots v_n,x_n;\tau,w,\rho) = \\
&\sum_{r\geq 0}\rho^r\sum_{u\in V_{[r]}}\mathrm{Res}_{z_1=0}z_1^{-1}\mathrm{Res}_{z_2=0}z_2^{-1}Z^{(1)}_V(\bar{u},w+z_2;v_1,x_1;\ldots v_n,x_n;u,z_1;\tau), \quad (71)
\end{aligned}$$

where the inner sum is taken over any basis for $V_{[r]}$. In particular, with the notation (62), the genus two partition function is

$$Z_V^{(2)}(\tau, w, \rho) = \sum_{n\geq 0} \rho^n \sum_{u\in V_{[n]}} Z_V^{(1)}(\overline{u}, u, w, \tau). \tag{72}$$

Next we consider $Z_V^{(2)}(\tau, w, \rho)$ in the two-tori degeneration limit. Define, much as in (64),

$$\chi = -\frac{\rho}{w^2}, \tag{73}$$

where w denotes a point on the torus and ρ is the genus two sewing parameter. Then one finds that the two-tori degeneration limit is given by $\rho, w \to 0$ for fixed χ, where

$$\Omega \to \begin{pmatrix} \tau & 0 \\ 0 & \frac{1}{2\pi \mathrm{i}} \log(f(\chi)) \end{pmatrix} \tag{74}$$

and $f(\chi)$ is the Catalan series (66) (cf. [18, Sect. 6.4]).

Theorem 7. *For fixed $|\chi| < \frac{1}{4}$, we have*

$$\lim_{w,\rho\to 0} Z_V^{(2)}(\tau, w, \rho) = f(\chi)^{c/24} Z_V^{(1)}(q) Z_V^{(1)}(f(\chi)).$$

Proof. By (62) we have

$$Z_V^{(1)}(\overline{u}, u, w, \tau) = \mathrm{Tr}_V \left(o(Y[\overline{u}, w]u)q^{L(0)}\right),$$

where $u \in V_{[n]}$. Using the non-degeneracy of the LiZ metric $\langle\ ,\ \rangle_{\mathrm{sq}}$ in the square bracket formalism we obtain

$$Y[\overline{u}, w]u = \sum_{m\geq 0} \sum_{v\in V_{[m]}} \langle \overline{v}, Y[\overline{u}, w]u\rangle_{\mathrm{sq}}\, v,$$

summing over any basis for $V_{[m]}$. Arguing much as in the first part of the proof of Theorem 6, we also find

$$\begin{aligned}
\langle \overline{v}, Y[\overline{u}, w]u\rangle_{\mathrm{sq}} &= (-w^{-2})^n \left\langle Y[\mathrm{e}^{wL[1]}\overline{u}, w^{-1}]\overline{v}, u\right\rangle_{\mathrm{sq}} \\
&= (-w^{-2})^n \left\langle \mathrm{e}^{w^{-1}L[-1]} Y[\overline{v}, -w^{-1}]\mathrm{e}^{wL[1]}\overline{u}, u\right\rangle_{\mathrm{sq}} \\
&= (-w^{-2})^n \left\langle E[\overline{v}, w]\overline{u}, u\right\rangle_{\mathrm{sq}},
\end{aligned}$$

where

$$E[\overline{v}, w] = \exp\left(w^{-1}L[-1]\right) Y[\overline{v}, -w^{-1}] \exp(wL[1]).$$

Hence

$$Z_V^{(2)}(\tau,w,\rho) = \sum_{m\geq 0}\sum_{v\in V_{[m]}}\sum_{n\geq 0}\chi^n \sum_{u\in V[n]} \langle E[\overline{v},w]\overline{u},u\rangle\, Z_V^{(1)}(v,q)$$
$$= \sum_{m\geq 0}\sum_{v\in V_{[m]}} \mathrm{Tr}_V\left(E[\overline{v},w]\chi^{L[0]}\right) Z_V^{(1)}(v,q).$$

Now consider

$$\mathrm{Tr}_V\left(E[\overline{v},w]\chi^{L[0]}\right) =$$
$$w^m \sum_{r,s\geq 0}(-1)^{r+m}\frac{1}{r!s!}\,\mathrm{Tr}_V\left(L[-1]^r\overline{v}[r-s-m-1]L[1]^s\chi^{L[0]}\right).$$

The leading term in w is w^0 (arising from $\overline{v} = \mathbf{1}$) and is given by

$$\mathrm{Tr}_V\left(E[\mathbf{1},w]\chi^{L[0]}\right) = f(\chi)^{c/24} Z_V^{(1)}(f(\chi)).$$

This follows from (69) and the isomorphism between the original and square bracket formalisms. Taking $w\to 0$ for fixed χ the result follows. □

5 The Heisenberg VOA

In this section we compute the genus two partition function in the ρ-formalism for the rank $l = 1$ Heisenberg VOA M. We also compute the genus two n-point function for n copies of the Heisenberg vector a and the genus two one-point function for the Virasoro vector ω. The main results mirror those obtained in the ϵ-formalism in Sect. 6 of [20].

5.1 *The Genus Two Partition Function* $Z_M^{(2)}(\tau,w,\rho)$

We begin by establishing a formula for $Z_M^{(2)}(\tau,w,\rho)$ in terms of the infinite matrix R (19). Recalling that the genus zero partition function is $Z_M^{(1)}(\tau) = 1/\eta(\tau)$ where $\eta(\tau) = q^{1/24}\prod_{n\geq 1}(1-q^n)$ is the Dedekind η-function, we find

Theorem 8. *We have*

$$Z_M^{(2)}(\tau,w,\rho) = \frac{Z_M^{(1)}(\tau)}{\det(1-R)^{1/2}}. \tag{75}$$

Remark 1. From Remark 2 of [20] it follows that the genus two partition function for l free bosons M^l is just the lth power of (75).

Proof. The proof is similar in structure to that of Theorem 5 of [20]. From (72) we have

$$Z_M^{(2)}(\tau, w, \rho) = \sum_{n \geq 0} \sum_{u \in M_{[n]}} Z_M^{(1)}(u, \overline{u}, w, \tau)\, \rho^n, \tag{76}$$

where u ranges over any basis of $M_{[n]}$ and $\overline{u}$ is the dual state with respect to the square-bracket LiZ metric. $Z_M^{(1)}(u, v, w, \tau)$ is a genus one Heisenberg 2-point function (62). We choose the square bracket Fock basis:

$$v = a[-1]^{e_1} \ldots a[-p]^{e_p}\mathbf{1}. \tag{77}$$

The Fock state v naturally corresponds to an unrestricted partition $\lambda = \{1^{e_1} \ldots p^{e_p}\}$ of $n = \sum_{1 \leq i \leq p} i e_i$. We write $v = v(\lambda)$ to indicate this correspondence. The Fock vectors form an orthogonal set from (54) with

$$\overline{v}(\lambda) = \frac{1}{\prod_{1 \leq i \leq p}(-i)^{e_i} e_i!} v(\lambda).$$

The 2-point function $Z_M^{(1)}(v(\lambda), v(\lambda), w, \tau)$ is given in Corollary 1 of [17] where it is denoted by $F_M(v, w_1, v, w_2; \tau)$. In order to describe this explicitly we introduce the set $\Phi_{\lambda,2}$ which is the disjoint union of two isomorphic label sets $\Phi_\lambda^{(1)}$, $\Phi_\lambda^{(2)}$ each with e_i elements labelled i determined by λ. Let $\iota : \Phi_\lambda^{(1)} \leftrightarrow \Phi_\lambda^{(2)}$ denote the canonical label identification. Then we have (loc. cit.)

$$Z_M^{(1)}(v(\lambda), v(\lambda), w, \tau) = Z_M^{(1)}(\tau) \sum_{\phi \in F(\Phi_{\lambda,2})} \Gamma(\phi), \tag{78}$$

where

$$\Gamma(\phi, w, \tau) = \Gamma(\phi) = \prod_{\{r,s\}} \xi(r, s, w, \tau), \tag{79}$$

and ϕ ranges over the elements of $F(\Phi_{\lambda,2})$, the fixed-point-free involutions in $\Sigma(\Phi_{\lambda,2})$ and where $\{r, s\}$ ranges over the orbits of ϕ on $\Phi_{\lambda,2}$. Finally

$$\xi(r, s) = \xi(r, s, w, \tau) = \begin{cases} C(r, s, \tau), & \text{if } \{r, s\} \subseteq \Phi_\lambda^{(a)}, a = 1 \text{ or } 2, \\ D(r, s, w_{ab}, \tau) & \text{if } r \in \Phi_\lambda^{(a)}, s \in \Phi_\lambda^{(b)}, a \neq b. \end{cases},$$

where $w_{12} = w_1 - w_2 = w$ and $w_{21} = w_2 - w_1 = -w$.

Remark 2. Note that ξ is well-defined since $D(r, s, w_{ab}, \tau) = D(s, r, w_{ba}, \tau)$.

Fig. 3 A doubly-indexed edge

Using the expression (78), it follows that the genus two partition function (76) can be expressed as

$$Z_M^{(2)}(\tau, w, \rho) = Z_M^{(1)}(\tau) \sum_{\lambda=\{i^{e_i}\}} \frac{E(\lambda)}{\prod_i (-i)^{e_i} e_i!} \rho^{\sum i e_i}, \tag{80}$$

where λ runs over all unrestricted partitions and

$$E(\lambda) = \sum_{\phi \in F(\Phi_{\lambda,2})} \Gamma(\phi). \tag{81}$$

We employ the doubly-indexed diagrams of Sect. 2.3. Consider the 'canonical' matching defined by ι as a fixed-point-free involution. We may then compose ι with each fixed-point-free involution $\phi \in F(\Phi_{\lambda,2})$ to define a 1-1 mapping $\iota\phi$ on the underlying labelled set $\Phi_{\lambda,2}$. For each ϕ we define a doubly-indexed diagram D whose nodes are labelled by k, a for an element $k \in \Phi_\lambda^{(a)}$ for $a = 1, 2$ and with cycles corresponding to the orbits of the cyclic group $\langle \iota\phi \rangle$. Thus, if $l = \phi(k)$ for $k \in \Phi_\lambda^{(a)}$ and $l \in \Phi_\lambda^{(\bar{b})}$ and $\iota : \bar{b} \mapsto b$ with convention (16) then the corresponding doubly-indexed diagram contains the edge (Fig. 3).
Consider the permutations of $\Phi_{\lambda,2}$ that commute with ι and preserve both $\Phi_\lambda^{(1)}$ and $\Phi_\lambda^{(2)}$. We denote this group, which is plainly isomorphic to $\Sigma(\Phi_\lambda)$, by Δ_λ. By definition, an automorphism of a doubly-indexed diagram D in the above sense is an element of Δ_λ which preserves edges and node labels.

For a doubly-indexed diagram D corresponding to the partition $\lambda = \{1^{e_1} \ldots p^{e_p}\}$ we set

$$\gamma(D) = \frac{\prod_{\{k,l\}} \xi(k, l, w, \tau)}{\prod (-i)^{e_i}} \rho^{\sum i e_i} \tag{82}$$

where $\{k, l\}$ ranges over the edges of D. We now have all the pieces assembled to copy the arguments used to prove Theorem 5 of [20]. First we find

$$\sum_{\lambda=\{i^{e_i}\}} \frac{E(\lambda)}{\prod_i (-i)^{e_i} e_i!} \rho^{\sum i e_i} = \sum_D \frac{\gamma(D)}{|\mathrm{Aut}(D)|}, \tag{83}$$

the sum ranging over all doubly-indexed diagrams.

We next introduce a weight function ζ as follows: for a doubly-indexed diagram D we set $\zeta(D) = \prod \zeta(E)$, the product running over all edges. Moreover for an edge E with nodes labelled (k, a) and (l, b) as in Fig. 3, we set

$$\zeta(E) = R_{ab}(k,l),$$

for R of (19). We then find

Lemma 4. $\zeta(D) = \gamma(D)$.

Proof. From (82) it follows that for a doubly-indexed diagram D we have

$$\gamma(D) = \prod_{\{k,l\}} -\frac{\xi(k,l,w,\tau)\rho^{(k+l)/2}}{\sqrt{kl}}, \tag{84}$$

the product ranging over the edges $\{k,l\}$ of D. So to prove the lemma it suffices to show that if k,l lie in $\Phi_\lambda^{(a)}, \Phi_\lambda^{(\bar{b})}$ respectively then the (a,b)-entry of $R(k,l)$ coincides with the corresponding factor of (84). This follows from our previous discussion together with Remark 2. □

From Lemma 4 and following similar arguments to the proof of Theorem 5 of [20] we find

$$\begin{aligned}\sum_D \frac{\gamma(D)}{|\mathrm{Aut}(D)|} &= \sum_D \frac{\zeta(D)}{|\mathrm{Aut}(D)|} \\ &= \exp\left(\sum_L \frac{\zeta(L)}{|\mathrm{Aut}(L)|}\right),\end{aligned}$$

where L denotes the set of non-isomorphic unoriented doubly indexed cycles. Orient these cycles, say in a clockwise direction. Let $\{M\}$ denote the set of non-isomorphic oriented doubly indexed cycles and $\{M_n\}$ the oriented cycles with n nodes. Then we find (cf. [20, Lemma 2]) that

$$\frac{1}{n}\mathrm{Tr}R^n = \sum_{M_n} \frac{\zeta(M_n)}{|\mathrm{Aut}(M_n)|}.$$

It follows that

$$\begin{aligned}\sum_L \frac{\zeta(L)}{|\mathrm{Aut}(L)|} &= \frac{1}{2}\sum_M \frac{\zeta(M)}{|\mathrm{Aut}(M)|} \\ &= \frac{1}{2}\mathrm{Tr}\left(\sum_{n\geq 1}\frac{1}{n}R^n\right) \\ &= -\frac{1}{2}\mathrm{Tr}\log(I-R) \\ &= -\frac{1}{2}\log\det(I-R).\end{aligned}$$

This completes the proof of Theorem 8. □

We may also find a product formula analogous to Theorem 6 of [20]. Let $\mathcal{R}$ denote the rotationless doubly-indexed oriented cycles i.e. cycles with trivial automorphism group. Then we find

Theorem 9.

$$Z_M^{(2)}(\tau, w, \rho) = \frac{Z_M^{(1)}(\tau)}{\prod_{\mathcal{R}}(1 - \zeta(N))^{1/2}}. \quad \square \tag{85}$$

5.2 *Holomorphic and Modular-Invariance Properties*

In Sect. 2.2 we reviewed the genus two ρ-sewing formalism and introduced the domain $\mathcal{D}^\rho$ which parametrizes the genus two surface. An immediate consequence of Theorem 1 is the following.

Theorem 10. *$Z_M^{(2)}(\tau, w, \rho)$ is holomorphic in $\mathcal{D}^\rho$.* $\square$

We next consider the invariance properties of the genus two partition function with respect to the action of the $\mathcal{D}^\rho$-preserving group Γ_1 reviewed in Sect. 2.2. Let χ be the character of $\mathrm{SL}(2,\mathbb{Z})$ defined by its action on $\eta(\tau)^{-2}$, i.e.

$$\eta(\gamma\tau)^{-2} = \chi(\gamma)\eta(\tau)^{-2}(c\tau + d)^{-1}, \tag{86}$$

where $\gamma = \begin{pmatrix} a & b \\ c & d \end{pmatrix} \in \mathrm{SL}(2,\mathbb{Z})$. Recall (e.g. [30]) that $\chi(\gamma)$ is a twelfth root of unity. For a function $f(\tau)$ on $\mathbb{H}_1, k \in \mathbb{Z}$ and $\gamma \in \mathrm{SL}(2,\mathbb{Z})$, we define

$$f(\tau)|_k\gamma = f(\gamma\tau)\,(c\tau + d)^{-k}, \tag{87}$$

so that

$$Z_{M^2}^{(1)}(\tau)|_{-1}\gamma = \chi(\gamma)Z_{M^2}^{(1)}(\tau). \tag{88}$$

At genus two, analogously to (87), we define

$$f(\tau, w, \rho)|_k\gamma = f(\gamma(\tau, w, \rho))\det(C\Omega + D)^{-k}. \tag{89}$$

Here, the action of γ on the right-hand-side is as in (18). We have abused notation by adopting the following conventions in (89), which we continue to use below:

$$\Omega = F^\rho(\tau, w, \rho),\ \gamma = \begin{pmatrix} A & B \\ C & D \end{pmatrix} \in \mathrm{Sp}(4,\mathbb{Z}) \tag{90}$$

where F^{ρ} is as in Theorem 4, and γ is identified with an element of $\mathrm{Sp}(4,\mathbb{Z})$ via (32) and (33). Note that (89) defines a right action of G on functions $f(\tau,w,\rho)$. We then have a natural analog of Theorem 8 of [20]

Theorem 11. *If* $\gamma \in \Gamma_1$ *then*

$$Z^{(2)}_{M^2}(\tau,w,\rho)|_{-1}\gamma = \chi(\gamma)Z^{(2)}_{M^2}(\tau,w,\rho).$$

Corollary 2. *If* $\gamma \in \Gamma_1$ *with* $Z^{(2)}_{M^{24}} = (Z^{(2)}_{M^2})^{12}$ *then*

$$Z^{(2)}_{M^{24}}(\tau,w,\rho)|_{-12}\gamma = Z^{(2)}_{M^{24}}(\tau,w,\rho).$$

Proof. The proof is similar to that of Theorem 8 of [20]. We have to show that

$$Z^{(2)}_{M^2}(\gamma.(\tau,w,\rho))\det(C\Omega + D) = \chi(\gamma)Z^{(2)}_{M^2}(\tau,w,\rho) \tag{91}$$

for $\gamma \in \Gamma_1$ where $\det(C\Omega_{11}+D) = c_1\Omega_{11}+d_1$. Consider the determinant formula (75). For $\gamma \in \Gamma_1$ define

$$R'_{ab}(k,l,\tau,w,\rho) = R_{ab}\left(k,l,\frac{a_1\tau+b_1}{c_1\tau+d_1},\frac{w}{c_1\tau+d_1},\frac{\rho}{(c_1\tau+d_1)^2}\right)$$

following (31). We find from Sect. 6.3 of [18] that

$$\begin{aligned} 1-R' &= 1-R-\kappa\Delta \\ &= (1-\kappa S).(1-R), \end{aligned}$$

where

$$\begin{aligned} \Delta_{ab}(k,l) &= \delta_{k1}\delta_{l1}, \\ \kappa &= \frac{\rho}{2\pi \mathrm{i}}\frac{c_1}{c_1\tau+d_1}, \\ S_{ab}(k,l) &= \delta_{k1}\sum_{c\in\{1,2\}}\left((1-R)^{-1}\right)_{cb}(1,l). \end{aligned}$$

Since $\det(1-R)$ and $\det(1-R')$ are convergent on $\mathcal{D}^{\rho}$ we find

$$\det(1-R') = \det(1-\kappa S).\det(1-R).$$

Indexing the columns and rows by $(a,k) = (1,1),(2,1),\ldots(1,k),(2,k)\ldots$ and noting that $S_{1b}(k,l) = S_{2b}(k,l)$ we find that

$$\det(1-\kappa S) = \begin{vmatrix} 1-\kappa S_{11}(1,1) & -\kappa S_{12}(1,1) & -\kappa S_{11}(1,2) & \cdots \\ -\kappa S_{11}(1,1) & 1-\kappa S_{12}(1,1) & -\kappa S_{11}(1,2) & \cdots \\ 0 & 0 & 1 & \cdots \\ \vdots & \vdots & \vdots & \ddots \end{vmatrix}$$

$$= 1-\kappa S_{11}(1,1) - \kappa S_{12}(1,1),$$

$$= 1-\kappa\sigma\left(\left(1-R\right)^{-1}\right)(1,1)),$$

where $\sigma(M)$ denotes the finite sum over the block labels for a 2×2 block matrix M. Applying (27), it is clear that

$$\det(1-\kappa S) = \frac{c_1\Omega_{11}+d_1}{c_1\tau+d_1}.$$

The theorem follows from (88). □

Remark 3. $Z^{(2)}_{M^2}(\tau,w,\rho)$ can be trivially considered as function on the covering space $\hat{\mathcal{D}}^\rho$ discussed in [18, Sect. 6.3]. Then $Z^{(2)}_{M^2}(\tau,w,\rho)$ is modular with respect to $L=\hat{H}\Gamma_1$ with trivial invariance under the action of the Heisenberg group $\hat{H}$ (loc. cit.).

5.3 Some Genus Two n-Point Functions

In this section we calculate some examples of genus two n-point functions for the rank one Heisenberg VOA M. We consider here the examples of the n-point function for the Heisenberg vector a and the 1-point function for the Virasoro vector $\tilde{\omega}$. We find that, up to an overall factor of the partition function, the formal differential form associated with the Heisenberg n-point function is described in terms of the global symmetric two form ω [33] whereas the Virasoro 1-point function is described by the genus two projective connection [11]. These results agree with those found in [20] in the ϵ-formalism up to an overall ϵ-formalism partition function factor.

The genus two Heisenberg vector 1-point function with the Heisenberg vector a inserted at x is $Z^{(2)}_M(a,x;\tau,w,\rho)=0$ since $Z^{(1)}_M(Y[a,x]Y[v,w]v,\tau)=0$ from [17]. The 2-point function for two Heisenberg vectors inserted at x_1,x_2 is

$$Z^{(2)}_M(a,x_1;a,x_2;\tau,w,\rho) = \sum_{r\geq 0}\rho^r\sum_{v\in M_{[r]}} Z^{(1)}_M(a,x_1;a,x_2;v,w_1,\overline{v},w_2;\tau). \quad (92)$$

We consider the associated formal differential form

$$\mathcal{F}^{(2)}_M(a,a;\tau,w,\rho) = Z^{(2)}_M(a,x_1;a,x_2;\tau,w,\rho)\,dx_1dx_2, \quad (93)$$

and find that it is determined by the bilinear form ω (12):

Theorem 12. *The genus two Heisenberg vector 2-point function is given by*

$$\mathcal{F}_M^{(2)}(a,a;\tau,w,\rho) = \omega(x_1,x_2) Z_M^{(2)}(\tau,w,\rho). \tag{94}$$

Proof. The proof proceeds along the same lines as Theorem 8. As before, we let $v(\lambda)$ denote a Heisenberg Fock vector (77) determined by an unrestricted partition $\lambda = \{1^{e_1} \ldots p^{e_p}\}$ with label set Φ_λ. Define a label set for the four vectors $a, a, v(\lambda), v(\lambda)$ given by $\Phi = \Phi_1 \cup \Phi_2 \cup \Phi_\lambda^{(1)} \cup \Phi_\lambda^{(2)}$ for $\Phi_1, \Phi_2 = \{1\}$ and let $F(\Phi)$ denote the set of fixed point free involutions on Φ. For $\phi = \ldots (rs) \ldots \in F(\Phi)$ let $\Gamma(x_1,x_2,\phi) = \prod_{(r,s)} \xi(r,s)$ as defined in (80) for $r,s \in \Phi_\lambda^{(2)} = \Phi_\lambda^{(1)} \cup \Phi_\lambda^{(2)}$ and

$$\xi(r,s) = \begin{cases} D(1,1,x_i - x_j,\tau) = P_2(\tau, x_i - x_j), & r,s \in \Phi_i,\ i \neq j, \\ D(1,s,x_i - w_a,\tau) = s P_{s+1}(\tau, x_i - w_a), & r \in \Phi_i,\ s \in \Phi_\lambda^{(a)}, \end{cases} \tag{95}$$

for $i, j, a \in \{1,2\}$ with D of (11). Then following Corollary 1 of [17] we have

$$Z_M^{(1)}(a,x_1;a,x_2;v(\lambda),w_1;v(\lambda),w_2;\tau) = Z_M^{(1)}(\tau) \sum_{\phi \in F(\Phi)} \Gamma(x_1,x_2,\phi).$$

We then obtain the following analog of (80)

$$\mathcal{F}^{(2)}(a,a;\tau,w,\rho) = Z_M^{(1)}(\tau) \sum_{\lambda=\{i^{e_i}\}} \frac{E(x_1,x_2,\lambda)}{\prod_i (-i)^{e_i} e_i!} \rho^{\sum i e_i} dx_1 dx_2, \tag{96}$$

where

$$E(x_1,x_2,\lambda) = \sum_{\phi \in F(\Phi)} \Gamma(x_1,x_2,\phi).$$

The sum in (96) can be re-expressed as the sum of weights $\zeta(D)$ for isomorphism classes of doubly-indexed configurations D where here D includes two distinguished valency one nodes labelled x_i (see Sect. 2.3) corresponding to the label sets $\Phi_1, \Phi_2 = \{1\}$. As before, $\zeta(D) = \prod_E \zeta(E)$ for standard doubly-indexed edges E augmented by the contributions from edges connected to the two valency one nodes with weights as in (37). Thus we find

$$\mathcal{F}^{(2)}(a,a;\tau,w,\rho) = Z_M^{(1)}(\tau) \sum_D \frac{\zeta(D)}{\prod_i e_i!} dx_1 dx_2,$$

Each D can be decomposed into *exactly* one necklace configuration N of type $\mathcal{N}(x;y)$ of (38) connecting the two distinguished nodes and a standard configuration $\hat{D}$ of the type appearing in the proof of Theorem 8 with $\zeta(D) = \zeta(N)\zeta(\hat{D})$. Since $|\mathrm{Aut}(N)| = 1$ we obtain

$$\begin{aligned}\mathcal{F}^{(2)}(a,a;\tau,w,\rho) &= Z_M^{(1)}(\tau)\sum_{\hat{D}}\frac{\zeta(\hat{D})}{|\mathrm{Aut}(\hat{D})|}\sum_{N\in\mathcal{N}(x;y)}\zeta(N)\\ &= Z_M^{(2)}(\tau,w,\rho)\zeta(x_1;x_2)\\ &= Z_M^{(2)}(\tau,w,\rho)\omega(x_1,x_2),\end{aligned}$$

using (41) of Proposition 2. □

Theorem 12 can be generalized to compute the n-point function corresponding to the insertion of n Heisenberg vectors. We find that it vanishes for n odd, and for n even is determined by the symmetric tensor

$$\mathrm{Sym}_n\,\omega = \sum_{\psi}\prod_{(r,s)}\omega(x_r,x_s), \tag{97}$$

where the sum is taken over the set of fixed point free involutions $\psi = \ldots(rs)\ldots$ of the labels $\{1,\ldots,n\}$. We then have

Theorem 13. *The genus two Heisenberg vector n-point function vanishes for odd n even; for even n it is given by the global symmetric meromorphic n-form:*

$$\frac{\mathcal{F}_M^{(2)}(a,\ldots,a;\tau,w,\rho)}{Z_M^{(2)}(\tau,w,\rho)} = \mathrm{Sym}_n\,\omega. \tag{98}$$

□

This agrees with the corresponding ratio in Theorem 10 of [20] in the ϵ-formalism, and also with earlier results in [33] which assume an analytic structure for the n-point function.

Using this result and the associativity of vertex operators, we can compute all n-point functions. In particular, the 1-point function for the Virasoro vector $\tilde{\omega} = \frac{1}{2}a[-1]a$ is as follows (cf. [20], Proposition 8):

Proposition 3. *The genus two 1-point function for the Virasoro vector $\tilde{\omega}$ inserted at x is given by*

$$\frac{\mathcal{F}_M^{(2)}(\tilde{\omega};\tau,w,\rho)}{Z_M^{(2)}(\tau,w,\rho)} = \frac{1}{12}s^{(2)}(x), \tag{99}$$

where $s^{(2)}(x) = 6\lim_{x\to y}\left(\omega(x,y) - \frac{dxdy}{(x-y)^2}\right)$ is the genus two projective connection [11]. □

6 Lattice VOAs

6.1 The Genus Two Partition Function $Z^{(2)}_{V_L}(\tau, w, \rho)$

Let L be an even lattice with V_L the corresponding lattice theory vertex operator algebra. The underlying Fock space is

$$V_L = M^l \otimes C[L] = \oplus_{\beta \in L} M^l \otimes \mathrm{e}^{\beta}, \tag{100}$$

where M^l is the corresponding Heisenberg free boson theory of rank $l = \dim L$ based on $H = C \otimes_Z L$. We follow Sect. 3.1 and [17] concerning further notation for lattice theories. We utilize the Fock basis $\{u \otimes \mathrm{e}^{\beta}\}$ where β ranges over L and u ranges over the usual orthogonal basis for M^l. From Lemma 1 and Corollary 1 we see that

$$Z^{(2)}_{V_L}(\tau, w, \rho) = \sum_{\alpha,\beta \in L} Z^{(2)}_{\alpha,\beta}(\tau, w, \rho), \tag{101}$$

$$Z^{(2)}_{\alpha,\beta}(\tau, w, \rho) = \sum_{n \geq 0} \sum_{u \in M^l_{[n]}} Z^{(1)}_{M^l \otimes \mathrm{e}^{\alpha}}\left(u \otimes \mathrm{e}^{\beta}, \overline{u} \otimes \mathrm{e}^{-\beta}, w, \tau\right) \rho^{n+(\beta,\beta)/2}. \tag{102}$$

The general shape of the 2-point function occurring in (102) is discussed extensively in [17]. By Proposition 1 (op. cit.) it splits as a product

$$\begin{aligned} &Z^{(1)}_{M^l \otimes \mathrm{e}^{\alpha}}\left(u \otimes \mathrm{e}^{\beta}, u \otimes \mathrm{e}^{-\beta}, w, \tau\right) = \\ &Q^{\beta}_{M^l \otimes \mathrm{e}^{\alpha}}(u, u, w, \tau)\, Z^{(1)}_{M^l \otimes \mathrm{e}^{\alpha}}\left(\mathrm{e}^{\beta}, \mathrm{e}^{-\beta}, w, \tau\right), \end{aligned} \tag{103}$$

where we have identified e^{β} with $\mathbf{1} \otimes \mathrm{e}^{\beta}$, and where $Q^{\beta}_{M^l \otimes \mathrm{e}^{\alpha}}$ is a function[6] that we will shortly discuss in greater detail. In [17, Corollary 5] (cf. the Appendix to the present paper) we established also that

$$Z^{(1)}_{M^l \otimes \mathrm{e}^{\alpha}}\left(\mathrm{e}^{\beta}, \mathrm{e}^{-\beta}, w, \tau\right) = \epsilon(\beta, -\beta) q^{(\alpha,\alpha)/2} \frac{\exp((\alpha, \beta) w)}{K(w, \tau)^{(\beta,\beta)}} Z^{(1)}_{M^l}(\tau), \tag{104}$$

where, as usual, we are taking w in place of $z_{12} = z_1 - z_2$. With cocycle choice $\epsilon(\beta, -\beta) = (-1)^{(\beta,\beta)/2}$ (cf. Appendix) we may then rewrite (102) as

[6] Note: in [17] the functional dependence on β, here denoted by a superscript, was omitted.

$$Z^{(2)}_{\alpha,\beta}(\tau,w,\rho) = Z^{(1)}_{M^l}(\tau)\exp\left\{\pi \mathrm{i}\left((\alpha,\alpha)\tau + 2(\alpha,\beta)\frac{w}{2\pi \mathrm{i}} + \frac{(\beta,\beta)}{2\pi \mathrm{i}}\log\left(\frac{-\rho}{K(w,\tau)^2}\right)\right)\right\}$$
$$\cdot \sum_{n\geq 0}\sum_{u\in M^l_{[n]}} Q^{\beta}_{M^l\otimes e^{\alpha}}(u,\overline{u},w,\tau)\rho^n. \tag{105}$$

We note that this expression is, as it should be, independent of the choice of branch for the logarithm function. We are going to establish the *precise* analog of Theorem 14 of [20] as follows:

Theorem 14. *We have*

$$Z^{(2)}_{V_L}(\tau,w,\rho) = Z^{(2)}_{M^l}(\tau,w,\rho)\theta^{(2)}_L(\Omega), \tag{106}$$

where $\theta^{(2)}_L(\Omega)$ is the genus two theta function of L [7].

Proof. We note that

$$\theta^{(2)}_L(\Omega) = \sum_{\alpha,\beta\in L}\exp\left(\pi \mathrm{i}((\alpha,\alpha)\Omega_{11} + 2(\alpha,\beta)\Omega_{12} + (\beta,\beta)\Omega_{22})\right). \tag{107}$$

We first handle the case of rank 1 lattices and then consider the general case. The inner double sum in (105) is the object which requires attention, and we can begin to deal with it along the lines of previous sections. Namely, arguments that we have already used several times show that the double sum may be written in the form

$$\sum_D \frac{\gamma(D)}{|\mathrm{Aut}(D)|} = \exp\left(\frac{1}{2}\sum_{N\in\mathcal{N}}\gamma(N)\right).$$

Here, D ranges over the oriented doubly indexed cycles of Sect. 5, while N ranges over oriented doubly-indexed necklaces $\mathcal{N} = \{\mathcal{N}(k,a;l,b)\}$ of (35). Leaving aside the definition of $\gamma(N)$ for now, we recognize as before that the piece involving only connected diagrams with no end nodes splits off as a factor. Apart from a $Z^{(1)}_M(\tau)$ term this factor is, of course, precisely the expression (83) for M. With these observations, we see from (105) that the following holds:

$$\frac{Z^{(2)}_{\alpha,\beta}(\tau,w,\rho)}{Z^{(2)}_M(\tau,w,\rho)} = \exp\left\{\mathrm{i}\pi\left((\alpha,\alpha)\tau + 2(\alpha,\beta)\frac{w}{2\pi \mathrm{i}} + \frac{(\beta,\beta)}{2\pi \mathrm{i}}\log\left(\frac{-\rho}{K(w,\tau)^2}\right)\right.\right.$$
$$\left.\left. + \frac{1}{2\pi \mathrm{i}}\sum_{N\in\mathcal{N}}\gamma(N)\right)\right\}. \tag{108}$$

To prove Theorem 14, we see from (107) and (108) that it is sufficient to establish that for each pair of lattice elements $\alpha,\beta\in L$, we have

$$\frac{Z^{(2)}_{\alpha,\beta}(\tau,w,\rho)}{Z^{(2)}_{M}(\tau,w,\rho)} = \exp\left(\pi \mathrm{i}\left((\alpha,\alpha)\Omega_{11} + 2(\alpha,\beta)\Omega_{12} + (\beta,\beta)\Omega_{22}\right)\right). \tag{109}$$

Recall the formula for Ω in Proposition 1. In order to reconcile (109) with the formula for Ω, we must carefully consider the expression $\sum_{N\in\mathcal{N}} \gamma(N)$. The function γ is essentially (82), except that we also get contributions from the end nodes which are now present. Suppose that an end node has label $k \in \Phi^{(a)}, a \in \{1,2\}$. Then according to Proposition 1 and display (45) of [17] (cf. (133) of the Appendix to the present paper), the contribution of the end node is equal to

$$\xi_{\alpha,\beta}(k,a) = \xi_{\alpha,\beta}(k,a,\tau,w,\rho) =$$
$$\begin{cases} \frac{\rho^{k/2}}{\sqrt{k}} \left(a, \delta_{k1}\alpha + C(k,0,\tau)\beta + D(k,0,w,\tau)(-\beta)\right), & a = 1 \\ \frac{\rho^{k/2}}{\sqrt{k}} \left(a, \delta_{k1}\alpha + C(k,0,\tau)(-\beta) + D(k,0,-w,\tau)\beta\right), & a = 2 \end{cases} \tag{110}$$

together with a contribution arising from the -1 in the denominator of (82) (we will come back to this point later). Using (cf. [17], displays (6), (11) and (12))

$$\begin{aligned} D(k,0,-w,\tau) &= (-1)^{k+1} P_k(-w,\tau) = -P_k(w,\tau), \\ C(k,0,\tau) &= (-1)^{k+1} E_k(\tau), \end{aligned}$$

we can combine the two possibilities in (110) as follows (recalling that $E_k = 0$ for odd k):

$$\xi_{\alpha,\beta}(k,a) = (a,\alpha)\rho^{1/2}\delta_{k1} + (a,\beta)d_{\overline{a}}(k), \tag{111}$$

where $d_a(k)$ is given by (25). We may then compute the weight for an oriented doubly-indexed necklace $N \in \mathcal{N}(k,a;l,b)$ (35). Let N' denote the oriented necklace from which the two end nodes and edges have been *removed* (we refer to these as *shortened* necklaces). From (111) we see that the total contribution to $\gamma(N)$ is

$$\begin{aligned} -\xi_{\alpha,\beta}(k,a)\xi_{\alpha,\beta}(l,b)\gamma(N') = -\big[&(\alpha,\alpha)\rho\delta_{k1}\delta_{l1} + (\beta,\beta)d_{\overline{a}}(k)d_{\overline{b}}(l) \\ &+(\alpha,\beta)\rho^{1/2}\left(d_{\overline{a}}(k)\delta_{l,1} + d_{\overline{b}}(l)\delta_{k,1}\right)\big]\gamma(N'), \end{aligned} \tag{112}$$

where we note that a sign -1 arises from each *pair* of nodes, as follows from (82).

We next consider the terms in (112) corresponding to (α,α), (α,β) and (β,β) separately, and show that they are precisely the corresponding terms on each side of (109). This will complete the proof of Theorem 14 in the case of rank 1 lattices. From (112), an (α,α) term arises only if the end node weights k,l are both equal to 1. Hence $\sum \gamma(N') = \zeta(1;1)$ (cf. (36)), where the sum ranges over shortened necklaces with end nodes of weight $1 \in \Phi^{(a)}$ and $1 \in \Phi^{(b)}$. Thus using Proposition 1, the total contribution to the right-hand-side of (109) is equal to

$$2\pi i\tau - \rho\zeta(1;1) = 2\pi i\Omega_{11}. \tag{113}$$

Next, from (112) we see that an (α, β)-contribution arises whenever at least one of the end nodes has label 1. If the labels of the end nodes are unequal then the shortened necklace with the *opposite* orientation makes an equal contribution. The upshot is that we may assume that the end node to the right of the shortened necklace has label $l = 1 \in \Phi^{(\overline{b})}$, as long as we count accordingly. We thus find $\sum \gamma(N') = \zeta(d;1)$ (cf. (36)), where the sum ranges over shortened necklaces with end nodes of weight $k \in \Phi^{(a)}$ and $1 \in \Phi^{(b)}$. Then Proposition 1 implies that the total contribution to the (α, β) term on the right-hand-side of (109) is

$$2w - 2\rho^{1/2}\zeta(d;1) = 2\Omega_{12},$$

as required.

It remains to deal with the (β, β) term, the details of which are very much along the lines as the case (α, β) just handled. A similar argument shows that the contribution to the (β, β)-term from (112) is equal to the expression $-\zeta(d;\overline{d})$ of (36). Thus the total contribution to the (β, β) term on the right-hand-side of (109) is

$$\log\left(\frac{-\rho}{K(w,\tau)^2}\right) - \zeta(d;\overline{d}) = \Omega_{22},$$

as in (29). This completes the proof of Theorem 14 in the rank 1 case.

As for the general case—we adopt the mercy rule and omit details! The reader who has progressed this far will have no difficulty in dealing with the general case, which follows by generalizing the calculations in the rank 1 case just considered. □

The analytic and automorphic properties of $Z^{(2)}_{V_L}(\tau, w, \rho)$ can be deduced from Theorem 14 using the known behaviour of $\theta^{(2)}_L(\Omega)$ and the analogous results for $Z^{(2)}_{M^l}(\tau, w, \rho)$ established in Sect. 5. We simply record

Theorem 15. *$Z^{(2)}_{V_L}(\tau, w, \rho)$ is holomorphic on the domain $\mathcal{D}^{\rho}$.* □

6.2 Some Genus Two n-Point Functions

In this section we consider the genus two n-point functions for n Heisenberg vectors and the 1-point function for the Virasoro vector $\tilde{\omega}$ for a rank l lattice VOA. The results are similar to those of Sect. 5.3 so that detailed proofs will not be given.

Consider the 1-point function for a Heisenberg vector a_i inserted at x. We define the differential 1-form

$$\mathcal{F}^{(2)}_{\alpha,\beta}(a_i;\tau,w,\rho)=\sum_{n\geq 0}\sum_{u\in M^l_{[n]}} Z^{(1)}_{M^l\otimes e^{\alpha}}\left(a_i,x;u\otimes e^{\beta},w;\overline{u}\otimes e^{-\beta},0,\tau\right)\rho^{n+(\beta,\beta)/2}dx. \tag{114}$$

This can be expressed in terms of the genus two holomorphic 1-forms ν_1,ν_2 of (24) in a similar way to Theorem 12 of [20]. Defining

$$\nu_{i,\alpha,\beta}(x)=(a_i,\alpha)\nu_1(x)+(a_i,\beta)\nu_2(x),$$

we find

Theorem 16.

$$\mathcal{F}^{(2)}_{\alpha,\beta}(a_i;\tau,w,\rho)=\nu_{i,\alpha,\beta}(x)Z^{(2)}_{\alpha,\beta}(\tau,w,\rho). \tag{115}$$

Proof. The proof proceeds along the same lines as Theorems 12 and 14 and Theorem 12 of (op. cit.). We find that

$$\mathcal{F}^{(2)}_{\alpha,\beta}(a_i;\tau,w,\rho)=Z^{(1)}_{M}(\tau)\sum_{D}\frac{\gamma(D)}{|\mathrm{Aut}(D)|}dx,$$

where the sum is taken over isomorphism classes of doubly-indexed configurations D where, in this case, each configuration includes one distinguished valence one node labelled by x as in (37). Each D can be decomposed into exactly one necklace configuration of type $\mathcal{N}(x;k,a)$ of (39), standard configurations of the type appearing in Theorem 12 and necklace contributions as in Theorem 106. The result then follows on applying (111) and the graphical expansion for $\nu_1(x),\nu_2(x)$ of (42) and (43). □

Summing over all lattice vectors, we find that the Heisenberg 1-point function vanishes for V_L. Similarly, one can generalize Theorem 13 concerning the n-point function for n Heisenberg vectors $a_{i_1},\ldots,a_{i_n}$. Defining

$$\mathcal{F}^{(2)}_{\alpha,\beta}(a_{i_1},\ldots,a_{i_n};\tau,w,\rho)=\prod_{t=1}^{n}dx_t\sum_{n\geq 0}\sum_{u\in M^l_{[n]}}\rho^{n+(\beta,\beta)/2}\,.$$
$$Z^{(1)}_{M^l\otimes e^{\alpha}}\left(a_{i_1},x_1;\ldots;a_{i_n},x_n;u\otimes e^{\beta},w;\overline{u}\otimes e^{-\beta},0,\tau\right),$$

we obtain the analogue of Theorem 13 of (op. cit.):

Theorem 17.

$$\mathcal{F}^{(2)}_{\alpha,\beta}(a_{i_1},\ldots,a_{i_n};\tau,w,\rho)=\mathrm{Sym}_n\left(\omega,\nu_{i_t,\alpha,\beta}\right)Z^{(2)}_{\alpha,\beta}(\tau,w,\rho), \tag{116}$$

the symmetric product of $\omega(x_r, x_s)$ *and* $\nu_{i_t,\alpha,\beta}(x_t)$ *defined by*

$$\mathrm{Sym}_n\left(\omega, \nu_{i_t,\alpha,\beta}\right) = \sum_{\psi} \prod_{(r,s)} \omega(x_s, x_s) \prod_{(t)} \nu_{i_t,\alpha,\beta}(x_t), \tag{117}$$

where the sum is taken over the set of involutions $\psi = \ldots(ij)\ldots(k)\ldots$ *of the labels* $\{1,\ldots,n\}$. □

We may also compute the genus two 1-point function for the Virasoro vector $\tilde{\omega} = \frac{1}{2}\sum_{i=1}^{l} a_i[-1]a_i$ using associativity of vertex operators as in Proposition 3. We find that for a rank l lattice,

$$\begin{aligned}\mathcal{F}^{(2)}_{\alpha,\beta}(\tilde{\omega};\tau,w,\rho) &\equiv \sum_{n\geq 0}\sum_{u\in M^l_{[n]}} Z^{(1)}_{M^l\otimes e^{\alpha}}\left(\tilde{\omega},x;u\otimes e^{\beta},w;\overline{u}\otimes e^{-\beta},0,\tau\right)\rho^{n+(\beta,\beta)/2}dx^2\\ &= \left(\frac{1}{2}\sum_i \nu_{i,\alpha,\beta}(x)^2 + \frac{l}{2}s^{(2)}(x)\right) Z^{(2)}_{\alpha,\beta}(\tau,w,\rho)\\ &= Z^{(2)}_{M^l}(\tau,w,\rho)\left(\mathcal{D}+\frac{l}{12}s^{(2)}\right) e^{i\pi((\alpha,\alpha)\Omega_{11}+2(\alpha,\beta)\Omega_{12}+(\beta,\beta)\Omega_{22})}.\end{aligned}$$

Here, we used (109) and the differential operator [5, 20, 35]

$$\mathcal{D} = \frac{1}{2\pi i}\sum_{1\leq a\leq b\leq 2} \nu_a\nu_b \frac{\partial}{\partial\Omega_{ab}}. \tag{118}$$

Defining the normalized Virasoro 1-point form

$$\hat{\mathcal{F}}^{(2)}_{V_L}(\tilde{\omega};\tau,w,\rho) = \frac{\mathcal{F}^{(2)}_{V_L}(\tilde{\omega};\tau,w,\rho)}{Z^{(2)}_{M^l}(\tau,w,\rho)}, \tag{119}$$

we obtain

Proposition 4. *The normalized Virasoro 1-point function for the lattice theory* V_L *satisfies*

$$\hat{\mathcal{F}}^{(2)}_{V_L}(\tilde{\omega};\tau,w,\rho) = \left(\mathcal{D}+\frac{l}{12}s^{(2)}\right)\theta^{(2)}_L(\Omega). \tag{120}$$

□

(The Ward identity (120) is similar to Proposition 11 in [20] in the ϵ-sewing formalism.)

Finally, we can obtain the analogue of Proposition 12 (op. cit.), where we find that $\hat{\mathcal{F}}^{(2)}_{V_L}$ enjoys the same modular properties as $\hat{Z}^{(2)}_{V_L} = \theta^{(2)}_L(\Omega(\tau,w,\rho))$. That is,

Proposition 5. *The normalized Virasoro 1-point function for a lattice VOA obeys*

$$\hat{\mathcal{F}}^{(2)}_{V_L}(\tilde{\omega};\tau,w,\rho)|_{l/2}\gamma = \left(\mathcal{D} + \frac{l}{12}s^{(2)}\right)\left(\hat{Z}^{(2)}_{V_L}(\tau,w,\rho)|_{l/2}\gamma\right), \qquad (121)$$

for $\gamma \in \Gamma_1$. □

7 Comparison Between the ϵ and ρ-Formalisms

In this section we consider the relationship between the genus two boson and lattice partition functions computed in the ϵ-formalism of [20] (based on a sewing construction with two separate tori with modular parameters τ_1, τ_2 and a sewing parameter ϵ) and the ρ-formalism developed in this paper. We write

$$Z^{(2)}_{V,\epsilon} = Z^{(2)}_{V,\epsilon}(\tau_1,\tau_2,\epsilon) = \sum_{n\geq 0}\epsilon^n \sum_{u\in V_{[n]}} Z^{(1)}_V(u,\tau_1)\, Z^{(1)}_V(\overline{u},\tau_2),$$

$$Z^{(2)}_{V,\rho} = Z^{(2)}_{V,\rho}(\tau,w,\rho) = \sum_{n\geq 0}\rho^n \sum_{u\in V_{[n]}} Z^{(1)}_V(\overline{u},u,w,\tau).$$

Although, for a given VOA V, the partition functions enjoy many similar properties, we show below that the partition functions are *not equal* in the two formalisms. This result follows from an explicit computation of the partition functions for two free bosons in the neighborhood of a two-tori degeneration points where $\Omega_{12} = 0$. It then follows that there is likewise no equality between the partition functions in the ϵ- and ρ-formalisms for a lattice VOA.

As shown in Theorem 12 of [18], we may relate the ϵ- and ρ-formalisms in certain open neighborhoods of the two-tori degeneration point, where $\Omega_{12} = 0$. In the ϵ-formalism, the genus two Riemann surface is parameterized by the domain

$$\mathcal{D}^{\epsilon} = \left\{(\tau_1,\tau_2,\epsilon) \in \mathbb{H}_1\times\mathbb{H}_1\times\mathbb{C} \mid |\epsilon| < \frac{1}{4}D(q_1)D(q_2)\right\}, \qquad (122)$$

with $q_a = \exp(2\pi i\tau_a)$ and $D(q)$ as in (15). In this case the two-tori degeneration is, by definition, given by $\epsilon \to 0$. In the ρ-formalism, the two torus degeneration is described by the limit (74). In order to understand this more precisely we introduce the domain [18]

$$\mathcal{D}^{\chi} = \left\{(\tau,w,\chi) \in \mathbb{H}_1\times\mathbb{C}\times\mathbb{C} \mid (\tau,w,-w^2\chi) \in \mathcal{D}^{\rho}, 0 < |\chi| < \frac{1}{4}\right\}, \qquad (123)$$

for $\mathcal{D}^{\rho}$ of (18) and $\chi = -\frac{\rho}{w^2}$ of (73). The period matrix is determined by a Γ_1-equivariant holomorphic map

$$F^{\chi} : \mathcal{D}^{\chi} \to \mathbb{H}_2,$$
$$(\tau, w, \chi) \mapsto \Omega^{(2)}(\tau, w, -w^2\chi). \quad (124)$$

Then

$$\mathcal{D}_0^{\chi} = \left\{ (\tau, 0, \chi) \in \mathbb{H}_1 \times \mathbb{C} \times \mathbb{C} \mid 0 < |\chi| < \frac{1}{4} \right\}, \quad (125)$$

is the space of two-tori degeneration limit points of the domain $\mathcal{D}^{\chi}$. We may compare the two parameterizations on certain Γ_1-invariant neighborhoods of a two-tori degeneration point in both parameterizations to obtain:

Theorem 18 (op. cit., Theorem 12). *There exists a 1-1 holomorphic mapping between Γ_1-invariant open domains $\mathcal{I}^{\chi} \subset (\mathcal{D}^{\chi} \cup \mathcal{D}_0^{\chi})$ and $\mathcal{I}^{\epsilon} \subset \mathcal{D}^{\epsilon}$ where $\mathcal{I}^{\chi}$ and $\mathcal{I}^{\epsilon}$ are open neighborhoods of a two-tori degeneration point.* □

We next describe the explicit relationship between $(\tau_1, \tau_2, \epsilon)$ and (τ, w, χ) in more detail. Firstly, from Theorem 4 of [18] we obtain

$$2\pi i \Omega_{11} = 2\pi i \tau_1 + E_2(\tau_2)\epsilon^2 + E_2(\tau_1) E_2(\tau_2)^2 \epsilon^4 + O(\epsilon^6),$$
$$2\pi i \Omega_{12} = -\epsilon - E_2(\tau_1) E_2(\tau_2)\epsilon^3 + O(\epsilon^5),$$
$$2\pi i \Omega_{22} = 2\pi i \tau_2 + E_2(\tau_1)\epsilon^2 + E_2(\tau_1)^2 E_2(\tau_2) \epsilon^4 + O(\epsilon^6).$$

Making use of the identity

$$\frac{1}{2\pi i}\frac{d}{d\tau} E_2(\tau) = 5E_4(\tau) - E_2(\tau)^2, \quad (126)$$

it is straightforward to invert $\Omega_{ij}(\tau_1, \tau_2, \epsilon)$ to find

Lemma 5. *In the neighborhood of the two-tori degeneration point $r = 2\pi i \Omega_{12} = 0$ of $\Omega \in \mathbb{H}_2$ we have*

$$2\pi i \tau_1 = 2\pi i \Omega_{11} - E_2(\Omega_{22}) r^2 + 5E_2(\Omega_{11}) E_4(\Omega_{22}) r^4 + O(r^6),$$
$$\epsilon = -r + E_2(\Omega_{11}) E_2(\Omega_{22}) r^3 + O(r^5),$$
$$2\pi i \tau_2 = 2\pi i \Omega_{22} - E_2(\Omega_{11}) r^2 + 5E_2(\Omega_{22}) E_4(\Omega_{11}) r^4 + O(r^6).$$

□

From Theorem 4 we may also determine $\Omega_{ij}(\tau, w, \chi)$ to $O(w^4)$ in a neighborhood of a two-tori degeneration point to find

Proposition 6. *For $(\tau, w, \chi) \in \mathcal{D}^{\chi} \cup \mathcal{D}_0^{\chi}$ we have*

$$2\pi i\Omega_{11} = 2\pi i\tau + G(\chi)\sigma^2 + G(\chi)^2 E_2(\tau)\sigma^4 + O(w^6),$$
$$2\pi i\Omega_{22} = \log f(\chi) + E_2(\tau)\sigma^2 + \left(G(\chi)E_2(\tau)^2 + \frac{1}{2}E_4(\tau)\right)\sigma^4 + O(w^6)$$
$$2\pi i\Omega_{12} = \sigma + G(\chi)E_2(\tau)\sigma^3 + O(w^5),$$

where $\sigma = w\sqrt{1-4\chi}$, $G(\chi) = \frac{1}{12} + E_2(q = f(\chi)) = O(\chi)$ *and* $f(\chi)$ *is the Catalan series (66).*

This result is an extension of [18, Proposition 13] and the general proof proceeds along the same lines. For our purposes, it is sufficient to expand the non-logarithmic terms to $O(w^4, \chi^0)$. Since $R(k,l) = O(\chi)$ and $d_a(k) = O(\chi^{1/2})$ then Theorem 4 implies

$$2\pi i\Omega_{11} = 2\pi i\tau + O(\chi), \tag{127}$$
$$2\pi i\Omega_{22} = \log\chi + 2\sum_{k\geq 2}\frac{1}{k}E_k(\tau)w^k + O(\chi), \tag{128}$$
$$2\pi i\Omega_{12} = w + O(\chi), \tag{129}$$

to all orders in w. In particular, we can readily confirm Proposition 6 to $O(w^4, \chi^0)$. Substituting (127)–(129) into Lemma 5 and using (126) and (136) we obtain

Proposition 7. *For* $(\tau, w, \chi) \in \mathcal{D}^\chi \cup \mathcal{D}_0^\chi$ *we have*

$$2\pi i\tau_1 = 2\pi i\tau + \frac{1}{12}w^2 + \frac{1}{144}E_2(\tau)w^4 + O(w^6, \chi),$$
$$2\pi i\tau_2 = \log(\chi) + \frac{1}{12}E_4(\tau)w^4 + O(w^6, \chi),$$
$$\epsilon = -w - \frac{1}{12}E_2(\tau)w^3 + O(w^5, \chi).$$

□

Define the ratio

$$T_{\epsilon,\rho}(\tau, w, \chi) = \frac{Z^{(2)}_{M^2,\epsilon}(\tau_1, \tau_2, \epsilon)}{Z^{(2)}_{M^2,\rho}(\tau, w, -w^2\chi)}, \tag{130}$$

for τ_1, τ_2, ϵ as given in Proposition 7. From Theorems 8 of [20] and Theorems 11 and 18 above we see that $T_{\epsilon,\rho}$ is Γ_1-invariant. From Theorem 7 for $V = M^2$, we find in the two tori degeneration limit that

$$\lim_{w\to 0} T_{\epsilon,\rho}(\tau, w, \chi) = f(\chi)^{-1/12},$$

i.e., the two partition functions do not even agree in this limit! The origin of this discrepancy may be thought to arise from the central charge dependent factors of $q^{-c/24}$ and $q_1^{-c/24}q_2^{-c/24}$ present in the definitions of $Z^{(2)}_{V,\rho}$ and $Z^{(2)}_{V,\epsilon}$ respectively (which, of course, are necessary for any modular invariance). One modification of the definition of the genus two partition functions compatible with the two tori degeneration limit might be:

$$Z^{\mathrm{new}(2)}_{V,\epsilon}(\tau_1,\tau_2,\epsilon) = \epsilon^{-c/12} Z^{(2)}_{V,\epsilon}(\tau_1,\tau_2,\epsilon), \quad Z^{\mathrm{new}(2)}_{V,\rho}(\tau,w,\rho) = \rho^{-c/24} Z^{(2)}_{V,\rho}(\tau,w,\rho).$$

However, for $V = M^2$, we immediately observe that the ratio cannot be unity due to the incompatible Γ_1 actions arising from

$$\epsilon \to \frac{\epsilon}{c_1\tau_1 + d_1}, \quad \rho \to \frac{\rho}{(c_1\tau + d_1)^2},$$

as given in Lemmas 8 and 15 of [18] (cf. (31)).

Consider instead a further Γ_1-invariant factor of $f(\chi)^{-c/24}$ in the definition of the genus two partition function in the ρ-formalism. Once again, we find that the partition functions do not agree in the neighborhood of a two-tori degeneration point:

Proposition 8.

$$f(\chi)^{1/12} T_{\epsilon,\rho}(\tau,w,\chi) = 1 - \frac{1}{288} E_4(\tau) w^4 + O(w^6,\chi).$$

Proof. As noted earlier, $R(k,l) = O(\chi)$ so that we immediately obtain

$$f(\chi)^{-1/12} Z^{(2)}_{M^2,\rho}(\tau,w,-w^2\chi) = \frac{1}{\eta(\tau)^2 \eta(f(\chi))^2} + O(\chi),$$

to all orders in w. On the other hand, $Z^{(2)}_{M^2,\epsilon}(\tau_1,\tau_2,\epsilon)$ of Theorem 5 of [20] to $O(\epsilon^4)$ is given by

$$\frac{1}{\eta(\tau_1)^2\eta(\tau_2)^2}\left[1 + E_2(\tau_1)E_2(\tau_2)\epsilon^2 + \left(E_2(\tau_1)^2 E_2(\tau_2)^2 + 15 E_4(\tau_1)E_4(\tau_2)\right)\epsilon^4\right] + O(\epsilon^6). \tag{131}$$

We expand this to $O(w^4,\chi)$ using Proposition 7, (126) and

$$\frac{1}{2\pi \mathrm{i}}\frac{d}{d\tau}\eta(\tau) = -\frac{1}{2}E_2(\tau)\eta(\tau),$$

to eventually find that

$$Z^{(2)}_{M^2,\epsilon}(\tau_1, \tau_2, \epsilon) = \frac{1}{\eta(q)^2\eta(f(\chi))^2}\left(1 - \frac{1}{288}E_4(\tau)w^4\right) + O(w^6, \chi). \quad \square$$

8 Final Remarks

Let us briefly and heuristically sketch how our results compare to some related ideas in the physics and mathematics literature. There is a wealth of literature concerning the bosonic string e.g. [10, 29]. In particular, the conformal anomaly implies that the physically defined path integral partition function Z_{string} cannot be reduced to an integral over the moduli space $\mathcal{M}_g$ of a Riemann surface of genus g except for the 26 dimensional critical string where the anomaly vanishes. Furthermore, for the critical string, Belavin and Knizhnik argue that

$$Z_{\text{string}} = \int_{\mathcal{M}_g} |F|^2 \, d\mu,$$

where $d\mu$ denotes a natural volume form on $\mathcal{M}_g$ and F is holomorphic and non-vanishing on $\mathcal{M}_g$ [1, 14]. They also claim that for $g \geq 2$, F is a global section for the line bundle $K \otimes \lambda^{-13}$ (where K is the canonical bundle and λ the Hodge bundle) on $\mathcal{M}_g$ which is trivial by Mumford's theorem [26]. In this identification, the λ^{-13} section is associated with 26 bosons, the K section with a $c = -26$ ghost system and the vanishing conformal anomaly to the vanishing first Chern class for $K \otimes \lambda^{-13}$ [28]. More recently, some of these ideas have also been rigorously proved for a zeta function regularized determinant of an appropriate Laplacian operator Δ_n [24]. The genus two partition functions $Z^{(2)}_{M^2,\epsilon}(\tau_1, \tau_2, \epsilon)$ and $Z^{(2)}_{M^2,\rho}(\tau, w, \rho)$ constructed in [20] and the present paper for a rank 2 Heisenberg VOA should correspond in these approaches to a local description of the holomorphic part of $\left(\frac{\det' \Delta_1}{\det N_1}\right)^{-1}$ of [14, 24], giving a local section of the line bundle λ^{-1}. Given these assumptions, it follows that $T_{\epsilon,\rho} = Z^{(2)}_{M^2,\epsilon}/Z^{(2)}_{M^2,\rho} \neq 1$ in the neighborhood of a two-tori degeneration point where the ratio of the two sections is a non-trivial transition function $T_{\epsilon,\rho}$.

In the case of a general rational conformal field theory, the conformal anomaly continues to obstruct the existence of a global partition function on moduli space for $g \geq 2$. However, *all* CFTs of a given central charge c are believed to share the same conformal anomaly e.g. [6]. Thus, the identification of the normalized lattice partition and n-point functions of Sect. 6 reflect the equality of the first Chern class of some bundle associated to a rank c lattice VOA to that for λ^{-c} with transition function $T^{c/2}_{\epsilon,\rho}$. It is interesting to note that even in the case of a unimodular lattice VOA with a unique conformal block [25, 33] the genus two partition function can therefore only be described locally. It would obviously be extremely valuable to find a rigorous description of the relationship between the VOA approach described here and these related ideas in conformal field theory and algebraic geometry.

Appendix

We list here some corrections to [17] and [18] that we needed above.

(a) Display (27) of [17] should read

$$\epsilon(\alpha, -\alpha) = \epsilon(\alpha, \alpha) = (-1)^{(\alpha,\alpha)/2}. \tag{132}$$

(b) Display (45) of [17] should read

$$\gamma(\Xi) = (a, \delta_{r,1}\beta + C(r, 0, \tau)\alpha_k + \sum_{l \neq k} D(r, 0, z_{kl}, \tau)\alpha_l). \tag{133}$$

(c) As a result of (a), displays (79) and (80) of [17] are modified and now read

$$F_N(\mathrm{e}^{\alpha}, z_1; \mathrm{e}^{-\alpha}, z_2; q) = \epsilon(\alpha, -\alpha)\frac{q^{(\beta,\beta)/2}}{\eta^l(\tau)} \frac{\exp((\beta, \alpha)z_{12})}{K(z_{12}, \tau)^{(\alpha,\alpha)}}, \tag{134}$$

$$F_{V_L}(\mathrm{e}^{\alpha}, z_1; \mathrm{e}^{-\alpha}, z_2; q) = \epsilon(\alpha, -\alpha)\frac{1}{\eta^l(\tau)} \frac{\Theta_{\alpha,L}(\tau, z_{12}/2\pi\mathrm{i})}{K(z_{12}, \tau)^{(\alpha,\alpha)}}. \tag{135}$$

(d) The expression for $\epsilon(\tau, w, \chi)$ of display (172) of [18] should read

$$\epsilon(\tau, w, \chi) = -w\sqrt{1 - 4\chi}\left(1 + \frac{1}{24}w^2 E_2(\tau)(1 - 4\chi) + O(w^4)\right) \tag{136}$$

Acknowledgements Geoffrey Mason was supported by the NSF and NSA.

References

1. Belavin, A.A., Knizhnik, V.G.: Algebraic geometry and the geometry of quantum strings. Phys. Lett. **168B**, 202–206 (1986)
2. Borcherds, R.E.: Vertex algebras. Kac-Moody algebras and the Monster. Proc. Natl. Acad. Sci. **83**, 3068–3071 (1986)
3. Dolan, L., Goddard, P., Montague, P.: Conformal field theories, representations and lattice constructions. Commun. Math. Phys. **179**, 61–120 (1996)
4. Farkas, H.M., Kra, I.: Riemann Surfaces. Springer, New York (1980)
5. Fay, J.: Theta Functions on Riemann Surfaces. Lecture Notes in Mathematics, vol. 352. Springer, Berlin/New York (1973)
6. Freidan, D., Shenker, S.: The analytic geometry of two dimensional conformal field theory. Nucl. Phys. **B281**, 509–545 (1987)
7. Freitag, E.: Siegelische Modulfunktionen. Springer, Berlin/New York (1983)
8. Frenkel, I., Lepowsky, J., Meurman, A.: Vertex Operator Algebras and the Monster. Academic, New York (1988)
9. Frenkel, I., Huang, Y., Lepowsky, J.: On axiomatic approaches to vertex operator algebras and modules. Mem. Am. Math. Soc. **104** (1993)

10. Green, M., Schwartz, J., Witten, E.: Superstring Theory I. Cambridge University Press, Cambridge (1987)
11. Gunning, R.C.: Lectures on Riemann Surfaces. Princeton University Press, Princeton (1966)
12. Hurley, D., Tuite, M.P.: On the torus degeneration of the genus two partition function. Int. J. Math. **24**, 1350056 (2013)
13. Kac, V.: Vertex Operator Algebras for Beginners. University Lecture Series, vol. 10. American Mathematical Society, Providence (1998)
14. Knizhnik, V.G.: Multiloop amplitudes in the theory of quantum strings and complex geometry. Sov. Phys. Usp. **32**, 945–971 (1989); Sov. Sci. Rev. A **10**, 1–76 (1989)
15. Lepowsky, J., Li, H.: Introduction to Vertex Operator Algebras and Their Representations. Birkhäuser, Boston (2004)
16. Li, H.: Symmetric invariant bilinear forms on vertex operator algebras. J. Pure Appl. Alg. **96**, 279–297 (1994)
17. Mason, G., Tuite, M.P.: Torus chiral n-point functions for free boson and lattice vertex operator algebras. Commun. Math. Phys. **235**, 47–68 (2003)
18. Mason, G., Tuite, M.P.: On genus two Riemann surfaces formed from sewn tori. Commun. Math. Phys. **270**, 587–634 (2007)
19. Mason, G., Tuite, M.P.: Partition functions and chiral algebras. Contemp. Math. **442**, 401–410 (2007)
20. Mason, G., Tuite, M.P.: Free bosonic vertex operator algebras on genus two Riemann surfaces I. Commun. Math. Phys. **300**, 673–713 (2010)
21. Mason, G., Tuite, M.P.: The genus two partition function for free bosonic and lattice vertex operator algebras. arXiv:0712.0628 (unpublished)
22. Mason, G., Tuite, M.P.: Vertex operators and modular forms. In: Kirsten, K., Williams, F. (eds.) A Window into Zeta and Modular Physics. MSRI Publications, vol. 57, pp. 183–278. Cambridge University Press, Cambridge (2010)
23. Matsuo, A., Nagatomo, K.: Axioms for a vertex algebra and the locality of quantum fields. Math. Soc. Jpn. Mem. **4** (1999)
24. McIntyre, A., Takhtajan, L.A.: Holomorphic factorization of determinants of Laplacians on Riemann surfaces and higher genus generalization of Kronecker's first limit formula. GAFA, Geom. Funct. Anal. **16**, 1291–1323 (2006)
25. Moore, G., Seiberg, N.: Classical and quantum conformal field theory. Commun. Math. Phys. **123**, 177–254 (1989)
26. Mumford, D.: Stability of projective varieties. L. Ens. Math. **23**, 39–110 (1977)
27. Mumford, D.: Tata Lectures on Theta I and II. Birkhäuser, Boston (1983)
28. Nelson, P.: Lectures on strings and moduli space. Phys. Rep. **149**, 337–375 (1987)
29. Polchinski, J.: String Theory I. Cambridge University Press, Cambridge (1998)
30. Serre, J.-P.: A Course in Arithmetic. Springer, Berlin (1978)
31. Sonoda, H.: Sewing conformal field theories I. Nucl. Phys. **B311**, 401–416 (1988)
32. Sonoda, H.: Sewing conformal field theories II. Nucl. Phys. **B311**, 417–432 (1988)
33. Tsuchiya, A., Ueno, K., Yamada, Y.: Conformal field theory on universal family of stable curves with gauge symmetries. Adv. Stud. Pure Math. **19**, 459–566 (1989)
34. Tuite, M.P: Genus two meromorphic conformal field theory. CRM Proc. Lect. Notes **30**, 231–251 (2001)
35. Ueno, K.: Introduction to conformal field theory with gauge symmetries. In: Geometry and Physics - Proceedings of the Conference at Aarhus University, Aaarhus. Marcel Dekker, New York (1997)
36. Yamada, A.: Precise variational formulas for abelian differentials. Kodai Math. J. **3**, 114–143 (1980)
37. Zhu, Y.: Modular invariance of characters of vertex operator algebras. J. Am. Math. Soc. **9**, 237–302 (1996)

Twisted Correlation Functions on Self-sewn Riemann Surfaces via Generalized Vertex Algebra of Intertwiners

Alexander Zuevsky

Abstract We review our recent results on computation of the partition and n-point "intertwined" functions for modules of vertex operator superalgebras with formal parameter associated to local parameters on Riemann surfaces obtained by self-sewing of a lower genus Riemann surface. We introduce the torus intertwined n-point functions containing two intertwining operators in the supertrace. Then we define the partition and n-point correlation functions for a vertex operator superalgebra on a genus two Riemann surface formed by self-sewing of the torus. For the free fermion vertex operator superalgebra we present a closed formula for the genus two continuous orbifold partition function in terms of an infinite dimensional determinant with entries arising from the original torus Szegő kernel. This partition function is holomorphic in the sewing parameters on a given suitable domain and possess natural modular properties. We describe modularity of the generating function for all n-point correlation functions in terms of a genus two Szegő kernel determinant.

1 Introduction

In this paper (based on the talk at the Conference "Conformal Field Theory, Automorphic Forms and Related Topics", Heidelberg Universität, Heidelberg, Germany, 2011) we review our recent result on construction and computation of correlation functions of vertex operator superalgebras with a formal parameter

To Proceeding of the Conference: "Conformal Field Theory, Automorphic Forms and Related Topics", Heidelberg Universität, Heidelberg, Germany, 2011

A. Zuevsky (✉)
Max–Planck Institut für Mathematik, Vivatsgasse 7, 53111 Bonn, Germany
e-mail: zuevsky@yahoo.com

W. Kohnen and R. Weissauer (eds.), *Conformal Field Theory, Automorphic Forms and Related Topics*, Contributions in Mathematical and Computational Sciences 8,
DOI 10.1007/978-3-662-43831-2_8, © Springer-Verlag Berlin Heidelberg 2014

associated to local coordinates on a self-sewn Riemann surface of genus g which forms a genus $g+1$ surface. In particular, we review result presented in the papers [33–37] accomplished in collaboration with M. P. Tuite (National University of Ireland, Galway).

1.1 Vertex Operator Super Algebras

A Vertex Operator Superalgebra (VOSA) [1,3,12,13,18] is a quadruple $(V, Y, \mathbf{1}, \omega)$: $V = V_{\bar{0}} \oplus V_{\bar{1}} = \bigoplus_{r \in \frac{1}{2}\mathbb{Z}} V_r$, $\dim V_r < \infty$, is a superspace, Y is a linear map $Y : V \to (\mathrm{End} V)[[z, z^{-1}]]$: so that for any vector (state) $u \in V$ we have $u(k)\mathbf{1} = \delta_{k,-1} u$, $k \geq -1$,

$$Y(u,z) = \sum_{n \in \mathbb{Z}} u(n) z^{-n-1},$$

$u(n)V_\alpha \subset V_{\alpha+p(u)}$, $p(u)$- parity. The linear operators (modes) $u(n) : V \to V$ satisfy creativity

$$Y(u,z)\mathbf{1} = u + O(z),$$

and lower truncation

$$u(n)v = 0,$$

conditions for $u, v \in V$ and $n \gg 0$.

These axioms identity imply locality, associativity, commutation and skew-symmetry:

$$(z_1 - z_2)^m Y(u, z_1) Y(v, z_2) = (-1)^{p(u,v)} (z_1 - z_2)^m Y(v, z_2) Y(u, z_1),$$
$$(z_0 + z_2)^n Y(u, z_0 + z_2) Y(v, z_2) w = (z_0 + z_2)^n Y(Y(u, z_0)v, z_2) w,$$
$$u(k)Y(v,z) - (-1)^{p(u,v)} Y(v,z) u(k) = \sum_{j \geq 0} \binom{k}{j} Y(u(j)v, z) z^{k-j},$$
$$Y(u,z)v = (-1)^{p(u,v)} e^{zL(-1)} Y(v, -z) u,$$

for $u, v, w \in V$ and integers $m, n \gg 0$, $p(u,v) = p(u)p(v)$.

The vacuum vector $\mathbf{1} \in V_{\bar{0},0}$ is such that, $Y(\mathbf{1}, z) = Id_V$, and $\omega \in V_{\bar{0},2}$ the conformal vector satisfies

$$Y(\omega, z) = \sum_{n \in \mathbb{Z}} L(n) z^{-n-2},$$

where $L(n)$ form a Virasoro algebra for a central charge C

$$[L(m), L(n)] = (m-n)L(m+n) + \frac{C}{12}(m^3 - m)\delta_{m,-n}.$$

$L(-1)$ satisfies the translation property

$$Y(L(-1)u, z) = \partial_z Y(u, z).$$

$L(0)$ describes a grading with $L(0)u = wt(u)u$, and $V_r = \{u \in V | wt(u) = r\}$.

1.2 VOSA Modules

Definition 1.1. A V-module for a VOSA V is a pair (W, Y_W), W is a $\mathbb{C}$-graded vector space $W = \bigoplus_{r\in\mathbb{C}} W_r$, $\dim W_r < \infty$, $W_{r+n} = 0$ for all r and $n \ll 0$. Y_W : $V \to \mathrm{End}(W)[[z, z^{-1}]]$

$$Y_W(u, z) = \sum_{n\in\mathbb{Z}} u_W(n) z^{-n-1},$$

for each $u \in V$ $u_W : W \to W$. $Y_W(\mathbf{1}, z) = \mathrm{Id}_W$, and for the conformal vector

$$Y_W(\omega, z) = \sum_{n\in\mathbb{Z}} L_W(n) z^{-n-2},$$

where $L_W(0)w = rw$, $w \in W_r$. The module vertex operators satisfy the Jacobi identity:

$$\begin{aligned} z_0^{-1}\delta\left(\frac{z_1 - z_2}{z_0}\right) & Y_W(u, z_1) Y_W(v, z_2) \\ -(-1)^{p(u,v)}\delta\left(\frac{z_2 - z_1}{-z_0}\right) & Y_W(v, z_2) Y_W(u, z_1) \\ = z_2^{-1}\delta\left(\frac{z_1 - z_0}{z_2}\right) & Y_W\left(Y(u, z_0)v, z_2\right). \end{aligned}$$

Recall that

$$\delta(z) = \sum_{n\in\mathbb{Z}} z^n.$$

The above axioms imply that $L_W(n)$ satisfies the Virasoro algebra for the same central charge C and that the translation property

$$Y_W(L(-1)u, z) = \partial_z Y_W(u, z).$$

1.3 Twisted Modules

We next define the notion of a twisted V-module [5,13]. Let g be a V-automorphism g, i.e., a linear map preserving $\mathbf{1}$ and ω such that

$$gY(v,z)g^{-1} = Y(gv,z),$$

for all $v \in V$. We assume that V can be decomposed into g-eigenspaces

$$V = \oplus_{\rho \in \mathbb{C}} V^{\rho},$$

where V^{ρ} denotes the eigenspace of g with eigenvalue $e^{2\pi i \rho}$.

Definition 1.2. A g-twisted V-module for a VOSA V is a pair (W^g, Y_g) $W^g = \bigoplus_{r \in \mathbb{C}} W_r^g$, $\dim W_r^g < \infty$, $W_{r+n}^g = 0$ for all r and $n \ll 0$. $Y_g : V \to \operatorname{End} W^g\{z\}$, the vector space of End W^g-valued formal series in z with arbitrary complex powers of z. For $v \in V^{\rho}$

$$Y_g(v,z) = \sum_{n \in \rho + \mathbb{Z}} v_g(n) z^{-n-1},$$

with $v_g(\rho + l)w = 0$, $w \in W^g$, $l \in \mathbb{Z}$ sufficiently large. $Y_g(\mathbf{1}, z) = \mathrm{Id}_{W^g}$,

$$Y_g(\omega, z) = \sum_{n \in \mathbb{Z}} L_g(n) z^{-n-2},$$

where $L_g(0)w = rw$, $w \in W_r^g$. The g-twisted vertex operators satisfy the twisted Jacobi identity:

$$\begin{aligned}
&z_0^{-1}\delta\left(\frac{z_1 - z_2}{z_0}\right) Y_g(u, z_1) Y_g(v, z_2) \\
&\quad -(-1)^{p(u,v)} z_0^{-1}\delta\left(\frac{z_2 - z_1}{-z_0}\right) Y_g(v, z_2) Y_g(u, z_1) \\
&\quad = z_2^{-1}\left(\frac{z_1 - z_0}{-z_2}\right)^{-\rho} \delta\left(\frac{z_1 - z_0}{-z_2}\right) Y_g(Y(u, z_0)v, z_2),
\end{aligned}$$

for $u \in V^{\rho}$.

1.4 Creative Intertwining Operators

We define the notion of creative intertwining operators in [36]. Suppose we have a VOA V with a V-module (W, Y_W).

Definition 1.3. A *Creative Intertwining Vertex Operator* $\mathcal{Y}$ for a VOA V-module (W, Y_W) is defined by a linear map

$$\mathcal{Y}(w, z) = \sum_{n\in\mathbb{Z}} w(n)z^{-n-1},$$

for $w \in W$ with modes $w(n) : V \to W$; satisfies creativity

$$\mathcal{Y}(w, z)\mathbf{1} = w + O(z),$$

for $w \in W$ and lower truncation

$$w(n)v = 0,$$

for $v \in V$, $w \in W$ and $n \gg 0$. The intertwining vertex operators satisfy the Jacobi identity:

$$\begin{aligned}
&z_0^{-1}\delta\left(\frac{z_1 - z_2}{z_0}\right) Y_W(u, z_1)\mathcal{Y}(w, z_2)\\
&-z_0^{-1}\delta\left(\frac{z_2 - z_1}{-z_0}\right) \mathcal{Y}(w, z_2)Y(u, z_1)\\
&= z_2^{-1}\delta\left(\frac{z_1 - z_0}{z_2}\right) \mathcal{Y}\left(Y_W(u, z_0)w, z_2\right),
\end{aligned}$$

for all $u \in V$ and $w \in W$.

These axioms imply that the intertwining vertex operators satisfy translation, locality, associativity, commutativity and skew-symmetry:

$$\mathcal{Y}(L_W(-1)w, z) = \partial_z\mathcal{Y}(w, z),$$

$$(z_1 - z_2)^m Y_W(u, z_1)\mathcal{Y}(w, z_2) = (z_1 - z_2)^m \mathcal{Y}(w, z_2)Y(u, z_1),$$

$$(z_0 + z_2)^n Y_W(u, z_0 + z_2)\mathcal{Y}(w, z_2)v = (z_0 + z_2)^n \mathcal{Y}(Y_W(u, z_0)w, z_2)v,$$

$$u_W(k)\mathcal{Y}(w, z) - \mathcal{Y}(w, z)u(k) = \sum_{j\geq 0} \binom{k}{j} \mathcal{Y}(u_W(j)w, z)z^{k-j},$$

$$\mathcal{Y}(w, z)v = e^{zL_W(-1)} Y_W(v, -z)w,$$

for $u, v \in V$, $w \in W$ and integers $m, n \gg 0$.

1.5 Example: Heisenberg Intertwiners

Consider the Heisenberg vertex operator algebra M, [18] generated by weight one normalized Heisenberg vector a with modes obeying

$$[a(n), a(m)] = n\delta_{n,-m},$$

$n, m \in \mathbb{Z}$. In [36] we consider an extension $\mathcal{M} = \cup_{\alpha \in \mathbb{C}} M_\alpha$ of M by its irreducible modules M_α generated by a $\mathbb{C}$-valued continuous parameter α automorphism $g = e^{2\pi i \alpha a(0)}$.

We introduce an extra operator q which is canonically conjugate to the zero mode $a(0)$, i.e.,

$$[a(n), q] = \delta_{n,0}.$$

The state $\mathbf{1} \otimes e^\alpha \in \mathcal{M}$ is created by the action of $e^{\alpha q}$ on the state $\mathbf{1} \otimes e^0$. Using q-conjugation and associativity properties, we explicitly construct in [36] the creative intertwining operators $\mathcal{Y}(u, z) : M \to M_\alpha$. We then prove

Theorem 1.4 (Tuite-Z). *The creative intertwining operators $\mathcal{Y}$ for $\mathcal{M}$ are generated by q-conjugation of vertex operators of M. For a Heisenberg state u,*

$$\mathcal{Y}(u \otimes e^\alpha, z) = e^{\alpha q}\, Y_-(e^\alpha, z)\, Y(u \otimes e^0)\, Y_+(e^\alpha, z)\, z^{\alpha\, a(0)},$$

$$Y_\pm(e^\alpha, z) \equiv \exp\left(\mp\alpha \sum_{n>0} a(\pm n)\, \frac{z^{\mp n}}{n} \right).$$

The operators $\mathcal{Y}$ with some extra cocycle structure satisfy a natural extension from rational to complex parameters of the notion of a *Generalized VOA* as described by Dong and Lepowsky [3,4]. We then prove in [36]:

Theorem 1.5 (Tuite-Z). *$\mathcal{Y}(u \otimes e^\alpha, z)$ satisfy the generalized Jacobi identity*

$$z_0^{-1} \left(\frac{z_1 - z_2}{z_0} \right)^{-\alpha\beta} \delta\left(\frac{z_1 - z_2}{z_0} \right) \mathcal{Y}(u \otimes e^\alpha, z_1)\, \mathcal{Y}(v \otimes e^\beta, z_2)$$
$$-C(\alpha, \beta) z_0^{-1} \left(\frac{z_2 - z_1}{z_0} \right)^{-\alpha\beta} \delta\left(\frac{z_2 - z_1}{-z_0} \right)$$
$$\mathcal{Y}(v \otimes e^\beta, z_2)\, \mathcal{Y}(u \otimes e^\alpha, z_1)$$
$$= z_2^{-1} \delta\left(\frac{z_1 - z_0}{z_2} \right) \mathcal{Y}(\mathcal{Y}(u \otimes e^\alpha, z_0)(v \otimes e^\beta), z_2) \left(\frac{z_1 - z_0}{z_2} \right)^{\alpha a(0)},$$

for all $u \otimes e^\alpha, v \otimes e^\beta \in \mathcal{M}$.

1.6 Invariant Form for Extended Heisenberg Algebra

The definitions of invariant forms [13, 20] for a VOSA and its g-twisted modules were given by Scheithauer [31] and in [34] correspondingly. A bilinear form $\langle \cdot, \cdot \rangle$ on $\mathcal{M}$ is said to be invariant if for all $u \otimes e^{\alpha}$, $v \otimes e^{\beta}$, $w \otimes e^{\gamma} \in \mathcal{M}$ we have

$$\langle \mathcal{Y}(u \otimes e^{\alpha}, z) v \otimes e^{\beta}, w \otimes e^{\gamma} \rangle = e^{i\pi\alpha\beta} \langle v \otimes e^{\beta}, \mathcal{Y}^{\dagger}(u \otimes e^{\alpha}, z) w \otimes e^{\gamma} \rangle,$$

$$\mathcal{Y}^{\dagger}(u \otimes e^{\alpha}, z) = \mathcal{Y}\left(e^{-z\lambda^{-2}L(1)} \left(-\frac{\lambda}{z} \right)^{2L(0)} (u \otimes e^{\alpha}), -\frac{\lambda^2}{z} \right).$$

We are interested in the Möbius map $z \mapsto w = \frac{\rho}{z}$ associated with the sewing condition so that $\lambda = -\xi \, \rho^{\frac{1}{2}}$, with $\xi \in \{\pm\sqrt{-1}\}$. We prove in [36]

Theorem 1.6 (Tuite-Z). *The invariant form $\langle ., . \rangle$ on $\mathcal{M}$ is symmetric, unique and invertible with*

$$\langle v \otimes e^{\alpha}, w \otimes e^{\beta} \rangle = \lambda^{-\alpha^2} \delta_{\alpha,-\beta} \langle v \otimes e^{0}, w \otimes e^{0} \rangle.$$

2 The Szegő Kernel

2.1 Torus Self-Sewing to Form a Genus Two Riemann Surface

In [22, 33] we describe procedures of sewing Riemann surfaces [8, 16]. Consider a self-sewing of the oriented torus $\Sigma^{(1)} = \mathbb{C}/\Lambda$, $\Lambda = 2\pi i(\mathbb{Z}\tau \oplus \mathbb{Z})$, $\tau \in \mathbb{H}_1$.

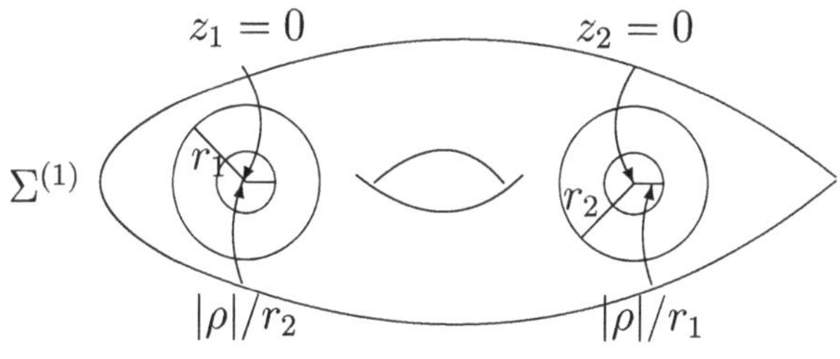

Define annuli $\mathcal{A}_a$, $a = 1, 2$ centered at $z = 0$ and $z = w$ of $\Sigma^{(1)}$ with local coordinates $z_1 = z$ and $z_2 = z - w$ respectively. We use the convention $\overline{1} = 2$, $\overline{2} = 1$. Take the outer radius of $\mathcal{A}_a$ to be $r_a < \frac{1}{2}D(q) = \min_{\lambda \in \Lambda, \lambda \neq 0} |\lambda|$. Introduce a complex parameter ρ, $|\rho| \leq r_1 r_2$. Take inner radius to be $|\rho|/r_{\overline{a}}$, with $|\rho| \leq r_1 r_2$. r_1, r_2 must be sufficiently small to ensure that the disks do not intersect. Excise the disks

$$\{z_a, |z_a| < |\rho| r_{\overline{a}}^{-1}\} \subset \Sigma^{(1)},$$

to form a twice-punctured surface

$$\hat{\Sigma}^{(1)} = \Sigma^{(1)} \backslash \bigcup_{a=1,2} \{z_a, |z_a| < |\rho| r_{\bar{a}}^{-1}\}.$$

Identify annular regions $\mathcal{A}_a \subset \hat{\Sigma}^{(1)}$, $\mathcal{A}_a = \{z_a, |\rho| r_{\bar{a}}^{-1} \leq |z_a| \leq r_a\}$ as a single region $\mathcal{A} = \mathcal{A}_1 \simeq \mathcal{A}_2$ via the sewing relation

$$z_1 z_2 = \rho,$$

to form a compact genus two Riemann surface $\Sigma^{(2)} = \hat{\Sigma}^{(1)} \backslash \{\mathcal{A}_1 \cup \mathcal{A}_2\} \cup \mathcal{A}$, parameterized by

$$\mathcal{D}^\rho = \{(\tau, w, \rho) \in \mathbb{H}_1 \times \mathbb{C} \times \mathbb{C}\,, |w - \lambda| > 2|\rho|^{\frac{1}{2}} > 0,\ \lambda \in \Lambda\}.$$

2.2 The Prime Form

Recall the prime form $E^{(g)}(z, z')$ [9, 10, 28]

$$E^{(g)}(z, z') = \frac{\vartheta \left[\begin{smallmatrix} \gamma \\ \delta \end{smallmatrix}\right] \left(\int_{z'}^{z} \nu | \Omega^{(g)}\right)}{\zeta(z)^{\frac{1}{2}} \zeta(z')^{\frac{1}{2}}} \sim (z - z')\, dz^{-\frac{1}{2}}\, dz'^{-\frac{1}{2}} \quad \text{for } z \sim z',$$

is a holomorphic differential form of weight $(-\frac{1}{2}, -\frac{1}{2})$ on $\tilde{\Sigma}^{(g)} \times \tilde{\Sigma}^{(g)}$,

$$E^{(g)}(z, z') = -E^{(g)}(z', z),$$

and has multipliers 1 and $e^{-i\pi\Omega_{jj}^{(g)} - \int_{z'}^{z} \nu_j}$ along the a_i and b_j cycles in z. Here

$$\zeta(z) = \sum_{i=1}^{g} \partial_{z_i} \vartheta \begin{bmatrix} \gamma \\ \delta \end{bmatrix} (0 | \Omega^{(g)}) \nu_i(z),$$

(a holomorphic one-form, and let $\zeta(z)^{\frac{1}{2}}$ denote the form of weight $\frac{1}{2}$ on the double cover $\tilde{\Sigma}^{(g)}$ of $\Sigma^{(g)}$).

In particular, the prime form on the torus is [28]

$$E^{(1)}(z, z') = K^{(1)}(z - z', \tau)\, dz^{-\frac{1}{2}}\, dz'^{-\frac{1}{2}},$$

$$K^{(1)}(z, \tau) = \frac{\vartheta_1(z, \tau)}{\partial_z \vartheta_1(0, \tau)},$$

for $z \in \mathbb{C}$ and $\tau \in \mathbb{H}_1$ and where $\vartheta_1(z, \tau) = \vartheta \begin{bmatrix} 1/2 \\ 1/2 \end{bmatrix} (z, \tau)$.

2.3 The Szegő Kernel

The Szegő Kernel [9, 10, 28] is defined by

$$S^{(g)}\begin{bmatrix}\theta\\ \phi\end{bmatrix}(z,z'|\Omega) = \frac{\vartheta\begin{bmatrix}\alpha\\ \beta\end{bmatrix}\left(\int_{z'}^{z}\nu\right)}{\vartheta\begin{bmatrix}\alpha\\ \beta\end{bmatrix}(0)E^{(g)}(z,z')} \sim \frac{dz^{\frac{1}{2}}dz'^{\frac{1}{2}}}{z-z'} \quad \text{for } z\sim z',$$

with $\vartheta\begin{bmatrix}\alpha\\ \beta\end{bmatrix}(0)\neq 0$,

$$\theta_j = -e^{-2\pi i\beta_j}, \quad \phi_j = -e^{2\pi i\alpha_j}, \quad j=1,\ldots,g,$$

where $E^{(g)}(z_1,z_2)$ is the genus g prime form. The Szegő kernel has multipliers along the a_i and b_j cycles in z given by $-\phi_i$ and $-\theta_j$ respectively and is a meromorphic $(\frac{1}{2},\frac{1}{2})$-form on $\tilde{\Sigma}^{(g)}\times\tilde{\Sigma}^{(g)}$.

$$S^{(g)}\begin{bmatrix}\theta\\ \phi\end{bmatrix}(z,z') = -S^{(g)}\begin{bmatrix}\theta^{-1}\\ \phi^{-1}\end{bmatrix}(z',z),$$

where $\theta^{-1}=(\theta_i^{-1})$ and $\phi^{-1}=(\phi_i^{-1})$.

Finally, we describe the modular invariance of the Szegő kernel under the symplectic group $Sp(2g,\mathbb{Z})$ where we find [Fay]

$$S^{(g)}\begin{bmatrix}\tilde{\theta}\\ \tilde{\phi}\end{bmatrix}(z,z'|\tilde{\Omega}^{(g)}) = S^{(g)}\begin{bmatrix}\theta\\ \phi\end{bmatrix}(z,z'|\Omega^{(g)}),$$

with $\tilde{\theta}_j=-e^{-2\pi i\tilde{\beta}_j}, \tilde{\phi}_j=-e^{2\pi i\tilde{\alpha}_j}$,

$$\begin{pmatrix}-\tilde{\beta}\\ \tilde{\alpha}\end{pmatrix} = \begin{pmatrix}A & B\\ C & D\end{pmatrix}\begin{pmatrix}-\beta\\ \alpha\end{pmatrix} + \frac{1}{2}\begin{pmatrix}-\mathrm{diag}(AB^T)\\ \mathrm{diag}(CD^T)\end{pmatrix},$$

$$\tilde{\Omega} = (A\Omega+B)(C\Omega+D)^{-1},$$

where $\mathrm{diag}(M)$ denotes the diagonal elements of a matrix M.

On the torus $\Sigma^{(1)}$ the Szegő kernel for $(\theta,\phi)\neq(1,1)$ is

$$S^{(1)}\begin{bmatrix}\theta\\ \phi\end{bmatrix}(z,z'|\tau) = P_1\begin{bmatrix}\theta\\ \phi\end{bmatrix}(z-z',\tau)\,dz^{\frac{1}{2}}\,dz'^{\frac{1}{2}},$$

where

$$P_1 \begin{bmatrix} \theta \\ \phi \end{bmatrix} (z,\tau) = \frac{\vartheta \begin{bmatrix} \alpha \\ \beta \end{bmatrix} (z,\tau)}{\vartheta \begin{bmatrix} \alpha \\ \beta \end{bmatrix} (0,\tau)} \frac{\partial_z \vartheta_1(0,\tau)}{\vartheta_1(z,\tau)}$$

$$= -\sum_{k\in\mathbb{Z}} \frac{q_z^{k+\lambda}}{1-\theta^{-1}q^{k+\lambda}},$$

for $\vartheta_1(z,\tau) = \vartheta \begin{bmatrix} \frac{1}{2} \\ \frac{1}{2} \end{bmatrix} (z,\tau)$, $q_z = e^z$, and $\phi = \exp(2\pi i\lambda)$ for $0 \le \lambda < 1$.

2.4 Genus Two Szegő Kernel in the ρ-Formalism

It is convenient to define $\kappa \in \left[-\frac{1}{2}, \frac{1}{2}\right)$ by $\phi_2 = -e^{2\pi i\kappa}$. Then we prove [33] the following

Theorem 2.1 (Tuite-Z). *$S^{(2)}$ is holomorphic in ρ for $|\rho| < r_1 r_2$ with*

$$S^{(2)}(x,y) = S_\kappa^{(1)}(x,y) + O(\rho),$$

for $x, y \in \hat{\Sigma}^{(1)}$ where $S_\kappa^{(1)}(x,y)$ is defined for $\kappa \neq -\frac{1}{2}$, by

$$S_\kappa^{(1)} \begin{bmatrix} \theta_1 \\ \phi_1 \end{bmatrix} (x,y|\tau,w) = \left(\frac{\vartheta_1(x-w,\tau)\vartheta_1(y,\tau)}{\vartheta_1(x,\tau)\vartheta_1(y-w,\tau)} \right)^\kappa$$

$$\cdot \frac{\vartheta^{(1)} \begin{bmatrix} \alpha_1 \\ \beta_1 \end{bmatrix} (x-y+\kappa w,\tau)}{\vartheta^{(1)} \begin{bmatrix} \alpha_1 \\ \beta_1 \end{bmatrix} (\kappa w,\tau)\, K^{(1)}(x-y,\tau)} dx^{\frac{1}{2}} dy^{\frac{1}{2}},$$

with similar expression for $S_{-\frac{1}{2}}^{(1)}(x,y)$ for $\kappa = -\frac{1}{2}$.

Let $k_a = k + (-1)^{\overline{a}}\kappa$, for $a = 1, 2$ and integer $k \ge 1$. We introduce the moments for $S_\kappa^{(1)}(x,y)$:

$$G_{ab}(k,l) = G_{ab} \begin{bmatrix} \theta^{(1)} \\ \phi^{(1)} \end{bmatrix} (\kappa; k, l)$$

$$= \frac{\rho^{\frac{1}{2}(k_a+l_b-1)}}{(2\pi i)^2} \oint_{\mathcal{C}_{\overline{a}}(x_{\overline{a}})} \oint_{\mathcal{C}_b(y_b)} (x_{\overline{a}})^{-k_a} (y_b)^{-l_b} S_\kappa^{(1)}(x_{\overline{a}}, y_b)\, dx_{\overline{a}}^{\frac{1}{2}}\, dy_b^{\frac{1}{2}},$$

with associated infinite matrix $G = (G_{ab}(k,l))$. We define also half-order differentials

$$h_a(k,x) = h_a \begin{bmatrix} \theta^{(1)} \\ \phi^{(1)} \end{bmatrix} (\kappa; k, x) = \frac{\rho^{\frac{1}{2}(k_a - \frac{1}{2})}}{2\pi i} \oint_{C_a(y_a)} y_a^{-k_a} S_\kappa^{(1)}(x, y_a)\, dy_a^{\frac{1}{2}},$$

$$\overline{h}_a(k,y) = \overline{h}_a \begin{bmatrix} \theta^{(1)} \\ \phi^{(1)} \end{bmatrix} (\kappa; k, y) = \frac{\rho^{\frac{1}{2}(k_a - \frac{1}{2})}}{2\pi i} \oint_{C_{\overline{a}}(x_{\overline{a}})} x_{\overline{a}}^{-k_a} S_\kappa^{(1)}(x_{\overline{a}}, y)\, dx_{\overline{a}}^{\frac{1}{2}},$$

and let $h(x) = (h_a(k,x))$ and $\overline{h}(y) = (\overline{h}_a(k,y))$ denote the infinite row vectors indexed by a, k. From the sewing relation $z_1 z_2 = \rho$ we have

$$dz_a^{\frac{1}{2}} = (-1)^{\overline{a}}\, \xi\, \rho^{\frac{1}{2}} \frac{dz_{\overline{a}}^{\frac{1}{2}}}{z_{\overline{a}}},$$

for $\xi \in \{\pm\sqrt{-1}\}$, depending on the branch of the double cover of $\Sigma^{(1)}$ chosen. It is convenient to define

$$T = \xi G D^\theta,$$

with an infinite diagonal matrix

$$D^\theta(k,l) = \begin{bmatrix} \theta^{-1} & 0 \\ 0 & -\theta \end{bmatrix} \delta(k,l).$$

Defining $\det(I - T)$ by the formal power series in ρ

$$\log\det(I - T) = Tr\log(I - T) = -\sum_{n\geq 1} \frac{1}{n} Tr(T^n),$$

we prove in [33]

Theorem 2.2 (Tuite-Z).

a.) $(I - T)^{-1} = \sum_{n\geq 0} T^n$ *is convergent for* $|\rho| < r_1 r_2$,
b.) $\det(I - T)$ *is non-vanishing and holomorphic in* ρ *on* $\mathcal{D}^\rho$.

Theorem 2.3 (Tuite-Z). $S^{(2)}(x,y)$ *is given by*

$$S^{(2)}(x,y) = S_\kappa^{(1)}(x,y) + \xi h(x) D^\theta (I - T)^{-1} \overline{h}^T(y).$$

3 Intertwined *n*-Point Functions

As in ordinary (non-intertwined) case [6, 17, 21, 23, 25–27, 34, 39] we construct in [37] the partition and n-point functions [2, 7, 11, 14, 15, 19, 29, 30, 32, 38] for vertex operator algebra modules.

3.1 Torus Intertwined n-Point Functions

Let g_i, f_i, $i = 1, 2$ be VOSA V automorphisms commuting with $\sigma v = (-1)^{p(v)} v$. For $u \in V_{\sigma g_2}$ and the states $v_1, \ldots, v_n \in V$ we define the *intertwined* n-point function [37] on the torus by

$$Z^{(1)} \begin{bmatrix} f_1 \\ g_1 \end{bmatrix} (u, z_2;\ v_1, x_1; \ldots; v_n, x_n; \overline{u}, z_1;\ \tau)$$

$$\equiv \mathrm{STr}_{V_{\sigma g_1}} \left(f_1\, \mathcal{Y}\left(q_{z_2}^{L_{\sigma g_2}(0)} u, q_{z_2} \right)\ Y(q_1^{L(0)} v_1, q_1) \right.$$

$$\left. \ldots\ Y(q_n^{L(0)} v_n, q_n)\, \mathcal{Y}\left(q_{z_1}^{L_{\sigma g_2^{-1}}(0)} \overline{u}, q_{z_1} \right)\ q^{L_{\sigma g_1}(0) - c/24} \right),$$

where $q = \exp(2\pi i \tau)$, $q_k = \exp(x_k)$, $q_{z_j} = \exp(z_j)$, $j = 1, 2$; $1 \le k \le n$, for variables $x_1, \ldots, x_n$ associated to the local coordinates on the torus, and $\overline{u}$ is dual for u with respect to the invariant form on $V_{\sigma g_2}$. The supertrace over a V-module N is defined by

$$\mathrm{STr}_N(X) = Tr_N(\sigma X).$$

For an element $u \in V_{\sigma g_2}$ of a VOSA g-twisted V-module we introduce also the differential form

$$\mathcal{F}^{(1)} \begin{bmatrix} f_1 \\ g_1 \end{bmatrix} (u, z_2;\ v_1, x_1;\ \ldots;\ v_n, x_n;\ \overline{u}, z_1;\ \tau)$$

$$\equiv Z^{(1)} \begin{bmatrix} f_1 \\ g_1 \end{bmatrix} (u, z_2;\ v_1, x_1; \ldots; v_n, x_n; \overline{u}, z_1;\ \tau)$$

$$\cdot dz_2^{wt[u]}\ dz_1^{wt[\overline{u}]}\ \prod_{i=1}^{n} dx_i^{wt\ [v_i]},$$

associated to the torus intertwined n-point function.

3.2 Genus Two Partition and n-Point Functions in ρ-Formalism

Let f_i, $i = 1, 2$ be automorphisms, and $V_{\sigma g_j}$ be twisted V-modules of a vertex operator superalgebra V. For $x_1, \ldots, x_n \in \Sigma^{(1)}$ with $|x_k| \ge |\rho|/r_2$ and $|x_k - w| \ge |\rho|/r_1$, $k = 1, \ldots, n$, we define the genus two n-point function [37] in the ρ-formalism by

$$Z^{(2)}\begin{bmatrix} f \\ g \end{bmatrix}(v_1, x_1; \ldots; v_n, x_n; \tau, w, \rho)$$
$$= \sum_{k\geq 0} \sum_{u\in V_{\sigma g_2}[k]} \rho^k Z^{(1)}\begin{bmatrix} f_1 \\ g_1 \end{bmatrix}(u, w + z_2; v_1, x_1; \ldots; v_n, x_n; f_2\overline{u}, z_1; \tau),$$

where $(f, g) = ((f_i), (g_i))$, where f (respectively g) denotes the pair f_1, f_2 (respectively g_1, g_2). The sum is taken over any $V_{\sigma g_2}$-basis.

In particular, introduce the genus two partition function

$$Z^{(2)}\begin{bmatrix} f \\ g \end{bmatrix}(\tau, w, \rho) = \sum_{u\in V_{\sigma g_2}} Z^{(1)}\begin{bmatrix} f_1 \\ g_1 \end{bmatrix}(u, w; f_2\, \overline{u}, 0;\ \tau),$$

where $Z^{(1)}\begin{bmatrix} f_1 \\ g_1 \end{bmatrix}(u, w; f_2\, \overline{u}, 0; \tau)$ is the genus one intertwined two point function.

Remark 3.1. We can generalize the genus two n-point function by introducing and computing the differential form associated to the torus n-point function containing several intertwining operators in the supertrace as well as corresponding genus two n-point functions.

Similar to the ordinary genus two case [34], we define the differential form [37] associated to the n-point function on a sewn genus two Riemann surface for $v_i \in V$ and $x_i \in \Sigma^{(2)}$, $i = 1, \ldots, n$ with $|x_i| \geq |\rho|/r_2$, $|x_i - w| \geq |\rho|/r_1$,

$$\mathcal{F}^{(2)}\begin{bmatrix} f \\ g \end{bmatrix}(v_1, \ldots, v_n; \tau, w, \rho)$$
$$\equiv Z^{(2)}\begin{bmatrix} f \\ g \end{bmatrix}(v_1, x_1; \ldots; v_n, x_n; \tau, w, \rho) \prod_{i=1}^{n} dx_i^{wt[v_i]}.$$

4 Free Fermion VOSA

4.1 Torus Intertwined Two-Point Function

The rank two free fermionic VOSA $V(H, \mathbb{Z} + \frac{1}{2})^{\otimes 2}$, [18] is generated by $\psi^{\pm}$ with

$$[\psi^{+}(m), \psi^{-}(n)] = \delta_{m,-n-1},\ [\psi^{+}(m), \psi^{+}(n)] = 0,\ [\psi^{-}(m), \psi^{-}(n)] = 0,$$

The rank two free fermion VOSA intertwined torus n-point function is parameterized by $\theta_1 = -e^{-2\pi i\beta_1}$, $\phi_1 = -e^{2\pi i\alpha_1}$, and $\phi_2 = -e^{-2\pi i\kappa}$, [34, 37] where

$$\sigma f_1 = e^{2\pi i \beta_1 a(0)}, \quad \sigma g_1 = e^{-2\pi i \alpha_1 a(0)}, \quad \sigma g_2 = e^{2\pi i \kappa a(0)},$$

for real valued $\alpha_1, \beta_1, \kappa, (\theta_1, \phi_1) \neq (1, 1)$.

For $u = \mathbf{1} \otimes e^\kappa \equiv e^\kappa \in V_{\sigma g_2}$ and $v_i = \mathbf{1}$, $i = 1, \ldots, n$ we obtain [37] the basic intertwined two-point function on the torus

$$Z^{(1)} \begin{bmatrix} f_1 \\ g_1 \end{bmatrix} (e^\kappa, z_2;\ e^{-\kappa}, z_1;\ \tau)$$
$$\equiv \mathrm{STr}_{V_{\sigma g_1}} \left(f_1 \mathcal{Y}\left(q_{z_2}^{L(0)} e^\kappa, q_{z_2}\right) \mathcal{Y}\left(q_{z_1}^{L(0)} e^{-\kappa}, q_{z_1}\right) q^{L_{\sigma g_1}(0)-c/24} \right).$$

We then consider the differential form

$$\mathcal{G}_n^{(1)} \begin{bmatrix} f_1 \\ g_1 \end{bmatrix} (x_1, y_1, \ldots, x_n, y_n)$$
$$\equiv \mathcal{F}^{(1)} \begin{bmatrix} f_1 \\ g_1 \end{bmatrix} (e^\kappa, w; \psi^+, x_1; \psi^-, y_1; \ldots; \psi^+, x_n; \psi^-, y_n; e^{-\kappa}, 0; \tau),$$

associated to the torus intertwined $2n$-point function

$$Z^{(1)} \begin{bmatrix} f_1 \\ g_1 \end{bmatrix} (\, e^\kappa, w;\ \psi^+, x_1; \psi^-, y_1; \ldots; \psi^+, x_n\ ; \psi^-, y_n; e^{-\kappa}, 0;\ \tau),$$

with alternatively inserted n states ψ^+ and n states ψ^- distributed on the resulting genus two Riemann surface $\Sigma^{(2)}$ at points $x_i, y_i \in \Sigma^{(2)}$, $i = 1, \ldots, n$.

We then prove in [37]

Theorem 4.1 (Tuite-Z). *For the rank two free fermion vertex operator superalgebra V and for $(\theta, \phi) \neq (1, 1)$ the generating form is given by*

$$\mathcal{G}_n^{(1)} \begin{bmatrix} f_1 \\ g_1 \end{bmatrix} (x_1, y_1, \ldots, x_n, y_n)$$
$$= Z^{(1)} \begin{bmatrix} f_1 \\ g_1 \end{bmatrix} (e^\kappa, w;\ e^{-\kappa}, 0;\ \tau)\ \det S_\kappa^{(1)},$$

$$Z^{(1)} \begin{bmatrix} f_1 \\ g_1 \end{bmatrix} (e^\kappa, w;\ e^{-\kappa}, 0;\ \tau) = \frac{1}{\eta(\tau)} \frac{\vartheta^{(1)} \begin{bmatrix} \alpha_1 \\ \beta_1 \end{bmatrix} (\kappa w, \tau)}{K^{(1)}(w, \tau)^{\kappa^2}},$$

is the basic intertwined two-point function on the torus, and $n \times n$-matrix $S_\kappa^{(1)} = \left[S_\kappa^{(1)} \begin{bmatrix} \theta_1 \\ \phi_1 \end{bmatrix} (x_i, y_j \mid \tau, w) \right]$, $i, j = 1 \ldots, n$, with elements given by parts of the Szegő kernel.

4.2 Genus Two Partition Function

In [37] we then prove:

Theorem 4.2 (Tuite-Z). *Let $V_{\sigma g_i}$, $i = 1, 2$ be σg_i-twisted V-modules for the rank two free fermion vertex operator superalgebra V. Let $(\theta, \phi) \neq (1, 1)$. Then the partition function on a genus two Riemann surface obtained in the ρ-self-sewing formalism of the torus is a non-vanishing holomorphic function on $\mathcal{D}^\rho$ given by*

$$Z^{(2)}\begin{bmatrix} f \\ g \end{bmatrix}(\tau, w, \rho) = Z^{(1)}\begin{bmatrix} f_1 \\ g_1 \end{bmatrix}(e^{\kappa}, w;\ e^{-\kappa}, 0;\ \tau)\ \det(1 - T),$$

where $Z^{(1)}\begin{bmatrix} f_1 \\ g_1 \end{bmatrix}(e^{\kappa}, w; e^{-\kappa}, 0; \tau)$ is the intertwined V-module $V_{\sigma g_1}$ torus basic two-point function.

We may similarly compute the genus two partition function in the ρ-formalism for the original rank one fermion VOSA $V\left(H, \mathbb{Z} + \frac{1}{2}\right)$ in which case we can only construct a σ-twisted module. Then we have [37] the following

Corollary 4.3 (Tuite-Z). *Let V be the rank one free fermion vertex operator superalgebra and f_1, $g_1 \in \{\sigma, 1\}$, $a = 1, 2$ be automorphisms. Then the partition function for V-module $V_{\sigma g_1}$ on a genus two Riemann surface obtained from ρ formalism of a self-sewn torus $\Sigma^{(1)}$ is given by*

$$Z^{(2)}_{\text{rank1}}\begin{bmatrix} f \\ g \end{bmatrix}(\tau, w, \rho) = Z^{(1)}_{\text{rank1}}\begin{bmatrix} f_1 \\ g_1 \end{bmatrix}(e^{\kappa}, w; e^{-\kappa}, 0; \tau)\det(I - T)^{1/2},$$

where $Z^{(1)}_{\text{rank 1}}\begin{bmatrix} f_1 \\ g_1 \end{bmatrix}(e^{\kappa}, w;\ e^{-\kappa}, 0;\ \tau)$ is the rank one fermion intertwined partition function on the original torus.

4.3 Genus Two Generating Form

In [37] we define matrices

$$S^{(2)} = \left(S^{(2)}(x_i, y_j)\right), \quad S^{(1)}_{\kappa} = \left(S^{(1)}_{\kappa}(x_i, y_j)\right),$$
$$H^{+} = \left((h(x_i))\,(k, a)\right), \quad H^{-} = \left(\left(\overline{h}(y_i)\right)(l, b)\right)^{T}.$$

$S^{(2)}$ and $S^{(1)}_{\kappa}$ are finite matrices indexed by x_i, y_j for $i, j = 1, \ldots, n$; H^{+} is semi-infinite with n rows indexed by x_i and columns indexed by $k \geq 1$ and $a = 1, 2$ and H^{-} is semi-infinite with rows indexed by $l \geq 1$ and $b = 1, 2$ and with n columns indexed by y_j. We then prove

Lemma 4.1 (Tuite-Z).

$$\det \begin{bmatrix} S_\kappa^{(1)} \ \xi \ H^+ \ D^{\theta_2} \\ H^- \quad I - T \end{bmatrix} = \det S^{(2)} \ \det(I - T),$$

with T, D^{θ_2}.

Introduce the differential form

$$\mathcal{G}_n^{(2)} \begin{bmatrix} f \\ g \end{bmatrix} (x_1, y_1, \ldots, x_n, y_n) = \mathcal{F}^{(2)} \begin{bmatrix} f \\ g \end{bmatrix} (\psi^+, \psi^-, \ldots, \psi^+, \psi^-; \tau, w, \rho),$$

associated to the rank two free fermion VOSA genus two $2n$-point function

$$Z^{(2)} \begin{bmatrix} f \\ g \end{bmatrix} (\psi^+, x_1; \psi^-, y_1; \ldots; \psi^+, x_n \ ; \psi^-, y_n; \tau, w, \rho),$$

with alternatively inserted n states ψ^+ and n states ψ^-. The states are distributed on the genus two Riemann surface $\Sigma^{(2)}$ at points x_i, $y_i \in \Sigma^{(2)}$, $i = 1, \ldots, n$. Then we have

Theorem 4.5 (Tuite-Z). *All* n*-point functions for rank two free fermion VOSA twisted modules* $V_{\sigma g}$ *on self-sewn torus are generated by the differential form*

$$\mathcal{G}_n^{(2)} \begin{bmatrix} f \\ g \end{bmatrix} (x_1, y_1, \ldots, x_n, y_n) = Z^{(2)} \begin{bmatrix} f \\ g \end{bmatrix} (\tau, w, \rho) \ \det S^{(2)},$$

where the elements of the matrix $S^{(2)} = \left[S^{(2)} \begin{bmatrix} \theta \\ \phi \end{bmatrix} (x_i, y_j \mid \tau, w) \right]$, $i, j = 1, \ldots, n$ *and* $Z^{(2)} \begin{bmatrix} f \\ g \end{bmatrix} (\tau, w, \rho)$ *is the genus two partition function.*

5 Modular Invariance Properties

Following the ordinary case [6,24,25] we would like to describe modular properties of genus two "intertwined" partition and n-point generating functions. As in [25], consider $\hat{H} \subset Sp(4, \mathbb{Z})$ with elements

$$\mu(a, b, c) = \begin{pmatrix} 1 & 0 & 0 & b \\ a & 1 & b & c \\ 0 & 0 & 1 & -a \\ 0 & 0 & 0 & 1 \end{pmatrix}.$$

$\hat{H}$ is generated by $A = \mu(1,0,0)$, $B = \mu(0,1,0)$ and $C = \mu(0,0,1)$ with relations $[A,B]C^{-2} = [A,C] = [B,C] = 1$. We also define $\Gamma_1 \subset Sp(4,\mathbb{Z})$ where $\Gamma_1 \cong SL(2,\mathbb{Z})$ with elements

$$\gamma_1 = \begin{pmatrix} a_1 & 0 & b_1 & 0 \\ 0 & 1 & 0 & 0 \\ c_1 & 0 & d_1 & 0 \\ 0 & 0 & 0 & 1 \end{pmatrix}, \quad a_1 d_1 - b_1 c_1 = 1.$$

Together these groups generate $L = \hat{H} \rtimes \Gamma_1 \subset Sp(4,\mathbb{Z})$. From [25] we find that L acts on the domain $\mathcal{D}^\rho$ of as follows:

$$\mu(a,b,c).(\tau,w,\rho) = (\tau, w + 2\pi i a\tau + 2\pi i b, \rho),$$
$$\gamma_1.(\tau,w,\rho) = \left(\frac{a_1\tau + b_1}{c_1\tau + d_1}, \frac{w}{c_1\tau + d_1}, \frac{\rho}{(c_1\tau + d_1)^2}\right).$$

We then define [37] a group action of $\gamma_1 \in SL(2,\mathbb{Z})$ on the torus intertwined two-point function $Z^{(1)}\begin{bmatrix} f_1 \\ g_1 \end{bmatrix}(u,w;\ v,0;\tau)$ for $u, v \in V_{\sigma g}$:

$$Z^{(1)}\begin{bmatrix} f_1 \\ g_1 \end{bmatrix}\Bigg|\gamma_1\ (u,w;\ v,0;\ \tau) = Z^{(1)}\left(\gamma_1.\begin{bmatrix} f_1 \\ g_1 \end{bmatrix}\right)(u,\gamma_1.w;\ v,0;\ \gamma_1.\tau),$$

with the standard action $\gamma_1.\tau$ and $\gamma_1.w$, and $\gamma_1.\begin{bmatrix} f_1 \\ g_1 \end{bmatrix} = \begin{bmatrix} f_1^{a_1} g_1^{b_1} \\ f_1^{c_1} g_1^{d_1} \end{bmatrix}$, and the torus multiplier $e^{(1)}_{\gamma_1}\begin{bmatrix} f_1 \\ g_1 \end{bmatrix} \in U(1)$, [26, 33]. Then we have [37]

Theorem 5.1 (Tuite-Z). *The torus intertwined two-point function for the rank two free fermion VOSA is a modular form (up to multiplier) with respect to L*

$$\begin{aligned} Z^{(1)}&\begin{bmatrix} f_1 \\ g_1 \end{bmatrix}\Bigg|\gamma_1\ (u,w;\ v,0;\ \tau) \\ &= e^{(1)}_{\gamma_1}\begin{bmatrix} f_1 \\ g_1 \end{bmatrix}(c_1\tau + d_1)^{wt u + wt v + \kappa^2} Z^{(1)}\begin{bmatrix} f_1 \\ g_1 \end{bmatrix}(u,w;v,0;\tau), \end{aligned}$$

where $u, v \in V_{\sigma g}$.

The action of the generators A, B and C is given by [33]

$$A\begin{bmatrix} f_1 \\ f_2 \\ g_1 \\ g_2 \end{bmatrix} = \begin{bmatrix} f_1 \\ f_1\, f_2\, \sigma \\ g_1\, g_2^{-1}\, \sigma \\ g_2 \end{bmatrix}, \quad B\begin{bmatrix} f_1 \\ f_2 \\ g_1 \\ g_2 \end{bmatrix} = \begin{bmatrix} f_1\, g_2\, \sigma \\ f_2\, g_1\, \sigma \\ g_1 \\ g_2 \end{bmatrix}, \quad C\begin{bmatrix} f_1 \\ f_2 \\ g_1 \\ g_2 \end{bmatrix} = \begin{bmatrix} f_1 \\ f_2\, g_2\, \sigma \\ g_1 \\ g_2 \end{bmatrix}.$$

In a similar way we may introduce the action of $\gamma \in L$ on the genus two partition function [37]

$$Z^{(2)}\begin{bmatrix} f \\ g \end{bmatrix}\Bigg| \gamma\ (\tau, w, \rho) = Z^{(2)}\left(\gamma.\begin{bmatrix} f \\ g \end{bmatrix}\right)\gamma.\ (\tau, w, \rho),$$

$$\gamma_1.\begin{bmatrix} f_1 \\ f_2 \\ g_1 \\ g_2 \end{bmatrix} = \begin{bmatrix} f_1^{a_1} g_1^{b_1} \\ f_2 \\ f_1^{c_1} g_1^{d_1} \\ g_2 \end{bmatrix}.$$

We may now describe the modular invariance of the genus two partition function for the rank two free fermion VOSA under the action of L. Define a genus two multiplier $e_\gamma^{(2)}\begin{bmatrix} f \\ g \end{bmatrix} \in U(1)$ for $\gamma \in L$ in terms of the genus one multiplier as follows

$$e_\gamma^{(2)}\begin{bmatrix} f \\ g \end{bmatrix} = e_{\gamma_1}^{(1)}\begin{bmatrix} f_1 \\ g_1 \end{bmatrix},$$

for the generator $\gamma_1 \in \Gamma_1$. We then find [37]

Theorem 5.2 (Tuite-Z). *The genus two partition function for the rank two VOSA is modular invariant with respect to L with the multiplier system, i.e.,*

$$Z^{(2)}\begin{bmatrix} f \\ g \end{bmatrix}\Bigg| \gamma\ (\tau, w, \rho) = e_\gamma^{(2)}\begin{bmatrix} f \\ g \end{bmatrix}\ Z^{(2)}\begin{bmatrix} f \\ g \end{bmatrix}(\tau, w, \rho).$$

Finally, we can also obtain modular invariance for the generating form

$$\mathcal{G}_n^{(2)}\begin{bmatrix} f \\ g \end{bmatrix}(x_1, y_1, \ldots, x_n, y_n),$$

for all genus two n-point functions [37].

Theorem 5.3. $\mathcal{G}_n^{(2)}\begin{bmatrix} f \\ g \end{bmatrix}(x_1, y_1, \ldots, x_n, y_n)$ *is modular invariant with respect to L with a multiplier.*

Acknowledgements The author would like to express his deep gratitude to the organizers of the Conference "Conformal Field Theory, Automorphic Forms and Related Topics", Heidelberg Universität, Heidelberg, Germany, 2011.

References

1. Borcherds, R.E.: Vertex algebras, Kac-Moody algebras and the Monster. Proc. Natl. Acad. Sci. **83**, 3068–3071 (1986)
2. di Vecchia, P., Hornfeck, K., Frau, M., Lerda, A., Sciuto, S.: N-string, g-loop vertex for the fermionic string. Phys. Lett. **B211**, 301–307 (1988)
3. Dong, C., Lepowsky, J.: Generalized Vertex Algebras and Relative Vertex Operators. Progress in Mathematics, vol. 112. Birkhäuser, Boston (1993)
4. Dong, C., Li, H., Mason, G.: Simple currents and extensions of vertex operator algebras. Commun. Math. Phys. **180**, 671–707 (1996)
5. Dong, C., Li, H., Mason, G.: Twisted representation of vertex operator algebras. Math. Ann. **310**, 571–600 (1998)
6. Dong, C., Li, H., Mason, G.: Modular-invariance of trace functions in orbifold theory and generalized moonshine. Commun. Math. Phys. **214**, 1–56 (2000)
7. Eguchi, T., Ooguri, H.: Chiral bosonization on a Riemann surface. Phys. Lett. **B187**, 127–134 (1987)
8. Farkas H.M., Kra, I: Theta Constants, Riemann Surfaces and the Modular Group. Graduate Studies in Mathematics, vol. 37, American Mathematical Society, Providence (2001)
9. Fay, J.D.: Theta functions on Riemann Surfaces. Lecture Notes in Mathematics. Springer, Berlin/New York (1973)
10. Fay, J.D.: Kernel functions, analytic torsion and moduli spaces. Mem. Am. Math. Soc. **96**, 123 (1992)
11. Freidan, D., Shenker, S.: The analytic geometry of two dimensional conformal field theory. Nucl. Phys. **B281**, 509–545 (1987)
12. Frenkel, I., Lepowsky J., Meurman A.: Vertex Operator Algebras and the Monster. Academic, New York (1988)
13. Frenkel, I., Huang, Y., Lepowsky, J.: On axiomatic approaches to vertex operator algebras and modules. Mem. Am. Math. Soc. **104**, 64 (1993)
14. Gaberdiel, M.R., Volpato, R.: Higher genus partition functions of meromorphic conformal field theories. J. High Energy Phys. **6**, 048 (2009)
15. Gaberdiel, M.R., Keller, Ch.A., Volpato, R.: Genus two partition functions of chiral conformal field theories. Commun. Number Theory Phys. **4**, 295–363 (2010)
16. Gunning, R.C.: Lectures on Riemann Surfaces. Princeton University Press, Princeton (1966)
17. Huang, Y.: Two-Dimensional Conformal Geometry and Vertex Operator Algebras. Progress in Mathematics, vol. 148. Birkhäuser, Boston (1997)
18. Kac, V.: Vertex Operator Algebras for Beginners. University Lecture Series, vol. 10. American Mathematical Society, Providence (1998)
19. Kawamoto, N., Namikawa, Y., Tsuchiya, A., Yamada, Y.: Geometric realization of conformal field theory on Riemann surfaces. Commun. Math. Phys. **116**, 247–308 (1988)
20. Li, H.: Symmetric invariant bilinear forms on vertex operator algebras. J. Pure. Appl. Alg. **96**, 279–297 (1994)
21. Mason, G., Tuite, M.P.: Chiral n-point functions for free boson and lattice vertex operator algebras. Commun. Math. Phys. **235**, 47–68 (2003)
22. Mason, G., Tuite, M.P.: On genus two Riemann surfaces formed from sewn tori. Commun. Math. Phys. **270**, 587–634 (2007)
23. Mason, G., Tuite, M.P.: Free bosonic vertex operator algebras on genus two Riemann surfaces I. Commun. Math. Phys. **300**, 673–713 (2010)
24. Mason, G., Tuite, M.P.: Vertex operators and modular forms. In: Kirsten, K., Williams, F. (eds.) A Window into Zeta and Modular Physics. MSRI Publications, vol. 57, pp. 183–278. Cambridge University Press, Cambridge (2010)
25. Mason, G., Tuite, M.P.: Free bosonic vertex operator algebras on genus two Riemann surfaces II. Comm. Math. Phys. **300**, 673–713 (2010)

26. Mason, G., Tuite, M.P., Zuevsky, A.: Torus n-point functions for $\mathbb{R}$-graded vertex operator superalgebras and continuous fermion orbifolds. Commun. Math. Phys. **283**, 305–342 (2008)
27. Matsuo, A., Nagatomo, K.: Axioms for a Vertex Algebra and the Locality of Quantum Fields. Math. Soc. Jpn. Mem. **4** (1999) 1–110
28. Mumford, D.: Tata Lectures on Theta I and II. Birkhäuser, Boston (1983)
29. Pezzella, F.: g-loop vertices for free fermions and bosons. Phys. Lett. **B220**, 544–550 (1989)
30. Raina, A.K.: Fay's trisecant identity and conformal field theory. Commun. Math. Phys. **122**, 625–641 (1989)
31. Scheithauer, N.: Vertex algebras, Lie algebras and superstrings. J. Alg. **200**, 363–403 (1998)
32. Tsuchiya, A, Ueno, K., Yamada, Y.: Conformal field theory on universal family of stable curves with gauge symmetries. Adv. Stud. Pure Math. **19**, 459–566 (1989)
33. Tuite, M.P., Zuevsky, A.: The Szegő kernel on a Sewn Riemann surface. Commun. Math. Phys. **306**(3), 617–645 (2011)
34. Tuite, M.P., Zuevsky, A.: Genus two partition and correlation functions for fermionic vertex operator superalgebras I. Commun. Math. Phys. **306**(2), 419–447 (2011)
35. Tuite, M.P., Zuevsky, A.: The bosonic vertex operator algebra on a genus g Riemann surface. RIMS Kokyuroko **1756**, 81–93 (2011)
36. Tuite, M.P., Zuevsky, A.: A generalized vertex operator algebra for Heisenberg intertwiners. J. Pure Appl. Alg. **216**(6), 1253–1492 (2012)
37. Tuite, M.P., Zuevsky, A.: Genus two partition and correlation functions for fermionic vertex 371 operator superalgebras II (2014, to appear). arxiv/1308.2441
38. Ueno, K.: Introduction to conformal field theory with gauge symmetries. In: Geometry and Physics - Proceedings of the Conference at Aarhus Univeristy, Aaarhus. Marcel Dekker, New York (1997)
39. Zhu, Y.: Modular invariance of characters of vertex operator algebras. J. Am. Math. Soc. **9**, 237–302 (1996)

The Theory of Vector-Valued Modular Forms for the Modular Group

Terry Gannon

Abstract We explain the basic ideas, describe with proofs the main results, and demonstrate the effectiveness, of an evolving theory of vector-valued modular forms (vvmf). To keep the exposition concrete, we restrict here to the special case of the modular group. Among other things, we construct vvmf for arbitrary multipliers, solve the Mittag-Leffler problem here, establish Serre duality and find a dimension formula for holomorphic vvmf, all in far greater generality than has been done elsewhere. More important, the new ideas involved are sufficiently simple and robust that this entire theory extends directly to any genus-0 Fuchsian group.

1 Introduction

Even the most classical modular forms (e.g. the Dedekind eta $\eta(\tau)$) need a multiplier, but this multiplier is typically a number (i.e. a 1-dimensional projective representation of some discrete group like $\Gamma = \mathrm{SL}_2(\mathbb{Z})$). Simple examples of *vector-valued* modular forms (vvmf) for $\mathrm{SL}_2(\mathbb{Z})$ are the weight-$\frac{1}{2}$ Jacobi theta functions $\Theta(\tau) = (\theta_2(\tau), \theta_3(\tau), \theta_4(\tau))^t$, which obey for instance

$$\Theta(-1/\tau) = \sqrt{\frac{\tau}{\mathrm{i}}}\begin{pmatrix} 0 & 0 & 1 \\ 0 & 1 & 0 \\ 1 & 0 & 0 \end{pmatrix} \Theta(\tau)\,, \tag{1}$$

T. Gannon (✉)
Department of Mathematics, University of Alberta, Edmonton, Alberta, Canada T6G 2G1
e-mail: tgannon@math.ualberta.ca

W. Kohnen and R. Weissauer (eds.), *Conformal Field Theory, Automorphic Forms and Related Topics*, Contributions in Mathematical and Computational Sciences 8,
DOI 10.1007/978-3-662-43831-2_9,

and $P(\tau) = (\tau, 1)^t$, which has weight -1 and obeys for instance

$$P(-1/\tau) = \tau^{-1} \begin{pmatrix} 0 & -1 \\ 1 & 0 \end{pmatrix} P(\tau) \,. \tag{2}$$

Back in the 1960s Selberg [31] called for the development of the theory of vvmf, as a way to study growth of coefficients of (scalar) modular forms for noncongruence groups. Since then, the relevance of vvmf has grown significantly, thanks largely to the work of Borcherds (see e.g. [9] and, in physics, the rise of rational conformal field theory (RCFT).

In particular, the characters of rational and logarithmic conformal field theories, or C_2-cofinite vertex operator algebras, form a weight-0 vvmf for Γ [29, 36]. Less appreciated is that the 4-point functions (conformal blocks) on the sphere in RCFT can naturally be interpreted as vvmf for $\Gamma(2)$, through the identification of the moduli space of 4-punctured spheres with $\Gamma(2)\backslash\mathbb{H}$. Moreover, the 1- and 2-point functions on a torus are naturally identified with vector-valued modular resp. Jacobi forms for Γ. Also, if the RCFT has additional structure, e.g. $N = 2$ supersymmetry or a Lie algebra symmetry, then the 1-point functions on the torus can be augmented, becoming vector-valued Jacobi (hence matrix-valued modular) forms for $\Gamma_0(2)$ or Γ [19, 28].

The impact of RCFT on mathematics makes it difficult to dismiss these as esoteric exotica. For example, 1-point torus functions for the orbifold of the Moonshine module $V^\natural$ by subgroups of the Monster contain as very special cases the Norton series of generalized Moonshine, so a study of them could lead to extensions of the Monstrous Moonshine conjectures. For all these less well-known applications to RCFT, the multiplier ρ will typically have infinite image, and the weights can be arbitrary rational numbers. Thus the typical classical assumptions that the weight be half-integral, and the modular form be fixed by some finite-index subgroup of Γ, will be violated by a plethora of potentially interesting examples. Hence in the following we do not make those classical assumptions (nor are they needed). In fact, there should be similar applications to sufficiently nice non-rational CFT, such as Liouville theory, where the weights w can be irrational.

In spite of its relevance, the general theory of vvmf has been slow in coming. Some effort has been made (c.f. [17, 33]) to lift to vvmf, classical results like dimension formulas and the 'elementary' growth estimates of Fourier coefficients. Moreover, differential equations have been recognised as valuable tools for studying vvmf, for many years (c.f. [1, 23, 25, 26] to name a few).

Now, an elementary observation is that a vvmf $\mathbb{X}(\tau)$ for a finite-index subgroup G of Γ can be lifted to one of Γ, by inducing the multiplier. This increases the rank of the vvmf by a factor equal to the index. This isomorphism tells us that developing a theory of vvmf for Γ gives for free that of any finite-index subgroup. But more important perhaps, it also shows that the theory of vvmf for Γ contains as a small subclass the scalar modular forms for noncongruence subgroups. This means that one can only be so successful in lifting results from the classical (=scalar) theory to vvmf. We should be looking for new ideas!

Our approach is somewhat different, and starts from the heuristic that a vvmf for a Fuchsian group Γ is a lift to $\mathbb{H}$ of a meromorphic section of a flat holomorphic vector bundle over the (singular) curve $\Gamma\backslash\mathbb{H}$ compactified if necessary by adjoining the cusps. The cusps mean we are not in the world of algebraic stacks. In place of the order-d ODE on $\mathbb{H}$ studied by other authors, we consider a first order Fuchsian DE on the sphere. *Fuchsian differential equations on compact curves, and vvmf for Fuchsian groups, are two sides of the same coin.* Another crucial ingredient of our theory is the behaviour at the elliptic fixed-points. This has been largely ignored in the literature. For simplicity, this paper restricts to the most familiar (and important) case: $\Gamma = \mathrm{SL}_2(\mathbb{Z})$, where $\Gamma\backslash\mathbb{H}^*$ is a sphere with three conical singularities (two at elliptic points and one at the cusp). The theory for other Fuchsian groups is developed elsewhere [6, 14].

There are two aspects to the theory: *holomorphic* (no poles anywhere) and *weakly holomorphic* (poles are allowed at the cusps, but only there). We address both. We start with weakly holomorphic not because it is more interesting, but because it is easier, and this makes it more fundamental. There is nothing particularly special about the cusps from this perspective—the poles could be allowed at any finitely many Γ-orbits, and the theory would be the same.

Section 3 is the heart of this paper. There we establish existence, using Röhrl's solution to the Riemann–Hilbert problem. We obtain analogues of the Birkhoff–Grothendieck and Riemann–Roch Theorems. We find a dimension formula for holomorphic vvmf, and are able to quantify the failure of exactness of the functor assigning to multipliers ρ, spaces of holomorphic vvmf. Our arguments are simpler and much more general than others in the literature. In Sect. 4 we give several illustrations of the effectiveness of the theory.

2 Elementary Remarks

2.1 The Geometry of the Modular Group

Fix $\xi_n = \exp(2\pi \mathrm{i}/n)$. Complex powers z^w throughout the paper are defined by $z^w = |z|^w e^{w\mathrm{i}\mathrm{Arg}(z)}$ for $-\pi \leq \mathrm{Arg}(z) < \pi$. We write $\mathbb{C}[x]$ for the space of polynomials, $\mathbb{C}[[x]]$ for power series $\sum_{n=0}^{\infty} a_n x^n$, and $\mathbb{C}[x^{-1}, x]]$ for Laurent expansions $\sum_{n=-N}^{\infty} a_n x^n$ for any N.

This paper restricts to the modular group $\Gamma := \mathrm{SL}_2(\mathbb{Z})$. Write $\overline{\Gamma} = \mathrm{PSL}_2(\mathbb{Z})$. Throughout this paper we use

$$S = \rho\begin{pmatrix} 0 & -1 \\ 1 & 0 \end{pmatrix}, \quad T = \rho\begin{pmatrix} 1 & 1 \\ 0 & 1 \end{pmatrix}, \quad U = \rho\begin{pmatrix} 0 & -1 \\ 1 & -1 \end{pmatrix} = ST^{-1}. \tag{3}$$

Write $\mathbb{H}^*$ for the *extended half-plane* $\mathbb{H} \cup \mathbb{Q} \cup \{\infty\}$. Then $\Gamma\backslash\mathbb{H}^*$ is topologically a sphere. As it is genus 0, it is uniformised by a Hauptmodul, which can be chosen to be

$$J(\tau) = q^{-1} + 744 + 196884\,q + 21493760\,q^2 + 864299970\,q^3 + \cdots\,, \qquad (4)$$

where as always $q = e^{2\pi i\tau}$. For $k \geq 2$ write $E_k(\tau)$ for the Eisenstein series, normalised so that $E_k(\tau) = 1 + \ldots$. For Γ, the ring of *weakly holomorphic* (i.e. poles allowed only at the cusp) modular functions is $\mathbb{C}[J]$ and the ring of *holomorphic* (i.e. holomorphic everywhere including the cusp) modular forms is $\mathrm{m} = \mathbb{C}[E_4, E_6]$.

In the differential structure induced by that of $\mathbb{H}$, $\Gamma\backslash\mathbb{H}^*$ will have a singularity for every orbit of $\overline{\Gamma}$-fixed-points. $J(\tau)$ smooths out these three singularities. The important q-expansion is the local expansion at one of those special points, but it is a mistake to completely ignore the other two, at $\tau = \mathrm{i}$ and $\tau = \xi_6$. These elliptic point expansions have been used for at least a century, even if they are largely ignored today. They play a crucial role in our analysis.

In particular, define $\tau_2 = \epsilon_2\,(\tau-\mathrm{i})/(\tau+\mathrm{i})$ and $\mathrm{j}_2(w;\tau) = E_4(\mathrm{i})^{-w/4}(1-\tau_2/\epsilon_2)^{-w}$, where $\epsilon_2 = \pi\sqrt{E_4(\mathrm{i})}$ and $E_4(\mathrm{i}) = 3\Gamma(1/4)^8/(2\pi)^6$, and define $\tau_3 = \epsilon_3\,(\tau - \xi_6)/(\tau - \xi_6^5)$ and $\mathrm{j}_3(w;\tau) = E_6(\xi_6)^{-w/6}(1-\tau_3/\epsilon_3)^{-w}$, where $\epsilon_3 = \pi\sqrt{3}E_6(\xi_6)^{1/3}$ and $E_6(\xi_6) = 27\Gamma(1/3)^{18}/(2^9\pi^{12})$. The transformations $\tau \mapsto \tau_i$ for $i = 2, 3$ map $\mathbb{H}$ onto the discs $|\tau_i| < \epsilon_i$, and send i and ξ_6 respectively to 0; j_2 and j_3 are proportional to the corresponding multipliers for a weight-w modular form. The elliptic fixed-point $\tau = \mathrm{i}$ is fixed by $\left(\begin{smallmatrix} 0 & -1 \\ 1 & 0 \end{smallmatrix}\right)$, which sends $\tau_2 \mapsto -\tau_2$. The elliptic fixed-point $\tau = \xi_6$ is fixed by $\left(\begin{smallmatrix} 0 & -1 \\ 1 & -1 \end{smallmatrix}\right)$; it sends $\tau_3 \mapsto \xi_3\tau_3$. If f is a scalar modular form of weight k, and we write $\tilde{f}_i(\tau) = \mathrm{j}_i(k;\tau)f(\tau_i)$ for $i = 2, 3$, we get $\tilde{f}_2(-\tau_2) = \mathrm{i}^k\,\tilde{f}(\tau_2)$ and $\tilde{f}_3(\xi_3\tau_3) = \xi_6^k\,\tilde{f}_3(\tau_3)$.

We have rescaled τ_2, τ_3 by ϵ_2, ϵ_3 to clean up the expansions, but this isn't used in the following. For instance, we have the rational expansions

$$\mathrm{j}_2(4;\tau)E_4(\tau) = 1 + 10\tau_2^2/9 + 5\tau_2^4/27 + 4\tau_2^6/81 + 19\tau_2^8/5103\cdots \qquad (5)$$

$$\mathrm{j}_2(6;\tau)E_6(\tau) = 2\tau_2 + 28\tau_2^3/27 + 56\tau_2^5/135 + 28\tau_2^7/405 + \cdots \qquad (6)$$

$$J(\tau) = 1728 + 6912\tau_2^2 + 11776\tau_2^4 + 1594112\tau_2^6/135 + \cdots \qquad (7)$$

Curiously, the expansion coefficients for $J(\tau)$ at i are all *positive* rationals, but infinitely many distinct primes divide the denominators. However, these denominators arise because of an $n!$ that appears implicitly in these coefficients (see Proposition 17 in [11]). Factoring that off, the sequence becomes:

1728, 10368, 158976, 3586752, 107057664, 4097780928,

193171879296, 10987178906592, 737967598470144,

57713234231210688, 5184724381875974016, ...

Is there a Moonshine connecting these numbers with representation theory?

The multiplier systems (see Definition 2.1 next subsection) of Γ at weight w are parametrised by the representations of $\overline{\Gamma}$. In dimension d, the $\overline{\Gamma}$-representations ρ—more precisely, the algebraic quotient of all group homomorphisms $\overline{\Gamma} \to GL_d(\mathbb{C})$ by the conjugate action of $GL_d(\mathbb{C})$—form a variety, and the completely reducible representations form an open subvariety. The connected components of this open subvariety correspond to ordered 5-tuples $(\alpha_i; \beta_j)$ [35], where α_i is the eigenvalue multiplicity of $(-1)^i$ for S, and β_j is that of ξ_3^j for U. Of course $\alpha_0 + \alpha_1 = \beta_0 + \beta_1 + \beta_2 = d$. When $d > 1$, the $(\alpha_i; \beta_j)$ component is nonempty iff

$$\max\{\beta_j\} \leq \min\{\alpha_i\}, \tag{8}$$

in which case its dimension is $d^2 + 1 - \sum_i \alpha_i^2 - \sum_j \beta_j^2$ [35]. The irreducible $\overline{\Gamma}$-representations of dimension $d < 6$ are explicitly described in [34]. For irreducible ρ, (8) is obtained using quiver-theoretic means in [35]; later we recover (8) within our theory.

2.2 Definitions

Write 1_d for the $d \times d$ identity matrix, and $e_j = (0, \ldots, 1, \ldots, 0)^t$ its jth column. See e.g. [27, 32] for the basics on scalar modular forms.

Definition 2.1. (a) An *admissible multiplier system* (ρ, w) consists of some $w \in \mathbb{C}$ called the *weight and* a map $\rho : \Gamma \to \mathrm{GL}_d(\mathbb{C})$ called the *multiplier*, for some positive integer d called the *rank*, such that:

(i) the associated automorphy factor

$$\tilde{\rho}_w(\gamma, \tau) = \rho(\gamma)\ (c\tau + d)^w$$

satisfies, for all $\gamma_1, \gamma_2 \in \Gamma$,

$$\tilde{\rho}_w(\gamma_1\gamma_2, \tau) = \tilde{\rho}_w(\gamma_1, \gamma_2\tau)\ \tilde{\rho}_w(\gamma_2, \tau)\,; \tag{9}$$

(ii) $\rho(1_2)$ and $e^{-\pi i w}\rho(-1_2)$ both equal the identity matrix.

The conditions on $\rho(\pm 1_2)$ in (ii) are necessary (and sufficient, as we'll see) for the existence of nontrivial vvmf at weight w. In practise modular forms are most important for their Fourier expansions. This is often true for vvmf, and this is why we take ρ to be matrix-valued in Definition 2.1. Note that the multiplier $\rho(\gamma)$ need not be unitary, and the weight w need not be real.

When $w \in \mathbb{Z}$, (ρ, w) is admissible iff ρ is a representation of Γ satisfying $\rho(-1_2) = e^{\pi i w}$. When $w \notin \mathbb{Z}$, ρ is only a *projective* representation of Γ, and is most elegantly described in terms of the braid group B_3. More precisely, $B_3 = \langle \sigma_1, \sigma_2 \,|\, \sigma_1\sigma_2\sigma_1 = \sigma_2\sigma_1\sigma_2 \rangle$ is a central extension by $\mathbb{Z}$ of Γ, where the surjection $B_3 \to \Gamma$ sends $\sigma_1 \mapsto \left(\begin{smallmatrix} 1 & 1 \\ 0 & 1 \end{smallmatrix}\right)$ and $\sigma_2 \mapsto \left(\begin{smallmatrix} 1 & 0 \\ -1 & 1 \end{smallmatrix}\right)$. The kernel is $\langle (\sigma_1\sigma_2\sigma_1)^4 \rangle$, half of the centre of B_3. Then (ρ, w) is admissible iff there is a representation $\hat{\rho}$ of B_3 (necessarily unique) satisfying

$$\hat{\rho}(\sigma_1) = T\,,\;\; \hat{\rho}((\sigma_1\sigma_2\sigma_1)^{-1}) = S\,,\;\; \hat{\rho}((\sigma_1\sigma_2)^{-2}) = U \tag{10}$$

and also $\hat{\rho}((\sigma_1\sigma_2\sigma_1)^2) = e^{\pi i w}$. Alternatively, we will see in Lemma 3.1 below that for any $w \in \mathbb{C}$, there is an admissible system (υ_w, w) of rank 1; then (ρ, w) is admissible iff $\overline{\upsilon_w} \otimes \rho$ is a $\overline{\Gamma}$-representation. From any of these descriptions, we see that a multiplier ρ determines the corresponding weight w modulo 2.

Definition 2.2. Let (ρ, w) be an admissible multiplier system of rank d. A map $\mathbb{X}\colon \mathbb{H} \to \mathbb{C}^d$ is called a *vector-valued modular form (vvmf)* provided

$$\mathbb{X}(\gamma\tau) = \tilde{\rho}_w(\gamma, \tau)\, \mathbb{X}(\tau) \tag{11}$$

for all $\gamma \in \Gamma$ and $\tau \in \mathbb{H}$, and each component $\mathbb{X}_i(\tau)$ is meromorphic throughout $\mathbb{H}^*$. We write $\mathcal{M}_w^!(\rho)$ for the space of all *weakly holomorphic* vvmf, i.e. those holomorphic throughout $\mathbb{H}$.

Meromorphicity at the cusps is defined as usual, e.g. by a growth condition or through the q-expansion given below.

Generic ρ will have T diagonalisable, which we can then insist is diagonal without loss of generality. For simplicity, we will assume throughout this paper that T is diagonal. The following theory generalises to T a direct sum of Jordan blocks (the so-called 'logarithmic' case) without difficulty, other than notational awkwardness [6, 14].

Assume then that T is diagonal. By an *exponent* λ for ρ, we mean any diagonal matrix such that $e^{2\pi i\lambda} = T$, i.e. $T_{jj} = e^{2\pi i\lambda_{jj}}$. An exponent is uniquely defined modulo 1. It is typical in the literature to fix the real part of λ to be between 0 and 1. But we learn in Theorem 3.2 below that often there will be better exponents to choose.

For any vvmf $\mathbb{X}$ and any exponent λ, $q^{-\lambda}\mathbb{X}(\tau)$ will be invariant under $\tau \mapsto \tau + 1$, where we write $q^\lambda = \mathrm{diag}(e^{2\pi i\tau\lambda_{11}}, \ldots, e^{2\pi i\tau\lambda_{dd}})$. This gives us a Fourier expansion

$$\mathbb{X}(\tau) = q^\lambda \sum_{n=-\infty}^{\infty} \mathbb{X}_{(n)}\, q^n\,, \tag{12}$$

where the coefficients $\mathbb{X}_{(n)}$ lie in $\mathbb{C}^d$. $\mathbb{X}(\tau)$ is meromorphic at the cusp $i\infty$ iff only finitely many coefficients $\mathbb{X}_{(n)}$, for $n < 0$, can be nonzero.

2.3 Local Expansions

We are now ready to describe the local expansions about any of the three special points $\mathrm{i}\infty, \mathrm{i}, \xi_6$ (indeed, the same method works for any point in $\mathbb{H}^*$).

Lemma 2.1. *Let (ρ, w) be admissible and T diagonal. Then any $\mathbb{X} \in \mathcal{M}_w^!(\rho)$ obeys*

$$\mathbb{X}(\tau) = q^{\lambda} \sum_{n=0}^{\infty} \mathbb{X}_{(n)}\, q^n = \mathrm{j}_2(w;\tau)^{-1} \sum_{n=0}^{\infty} \mathbb{X}_{[n]}\, \tau_2^n = \mathrm{j}_3(w;\tau)^{-1} \sum_{n=0}^{\infty} \mathbb{X}_{\langle n\rangle}\, \tau_3^n\,, \quad (13)$$

for some exponent λ. These converge for $0 < |q| < 1$ and $|\tau_i| < \epsilon_i$. Also,

$$e^{\pi \mathrm{i} w/2} S\mathbb{X}_{[n]} = (-1)^n \mathbb{X}_{[n]} \qquad \text{and} \qquad e^{2\pi \mathrm{i} w/3} U \mathbb{X}_{\langle n\rangle} = \xi_3^n \mathbb{X}_{\langle n\rangle}\,. \quad (14)$$

The existence of the q-series follows from the explanation at the end of last subsection. For $\tau = \mathrm{i}$ ($\tau = \xi_6$ is identical), $\mathrm{j}_2(w;\tau)\,\mathbb{X}(\tau)$ is holomorphic in the disc $|\tau_2| < \epsilon_2$ and so has a Taylor expansion. The transformation (14) can be seen by direct calculation, but will be trivial once we know Lemma 3.1 below.

We label components by $\mathbb{X}_{(n)\,i}$ etc. A more uniform notation would have been to define e.g. $q_2 = \tau_2^2$, find a matrix P_2 and an 'exponent matrix' λ_2 whose diagonal entries lie in $\frac{1}{2}\mathbb{Z}$, such that $P_2 S P_2^{-1} = e^{2\pi \mathrm{i}\lambda_2}$ and $\mathbb{X} = \mathrm{j}_2(w;\tau)^{-1} P_2^{-1} q_2^{\lambda_2} \sum_{n=0}^{\infty} \tilde{\mathbb{X}}_{[n]} q_2^n$. For most purposes the simpler (13) is adequate, but see (35) below.

2.4 Differential Operators

Differential equations have played a large role in the theory of vvmf. The starting point is the modular derivative

$$\mathrm{D}_w f = \frac{1}{2\pi \mathrm{i}} \frac{\mathrm{d}}{\mathrm{d}\tau} - \frac{w}{12} E_2 = q \frac{\mathrm{d}}{\mathrm{d}q} - \frac{w}{12} E_2\,, \quad (15)$$

where $E_2(\tau) = 1 - 24q - 72q^2 - \cdots$ is the quasi-modular Eisenstein function. Note that D_{12} kills the discriminant form $\Delta(\tau) = \eta(\tau)^{24}$. This D_w maps $\mathcal{M}_w^!(\rho)$ to $\mathcal{M}_{w+2}^!(\rho)$. It is a derivation in the sense that if $f \in \mathfrak{m}$ is weight k and $\mathbb{X} \in \mathcal{M}_w^!(\rho)$, then $\mathrm{D}_{k+w}(f\mathbb{X}) = \mathrm{D}_k(f)\,\mathbb{X} + f\,\mathrm{D}_w(\mathbb{X})$. We write $\mathrm{D}_w^j = \mathrm{D}_{w+2j-2} \circ \cdots \circ \mathrm{D}_{w+2} \circ \mathrm{D}_w$.

There are several different applications of differential equations to modular forms—some are reviewed in [11]. But outside of our work, the most influential for the theory of vvmf (see e.g. [1, 21, 23, 26]) has been the differential equation coming from the Wronskian (see e.g. [23] for the straightforward proof):

Lemma 2.2. **(a)** *Let (ρ, w) be admissible. For $\mathbb{X} \in \mathcal{M}^!_w(\rho)$, define*

$$\mathrm{Wr}(\mathbb{X}) := \det \begin{pmatrix} \mathbb{X}_1 & \frac{\mathrm{d}}{2\pi \mathrm{i}\mathrm{d}\tau}\mathbb{X}_1 & \cdots & (\frac{\mathrm{d}}{2\pi \mathrm{i}\mathrm{d}\tau})^{d-1}\mathbb{X}_1 \\ \vdots & \vdots & & \vdots \\ \mathbb{X}_d & \frac{\mathrm{d}}{2\pi \mathrm{i}\mathrm{d}\tau}\mathbb{X}_d & \cdots & (\frac{\mathrm{d}}{2\pi \mathrm{i}\mathrm{d}\tau})^{d-1}\mathbb{X}_d \end{pmatrix} = \det \begin{pmatrix} \mathbb{X}_1 & \mathrm{D}_w\mathbb{X}_1 & \cdots & \mathrm{D}_w^{d-1}\mathbb{X}_1 \\ \vdots & \vdots & & \vdots \\ \mathbb{X}_d & \mathrm{D}_w\mathbb{X}_d & \cdots & \mathrm{D}_w^{d-1}\mathbb{X}_d \end{pmatrix}.$$

Then $Wr(\mathbb{X})(\tau) \in \mathcal{M}^!_{(d+w-1)d}(\det\rho)$. If the coefficients of $\mathbb{X}$ are linearly independent over $\mathbb{C}$, then the function $Wr(\mathbb{X})(\tau)$ is nonzero.

(b) *Given admissible (ρ, w) and $\mathbb{X} \in \mathcal{M}^!_w(\rho)$, define an operator $L_{\mathbb{X}}$ on the space of all functions y meromorphic on $\mathbb{H}^*$, by*

$$L_{\mathbb{X}} = \det \begin{pmatrix} y & \mathrm{D}_w y & \cdots & \mathrm{D}_w^d y \\ \mathbb{X}_1 & \mathrm{D}_w\mathbb{X}_1 & \cdots & \mathrm{D}_w^d\mathbb{X}_1 \\ \vdots & \vdots & & \vdots \\ \mathbb{X}_d & \mathrm{D}_w\mathbb{X}_d & \cdots & \mathrm{D}_w^d\mathbb{X}_d \end{pmatrix} = \sum_{l=0}^{d} h_l(\tau)\,\mathrm{D}_w^l y\,, \tag{16}$$

where $h_d = \mathrm{Wr}(\mathbb{X})$ and each h_l is a (meromorphic scalar) modular form of weight $(d+w+1)d-2l$ with multiplier $\det\rho$. Then $L_{\mathbb{X}}\mathbb{X}_i = 0$ for all components $\mathbb{X}_i$ of $\mathbb{X}$. Conversely, when the components of $\mathbb{X}$ are linearly independent, the solution space to $L_{\mathbb{X}}y = 0$ is $\mathbb{C}-\mathrm{Span}\{\mathbb{X}_i\}$.

In our theory, the differential equation (16) plays a minor role. More important are the differential operators which don't change the weight:

$$\nabla_{1,w} = \frac{E_4E_6}{\Delta}\mathrm{D}_w\,, \quad \nabla_{2,w} = \frac{E_4^2}{\Delta}\mathrm{D}_w^2\,, \quad \nabla_{3,w} = \frac{E_6}{\Delta}\mathrm{D}_w^3\,. \tag{17}$$

Each $\nabla_{i,w}$ operates on $\mathcal{M}^!_w(\rho)$, and an easy calculation shows that any $f\mathrm{D}_w^j$ for $f \in \mathcal{M}^!_{-2j}(1)$ is a polynomial in these three $\nabla_{i,w}$ with coefficients in $\mathbb{C}[J]$. Conversely, $\nabla_{3,w}$ is not in $\mathbb{C}[J, \nabla_{1,w}, \nabla_{2,w}]$ and $\nabla_{2,w}$ is not in $\mathbb{C}[J, \nabla_{1,w}]$. The reason $\nabla_{2,w}$ and $\nabla_{3,w}$ are needed is because Γ has elliptic points of order 2 and 3—this is made explicit in the proof of Proposition 3.2. It is crucial to our theory that $\mathcal{M}^!_w(\rho)$ is a module over $\mathbb{C}[J, \nabla_{1,w}, \nabla_{2,w}, \nabla_{3,w}]$. Background on Fuchsian differential equations is provided in e.g. [15].

We sometimes drop the subscript w on D_w and $\nabla_{i,w}$ for readability.

3 Our Main Results

3.1 Existence of vvmf

The main result (Theorem 3.1) of this subsection is the existence proof of vvmf for any (ρ, w). As a warm-up, let us show there is a (scalar) modular form of every complex weight, and compute its multiplier.

Lemma 3.1. *For any $w \in \mathbb{C}$, there is a weakly holomorphic modular form $\Delta^w(\tau) = q^w(1 - 24wq + \cdots)$ of weight $12w$, nonvanishing everywhere except at the cusps. The multiplier υ_{12w}, in terms of the braid group B_3 (recall* (10)*), is*

$$\hat{\upsilon}_{12w}(\sigma_1) = \hat{\upsilon}_{12w}(\sigma_2) = \exp(2\pi \mathrm{i} w)\,. \tag{18}$$

Proof. First note that the discriminant form $\Delta(\tau)$ is holomorphic and nonzero in the simply-connected domain $\mathbb{H}$, and so has a logarithmic derivative $\mathrm{Log}\,\Delta(\tau)$ there. Hence $\Delta^w(\tau) := \exp(w\mathrm{Log}\,\Delta(\tau))$ is well-defined and holomorphic throughout $\mathbb{H}$. It is easy to verify that Δ^w satisfies the differential equation

$$\frac{1}{2\pi \mathrm{i}}\frac{\mathrm{d}f}{\mathrm{d}\tau} = wE_2\, f \tag{19}$$

—indeed, this simply reduces to the statement that E_2 is the logarithmic derivative of Δ. Therefore any other solution to (19) will be a scalar multiple of Δ^w.

Now, fix $\gamma \in \Gamma$. Then $f(\tau) = (c\tau + d)^{-12w}\Delta^w(\gamma\tau)$ exists and is holomorphic throughout $\mathbb{H}$ for the same reason. Note that $f(\tau)$ also satisfies the differential equation (19):

$$\begin{aligned}\frac{1}{2\pi \mathrm{i}}\frac{\mathrm{d}}{\mathrm{d}\tau} f(\tau) =& \frac{-6cw}{\pi \mathrm{i}}(c\tau + d)^{-12w}(c\tau + d)^{-1}\Delta^w(\gamma\tau)\\ &+ (c\tau + d)^{-12w}(c\tau + d)^{-2}wE_2(\gamma\tau)\Delta^w(\gamma\tau) = wE_2(\tau)\, f(\tau)\,,\end{aligned}$$

using quasi-modularity of E_2, and thus $f(\tau) = \upsilon\Delta^w(\tau)$ throughout $\mathbb{H}$, for some constant $\upsilon = \upsilon(\gamma) \in \mathbb{C}$.

The final step needed to verify that Δ^w is a modular form of weight $12w$ is that it behaves well at the cusps. By the previous paragraph it suffices to consider $\mathrm{i}\infty$. But there Δ^w has the expansion $\Delta^w = q^w(1 - 24wq + \cdots)$, up to a constant factor which we can take to be 1. As all cusps lie in the Γ-orbit of $\mathrm{i}\infty$, we see that $\Delta^w(\tau)$ is indeed holomorphic at all cusps.

Because the weight is nonintegral in general, ν_{12w} will be a representation of B_3. This expansion tells us that $\Delta^w(\tau + 1) = \exp(2\pi \mathrm{i} w)\Delta^w(\tau)$, so we have $\hat{\upsilon}_{12w}(\sigma_1) = \exp(2\pi \mathrm{i} w)$. From the familiar B_3 presentation we see that any one-dimensional representation of B_3 takes the same value on both generators σ_i, so $\hat{\upsilon}_{12w}$ is determined. □

For admissible (ρ, w), Lemma 3.1 says the matrices $e^{\pi \mathrm{i} w/2}S$ and $e^{2\pi \mathrm{i} w/3}U$ have order 2 and 3 respectively. Write $\alpha_j(\rho, w)$ for the multiplicity of $(-1)^j$ as an eigenvalue of $e^{\pi \mathrm{i} w/2}S$, and $\beta_j(\rho, w)$ for the eigenvalue multiplicity of ξ_3^j for $e^{2\pi \mathrm{i} w/3}U$.

Theorem 3.1. *Let (ρ, w) be admissible and T diagonal. Then there is a $d \times d$ matrix $\Phi(\tau)$ and exponent λ such that the columns of Φ lie in $\mathcal{M}_w^!(\rho)$, and*

$$\Phi(\tau) = q^{\lambda}\,\mathrm{F}(q) = q^{\lambda}\sum_{n=0}^{\infty} F_{(n)} q^{n}\,, \tag{20}$$

where the matrix $\mathrm{F}(q)$ *is holomorphic and invertible in a neighbourhood of* $q = 0$.

Proof. Consider any representation ρ' of $\pi_1(\mathbb{P}^1 \setminus \{0, 1, \infty\}) \cong F_2$, the free group with 2 generators. Rörhl's solution [30] to the *Riemann–Hilbert problem* [7,8] (see also [16]) says that there exists a Fuchsian differential equation

$$\frac{\mathrm{d}}{\mathrm{d}z}\Psi(z) = \Psi\left(\frac{A_1}{z} + \frac{A_2}{z-1} + \frac{B}{z-b}\right) \tag{21}$$

on the Riemann sphere $\mathbb{P}^1$ whose monodromy is given by ρ'—i.e. the monodromy corresponding to a small circle about 0 and 1, respectively, equals the value of ρ' at the corresponding loops in the fundamental group. There will also be a simple pole at ∞, with residue $A_\infty := -A_0 - A_1 - B$. The B term in (21) corresponds to an apparent singularity; it can be dropped if the monodromies about 0 or 1 or ∞ have finite order [7,8] (which will happen for the ρ' of interest to us). About any of these 3 or 4 singular points $c \in \{0, 1, \infty, b\}$, Levelt [20] proved that a solution $\Psi(z)$ to such a differential equation has the form

$$\Psi(z) = P_c^{-1}\tilde{z}^{N_c}\tilde{z}^{\lambda_c}\mathrm{F}_c(z) \tag{22}$$

where N_c is nilpotent, λ_c is diagonal, F_c is holomorphic and holomorphically invertible about $z = c$, and $\tilde{z}$ is a coordinate on the universal cover of a small disc punctured at c. By [16], we may take $N_c + \lambda_c$ to be conjugate to A_c.

Now suppose we are given a representation of the free product $\overline{\Gamma} \cong \mathbb{Z}_2 * \mathbb{Z}_3$. This is a homomorphic image of the free group F_2, so we can lift ρ' to $\pi_1(\mathbb{P}^1 \setminus \{0, 1, \infty\})$. For us, $\mathbb{P}^1$ is $\Gamma\backslash\mathbb{H}^*$, with (smooth) global coordinate $z = J/1728$. The monodromy at $z = 0$ and $z = 1$ has finite order 2 and 3 respectively, so we won't need the apparent singularity b. The point ∞ corresponds to $\mathrm{i}\infty$ (or rather its Γ-orbit), where $q_\infty = q$, and we have $N_\infty = 0$ and $P_\infty = 1_d$ since T is diagonal; in this case Levelt's equation (22) reduces to (20). The singularities 0 and 1 correspond to the order 2 and 3 elliptic points i and ξ_6; for them, $N = 0$, z locally looks like $(\tau - \mathrm{i})^2$ and $(\tau - \xi_6)^3$, and the diagonal elements of λ lie in $\frac{1}{2}\mathbb{Z}$ and $\frac{1}{3}\mathbb{Z}$. Then Levelt's equation (22) says $\Psi(J(\tau))$ is meromorphic at i and ξ_6 (hence their Γ-orbits). It will be automatically holomorphic at all other points.

The desired matrix is $\Phi(\tau) = \Psi(J(\tau)/1728)\Delta^{w/12}(\tau)J(\tau)^m(J(\tau)/1728 - 1)^n$, where $m, n \in \mathbb{Z}_{\geq 0}$ are taken large enough to kill any poles at $z = 0$ and 1 (i.e. at the elliptic points), and Ψ corresponds to the $\overline{\Gamma}$-representation $\rho' = \upsilon_{-w},\otimes\rho$. □

The proof generalises without change to nondiagonalisable T, and to any other genus-0 Fuchsian group of the first kind [14]. Theorem 3.1 is vastly more general than previous vvmf existence proofs. Previously (see [17, 18]), existence of vvmf was only established for Γ with real weight w, and requiring in addition that the eigenvalues of T all have modulus 1. Their proof used Poincaré series; the difficult step there is to establish convergence, and that is what has prevented their methods to be generalised. That analytic complexity was handled here by Röhrl's argument.

3.2 Mittag-Leffler

Let (ρ, w) be admissible and T diagonal. In this subsection we study the principal part map and calculate its index. This is fundamental to our theory. As always in this paper, the generalisation to nondiagonalisable T and to other genus-0 Fuchsian groups is straightforward [6].

Given any exponent λ and any $\mathbb{X}(\tau) \in \mathcal{M}^!_w(\rho)$, we have the q-expansion (12). Define the *principal part map* $\mathcal{P}_\lambda : \mathcal{M}^!_w(\rho) \to \mathbb{C}^d[q^{-1}]$ by

$$\mathcal{P}_\lambda(\mathbb{X}) = \sum_{n \leq 0} \mathbb{X}_{(n)} q^n . \tag{23}$$

When we want to emphasise the domain, we'll write this $\mathcal{P}_{\lambda;(\rho,w)}$.

Theorem 3.2 ([6]). *Assume (ρ, w) is admissible, and T is diagonal. Recall the eigenvalue multiplicities $\alpha_j = \alpha_j(\rho, w)$ and $\beta_j = \beta_j(\rho, w)$ from Sect. 3.1.*

(a) *For any exponent λ, $\mathcal{P}_\lambda : \mathcal{M}^!_w(\rho) \to \mathbb{C}^d[q^{-1}]$ has finite-dimensional kernel and cokernel, and the index is*

$$\dim\ker \mathcal{P}_\lambda - \dim\operatorname{coker} \mathcal{P}_\lambda = -\operatorname{Tr}\lambda + c_{(\rho,w)} , \tag{24}$$

for

$$\begin{aligned} c_{(\rho,w)} &= \frac{(w-7)d}{12} + \frac{e^{\pi \mathrm{i} w/2}}{4}\operatorname{Tr} S + \frac{2}{3\sqrt{3}}\operatorname{Re}\left(e^{\frac{-\pi \mathrm{i}}{6} - \frac{2\pi \mathrm{i} w}{3}}\operatorname{Tr} U\right) \\ &= \frac{wd}{12} - \frac{\alpha_1}{2} - \frac{\beta_1 + 2\beta_2}{3} . \end{aligned} \tag{25}$$

(b) *There exist exponents λ for which $\mathcal{P}_\lambda : \mathcal{M}^!_w(\rho) \to \mathbb{C}^d[q^{-1}]$ is a vector space isomorphism.*

By a *bijective exponent* we mean any exponent λ for which $\mathcal{P}_\lambda : \mathcal{M}^!_w(\rho) \to \mathbb{C}^d[q^{-1}]$ is an isomorphism. Of course by (24) their trace $\sum_j \lambda_{jj}$ must equal $c_{(\rho,w)}$, but the converse is not true as we will see.

For example, for the trivial 1-dimensional representation, $T = 1$ so an exponent is just an integer. Here, $\mathcal{M}^!_0(1) = \mathbb{C}[J]$ and $c_{(1,0)} = 0$. The map $\mathcal{P}_1$ is injective but not surjective (nothing has principal part 1). On the other hand, the map $\mathcal{P}_{-1}$ is surjective but not injective (it kills all constants). For another example, taking $\rho = \upsilon_{-2}$ (the multiplier of η^{-4}), we have $c_{(\upsilon_{-2},0)} = -7/6$.

It is standard in the literature to restrict from the start to exponents satisfying $0 \leq \lambda_{jj} < 1$. However, such λ are seldom bijective. It is rarely wise to casually throw away a freedom.

Theorem 3.2(b) first appeared in [5], though for restricted (ρ, w), and with the erroneous claim that λ is bijective iff $\mathrm{Tr}\,\lambda = c_{(\rho,w)}$. The deeper part of Theorem 3.2 is the index formula, which is new. We interpret it later as Riemann–Roch, and obtain from it dimensions of spaces of holomorphic vvmf.

The right-side of (25) is always integral. To see that, take $w = 0$ and note

$$\exp(2\pi \mathrm{i}\, \mathrm{Tr}\,\lambda) = \det T = \det S \, \det U^{-1} = (-1)^{\alpha_1} \xi_3^{-\beta_1+\beta_2}. \tag{26}$$

Fix an admissible (ρ, w) with diagonal T, and a bijective exponent Λ. As a vector space over $\mathbb{C}$, $\mathscr{M}_w^!(\rho)$ has a basis

$$\mathbb{X}^{(j;n)}(\tau) = \mathscr{P}_\Lambda^{-1}(q^{-n} e_j) = q^\Lambda \left(q^{-n} e_j + \sum_{m=1}^{\infty} \mathbb{X}_{(m)}^{(j;n)} q^m \right), \tag{27}$$

where $e_j = (0, \ldots, 1, \ldots, 0)^t$ and $n \in \mathbb{Z}_{\geq 0}$. We'll describe next subsection an effective way to find all these $\mathbb{X}^{(j;n)}(\tau)$, given the d^2 coefficients

$$\chi_{ij} := \mathbb{X}_{(1)i}^{(j;0)} \in \mathbb{C}. \tag{28}$$

We learn there that $\mathscr{M}_w^!(\rho)$ is under total control once a bijective exponent Λ and its corresponding matrix $\chi = \chi(\Lambda)$ are found.

Recall that for fixed ρ, the weight w is only determined mod 2, i.e. (ρ, w) is admissible iff $(\rho, w + 2k)$ is, for any $k \in \mathbb{Z}$. We find from the definition of the α_i and β_j that

$$\alpha_j(\rho, w + 2k) = \alpha_{j+k}(\rho, w), \qquad \beta_j(\rho, w + 2k) = \beta_{j+k}(\rho, w). \tag{29}$$

Plugging this into (25), we obtain the trace of a bijective exponent for $(\rho, w + 2k)$:

$$c_{(\rho,w+2k+12l)} = c_{(\rho,w)} + ld + \begin{cases} 0 & \text{if } k = 0 \\ \alpha_1 - \beta_0 & \text{if } k = 1 \\ \beta_2 & \text{if } k = 2 \\ \alpha_1 & \text{if } k = 3 \\ \beta_1 + \beta_2 & \text{if } k = 4 \\ \alpha_1 + \beta_2 & \text{if } k = 5 \end{cases}. \tag{30}$$

Sketch of proof of Theorem 3.2. (See [6] for the complete proof and its generalisation) First of all, it is easy to show that if the real part of an exponent λ is sufficiently large, then $\mathscr{P}_\lambda$ is necessarily injective. This implies that the kernel of any $\mathscr{P}_{\lambda'}$ will be finite-dimensional, in fact $\dim \ker \mathscr{P}_{\lambda'} \leq \sum_j \max\{\lambda_{jj} - \lambda'_{jj}, 0\}$ for any λ with $\mathscr{P}_\lambda$ injective.

Let M be the $\mathbb{C}[J]$-span of the columns of the matrix Φ of Theorem 3.1. Let λ_M be the corresponding exponent appearing in (20). Then invertibility of F(q) implies invertibility of $F_{(0)}$, which in turn implies $\mathscr{P}_{\lambda_M} : M \to \mathbb{C}^d[q^{-1}]$ is surjective, and hence so is $\mathscr{P}_{\lambda_M} : \mathscr{M}^!_w(\rho) \to \mathbb{C}^d[q^{-1}]$. This implies any cokernel is also finite-dimensional.

A bijective exponent can be obtained by starting from λ_M, and recursively increasing one of its entries by $+1$ to kill something in the kernel, which also means nothing gets lost from the range. This process must terminate, by finiteness of the kernel.

The index formula (24), for some value of $c_{(\rho,w)}$, is now computed by showing that, whenever $\lambda' \geq \lambda$, the index of $\mathscr{P}_\lambda$ minus that of $\mathscr{P}_{\lambda'}$ equals $\mathrm{Tr}\,(\lambda' - \lambda)$. The constant $c_{(\rho,w)}$ is computed next subsection. □

Corollary 3.1. *Suppose (ρ, w) is admissible, T is diagonal, and ρ has no 1-dimensional direct summand. Then setting $\epsilon = 0, 1$ for d even, odd respectively, we obtain the bounds*

$$\frac{wd}{12} - d + \frac{\epsilon}{4} \leq c_{(\rho,w)} \leq \frac{wd}{12} - \frac{5d}{12} - \frac{\epsilon}{4}. \tag{31}$$

Proof. Assume without loss of generality that weight $w = 0$. Then (25) tells us that $c_{(\rho,w)} = -\alpha_1/2 - (\beta_1 + 2\beta_2)/3$, where $\alpha_i = \alpha_i(\rho, w), \beta_j = \beta_j(\rho, w)$.

First, let's try to maximise $c_{(\rho,w)}$, subject to the inequalities (8). Clearly, $c_{(\rho,w)}$ is largest when $\alpha_0 \geq \alpha_1 \geq \beta_0 \geq \beta_1 \geq \beta_2$. In fact, we should take β_0 as large as possible, i.e. $\alpha_1 = \beta_0$. Then our formula for $c_{(\rho,w)}$ simplifies to $-\frac{d}{2} + \frac{\beta_1 - \beta_2}{6}$. It is now clear this is maximised by $\beta_2 = \epsilon$ and $\beta_0 = \beta_1 = \frac{d-\epsilon}{2}$, which recovers the upper bound in (31). Similarly, the lower bound in (31) is realised by $\alpha_0 = \beta_2 = \beta_1 = (d - \epsilon)/2$. □

The question of which exponents λ with the correct trace are bijective, can be subtle, though we see next that for generic ρ the trace condition $\mathrm{Tr}\,\lambda = c_{(\rho,w)}$ is also sufficient. With this in mind, define the $\infty \times \infty$ complex matrix $\mathscr{X} = (\mathscr{X}^{(j;n)}_{(m)i})$ built from the mth coefficient of the ith component of the basis (27). For any $\ell \in \mathbb{Z}^d$ define a $\left(\sum_i \max\{\ell_i, 0\}\right) \times \left(\sum_j \max\{-\ell_j, 0\}\right)$ submatrix $\mathscr{X}(\ell)$ of $\mathscr{X}$ by restricting to the rows $(i; m)$ with $0 \leq m < \ell_i$ and columns $(j; n)$ with $1 \leq n \leq -\ell_j$.

Proposition 3.1. *Let (ρ, w) be admissible, T diagonal, and Λ bijective. Then an exponent λ is also bijective iff the matrix $\mathscr{X}(\lambda - \Lambda)$ is invertible.*

The effectiveness of this test will be clear next subsection, where we explain how to compute the $\mathbb{X}^{(j;n)}$ and hence the submatrices $\mathscr{X}(\ell)$. Of course, invertibility forces $\mathscr{X}(\lambda - \Lambda)$ to be square, i.e. $\sum_i \ell_i = 0$, i.e. that $\mathrm{Tr}\,\lambda = \mathrm{Tr}\,\Lambda$.

To prove Proposition 3.1, observe that the spaces $\ker \mathscr{P}_\lambda$ and $\mathrm{null}\,\mathscr{X}(\ell)$ are isomorphic, with $v \in \mathrm{null}\,\mathscr{X}(\ell)$ identified with $\sum_j \sum_{n=0}^{-\ell_j - 1} v_{(j;n)} \mathbb{X}^{(j;n)}$ (the nullspace $\mathrm{null}\,M$ of a $m \times n$ matrix M is all $v \in \mathbb{C}^n$ such that $Mv = 0$). Similarly, the spaces $\mathrm{coker}\,\mathscr{P}_\lambda$ and $\mathrm{null}\,\mathscr{X}(\ell)^t$ are isomorphic.

For example, given a bijective exponent Λ, Proposition 3.1 says that $\chi_{ij} \neq 0$ iff $\Lambda + e_i - e_j$ is bijective, where χ is defined in (28).

3.3 Birkhoff–Grothendieck and Fuchsian Differential Equations

As mentioned in the introduction, one of our (vague) heuristics is to think of vvmf as meromorphic sections of holomorphic bundles over $\mathbb{P}^1$. But the Birkhoff–Grothendieck Theorem says that such a bundle is a direct sum of line bundles. If taken literally, it would say that, up to a change P of basis, each component $\mathbb{X}_i(\tau)$ of any $\mathbb{X} \in \mathcal{M}_w^!(\rho)$ would be a *scalar* modular form of weight w (and some multiplier) for Γ. This is absurd, as it only happens when ρ is equivalent to a sum of 1-dimensional projective representations. The reason Birkhoff–Grothendieck cannot be applied here is that we do *not* have a bundle (in the usual sense) over $\mathbb{P}^1$—indeed, our space $\mathbb{H}^*/\Gamma$ has three singularities.

Nevertheless, Theorem 3.3(a) below says Birkhoff–Grothendieck still holds in spirit.

Theorem 3.3. *Let (ρ, w) be admissible, T diagonal, and Λ bijective.*

(a) *$\mathcal{M}_w^!(\rho)$ is a free $\mathbb{C}[J]$-module of rank $d = \mathrm{rank}(\rho)$. Free generators are $\mathbb{X}^{(j;0)}$ (see (27)).*

(b) *Let $\Xi(\tau) = q^\Lambda(1_d + \chi q + \sum_{n=2}^\infty \Xi_{(n)} q^n)$ be the $d \times d$ matrix whose columns are the $\mathbb{X}^{(j;0)}$. Then*

$$\frac{E_4(\tau)\,E_6(\tau)}{\Delta(\tau)} \mathrm{D}_w \Xi(\tau) = \Xi(\tau)\{(J(\tau) - 984)\Lambda_w + \chi_w + [\Lambda_w, \chi_w]\}, \quad (32)$$

where $\Lambda_w := \Lambda - \frac{w}{12} 1_d$, $\chi_w := \chi + 2w 1_d$, and $[\cdot,\cdot]$ denotes the usual bracket.

(c) *Assume weight $w = 0$. The multi-valued function $\tilde{\Xi}(z) := \Xi(\tau(z))$, where $z(\tau) = J(\tau)/1728$, obeys the Fuchsian differential equation*

$$\frac{\mathrm{d}}{\mathrm{d}z} \tilde{\Xi}(z) = \tilde{\Xi}(z) \left(\frac{\mathscr{A}_2}{z-1} + \frac{\mathscr{A}_3}{z} \right), \quad (33)$$

$$\mathscr{A}_2 = -\frac{31}{72}\,\Lambda - \frac{1}{1728}(\chi + [\Lambda, \chi]), \quad \mathscr{A}_3 = -\frac{41}{72}\Lambda + \frac{1}{1728}(\chi + [\Lambda, \chi]) \quad (34)$$

(recall (21)). Moreover, $\mathscr{A}_2, \mathscr{A}_3$ are diagonalisable, with eigenvalues in $\{0, \frac{1}{2}\}$ and $\{0, \frac{1}{3}, \frac{2}{3}\}$, respectively.

Sketch of proof (see [6] for the complete proof and generalisation). The basis vvmf $\mathbb{X}^{(j;n)}$ exist by surjectivity of $\mathscr{P}_\Lambda$. To show $\mathcal{M}_w^!(\rho)$ is generated over $\mathbb{C}[J]$ by the $\mathbb{X}^{(i;0)}$, follows from an elementary induction on n: if the $\mathbb{X}^{(i;m)}$ all lie in $\oplus_l \mathbb{C}[J]\mathbb{X}^{(l;0)}$ for all i and all $m < n$, then $\mathbb{X}^{(j;n)} \in J\mathbb{X}^{(j;n-1)} + \oplus_l \mathbb{C}[J]\mathbb{X}^{(l;0)} \subseteq \oplus_l \mathbb{C}[J]\mathbb{X}^{(l;0)}$, using the fact that $\mathscr{P}_\Lambda$ is injective. That these generators are free, follows by noting the determinant of Ξ has a nontrivial leading term (namely q^{λ}) and so is nonzero.

The columns of $\nabla_{1,w}\Xi$ also lie in $\mathcal{M}^!_w(\rho)$, and so $\nabla_{1,w}\Xi = \Xi\, D(J)$ for some $d \times d$ matrix-valued polynomial D. That polynomial $D(J)$ is determined by Theorem 3.2 by comparing principal parts. (33) follows directly from (32), by changing variables. Because $e^{2\pi \mathrm{i}\mathscr{A}_2}$ and $e^{2\pi \mathrm{i}\mathscr{A}_3}$ must be conjugate to S and U, respectively, then $\mathscr{A}_2$ and $\mathscr{A}_3$ are diagonalisable with eigenvalues in $\frac{1}{2}\mathbb{Z}$ and $\frac{1}{3}\mathbb{Z}$. But (22) implies that none of these eigenvalues can be negative (otherwise holomorphicity at $\tau = \mathrm{i}$ or $\tau = \xi_6$ would be lost). And none of these eigenvalues can be ≥ 1, as otherwise the corresponding column of Ξ can be divided by $J-1728$ or J, retaining holomorphicity in $\mathbb{H}$ but spanning over $\mathbb{C}[J]$ a strictly larger space of weakly holomorphic vvmf. Since by (34) $\mathscr{A}_2 + \mathscr{A}_3 = \Lambda$, we can obtain the trace of Λ by summing the eigenvalues of $\mathscr{A}_2$ and $\mathscr{A}_3$, and thus finally obtain (25). □

Theorem 3.3 first appeared in [5], though for restricted (ρ, w). It is generalised to arbitrary T and arbitrary genus-0 groups in [6].

Theorem 3.3(c) and (22) say that

$$\Xi(\tau) = \mathrm{j}_2(w;\tau)^{-1} P_2^{-1} q_2^{\lambda_2} \sum_{n=0}^{\infty} \Xi_{[n]} q_2^n = \mathrm{j}_3(w;\tau)^{-1} P_3^{-1} q_3^{\lambda_3} \sum_{n=0}^{\infty} \Xi_{\langle n\rangle} q_3^n \tag{35}$$

where $\Xi_{[0]}$ and $\Xi_{\langle 0\rangle}$ are invertible, λ_2, λ_3 are diagonal, λ_2 has α_i diagonal values equal to $i/2$ for $i = 0, 1$, λ_3 has β_j diagonal entries equal to $j/3$ for $j = 0, 1, 2$, $P_2 S P_2^{-1} = e^{2\pi \mathrm{i}\lambda_2}$, and $P_3 U P_3^{-1} = e^{2\pi \mathrm{i}\lambda_3}$. The key properties here are the bounds $0 \leq (\lambda_2)_{ii} < 1$ and $0 \leq (\lambda_3)_{jj} < 1$, and the invertibility of $\Xi_{[0]}$ and $\Xi_{\langle 0\rangle}$.

Given Λ and χ, it is easy to solve (32) recursively:

$$[\Lambda_w, \Xi_{(n)}] + n\Xi_{(n)} = \sum_{l=0}^{n-1} \Xi_{(l)} \left(f_{n-l}\Lambda_w + \frac{w}{12} t_{n-l} + g_{n-l}(\chi_w + [\Lambda_w, \chi_w]) \right) \tag{36}$$

for $n \geq 2$, where we write $E_2(\tau) = \sum_{n=0}^{\infty} t_n q^n = 1 - 24q - \cdots$, $(J(\tau) - 984)\Delta(\tau)/E_{10}(\tau) = \sum_{n=0}^{\infty} f_n q^n = 1 + 0q + \cdots$ and $\Delta(\tau)/E_{10}(\tau) = \sum_{n=0}^{\infty} g_n q^n = q + \cdots$. We require $\Xi_{(0)} = 1_d$. Note that the ij-entry on the left-side of (36) is $(\Lambda_{ii} - \Lambda_{jj} + n)\, \Xi_{(n)\, ij}$, so (36) allows us to recursively identify all entries of $\Xi_{(n)}$, at least when all $|\Lambda_{jj} - \Lambda_{ii}| \neq n$. Indeed, $\Lambda_{jj} - \Lambda_{ii}$ can never lie in $\mathbb{Z}_{\geq 2}$, thanks to this recursion, since then the value of $\Xi_{(n)\,ij}$ would be unconstrained, contradicting uniqueness of the solution to (32) with $\Xi_{(0)} = 1_d$.

Theorem 3.3 tells us that $\mathbb{X} \in \mathcal{M}^!_w(\rho)$ iff $\mathbb{X} = \Xi P(J)$, where $P(J) \in \mathbb{C}^d[q^{-1}]$. The basis vvmf $\mathbb{X}^{(j;n)}$ of (27) can be easily found recursively from this [5]. They can also be found as follows. Define the generating function

$$\mathscr{X}_{ij}(\tau, \sigma) := q^{-\Lambda_{ii}-1} \sum_{n=0}^{\infty} \left[\mathbb{X}^{(j;n)}(\tau)\right]_i z^n = \frac{\delta_{ij}}{q-z} + \sum_{m=1}^{\infty}\sum_{n=0}^{\infty} \mathscr{X}^{(j;n)}_{(m)\,i} q^{m-1} z^n\,, \tag{37}$$

where we write $z = e^{2\pi i\sigma}$. Then, writing $J' = D_0 J = -E_4^2 E_6/\Delta$, we have [5]

$$\mathscr{X}(\tau,\sigma) = \frac{J'(\sigma)\, q^{-\Lambda-1_d}}{J(\tau) - J(\sigma)}\, \Xi(\tau)\, \Xi(\sigma)^{-1} z^{\Lambda}\,. \tag{38}$$

We call $\Xi(\tau)$ in Theorem 3.3(b) the *fundamental matrix* associated to Λ. A $d \times d$ matrix $\Xi(\tau)$ is a fundamental matrix for (ρ, w) iff all columns lie in $\mathscr{M}_w^!(\rho)$, and $\Xi(\tau) = q^{\Lambda}(1_d + \sum_{n=1}^{\infty} \Xi_{(n)} q^n)$ where $\operatorname{Tr}\Lambda = c_{(\rho,w)}$. The reason is that $\mathscr{P}_\Lambda : \mathscr{M}_w^!(\rho) \to \mathbb{C}^d[q^{-1}]$ is then surjective, so by the index formula it must also be injective.

The determinant of any fundamental matrix is now easy to compute [5]:

$$\det \Xi(\tau) = E_4(\tau)^{\beta_1+2\beta_2} E_6(\tau)^{\alpha_1} \Delta(\tau)^{(dw-4\beta_1-8\beta_2-6\alpha_1)/12}\,, \tag{39}$$

where $\alpha_i = \alpha_i(\rho, w)$, $\beta_j = \beta_j(\rho, w)$ are the eigenvalue multiplicities of Sect. 3.1. Indeed, the determinant is a scalar modular form (with multiplier); use E_4 and E_6 to factor off the zeros at the elliptic points (which we can read off from (35)), and note that the resulting modular form will have no zeros in $\mathbb{H}$ and hence must be a power of Δ, where the power is determined by the weight.

A very practical way to obtain bijective Λ and χ, and hence a fundamental matrix $\Xi(\tau)$, is through cyclicity:

Proposition 3.2. *Suppose (ρ, w) is admissible, and T is diagonal. Suppose $\mathbb{X} \in \mathscr{M}_w^!(\rho)$, and the components of $\mathbb{X}$ are linearly independent over $\mathbb{C}$. Then $\mathscr{M}_w^!(\rho) = \mathbb{C}[J, \nabla_{1,w}, \nabla_{2,w}, \nabla_{3,w}]\mathbb{X}$.*

Proof. Let $\mathscr{M}_{\mathbb{X}} := \mathbb{C}[J, \nabla_{1,w}]\mathbb{X}$. Since $\mathscr{M}_{\mathbb{X}}$ is a module of a PID $\mathbb{C}[J]$, it is a sum of cyclic submodules $\mathbb{C}[J]\mathbb{Y}^{(i)}$. Each $\mathbb{C}[J]\mathbb{Y}^{(i)}$ is torsion-free (by looking at leading powers of q). So $\mathscr{M}_{\mathbb{X}}$ must be free of some rank d'. Because it is a submodule of the rank d module $\mathscr{M}_w^!(\rho)$ (and again using the fact that $\mathbb{C}[J]$ is a PID), $d' \le d$. That $d' = d$ follows by computing the determinant of the $d \times d$ matrix with columns $\nabla_{1,w}^{i-1}\mathbb{X}$: that determinant equals $(2\pi i)^{1-d}(E_4E_6/\Delta)^{(d-1)d/2}$ times the Wronskian of $\mathbb{X}$, which is nonzero by Lemma 2.2(a).

Let $\Xi_{\mathbb{X}}$ be the matrix formed by those d generators of $\mathscr{M}_{\mathbb{X}}$. Because $\nabla_{1,w}\mathscr{M}_{\mathbb{X}} \subseteq \mathscr{M}_{\mathbb{X}}$, the argument of Theorem 3.3 applies and $\Xi_{\mathbb{X}}$ satisfies analogues of (32) and hence (33). This means $\mathscr{M}_{\mathbb{X}}$ will have its own analogues $\Lambda_{\mathbb{X}}, \mathscr{A}_{\mathbb{X}2}, \mathscr{A}_{\mathbb{X}3}$ (their exponentials $e^{2\pi i\Lambda_{\mathbb{X}}}$ etc will be conjugate to $e^{2\pi i\Lambda}$ etc). The trace of $\Lambda_{\mathbb{X}}$ will equal the trace of $\mathscr{A}_{\mathbb{X}2} + \mathscr{A}_{\mathbb{X}3}$, for the same reason it did in Theorem 3.3. Now, $\mathscr{P}_{\Lambda_{\mathbb{X}}}$ is an isomorphism when restricted to $\mathscr{M}_{\mathbb{X}}$, so when extended to $\mathscr{M}_w^!(\rho)$ will also have trivial cokernel. Thus the dimension of $\mathscr{M}_w^!(\rho)/\mathscr{M}_{\mathbb{X}}$ will equal the dimension of $\ker \mathscr{P}_{\Lambda_{\mathbb{X}}}$ which, by the index formula (24), equals

$$\operatorname{Tr}\Lambda_{\mathbb{X}} - \operatorname{Tr}\Lambda = (\operatorname{Tr}\mathscr{A}_{\mathbb{X}2} - \operatorname{Tr}\mathscr{A}_2) + (\operatorname{Tr}\mathscr{A}_{\mathbb{X}3} - \mathscr{A}_3)\,.$$

This means that if $\mathscr{M}_{\mathbb{X}} \neq \mathscr{M}_w^!(\rho)$, then at least one eigenvalue of $\mathscr{A}_{\mathbb{X}2}$ or $\mathscr{A}_{\mathbb{X}3}$ is ≥ 1. Suppose one of $\mathscr{A}_{\mathbb{X}2}$ is. Then by (35) some row of $\Xi_{\mathbb{X}}$ will have order ≥ 2 at

$\tau = \mathrm{i}$. This means every vvmf in $\mathcal{M}_{\mathbb{X}}$ has some component with a zero at $\tau = \mathrm{i}$ of order ≥ 2. Hit each column of $\Xi_{\mathbb{X}}$ with $\nabla_{2,w} = \Delta^{-1}E_4^2\mathrm{D}_w^2$: it will reduce the order of that zero everywhere by 2 (because at $\tau = \mathrm{i}$ $\nabla_{2,w}$ looks like $a\frac{\mathrm{d}^2}{\mathrm{d}\tau_2^2} + b\frac{\mathrm{d}}{\mathrm{d}\tau_2} + c$ for $a \neq 0$). This means the module generated over $\mathbb{C}[J, \nabla_{1,w}]$ by the columns of $\Xi_{\mathbb{X}}$ and $\nabla_{2,w}\Xi_{\mathbb{X}}$ will be strictly greater than $\mathcal{M}_{\mathbb{X}}$, as some vvmf in it will have a smaller order at that component than all vvmf in $\mathcal{M}_{\mathbb{X}}$. So repeat this argument with $\mathcal{M}_{\mathbb{X}}$ replaced with this extension. If instead $\mathcal{A}_{\mathbb{X}\,3}$ has an eigenvalue ≥ 1, then some component of every vvmf in $\mathcal{M}_{\mathbb{X}}$ will have a zero at $\tau = \xi_6$ of order ≥ 3, so use $\nabla_{3,w}$, which will reduce the order of that zero by 3. Eventually all eigenvalues will be < 1, in which case the dimension of $\mathcal{M}_w^!(\rho)/\mathcal{M}$ will be 0. □

To indicate the nontriviality of our theory, we get a 1-line proof of the solution (8) to the Deligne–Simpson problem for $\overline{\Gamma}$, at least for most ρ (the general case requires slightly more work). Let ρ be any $\overline{\Gamma}$-representation with T diagonal, and let Λ be bijective for $(\rho, 0)$ and Ξ a corresponding fundamental matrix. Then as long as all $\Lambda_{ii} \neq 0$, the columns of the derivative $\mathrm{D}_0\Xi$ will span a free rank-d submodule of $\mathcal{M}_2^!(\rho)$ over $\mathbb{C}[J]$, on which $\mathcal{P}_\Lambda$ is surjective. Thus $c_{(\rho,0)} \leq c_{(\rho,2)}$ and so by (30) we obtain $\alpha_1 \geq \beta_0$. (It is clear that things are more subtle when some $\Lambda_{ii} = 0$, as this inequality fails for $\rho = 1$!) The other inequalities $\alpha_i \geq \beta_j$ follow by comparing $c_{(\rho,2k)}$ and $c_{(\rho,2k+2)}$ in the identical way.

As mentioned earlier, S and U are conjugate to $e^{2\pi\mathrm{i}\mathcal{A}_2}$ and $e^{2\pi\mathrm{i}\mathcal{A}_3}$, respectively, but identifying precisely *which* conjugate is a transcendental and subtle question. For example, we see in Sect. 4.2 below that when $d = 2$, they are related by Gamma function values.

3.4 Holomorphic vvmf

Until this point in the paper, our focus has been on weakly holomorphic vvmf of fixed weight, i.e. vvmf holomorphic everywhere except at the Γ-orbit of $\mathrm{i}\infty$. The reason is that structurally it is the simplest and most fundamental. For example, it is acted on by the ring of (scalar) modular functions holomorphic away from $\Gamma\mathrm{i}\infty$, which for any genus-0 Fuchsian group is a PID. By contrast, the holomorphic vvmf (say of arbitrary even integral weight) is a module over the ring of holomorphic modular forms, which in genus-0 is usually not even polynomial. However, there is probably more interest in holomorphic vvmf, so it is to these we now turn. The two main questions we address are the algebraic structure (see Theorem 3.4 below), and dimensions (see Theorem 3.5 next subsection).

Definition 3.1. Let (ρ, w) be admissible, T diagonal, and λ any exponent. Define

$$\mathcal{M}_w^\lambda(\rho) := \ker \mathcal{P}_{\lambda - 1_d} = \left\{ \mathbb{X} \in \mathcal{M}_w^!(\rho) \mid \mathbb{X}(\tau) = q^\lambda \sum_{n=0}^{\infty} \mathbb{X}_{(n)} q^n \right\} \tag{40}$$

and $\mathcal{M}^\lambda(\rho) = \coprod_{k\in\mathbb{Z}} \mathcal{M}_{w+2k}^\lambda(\rho)$. We call any $\mathbb{X} \in \mathcal{M}^\lambda(\rho)$, *$\lambda$-holomorphic*.

For example, for the trivial representation, $\mathcal{M}^0(1)$ are the modular forms $\mathfrak{m} = \mathbb{C}[E_4, E_6]$, while $\mathcal{M}^1(1)$ are the cusp forms $\mathfrak{m}\Delta$. More generally, define λ^{hol} to be the unique exponent with $0 \leq \mathrm{Re}\,\lambda_{ii} < 1$ for all i. Then $\mathcal{M}^{\lambda^{hol}}(\rho)$ coincides with the usual definition of holomorphic vvmf. Choosing $0 < \mathrm{Re}\,\lambda_{ii} \leq 1$ would give the vector-valued cusp forms.

Theorem 3.2 computes $\dim \mathcal{M}^{\lambda}_{w+2k}(\rho)$ for all sufficiently large $|k|$:

Lemma 3.2. *Let (ρ, w) be admissible and T diagonal. Let λ be any exponent, and Λ any bijective exponent.*

(a) *For any $k \in \mathbb{Z}$, $\mathcal{M}^{\lambda}_{w+2k}(\rho)$ is finite-dimensional, and obeys the bound*

$$\dim \mathcal{M}^{\lambda}_{w+2k}(\rho) \geq \max\left\{0, \frac{w+2k+2}{12}d + \frac{\alpha_k}{2} + \frac{\beta_k - \beta_{k+2}}{3} - \mathrm{Tr}\,\lambda\right\}. \quad (41)$$

(b) *Choose any $m, n \in \mathbb{Z}$ satisfying $\mathrm{Re}\,(\Lambda + m1_d) \geq \mathrm{Re}\,\lambda \geq \mathrm{Re}\,(\Lambda - n1_d)$ entrywise. Then $\mathcal{M}^{\lambda}_{w-2k}(\rho) = 0$ when $k = 6n+6$ or $k \geq 6n+8$, and equality holds in (41) whenever $k = 6m-6$ or $k \geq 6m-4$.*

(c) *Let w_0 be the weight with smallest real part for which $\mathcal{M}^{\lambda}_{w_0}(\rho) \neq 0$ and write $\epsilon = 0, 1$ for d even, odd respectively. Suppose ρ is irreducible and not 1-dimensional. Then w_0 satisfies the bounds*

$$\frac{12}{d}\mathrm{Tr}\,\lambda + 1 - d \leq w_0 \leq \frac{12}{d}\mathrm{Tr}\,\lambda - \frac{3\epsilon}{d}. \quad (42)$$

Proof. Theorem 3.2(a) gives finite-dimensionality. The bound (41) follows from $\dim\ker \mathcal{P}_{\lambda-1_d} \geq \mathrm{index}\,\mathcal{P}_{\lambda-1_d}$, the index formula in Theorem 3.2(a), and (29). Note that $\Delta^k \mathcal{M}^!_w(\rho) = \mathcal{M}^!_{w+2k}(\rho)$ for any $k \in \mathbb{Z}$ so $\Lambda + k1_d$ is bijective for $(\rho, w+12k)$. Hence $\mathcal{P}_{\lambda-1_d}$ is injective on $\mathcal{M}^!_{w-12(n+1)}(\rho)$ because $\mathcal{P}_{\Lambda-(n+1)1_d}$ is, while $\mathcal{P}_{\lambda-1_d}$ is surjective on $\mathcal{M}^!_{w+12(m-1)}(\rho)$ because $\mathcal{P}_{\Lambda+(m-1)1_d}$ is. This proves (b) for those weights. Now, for any $k \geq 2$ there is a scalar modular form $f \in \mathfrak{m}$ of weight $2k$ with nonzero constant term, so $f(\tau)^{-1} \in \mathbb{C}[[q]]$ and the surjectivity of $\mathcal{P}_{\lambda-1_d}$ on $\mathcal{M}^!_{w+12(m-1)}(\rho)$ implies that on $\mathcal{M}^!_{w+12(m-1)+2k}(\rho)$. More directly, injectivity of $\mathcal{P}_{\lambda-1_d}$ on $\mathcal{M}^!_{w-12(n+1)}(\rho)$ implies that on $\mathcal{M}^!_{w-12(n+1)+2k}(\rho)$.

Now turn to (c). We know that $\dim \mathcal{M}^{\lambda}_w(\rho) > 0$ for any w for which $\mathrm{Tr}\,\Lambda > \mathrm{Tr}\,\lambda - d$, since $\dim\ker \mathcal{P}_{\lambda-1_d} \geq \mathrm{Index}\,\mathcal{P}_{\lambda-1_d} > 0$. Using (31), we obtain the upper bound of (42).

Choose any nonzero $\mathbb{X} \in \mathcal{M}^{\lambda}_w(\rho)$. Then its components must be linearly independent over $\mathbb{C}$, because they span a subrepresentation of the irreducible ρ. Therefore, Lemma 2.2(a) says $\mathrm{Wr}(\mathbb{X})(\tau) \in \mathcal{M}^!_{d(w+d-1)}(\det\rho)$ is nonzero, with leading power of q in $\mathrm{Tr}\,\lambda + \mathbb{Z}_{\geq 0}$. This implies $\mathrm{Wr}(\mathbb{X})\tau)/\Delta^{\mathrm{Tr}\,\lambda}(\tau)$ lies in $\mathcal{M}^0_{dw+d(d-1)-12\mathrm{Tr}\,\lambda}(\upsilon_u)$ for some $u \in 2\mathbb{Z}$, from which follows the lower bound of (42). □

The lower bound in (42) is due to Mason [23] (he proved it for $\lambda = \lambda^{hol}$ but the generalisation given here is trivial). The $d \neq 1$ assumption in (c) is only needed for the lower bound.

Theorem 3.3(a) tells us the space $\mathcal{M}^!_w(\rho)$ of weakly holomorphic vvmf is a free module of rank d over $\mathbb{C}[J]$. The analogous statement for holomorphic vvmf, namely that $\mathcal{M}^{\lambda^{hol}}(\rho)$ is free of rank d over m, is implicit in [13] (see the Remark there on page 98). It was also proved independently in [22], and independently but simultaneously we obtained the following generalisation. The proof of freeness given here is far simpler than in [22], is more general (as it applies to arbitrary λ), gives more information (see (b) below), and generalises directly to arbitrary T and arbitrary genus-0 groups [14].

Theorem 3.4. *Let (ρ, w) be admissible and T diagonal. Choose any exponent λ. Let $\alpha_j = \alpha_j(\rho, w)$ and $\beta_j = \beta_j(\rho, w)$.*

(a) *$\mathcal{M}^\lambda(\rho)$ is a free module over $\mathrm{m} = \mathbb{C}[E_4, E_6]$, of rank d.*

(b) *Let $w_0 = w^{(1)} \le w^{(2)} \le \cdots \le w^{(d)}$ be the weights of the free generators. Then precisely α_i of the $w^{(j)}$ will be congruent mod 4 to $w + 2i$, and precisely β_i of them will be congruent mod 6 to $w - 2i$. Moreover, $\sum_j w^{(j)} = 12\,\mathrm{Tr}\,\lambda$. Let Ξ^λ be the matrix obtained by putting these d generators into d columns. Then (up to an irrelevant nonzero constant)*

$$\det \Xi^\lambda = \Delta^{\mathrm{Tr}\,\lambda}\,.$$

Proof. Let $w_0 \in w + 2\mathbb{Z}$ be the weight of minimal real part with $\mathcal{M}^\lambda_{w_0}(\rho) \neq 0$—this exists by Lemma 3.2(b). Write $\mathcal{M} = \mathcal{M}^\lambda(\rho)$, $w_k = w_0 + 2k$ and $\mathcal{M}_l = \mathcal{M}^\lambda_{w_l}(\rho)$. For $\mathbb{X}(\tau) \in \mathcal{M}_w(\rho)$, recall from (13) that the constant term $\mathbb{X}_{[0]}$ at $\tau = \mathrm{i}$ is $\mathbb{X}(\mathrm{i})/E_4(\mathrm{i})^{w/4}$. Fix $S_0 = e^{\pi \mathrm{i} w_0/2} S$; then $e^{\pi \mathrm{i} w_k/2} S = (-1)^k S_0$ and so for any $\mathbb{X} \in \mathcal{M}_k$, its constant term satisfies $S_0 \mathbb{X}_{[0]} = (-1)^k \mathbb{X}_{[0]}$ thanks to (14).

Find $\mathbb{X}^{(i)}(\tau) \in \mathcal{M}_{l_i}$ with the property that, for any $k \geq 0$, the space of constant terms $\mathbb{X}_{[0]}$, as $\mathbb{X}(\tau)$ runs over all $\cup_{l=0}^k \mathcal{M}_l$, has a basis given by the constant terms $\mathbb{X}^{(i)}_{[0]}$ for those $\mathbb{X}^{(i)}(\tau)$ in $\cup_{l=0}^k \mathcal{M}_l$ (i.e. for those i with $l_i \leq k$). This is done recursively with k. We will show that these $\mathbb{X}^{(i)}(\tau)$ are the desired free generators for $\mathcal{M}$.

The key observation is the following. Consider any $\mathbb{X}(\tau) \in \mathcal{M}_k$. Then by definition of the $\mathbb{X}^{(i)}(\tau)$, $\mathbb{X}_{[0]} = \sum_i c_i \mathbb{X}^{(i)}_{[0]}$ where $c_i = 0$ unless $l_i \leq k$ and $l_i \equiv k \pmod 2$. The key observation is that the constant term $\mathbb{X}'_{[0]}$ of $\mathbb{X}'(\tau) = \mathbb{X}(\tau) - \sum_i c_i\, E_4(\tau)^{(k-l_i)/2} \mathbb{X}^{(i)}(\tau)$ is 0, so $\mathbb{X}'(\tau)/E_6(\tau) \in \mathcal{M}_{k-3}$.

One consequence of this observation is that, by an easy induction on k, any $\mathbb{X}(\tau) \in \mathcal{M}$ must lie in $\oplus_i \mathrm{m}\mathbb{X}^{(i)}(\tau)$. Another consequence is that there are exactly d of these $\mathbb{X}^{(i)}(\tau)$, in particular their constant terms form a basis for $\mathbb{C}^d$. To see this, take any fundamental matrix $\Xi(\tau)$ at weight w_0. Then for sufficiently large l, each column of $\Delta(\tau)^l\, \Xi(\tau)$ is in $\mathcal{M}_{12l}$. The constant terms of $\Delta(\tau)^l\, \Xi(\tau)$ (which have S_0-eigenvalues +1) and of $\Delta(\tau)^l \mathrm{D}_{w_0} \Xi(\tau)$ (which have S_0-eigenvalues -1) must have rank $\alpha_0(\rho, w_0)$ and $\alpha_1(\rho, w_0)$, respectively, as otherwise the columns of $\Delta(\tau)^l\, \Xi(\tau)$ would be linearly dependent over m, contradicting their linear independence over $\mathbb{C}[J]$. This means of course that exactly $\alpha_0(\rho, w_0)$ of these $\mathbb{X}^{(i)}(\tau)$ have l_i even, and exactly $\alpha_1(\rho, w_0)$ have l_i odd.

Thus these d $\mathbb{X}^{(i)}(\tau)$ generate over m all of $\mathcal{M}$. To see they are linearly independent over m, suppose we have a relation $\sum_i p_i(\tau)\,\mathbb{X}^{(i)}(\tau) = 0$, for modular forms $p_i(\tau) \in$ m which do not share a (nontrivial) common divisor. The constant term at i of that relation reads $\sum_i p_i(\mathrm{i})\,\mathbb{X}^{(i)}_{[0]} = 0$ and hence each $p_i(\mathrm{i}) = 0$, since the $\mathbb{X}^{(i)}_{[0]}$ are linearly independent by construction. This forces all $p_i(\tau)$ to be 0, since otherwise we could divide them all by E_6, which would contradict the hypothesis that they share no common divisor.

The identical argument applies to the constant terms $\mathbb{X}_{\langle 0\rangle} = \mathbb{X}(\xi_6)/E_6(\xi_6)^{w/6}$ at $\tau = \xi_6$; this implies that exactly $\beta_i(\rho, w_0)$ generators $\mathbb{X}^{(j)}(\tau)$ have $l_j \equiv i \pmod 3$.

Form the $d \times d$ matrix $\Xi^\lambda(\tau)$ from these d generators $\mathbb{X}^{(i)}(\tau)$ and call the determinant $\delta(\tau)$. The linear independence of the constant terms of the generators at the elliptic fixed points, says that δ cannot vanish at any elliptic fixed point. δ also can't have a zero anywhere else in $\mathbb{H}$. To see this, first note that a zero at $\tau^* \in \mathbb{H}$ implies there is nonzero row vector $v \in \mathbb{C}^d$ such that $v\Xi^\lambda(\tau) = 0$ and hence $v\mathbb{X}(\tau^*) = 0$ for any $\mathbb{X}(\tau) \in \mathcal{M}^\lambda(\rho)$. But (39) says the determinant of any fundamental matrix $\Xi(\tau)$ for (ρ, w_0) can only vanish at elliptic fixed points and cusps, and so $v\Xi(\tau^*) \neq 0$ for any fundamental matrix, and hence $v\mathbb{Y}(\tau^*) \neq 0$ for some $\mathbb{Y}(\tau) \in \mathcal{M}^!_{w_0}(\rho)$. To get a contradiction, choose N big enough so that $\Delta(\tau)^N\mathbb{Y}(\tau) \in \mathcal{M}$. This means $\delta(\tau)$ is a scalar modular form (with multiplier) which doesn't vanish anywhere in $\mathbb{H}$. Hence $\delta(\tau)$ must be a power of $\Delta(\tau)$, and considering weights we see this must be $\delta(\tau) = \Delta^{\sum_i w_{l_i}/12}(\tau)$. We compute that sum over i, shortly.

Find the smallest ℓ such that $d + c_{(\rho, w_0+2\ell)} > \mathrm{Tr}\,\lambda$ (recall (30)) and put $w_0' := w_\ell$. Define $n_k' = 0$ for $k < 0$, and

$$n_k' = \frac{w_0' + 2k + 2}{12}d + \frac{\alpha_k}{2} + \frac{\beta_k - \beta_{2+k}}{3} - \mathrm{Tr}\,\lambda \tag{43}$$

for $k \geq 0$. Using (29), the numbers n_k' are the values $\max\{c_{(\rho, w_0'+2k)}, 0\}$. From Lemma 3.2 we obtain the equality $n_k = n_{k-\ell}'$ for $|k|$ sufficiently large. The 'tight' Hilbert–Poincaré series $H_{tt}^\lambda(\mathcal{M}; x) := \sum_k n_k' x^{w_{k+\ell}}$ equals

$$H_{tt}^\lambda(\mathcal{M}; x) = x^{w_0'} \frac{n_0' + n_1'x^2 + (n_2' - n_0')x^4 + (n_3' - n_1' - n_0')x^6 + (n_4' - n_2' - n_1')x^8}{(1 - x^4)(1 - x^6)}, \tag{44}$$

by a simple calculation (the significance of $H_{tt}^\lambda(x)$ is explained in Proposition 3.3 below). From the numerator we read off the weights of the 'tight' generators: write $w'^{(i)} = w_0'$ for $1 \leq i \leq n_0'$, …, $w'^{(i)} = w_0' + 8$ for $d - n_4' + n_2' + n_1' < i \leq d$. We know that the actual Hilbert–Poincaré series, $H^\lambda(\mathcal{M}; x) = \sum_k n_k x^{w_k}$, minus the tight one, equals a finite sum of terms, since $n_k = n'_{k-\ell}$ for $|k|$ large. Therefore $(1-x^4)(1-x^6)(H^\lambda(x) - H^\lambda_{tt}(x)) = \sum x^{w^{(i)}} - \sum x^{w'^{(i)}}$ is simply a polynomial identity. Differentiating with respect to x and setting $x = 1$ gives $\sum w^{(i)} = \sum w'^{(i)}$, and the latter is readily computed to be $12\,\mathrm{Tr}\,\lambda$. □

This freeness doesn't seem directly related to that of Theorem 3.3(a). The reason it is natural to look at local expansions about $\tau = \mathrm{i}$ and $\tau = \zeta_6$ in the proof of (a) is because if $\mathbb{X}(\tau)$ is holomorphic and $\mathbb{X}(\mathrm{i}) = 0$, then $\mathbb{X}(\tau)/E_6(\tau)$ is also holomorphic (similarly for $\mathbb{X}(\zeta_6) = 0$ and $\mathbb{X}(\tau)/E_4(\tau)$).

Call an admissible (ρ, w) *tight* if it has the property that for all $k \in \mathbb{Z}$, an exponent λ is bijective for $w + 2k$ iff $\mathrm{Tr}\,\lambda = c_{(\rho,w+2k)}$. In this case we also say ρ is tight. Generic ρ are tight. We learn in Theorem 4.1 that all irreducible ρ in dimension $d < 6$ are tight. For tight ρ, most quantities can be easily determined:

Proposition 3.3. *Suppose (ρ, w) is admissible and tight, and T is diagonal.*

(a) *Then $\upsilon_u \otimes \rho$ is also tight for any $u \in \mathbb{C}$, as is the contragredient $\rho^* = (\rho^t)^{-1}$.*
(b) *For any exponent λ (not necessarily bijective), either $\ker \mathcal{P}_\lambda = 0$ (if $\mathrm{Tr}\,\lambda \geq c_{(\rho,w)}$) or $\mathrm{coker}\, \mathcal{P}_\lambda = 0$ (if $\mathrm{Tr}\,\lambda \leq c_{(\rho,w)}$). Moreover,*

$$\dim \mathcal{M}^\lambda_w(\rho) = \max\{0, c_{(\rho,w)} + d - \mathrm{Tr}\,\lambda\}\,, \tag{45}$$

and the Hilbert–Poincaré series $H^\lambda(x)$ of $\mathcal{M}^\lambda(\rho)$ equals the tight Hilbert–Poincaré series $H^\lambda_{tt}(x)$ of (44).

The proof of (a) uses the equality $\mathcal{M}^!_{w+u}(\upsilon_u \otimes \rho) = \Delta^{u/12}(\tau)\mathcal{M}^!_w(\rho)$, as well as the duality in Proposition 3.4 below. To prove (b), let Λ be the unique bijective exponent matrix for (ρ, w) satisfying $\Lambda_{ii} = \lambda_{ii} - 1$ for all $i \neq 1$. Write $n = \Lambda_{11} + 1 - \lambda_{11} = c_{(\rho,w)} + d - \mathrm{Tr}\,\lambda \in \mathbb{Z}$. If $n \geq 0$ $\mathcal{P}_\lambda$ inherits the surjectivity of $\mathcal{P}_\Lambda$, while if $n \leq 0$ it inherits the injectivity. The index formula (24) gives the dimension, which by (29) equals $n'_{k-\ell}$. This is why H^λ_{tt} arises.

As long as (ρ, w) is tight, this argument tells us how to find a basis for $\mathcal{M}^\lambda_w(\rho)$. Let Λ and n be as above. For $n > 0$ a basis for $\mathcal{M}^\lambda_w(\rho)$ consists of the basis vectors $\mathbb{X}^{(1;i)}$ (see Sect. 3.2) for $0 \leq i < n$. Incidentally, the name 'tight' refers to the fact that the numerator of H^λ_{tt} is maximally bundled together.

There are other constraints on the possible weights $w^{(i)}$ of Theorem 3.4(b). A useful observation in practice is that if ρ is irreducible, then the set $\{w^{(1)}, \ldots, w^{(d)}\}$ can't have gaps, i.e. for $n = (w^{(d)} - w^{(1)})/2$,

$$\{w^{(1)}, \ldots, w^{(d)}\} = \{w^{(1)}, w^{(1)} + 2, w^{(1)} + 4, \ldots, w^{(1)} + 2n\}\,. \tag{46}$$

The reason is that when $w^{(1)}+2k$ doesn't equal any $w^{(i)}$, then when $w^{(j)} < w^{(1)}+2k$, $\mathrm{D}^{k-(w^{(j)}-w^{(1)})/2}\mathbb{X}^{(j)} = \sum_{w^{(l)}<w^{(1)}+2k} f_{jl}\mathbb{X}^{(l)}$ for $f_{jl} \in \mathrm{m}$, where the sum is over all l with $w^{(l)} < w^{(1)} + 2k$. If in addition $w^{(1)} + 2k < w^{(d)}$ (i.e. $w^{(1)} + 2k$ is a *gap*), then the Wronskian $\mathrm{Wr}(\mathbb{X}^{(1)})$ would have to vanish, contradicting irreducibility.

Several papers (e.g. [21, 23]) consider a 'cyclic' class of vvmf where the components of $\mathbb{X}$ span the solution space to a monic modular differential equation $\mathrm{D}_k^d + \sum_{l=0}^{d-1} f_l D_k^l = 0$, where each $f_l \in \mathfrak{m}$ is of weight $2d - 2l$. In this accessible case the free generators $\mathbb{X}^{(i)}$ can be taken to be $\mathbb{X}, \mathrm{D}_k\mathbb{X}, \ldots, \mathrm{D}_k^{d-1}\mathbb{X}$, i.e. the weights are $\alpha^{(i)} = \alpha^{(1)} + (i-1)2$. This means the corresponding ρ *cannot* be tight, when $d \geq 6$. Indeed, the multipliers of such vvmf are exceptional, requiring the multiplicities of ρ to satisfy $|\alpha_i - \alpha_j| \leq 1$ and $|\beta_i - \beta_j| \leq 1$. Recall that the connected components of the moduli space of $\overline{\Gamma}$-representations are parametrised by these multiplicities; these particular components are of maximal dimension. For example, when $d = 6$, such a representation ρ can lie in only 1 of the 12 possible connected components, and these ρ define a 6-dimensional subspace inside that 7-dimensional component.

3.5 Serre Duality and the Dimension Formula

A crucial symmetry of the theory is called the *adjoint* in the language of Fuchsian equations, and shortly we reinterpret this as Serre duality.

Proposition 3.4. *Let* (ρ, w) *be admissible,* T *diagonal, and* Λ *bijective, and let* $\Xi(\tau)$ *be the associated fundamental matrix of* $\mathcal{M} = \mathcal{M}_w^!(\rho)$. *Let* ρ^* *denote the contragredient* $(\rho^{-1})^t$ *of* ρ. *Then* $\mathcal{M}_w^!(\rho)$ *and* $\mathcal{M}_{2-w}^!(\rho^*)$ *are naturally isomorphic as* $\mathbb{C}[J, \nabla_1, \nabla_2, \nabla_3]$*-modules. Moreover,*

$$\Xi^*(\tau) = E_4(\tau)^2 E_6(\tau)\, \Delta(\tau)^{-1} \left(\Xi(\tau)^t\right)^{-1}, \tag{47}$$

$$\mathcal{X}^*(z,q) = -\,\mathcal{X}(q,z)^t, \tag{48}$$

where Ξ^* *is the fundamental matrix of* $\mathcal{M}_{2-w}^!(\rho^*)$ *corresponding to the bijective exponent* $\Lambda^* = -1_d - \Lambda$, *and* $\mathcal{X}, \mathcal{X}^*$ *are the generating functions* (37) *for* (ρ, w) *and* $(\rho^*, 2-w)$ *respectively. In other words the* $q^{m+\Lambda_{ii}}$ *coefficient* $\mathcal{X}_{(m)i}^{(j;n-1)}$ *of the basis vector* $\mathbb{X}_i^{(j;n-1)}(\tau)$ *is the negative of the* $q^{n+\Lambda_{jj}^*}$ *coefficient* $\mathcal{X}_{(n)j}^{*(i;m-1)}$ *of the basis vector* $\mathbb{X}_j^{*(i;m-1)}(\tau)$, *for all* $m, n \geq 1$.

Proof. Define $\Xi^*(\tau)$ by (47); to show it is the fundamental matrix associated to the bijective exponent $\Lambda^* = -1_d - \Lambda$, we need to show that $\Xi^*(\tau) = q^{-1_d - \Lambda}(1_d + \sum_{n=1}^{\infty} \Xi_{(n)}^* q^n)$ (this is clear), that $-1_d - \Lambda$ has trace $c_{(\rho^*, 2-w)}$ (we'll do this next), and that the columns of $\Xi^*(\tau)$ are in $\mathcal{M}_{2-w}^!(\rho^*)$. Using (29), we find $\alpha_i(\rho^*, 2-w) = \alpha_{i+1}(\rho, w)$, and $\beta_j(\rho^*, 2-w) = \beta_{2-j}(\rho, w)$, so we compute from (25) that $c_{(\rho^*, 2-w)} = \mathrm{Tr}\,(-1_d - \Lambda)$.

From $\Xi(\gamma\tau) = \tilde{\rho}_w(\gamma, \tau)\, \Xi(\tau)$, we get $(\Xi(\gamma\tau)^t)^{-1} = \tilde{\rho}_{-w}^*(\gamma, \tau)(\Xi(\tau)^t)^{-1}$. It thus suffices to show $\Xi^*(\tau)$ is holomorphic in $\mathbb{H}$. We see from (39) that $\Xi(\tau)^{-1}$ is meromorphic everywhere in $\mathbb{H}^*$, with poles possible only at the elliptic points and the cusp. Locally about $\tau = \mathrm{i}$, (35) tells us

$$(\Xi(\tau)^t)^{-1} = \mathrm{j}(-w;\tau)\, P_2^t q_2^{-\lambda_2} \left(1_d + \sum_{n=1}^{\infty} (\Xi_{[0]}^t)^{-1} \Xi_{[n]}^t q^n\right)^{-1} (\Xi_{[0]}^t)^{-1}. \quad (49)$$

The series in the middle bracketed factor is invertible at $\tau = \mathrm{i}$, because its determinant equals 1 there. So every entry of $(\Xi(\tau)^t)^{-1}$ at $\tau = \mathrm{i}$ has at worst a simple pole (coming from $q_2^{-\lambda_2} = \tau_2^{-2\lambda_2}$). But $J' = -E_4^2 E_6/\Delta$ has a simple pole at $\tau = \mathrm{i}$. Therefore $\Xi^*(\tau)$ is holomorphic at $\tau = \mathrm{i}$. Likewise, at $\tau = \xi_6$, the entries of $(\Xi(\tau)^t)^{-1}$ have at worse an order 2 pole (coming from $q_3^{-\lambda_3} = \tau_3^{-3\lambda_3}$), but J' has an order 2 zero at $\tau = \xi_6$, so $\Xi^*(\tau)$ is also holomorphic at $\tau = \xi_6$. Thus the columns of $\Xi^*(\tau)$ lie in $\mathcal{M}_{2-w}^!(\rho^*)$ (the 2 comes from J').

This concludes the proof that Ξ^* is a fundamental matrix for $(\rho^*, 2-w)$. Equation (48) now follows directly from (38). □

A special case of Eq. (47) was found in [5]. An interesting special case of (48) is that the constant term of any weakly holomorphic modular form $f \in \mathcal{M}_2^!(1)$ is 0. To see this, recall first that $(\rho, w) = (1, 0)$ has $\Lambda = 0$ and $\Xi(\tau) = \mathbb{X}^{(1;0)}(\tau) = 1$, i.e. the q^{m+0}-coefficient of $\mathbb{X}_1^{(1;0)}$ vanishes for all $m \geq 1$. Then (48) says the q^{1-1}-coefficient of all $\mathbb{X}_1^{(1;m-1)}$ must also vanish. Since the $\mathbb{X}^{(1;k)}$ span (over $\mathbb{C}$) all of $\mathcal{M}_2^!(1)$, the constant term of any $f \in \mathcal{M}_2^!(1)$ must vanish. The same result holds for any genus-0 group.

The final fundamental ingredient of our theory connects the index formula (24) to this duality:

Theorem 3.5. *Let (ρ, w) be admissible and T diagonal, and recall the quantity $c_{(\rho,w)}$ computed in* (25). *Then for any exponent λ,*

$$\operatorname{coker} \mathcal{P}_{\lambda;(\rho,w)} \cong \left(\mathcal{M}_{2-w}^{-\lambda}(\rho^*)\right)^*, \quad (50)$$

$$\dim \mathcal{M}_w^{\lambda}(\rho) - \dim \mathcal{M}_{2-w}^{1_d-\lambda}(\rho^*) = c_{(\rho,w)} + d - \operatorname{Tr}\lambda. \quad (51)$$

Proof. Let $\mathbb{X}(\tau) \in q^{\lambda}\mathbb{C}^d[q^{-1}, q]]$, i.e. $\mathbb{X}(\tau) = q^{\lambda} \sum_{n=-N}^{\infty} \mathbb{X}_{(n)} q^n$ for some $N = N(\mathbb{X})$, and let $\mathbb{Y}(\tau) \in \mathcal{M}_{2-w}^{-\lambda}(\rho^*)$, and define a pairing $\langle \mathbb{X}, \mathbb{Y}\rangle$ to be the q^0-coefficient f_0 of $\mathbb{X}(\tau)^t \mathbb{Y}(\tau) = \sum_{i=1}^d \mathbb{X}_i(\tau)\, \mathbb{Y}_i(\tau) = \sum_{n=-N}^{\infty} f_n q^n$. Note that $\langle \mathbb{X}, \mathbb{Y}\rangle = \sum_{-N\leq n\leq 0} \mathbb{X}_{(n)} \mathbb{Y}_{(-n)}$ depends only on the coefficients $\mathbb{X}_{(n)}$ for $n \leq 0$, since by hypothesis $\mathbb{Y}_{(k)} = 0$ for $k < 0$. In other words, the pairing $\langle \mathbb{X}, \mathbb{Y}\rangle$ depends only on $\mathbb{Y}(\tau)$ and the principal part $\mathcal{P}_{\lambda}\mathbb{X}(q)$ of $\mathbb{X}(\tau)$.

If $\mathbb{X}(\tau) \in \mathcal{M}_w^!(\rho)$, then $\mathbb{X}(\tau)^t \mathbb{Y}(\tau)$ will lie in $\mathcal{M}_2^!(1)$. Hence from the observation after Proposition 3.4, in that case $\langle \mathbb{X}, \mathbb{Y}\rangle$ will vanish. This means the pairing $\langle \mathbb{X}, \mathbb{Y}\rangle$ is a well-defined pairing between the cokernel of $\mathcal{P}_{\lambda;(\rho,w)}$, and $\mathcal{M}_{2-w}^{-\lambda}(\rho^*)$.

Let $\mathbb{Y}^1(\tau), \ldots, \mathbb{Y}^m(\tau)$ be a basis of $\mathcal{M}_{2-w}^{-\lambda}(\rho^*)$. We can require that this basis be triangular in the sense that for each $1 \leq i \leq m$ there is an n_i, k_i so that the coefficient $\mathbb{Y}^j_{(n_i)\,k_i} = \delta_{ji}$ for all i, j. Indeed, choose any n_1, k_1 such that $\mathbb{Y}^1_{(n_1)\,k_1} \neq 0$, and rescale $\mathbb{Y}^1(\tau)$ so that coefficient equals 1. Subtract if necessary a multiple of $\mathbb{Y}^1(\tau)$ from the other $\mathbb{Y}^i(\tau)$ so that $\mathbb{Y}^i_{(n_1)\,k_1} = 0$. Now, repeat: choose any n_2, k_2

such that $\mathbb{Y}^2_{(n_2)k_2} \neq 0$, etc. If we take $\mathbb{X}^i(\tau)$ to be $q^{-n_i}e_{k_i}$ (i.e. all coefficients vanish except one coefficient in one component), then $\langle \mathbb{X}^i, \mathbb{Y}^j \rangle = \delta_{ij}$. This form of nondegeneracy means that $\dim \mathcal{M}^{-\lambda}_{2-w}(\rho^*) \leq \dim \operatorname{coker} \mathcal{P}_{\lambda;(\rho,w)}$.

Repeating this argument in the dual direction, more precisely replacing ρ, w, λ with $\rho^*, 2 - w, -1_d - \lambda$ respectively, gives us the dual inequality $\dim \mathcal{M}^{\lambda+1_d}_w(\rho) \leq \dim \operatorname{coker} \mathcal{P}_{-1_d-\lambda;(\rho^*,2-w)}$. However, from Proposition 3.4 we know that $c_{(\rho^*,2-w)} = \operatorname{Tr}(-1_d - \Lambda) = -d - c_{(\rho,w)}$, so the index formula (24) gives us both

$$\dim \mathcal{M}^{\lambda+1_d}_w(\rho) - \dim \operatorname{coker} \mathcal{P}_{\lambda;(\rho,w)} = c_{(\rho,w)} - \operatorname{Tr} \lambda \,, \tag{52}$$

$$\dim \mathcal{M}^{-\lambda}_{2-w}(\rho^*) - \dim \operatorname{coker} \mathcal{P}_{-1_d-\lambda;(\rho^*,2-w)} = -d - c_{(\rho,w)} + d + \operatorname{Tr} \lambda \,. \tag{53}$$

Adding these gives

$$\dim \mathcal{M}^{\lambda+1_d}_w(\rho) + \dim \mathcal{M}^{-\lambda}_{2-w}(\rho^*) = \dim \operatorname{coker} \mathcal{P}_{\lambda;(\rho,w)} + \dim \operatorname{coker} \mathcal{P}_{-1_d-\lambda;(\rho^*,2-w)} \,.$$

Together with the two inequalities, this shows that the dimensions of $\mathcal{M}^{-\lambda}_{2-w}(\rho^*)$ and $\operatorname{coker} \mathcal{P}_{\lambda;(\rho,w)}$ match. Hence the pairing $\langle \mathbb{X}, \mathbb{Y} \rangle$ is nondegenerate and establishes the isomorphism (50). Equation (51) now follows immediately from the index formula (24). □

We suggest calling part (a) *Serre duality* because $\ker \mathcal{P}_\lambda$ has an interpretation as $H^0(B_\Gamma; \mathcal{A})$ for some space $\mathcal{A} = \mathcal{A}_{\lambda;(\rho,w)}$ of meromorphic functions on which Γ acts by (ρ, w), while $\operatorname{coker} \mathcal{P}_\lambda$, being the obstruction to finding meromorphic sections of our (ρ, w)-vector bundle, should have an interpretation as some H^1. The shifts by 1_d and 2 would be associated to the canonical line bundle. This interpretation is at this point merely a heuristic, however.

Compare (50) to Theorem 3.1 in [10]. There, Borcherds restricts attention to weight $w = k$ a half-integer, groups commensurable to $\mathrm{SL}_2(\mathbb{Z})$, representations ρ with finite image, and $\lambda = \lambda^{hol}$. The assumption of finite image was essential to his proof. Our proof extends to arbitrary T and arbitrary genus-0 groups [14] (so by inducing the representation, it also applies to any finite-index subgroup of a genus-0 group).

The most important special case of the dimension formula (51) is $\lambda = \lambda^{hol}$, which relates the dimensions of holomorphic vvmf with those of vector-valued cusp forms. Compare (51) to [33], where Skoruppa obtained the formula assuming $2w \in \mathbb{Z}$, and that ρ has finite image. His proof used the Eichler-Selberg trace formula. Once again we see that these results hold in much greater generality. Of course thanks to the induction trick, (51) also gives the dimension formulas for spaces of e.g. holomorphic and cusp forms, for any finite index subgroup of Γ.

Incidentally, it is possible to have nonconstant modular functions, holomorphic everywhere in $\mathbb{H}^*$, for some multipliers with infinite image (see e.g. Remark 2 in Sect. 3 of [17]). On the other hand, Lemma 2.4 of [17] prove that for *unitary* multipliers ρ, there are no nonzero holomorphic vvmf $\mathbb{X} \in \mathcal{M}^{\lambda^{hol}}_w(\rho)$ of weight

$w < 0$. This implies there are no vector-valued cusp forms of weight $w \le 0$, for unitary ρ.

3.6 The q-Expansion Coefficients

Modular forms—vector-valued or otherwise—are most important for their q-expansion coefficients. In this subsection we study these. A special case of Proposition 3.5(b) (namely, $w = 0$ and $\mathbb{Q}_{\mathbb{X}} = \mathbb{Q}$) is given in [1], but our proof generalises without change to T nondiagonalisable and to any genus-0 group [14].

As a quick remark, note from Theorem 3.3(b) that the coefficients $\mathbb{X}_{(n)i}$ of a vvmf $\mathbb{X} \in \mathcal{M}_w^!(\rho)$ will all lie in the field generated over $\mathbb{Q}$ by the entries of Λ_w and χ_w, as well as the coefficients $\mathbb{X}_{(n)i}$ of the principal part $\mathcal{P}_\Lambda(\mathbb{X})$, where Λ is bijective and $\chi = \Xi_{(1)}$.

Let $\mathcal{F}$ denote the span of all $p(\tau)\,q^u h(q)$, where $p(\tau) \in \mathbb{C}[\tau]$, $u \in \mathbb{C}$, and $h \in \mathbb{C}[[q]]$ is holomorphic at $q = 0$. The components of any vvmf (including 'logarithmic', where T is nondiagonalisable) must lie in $\mathcal{F}$. $\mathcal{F}$ is a ring, where terms simplify in the obvious way (thanks to the series fs converging absolutely). $\mathcal{F}$ is closed under differentiation and $\tau \mapsto \tau + 1$. The key to analysing functions in $\mathcal{F}$ is the fact that they are equal only when it is obvious that they are equal:

Lemma 3.2. *Suppose $\sum_{i=1}^n p_i(\tau)\,q^{u_i} h_i(q) = 0$ for all q in some sufficiently small disc about $q = 0$, where each $p_i(\tau) \in \mathbb{C}[\tau]$, $u_i \in \mathbb{C}$, $0 \le \mathrm{Re}\, u_i < 1$, and $h_i \in \mathbb{C}[[q]]$ is holomorphic at $q = 0$, and all u_i are pairwise distinct. Then for each i, either $p_i(\tau)$ is identically 0 or $h_i(q)$ is identically 0 (i.e. all coefficients of $h_i(q)$ vanish).*

Proof. For each $v \in \mathbb{C}$, define an operator $\mathcal{T}_v$ on $\mathcal{F}$ by $(\mathcal{T}_v g)(\tau) = g(\tau+1) - e^{2\pi i v} g(\tau)$. Note that $p(\tau)q^u h(q)$ lies in the kernel of $\mathcal{T}_u^k$ if (and we will see only if) the degree of the polynomial $p(\tau)$ is $< k$.

Assume for contradiction that no $p_i(\tau)$ nor $h_i(q)$ are identically 0. Let k_i be the degree of $p_i(\tau)$. Apply $\mathcal{T}_{u_d}^{k_d+1} \circ \cdots \circ \mathcal{T}_{u_2}^{k_2+1} \circ \mathcal{T}_{u_1}^{k_1}$ to $\sum_{i=1}^n p_i(\tau)\,q^{u_i} h_i(q) = 0$ to obtain $a q^{u_1} h_1(q) = 0$ for some nonzero $a \in \mathbb{C}$ and all q in that disc. This forces $h_1(q) \equiv 0$, a contradiction. □

Proposition 3.5. *Let (ρ, w) be admissible and T diagonal, and choose any vvmf $\mathbb{X} \in \mathcal{M}_w^!(\rho)$. Write $\mathbb{X}(\tau) = q^\lambda \sum_{n=0}^\infty \mathbb{X}_{(n)} q^n$.*

(a) *Let σ be any field automorphism of $\mathbb{C}$, and for each $1 \le i \le d$ define $\mathbb{X}_i^\sigma(\tau) = q^{\sigma(\lambda_{ii})} \sum_{n=0}^\infty \sigma(\mathbb{X}_{(n)i}) q^n$. Then $\mathbb{X}^\sigma(\tau) \in \mathcal{M}_{\sigma w}^!(\rho^\sigma)$, where $(\rho^\sigma, \sigma w)$ is admissible, $T^\sigma = e^{2\pi i \sigma\lambda}$, $e^{\pi i \sigma w/2} S^\sigma$ is conjugate to $e^{\pi i w/2} S$ and $e^{2\pi i \sigma w/3} U^\sigma$ is conjugate to $e^{2\pi i w/3} U$. $\mathbb{X}^\sigma(\tau)$ will be $\sigma\lambda$-holomorphic iff $\mathbb{X}(\tau)$ is λ-holomorphic. Choose any bijective exponent Λ and fundamental matrix $\Xi(\tau)$ for (ρ, w); then $(\rho^\sigma, \sigma w)$ will have bijective exponent $\sigma\Lambda$ and fundamental matrix $\Xi^\sigma(\tau)$, where σ acts on $\Xi(\tau)$ column-wise.*

(b) *Let $\mathbb{Q}_{\mathbb{X}}$ be the field generated over $\mathbb{Q}$ by all Fourier coefficients $\mathbb{X}_{(n)}$ of $\mathbb{X}(\tau)$. Assume the components $\mathbb{X}_i(\tau)$ of $\mathbb{X}(\tau)$ are linearly independent over $\mathbb{C}$. Then both the weight w and all exponents λ_i lie in $\mathbb{Q}_{\mathbb{X}}$.*

Proof. Start with (a), and consider first $w = 0$. Write $\chi = \Xi_{(1)}$. Then $\Xi(\tau)$ obeys the differential equation (32). Recall that J, E_2, E_4, E_6, Δ all have coefficients in $\mathbb{Z}$ and hence are fixed by σ. Then $\Xi^\sigma(\tau)$ formally satisfies (32) with Λ_0, χ_0 replaced with $\sigma(\Lambda_0), \sigma(\chi_0)$ (recall that (32) is equivalent to the recursions (36)). Therefore $\Xi^\sigma(\tau)$ is a fundamental solution of that differential equation, with entries meromorphic in $\mathbb{H}^*$.

In fact, thanks to the Fuchsian equation (32), the only possible poles of $\Xi^\sigma(\tau)$ are at the cusps or elliptic fixed points. The behaviour at the elliptic points is easiest to see from (33); in particular, S^σ and U^σ will be conjugate to $e^{2\pi \mathrm{i}\mathscr{A}_2^\sigma}$ and $e^{2\pi \mathrm{i}\mathscr{A}_3^\sigma}$ respectively. But $\mathscr{A}_j^\sigma = \sigma\mathscr{A}_j$ entry-wise and both $\mathscr{A}_j$ are diagonalisable with rational eigenvalues, so $\mathscr{A}_j^\sigma$ is conjugate to $\mathscr{A}_j$ (as it has the identical eigenvalue multiplicities).

To generalise to arbitrary weight w, it is clear from Lemma 3.1 that $(\Delta^{w/12})^\sigma(\tau) = \Delta^{\sigma w/12}(\tau)$. Any fundamental matrix in weight w is $\Delta^{w/12}(\tau)$ times one in weight 0.

Finally, if $\mathbb{X}(\tau) \in \mathscr{M}_w^!(\rho)$, then there exists a polynomial $p(J) \in \mathbb{C}^d[J]$ such that $\mathbb{X}(\tau) = \Xi(\tau)p(J(\tau))$, so $\mathbb{X}^\sigma(\tau) = \Xi^\sigma(\tau)p(J(\tau))^\sigma$. Since σ fixes the coefficients of $J(\tau)$, $\mathbb{X}^\sigma(\tau)$ manifestly lies in $\mathscr{M}_{\sigma w}^!(\rho^\sigma)$.

Now turn to (b). Suppose first that there is some entry $\Lambda_{ii} \notin \mathbb{Q}_{\mathbb{X}}$. Then there exists some field automorphism σ of $\mathbb{C}$ fixing $\mathbb{Q}_{\mathbb{X}}$ but with $\delta := \sigma\Lambda_{ii} - \Lambda_{ii}$ nonzero (see e.g. [1] for a proof of why such a σ exists).

From part (a), $\mathbb{X}^\sigma(\tau) \in \mathscr{M}_{\sigma w}^!(\rho^\sigma)$. Then $\mathbb{X}^\sigma(\tau) = q^{\sigma\Lambda-\Lambda}\mathbb{X}(\tau)$. Choose any $\gamma = \begin{pmatrix} a & b \\ c & d' \end{pmatrix} \in \Gamma$ with $c \neq 0$ (because Γ is Fuchsian of the first kind it will have many such γ). Then $\tau \mapsto -1/\tau$ gives

$$S^\sigma \mathbb{X}^\sigma(\tau) = \tau^{w-\sigma w} \exp(-2\pi \mathrm{i}(\sigma\Lambda - \Lambda)/\tau)\, S\, \mathbb{X}(\tau)\,. \tag{54}$$

Write $f(\tau), g(\tau) \in \mathscr{F}$ for the ith entry of $S\mathbb{X}$ and of $S^\sigma\mathbb{X}^\sigma(\tau)$, respectively. Because the entries of $\mathbb{X}$ are linearly independent, f (and also g) must be nonzero. Write $w' = w - \sigma w$. By (54), $g(\tau) = \tau^{w'} \exp(-2\pi \mathrm{i}\delta/\tau)\, g(\tau)$. Then like all entries of $S^\sigma\mathbb{X}^\sigma(\tau)$, $g(\tau)$ is killed by the order-d differential operator $L_{S^\sigma\mathbb{X}^\sigma} = \sum \tilde{h}_l(\tau)\left(\frac{\mathrm{d}}{2\pi \mathrm{i}\mathrm{d}\tau}\right)^l$ obtained from Lemma 2.2(b) by expanding out each $\mathrm{D}_{\sigma w}^l$; note that each $\tilde{h}_l(\tau) \in q^v\mathbb{C}[[q]]$ for some $v \in \mathbb{C}$, being a combination of the modular forms $h_l(\tau)$ of Lemma 2.2(b) and various derivatives of $E_2(\tau)$. The product rule and induction on l gives

$$\left(\frac{\mathrm{d}}{2\pi \mathrm{i}\mathrm{d}\tau}\right)^l g(\tau) = \tau^{w'} \exp(-2\pi \mathrm{i}\,\delta/\tau) \sum_{k=0}^{l} \frac{p_{l,k}(\tau)}{\tau^{2l-2k}} \left(\frac{\mathrm{d}}{2\pi \mathrm{i}\mathrm{d}\tau}\right)^k f(\tau)\,, \tag{55}$$

where $p_{l,k}(\tau)$ is a polynomial in τ of degree $\leq l-k$ and $p_{l,0}(\tau)$ has nonzero constant term $(2\pi i\delta)^{l-k}$. Multiplying $L_{S^\sigma \mathbb{X}^\sigma} g = 0$ by $\tau^{2d}\tau^{-w'}\exp(2\pi i\delta/\tau)$, we obtain

$$\sum_{l=0}^{d} \tilde{h}_l(\tau) \sum_{k=0}^{l} \tau^{2k} p_{l,k}(\tau) \left(\frac{\mathrm{d}}{2\pi i \mathrm{d}\tau}\right)^k f(\tau) = 0\,. \tag{56}$$

Now, the derivatives $\left(\frac{\mathrm{d}}{2\pi i \mathrm{d}\tau}\right)^k f(\tau)$ are manifestly q-series (i.e. functions in $\mathscr{F}$ where the polynomial parts $p_i(\tau)$ are all constant), so we see from Lemma 3.2 that (regarding (56) as a polynomial in τ with q-series coefficients) the τ^0 coefficient of (56) must itself vanish. This is simply $\tilde{h}_d(\tau)\,(2\pi i\delta)^d f(\tau) = 0$, where $\tilde{h}_d(\tau) = \mathrm{Wr}(\mathbb{X}^\sigma)(\tau) \neq 0$ by Lemma 2.2(a). This forces $\delta = 0$, i.e. $\sigma\Lambda_{ii} = \Lambda_{ii}$.

Therefore all entries Λ_{ii} must lie in $\mathbb{Q}_{\mathbb{X}}$. Suppose now that the weight w doesn't lie in $\mathbb{Q}_{\mathbb{X}}$, and as above choose a field automorphism σ of $\mathbb{C}$ fixing $\mathbb{Q}_{\mathbb{X}}$ but with $\sigma w - w \notin \mathbb{Z}$ (if w lies in an algebraic extension of $\mathbb{Q}_{\mathbb{X}}$ then this is automatic, while if w lies in a transcendental extension we can select σw to likewise be an arbitrary transcendental). The remainder of the argument is as above: $\frac{p_{l,k}(\tau)}{\tau^{2l-2k}}$ in (55) is now replaced with $\frac{c_{l,k}}{\tau^{l-k}}$ for some $c_{l,k} \in \mathbb{C}$, and $c_{l,0} = w'(w'-1)\cdots(w'-l+1)) \neq 0$. Multiplying $L_{S^\sigma \mathbb{X}^\sigma} g = 0$ by $\tau^d\tau^{-w'}$, we see that $\tilde{h}_d(\tau)c_{d,0}f(\tau) = 0$, likewise impossible unless $w \in \mathbb{Q}_{\mathbb{X}}$. □

The obvious Galois action on representations, namely $(\sigma\rho)(\gamma)_{ij} = \sigma(\rho(\gamma)_{ij})$, is unrelated to this ρ^σ. It would be interesting though to understand the relation between the vvmf of $\sigma\rho$ and those of ρ.

For vvmf with rank-1 multipliers, we have $T^\sigma = e^{2\pi i\sigma\lambda}$, $e^{\pi i\sigma w/2}S^\sigma = e^{\pi i w/3}S$, and $e^{2\pi i\sigma w/3}U^\sigma = e^{2\pi i w/3}U$. When the rank d is greater than 1, however, the precise formula for S^σ and U^σ is delicate. For example, we learn in Sect. 4.2 that when $d = 2$, the explicit relation between S^σ and S involves the relation between the Gamma function values $\Gamma(\sigma\Lambda_{11})$ and $\Gamma(\Lambda_{11})$.

The next result is formulated in terms of certain modules $\mathscr{K}$. One important example is $\mathscr{K} = \mathbb{K}[q^{-1}, q]]$ for any subfield $\mathbb{K}$ of $\mathbb{C}$. Another example is the subset $\mathscr{K}$ of $f \in \mathbb{Q}[[q]]$ with bounded denominator, i.e. for which there is an $N \in \mathbb{Z}_{>0}$ such that $Nf \in \mathbb{Z}[[q]]$. Both examples satisfy all conditions of Proposition 3.6. This latter example can be refined in several ways, e.g. by fixing from the start a set P of primes and requiring that the powers of the primes in P appearing in the denominators be bounded, but primes $p \notin P$ be unconstrained.

Proposition 3.6. *Suppose (ρ, w) is admissible, ρ is irreducible and T is diagonal, and choose any exponent λ. Let $\mathbb{K}$ be any subfield of $\mathbb{C}$ and let $\mathscr{K} \subset \mathbb{K}[q^{-1}, q]]$ be any module over both $\mathbb{K}$ and $\mathbb{Z}[q^{-1}, q]]$ (where both of these act by multiplication), such that $\frac{\mathrm{d}}{\mathrm{d}q}\mathscr{K} \subseteq \mathscr{K}$. Let $\mathscr{M}_w^{!\,\mathscr{K}}(\rho)$ denote the intersection $\mathscr{M}_w^!(\rho) \cap q^\lambda\mathscr{K}$. Then the following are equivalent:*

(i) $\mathscr{M}_w^{!\,\mathbb{K}}(\rho) \neq 0$;
(ii) $\mathscr{M}_{w+2k}^{!\,\mathbb{K}}(\rho) \neq 0$ *for all* $k \in \mathbb{Z}$;

(iii) $\mathrm{span}_{\mathbb{C}}\, \mathcal{M}_w^{!\,\mathbb{K}}(\rho) = \mathcal{M}_w^!(\rho)$;
(iv) *for any bijective exponent Λ, all entries $\Xi_{ij}(\tau)$ of the corresponding fundamental matrix $\Xi(\tau)$ lie in $q^{\Lambda_{ii}}\mathcal{K}$;*
(v) *for any exponent λ, the space* $\mathrm{span}_{\mathbb{C}}\left(\mathcal{M}^{\lambda}(\rho) \cap q^{\lambda}\mathcal{K}\right) = \mathcal{M}^{\lambda}(\rho)$.

Proof. Assume (i) holds, and choose any nonzero $\mathbb{X}(\tau) \in \mathcal{M}_w^{!\,\mathcal{K}}(\rho)$. Then $\Delta^h E_4^i E_6^j \mathbb{X} \in q^{\lambda}\mathcal{K}$ for any $h \in \mathbb{Z}$ and $i, j \in \mathbb{Z}_{\geq 0}$, which gives us (ii). That ρ is irreducible forces the components of $\mathbb{X}(\tau)$ to be linearly independent over $\mathbb{C}$. Then Proposition 3.5 would require all $w, \lambda_{ii} \in \mathbb{K}$, and hence $q^{\lambda}\mathcal{K}^d$ is mapped into itself by $\nabla_{w',i}$ for any $w' \in w + 2\mathbb{Z}$. Moreover, Proposition 3.2 applies and $\mathcal{M}_w^!(\rho) = \mathbb{C}[J, \nabla_1, \nabla_2, \nabla_3]\mathbb{X}$, which gives us (iii). Let Λ be any bijective exponent. Then each column of $\Xi(\tau)$ will be a linear combination over $\mathbb{C}$ of finitely many vvmf in $\mathcal{M}_w^{!\,\mathcal{K}}(\rho)$; but all of these vectors are uniquely determined by their principal parts $\mathcal{P}_{\Lambda}$, which are all in $\mathbb{K}^d[q^{-1}]$, so that linear combination must have a solution over the field $\mathbb{K}$. This gives us (iv). To get (v), note that in the proof of Theorem 3.4(a) we may choose our $\mathbb{X}^{(i)}(\tau)$ to lie in $q^{\lambda}\mathcal{K}$, since all we require of them is a linear independence condition. □

This proposition is relevant to the study of modular forms for noncongruence subgroups of Γ. A conjecture attributed to Atkin–Swinnerton-Dyer [2] states that a (scalar) modular form for some subgroup of Γ will have bounded denominator only if it is a modular form for some congruence subgroup. More generally, it is expected that a vvmf $\mathbb{X}(\tau)$ for Γ, with entries $\mathbb{X}_i(\tau)$ linearly independent over $\mathbb{C}$ and with coefficients $\mathbb{X}_{(n)\,i}$ all in $\mathbb{Q}$, will have bounded denominators only if its weight w lies in $\frac{1}{2}\mathbb{Z}$ and the kernel of $\upsilon_{-w} \otimes \rho$ is a congruence subgroup. If the kernel is of infinite index, then it is expected that infinitely many distinct primes will appear in denominators of coefficients.

3.7 Extensions of Modules and Exactness

If ρ is a direct sum $\rho' \oplus \rho''$, then trivially its vvmf are comprised of those of ρ' and ρ''. But what if ρ is a semi-direct sum or a more general extension?

Consider a short exact sequence

$$0 \to U \xrightarrow{\iota} V \xrightarrow{\pi} W \to 0 \tag{57}$$

of finite-dimensional $\overline{\Gamma}$-modules. We can consider an action at arbitrary weight by tensoring with υ_w. Choose a basis for V in which $\rho_V = \begin{pmatrix} \rho_U & * \\ 0 & \rho_W \end{pmatrix}$. In terms of this basis, $\mathbb{X}_V \in \mathcal{M}_w^!(\rho_V)$ implies $\mathbb{X}_V = \begin{pmatrix} \mathbb{X}_U \\ \mathbb{X}_W \end{pmatrix}$ where $\mathbb{X}_W \in \mathcal{M}_w^!(\rho_W)$, and $\begin{pmatrix} \mathbb{X}_U \\ 0 \end{pmatrix} \in \mathcal{M}_w^!(\rho_V)$ iff $\mathbb{X}_U \in \mathcal{M}_w^!(\rho_U)$. In terms of this basis we have the natural embedding $\iota'(\mathbb{X}_U) = \begin{pmatrix} \mathbb{X}_U \\ 0 \end{pmatrix}$ and projection $\pi'\begin{pmatrix} \mathbb{X}_U \\ \mathbb{X}_W \end{pmatrix} = \mathbb{X}_W$. We write (57) as $\rho_V = \rho_U \ni \rho_W$ even when it's not semi-direct.

The fundamental objects in the theory of vvmf are the functors $(\rho, w) \mapsto \mathcal{M}^!_w(\rho)$, $(\rho, w; \lambda) \mapsto \mathcal{M}^\lambda_w(\rho)$ and $(\rho, \lambda) \mapsto \mathcal{M}^\lambda(\rho)$, attaching spaces of vvmf to multipliers, weights and exponents. Marks–Mason [22] suggest considering the effects of these functors on (57). It is elementary that they are left-exact, i.e. that $\rho_V = \rho_U \oplus \rho_W$ trivially implies

$$0 \to \mathcal{M}^{\lambda_U}_w(\rho_U) \stackrel{\iota'}{\to} \mathcal{M}^{\lambda_V}_w(\rho_V) \stackrel{\pi'}{\to} \mathcal{M}^{\lambda_W}_w(\rho_W) \tag{58}$$

and similarly for $\mathcal{M}^!_w$ etc. However [22] found 1-dimensional U and W such that the functor $\rho \mapsto \mathcal{M}^{\lambda^{hol}}(\rho)$ is not right-exact.

Thanks to Theorem 3.5, we can generalise and quantify this discrepancy. We thank Geoff Mason for suggesting the naturalness of explaining the failure of right-exactness with a long exact sequence.

Theorem 3.6. *Write $\rho_V = \rho_U \oplus \rho_W$ as in (57), and suppose (ρ_V, w) is admissible and T_V diagonal. Choose any exponent $\lambda_V = \mathrm{diag}(\lambda_U, \lambda_W)$. Then we obtain the exact sequences of $\mathbb{C}$-spaces*

$$0 \to \mathcal{M}^!_w(\rho_U) \stackrel{\iota'}{\to} \mathcal{M}^!_w(\rho_V) \stackrel{\pi'}{\to} \mathcal{M}^!_w(\rho_W) \to 0\,, \tag{59}$$

$$0 \to \mathcal{M}_U \stackrel{\iota'}{\to} \mathcal{M}_V \stackrel{\pi'}{\to} \mathcal{M}_W \stackrel{\delta}{\to} \mathrm{coker}\, \mathcal{P}_U \cong (\mathcal{M}_{U*})^* \stackrel{\pi''*}{\to} (\mathcal{M}_{V\star})^* \stackrel{\iota''*}{\to} (\mathcal{M}_{W*})^* \to 0 \tag{60}$$

where we write $\mathcal{P}_U = \mathcal{P}_{\lambda_U - 1_{d_U};(\rho_U, w)}$, $\mathcal{M}_U = \mathcal{M}^{\lambda_U}_w(\rho_U)$, $\mathcal{M}_{U*} = \mathcal{M}^{1_{d_U}-\lambda_U}_{2-w}(\rho^*_U)$ etc. ι''^* and π''^* are restrictions of the dual (transpose) maps. The isomorphism in (60) *is* (50), *and the connecting map δ is defined in the proof. Moreover, for any bijective exponents Λ_U of (ρ_U, w) and Λ_W of (ρ_W, w), diag(Λ_U, Λ_W) is bijective for (ρ_V, w).*

Proof. First let's prove (59). Let $\mathcal{M}^!_V, \mathcal{M}^!_U, \mathcal{M}^!_W$ denote $\mathcal{M}^!_w(\rho_V), \mathcal{M}^!_w(\rho_U), \mathcal{M}^!_w(\rho_W)$ respectively. Choose any bijective exponents Λ_U, Λ_W and define $\Lambda_V = \mathrm{diag}(\Lambda_U, \Lambda_W)$, $\mathcal{P}'_V, \mathcal{P}'_U, \mathcal{P}'_W$ for $\mathcal{P}_{\Lambda_V}$ etc. If $\begin{pmatrix} \mathbb{X}_U \\ \mathbb{X}_W \end{pmatrix} \in \ker \mathcal{P}'_V$, then $\mathbb{X}_W(\tau) \in \ker \mathcal{P}'_W$ and hence $\mathbb{X}_W(\tau) = 0$ because Λ_W is bijective. This means $\mathbb{X}_U(\tau) \in \mathcal{M}^!_U$, so also $\mathbb{X}_U(\tau) \in \ker \mathcal{P}'_U$ and $\mathbb{X}_U(\tau) = 0$ because Λ_U is bijective. Therefore $\mathcal{P}'_V$ is injective. From the first line of (25) we know $c_{(\rho_V, w)} = c_{(\rho_U, w)} + c_{(\rho_W, w)}$ (since $\mathrm{Tr}\, S_V = \mathrm{Tr}\, S_U + \mathrm{Tr}\, S_W$ etc); the index formula (24) then implies coker $\mathcal{P}'_V = 0$ and hence Λ_V is bijective. In order to establish (59), only the surjectivity of π' needs to be shown, but this follows from the surjectivity of $\mathcal{P}'_V$ together with the injectivity of $\mathcal{P}'_W$.

Now let's turn to (60). Most of this exactness again comes from (58): the dual of $0 \to \mathcal{M}_{W*} \to \mathcal{M}_{V*} \to \mathcal{M}_{U*}$ gives the second half of (60). The connecting map δ is defined as follows. Given any $\mathbb{X}_W \in \mathcal{M}_W$, exactness of (59) says that there is an $\mathbb{X}_U$, unique mod $\mathcal{M}_U^!$, such that $\binom{\mathbb{X}_U}{\mathbb{X}_W} \in \mathcal{M}_V^!$. The connecting map $\delta : \mathcal{M}_W \to \operatorname{coker} \mathcal{P}_U$ sends $\mathbb{X}_W$ to $\mathcal{P}_U(\mathbb{X}_U) + \operatorname{Im} \mathcal{P}_U$. Exactness at $\mathcal{M}_W$ is now clear: if $\mathbb{X}_W \in \ker \delta$ then $\mathcal{P}_U(\mathbb{X}_U) = \mathcal{P}_U(\mathbb{X}'_U)$ for some $\mathbb{X}'_U \in \mathcal{M}_U$, so $\binom{\mathbb{X}'_U}{0}$ and hence $\binom{\mathbb{X}_U - \mathbb{X}'_U}{\mathbb{X}_W}$ both lie in $\mathcal{M}_V^!$. However the latter manifestly lies in the kernel of $\mathcal{P}_V$, so $\mathbb{X}_W = \pi' \binom{\mathbb{X}_U - \mathbb{X}'_U}{\mathbb{X}_W}$ as desired. To see exactness at π''^*, suppose $f_U \in \ker \pi''^*$. Then f_U is a functional on $\mathcal{M}_{U*}$ and the associated functional $\binom{\mathbb{Y}_U}{\mathbb{Y}_W} \mapsto f_U(\mathbb{Y}_U)$ on $\mathcal{M}_{V*}$ is 0. Thanks to (50), these functionals can be expressed as $\mathbb{Y}_U \mapsto \langle \mathbb{P}'_U, \mathbb{Y}_U \rangle$ and $\binom{\mathbb{Y}_U}{\mathbb{Y}_W} \mapsto \langle \binom{\mathbb{P}_U}{\mathbb{P}_W}, \binom{\mathbb{Y}_U}{\mathbb{Y}_W} \rangle$ for some $\mathbb{P}'_U(\tau) \in \mathbb{C}^{d_U}[q^{-1}]$ and $\binom{\mathbb{P}_U}{\mathbb{P}_W} \in \mathbb{C}^{d_U + d_W}[q^{-1}]$. The independence of the functional on $\mathbb{Y}_W$ means $\mathbb{P}_W(\tau) = 0$, so we can take $\mathbb{P}_U(\tau) = \mathbb{P}'_U(\tau)$. That the functional on $\mathcal{M}_{V*}$ is 0 means (again from (50)) that $\binom{\mathbb{P}_U}{0} = \mathcal{P}_V \binom{\mathbb{X}_U}{\mathbb{X}_W}$ for some $\binom{\mathbb{X}_U}{\mathbb{X}_W} \in \mathcal{M}_V^!$. Moreover, $\mathbb{X}_W(\tau) \in \mathcal{M}_W$, so $\mathbb{P}_U = \delta(\mathbb{X}_W)$, as desired. □

Theorem 3.6 allows us to classify all bijective Λ_V. These are given by all exponents $\operatorname{diag}(\lambda_U, \lambda_W)$ such that $\mathcal{P}_{\lambda_U}$ is injective, $\mathcal{P}_{\lambda_W}$ is surjective, and the connecting map $\delta : \ker \mathcal{P}_{\lambda_W} \to \operatorname{coker} \mathcal{P}_{\lambda_U}$ is an isomorphism.

We can now quantify the failure of $\mathcal{M}^\lambda(\rho)$ to be exact. For each fixed w, the discrepancy is

$$\dim \mathcal{M}_U + \dim \mathcal{M}_W - \dim \mathcal{M}_V = \dim \operatorname{Im} \delta\,. \tag{61}$$

For $w \ll 0$, $\mathcal{M}_w^\lambda(\rho_W) = 0$ so $\delta = 0$ and the discrepancy is 0, while for $w \gg 2$, then $\mathcal{M}_{2-w}^{1_{d_U} - \lambda^{hol}}(\rho_U^*) = 0$ so again $\delta = 0$ and the discrepancy is 0. Thus the total discrepancy, summed over all w, is finite.

Let us recover in our picture the calculation in Theorem 4 of [22]. Take $\lambda = \lambda^{hol}$. Consider ρ_V of the form $\upsilon_{2a} \ni \upsilon_{2b}$, where as always υ_j has $T = e^{2\pi i j/12}$. Then Theorem 4 of [22] says ρ_V can be indecomposable iff $|a - b| = 1$. As above, if $w < 0$ then $\mathcal{M}_W = 0$ while if $w \geq 2$ then $\mathcal{M}_{U*} = 0$, so only at $w = 0$ can $\delta \neq 0$. We find that $w = 0$ and $\mathcal{M}_W, \mathcal{M}_{U*} \neq 0$ forces $b = 0$ and $a = 5$, in which case $\delta : \mathbb{C} \to \mathbb{C}$. Now, a bijective exponent for $(\rho_V, 0)$ is $\operatorname{diag}(-\frac{7}{6}, 0)$ by Theorem 3.6 (and Sect. 4.1 below), so there must be a $\mathbb{X}_U(\tau) \in q^{-1/6} \sum_{n=0}^{\infty} c_n q^n$ such that $\binom{\mathbb{X}_U}{1} \in \mathcal{M}_V^!$, by surjectivity. Indeed, $\delta(1) = \mathcal{P}_U(\mathbb{X}_U) = c_0$. If $c_0 = 0$ then the Wronskian of $\binom{c_1 q^{5/6} + \cdots}{1}$ will be a nonzero holomorphic modular form $cq^{5/6} + \cdots$ of weight 2 (the Wronskian is nonzero because ρ_V is indecomposable). This is impossible (e.g. η^{-20} times it would also be holomorphic but with trivial multiplier and weight -8). Therefore $c_0 \neq 0$, so $\delta \neq 0$ and the total discrepancy is 1-dimensional.

4 Effectiveness of the Theory

Explicit computations within our theory are completely feasible. Recall from Theorem 3.3 that we have complete and explicit knowledge of the space $\mathcal{M}^!_w(\rho)$ of weakly holomorphic vvmf, if we know the diagonal matrix Λ and the complex matrix $\mathcal{X}$. We know Tr Λ, and generically any matrix with the right trace and with $e^{2\pi i \Lambda} = T$ is a bijective exponent. This $\mathcal{X}$ can be obtained in principle from ρ and Λ using the Rademacher expansion (see e.g. [12]). However this series expansion for $\mathcal{X}$ converges notoriously slowly, and obscures any properties of $\mathcal{X}$ that may be present (e.g. integrality). Hence other methods are needed for identifying $\mathcal{X}$. In this section we provide several examples, illustrating some of the ideas available. See [4, 5] for further techniques and examples.

4.1 One Dimension

It is trivial to solve the $d = 1$ case [5]. Here, $\rho = \upsilon_u$, and $u \in \mathbb{C}$, $0 \leq \operatorname{Re} u < 1$, and the weight w is required to be $w \in 12u + 2\mathbb{Z}$. Write $w = 12u + 2j + 12n$ where $0 \leq j \leq 5$ and $n \in \mathbb{Z}$; then for $j = 0, 1, 2, 3, 4, 5$ resp., the fundamental matrix $\Xi(\tau)$ for $\mathcal{M}^!_{12u+2j+12n}(\upsilon_u)$ is (recall Lemma 3.1):

$$\Delta^{u+n}(\tau), \qquad \Delta^{u+n-1}(\tau)\, E_{14}(\tau), \qquad \Delta^{u+n}(\tau)\, E_4(\tau),$$
$$\Delta^{u+n}(\tau)\, E_6(\tau), \qquad \Delta^{u+n}(\tau)\, E_8(\tau), \qquad \Delta^{u+n}(\tau)\, E_{10}(\tau),$$

respectively, where $E_8 = E_4^2$, $E_{10} = E_4 E_6$ and $E_{14} = E_4^2 E_6$. The unique free generator of $\mathcal{M}^{\lambda^{hol}}(\upsilon_u)$ is $\Delta^u(\tau)$. This means that $\dim \mathcal{M}^{\lambda^{hol}}_w(\upsilon_u)$ equals the dimension of the weight $w - 12u$ subspace of m, assuming as above that $0 \leq \operatorname{Re} u < 1$.

4.2 Two Dimensions

Much more interesting is $d = 2$ (see e.g. [24, 34]). Let's start with weakly holomorphic. As usual, it suffices to consider weight $w = 0$. We continue to require that T be diagonal, although we give a 'logarithmic' example shortly.

The moduli space of equivalence classes of 2-dimensional representations of $\overline{\Gamma}$ consists of 15 isolated points, together with 3 half-planes. Each half-plane has four singularities: two conical singularities and two triple points. The equivalence classes of irreducible representations with T diagonalisable correspond bijectively with the regular points on the three half-planes. The 15 isolated points are all direct sums $\rho_1 \oplus \rho_2$ of 1-dimensional representations which violate the inequalities (8), and can be recovered from the 1-dimensional solution. Each conical singularity corresponds to an irreducible 'logarithmic' representation with T nondiagonalisable. Each triple point consists of a direct sum $\rho \oplus \rho'$ as well as the semi-direct sums $\rho \ni \rho'$ and $\rho' \ni \rho$. In all cases except the six triple points, the set of eigenvalues of T uniquely determine the representation. The three half-planes correspond to the three different choices of (α_i, β_j) possible at $d = 2$.

More explicitly, one half-plane has $\det T = \xi_6 =: \zeta$:

$$T = \begin{pmatrix} z & 0 \\ 0 & \zeta z^{-1} \end{pmatrix}, \qquad S = \begin{pmatrix} \frac{\bar{\xi} z}{z^2-\zeta} & y \\ \frac{(z^2-1)(z^2-\zeta^2)}{y(z^2-\zeta)^2} & -\frac{\bar{\xi} z}{z^2-\zeta} \end{pmatrix}, \tag{62}$$

for arbitrary $y, z \in \mathbb{C}$ provided $y \neq 0$ and $z \notin \{0, \pm\xi_{12}\}$. This will be irreducible iff $z \notin \{\pm 1, \pm\zeta\}$. The redundant parameter y is introduced for later convenience. *Irreducible* ρ, ρ' with z, z' related by $(zz')^2 = \zeta$ are naturally isomorphic. Here, $\alpha_1 = \beta_0 = \beta_1 = 1$, so $\operatorname{Tr} \Lambda = -\frac{5}{6}$ and $\Lambda = \operatorname{diag}(t, -\frac{5}{6} - t)$ where $z = e^{2\pi i t}$ for some $t \not\equiv \frac{1}{12}, \frac{7}{12}$ (mod 1). Then Theorem 3.3(c) tells us

$$\mathscr{X} = \begin{pmatrix} 24\frac{t(60t-11)}{12t+5} & 10368x\frac{t(2t+1)(3t+1)(6t+5)}{(12t+11)(12t+5)^2} \\ \frac{10368}{x(12t-1)} & -4\frac{(6t+5)(60t+61)}{12t+5} \end{pmatrix}, \tag{63}$$

for some $x \neq 0$ to be determined shortly. Equation (33) is ideally suited to relate x and y, since at $d = 2$ it reduces to the classical hypergeometric equation. We read off from it the fundamental matrix

$$\Xi(\mathrm{z}(\tau)) = \begin{pmatrix} f(t; \frac{5}{6}; \mathrm{z}(\tau)) & \mathscr{X}_{12} f(t+1; \frac{5}{6}; \mathrm{z}(\tau)) \\ \mathscr{X}_{21} f(\frac{1}{6} - t; \frac{5}{6}; \mathrm{z}(\tau)) & f(-\frac{5}{6} - t; \frac{5}{6}; \mathrm{z}) \end{pmatrix} \tag{64}$$

for $\mathrm{z}(\tau) = J(\tau)/1728$, where we write

$$f(a; c; \mathrm{z}) = (-1728\mathrm{z})^{-a} F(a, a + \frac{1}{2}; 2a + c; \mathrm{z}^{-1}) \tag{65}$$

for $F(a, b; c; z) = 1 + \frac{ab}{c} z + \cdots$ the hypergeometric series. Substituting $\mathrm{z}(\tau) = J(\tau)/1{,}728$ directly into (65) and (64) gives the q-expansion of Ξ. The parameters x, y appearing in S and $\mathscr{X}$ can be related by the standard analytic continuation of $F(a, b; c; z)$ from $z \approx 0$ to $z \approx \infty$, which implies

$$f(a; c; \mathrm{z}) = (1728)^{-a} \left\{ \frac{\sqrt{\pi}\,\Gamma(2a + c)}{\Gamma(a + \frac{1}{2})\,\Gamma(a + c)} - \frac{2\sqrt{\pi}\,\Gamma(2a + c)}{\Gamma(a)\,\Gamma(a + c - \frac{1}{2})} \mathrm{z}^{1/2} + \cdots \right\} \tag{66}$$

for small $|z|$. Hence

$$y = \frac{\sqrt{3}x}{1728}\,\frac{2^{2/3}}{432^{2t}}\,\frac{\Gamma(2t+\frac{5}{6})^2}{\Gamma(2t)\,\Gamma(2t+\frac{2}{3})}\,. \tag{67}$$

In particular, we see that when ρ is irreducible, Λ is bijective for a $\mathbb{Z}$ worth of t's (i.e. the necessary conditions $e^{2\pi i\Lambda} = T$ and $\mathrm{Tr}\,\Lambda = -\frac{5}{6}$ are also sufficient)—at the end of Sect. 3.4 we called such ρ *tight*. This fact is generalised in Theorem 4.1 below. There are four indecomposable but reducible ρ here:

$z = 1$: this ρ is the extension $1 \,\text{Ð}\, \upsilon_2$ of 1 (the subrepresentation) with υ_2 (the quotient); in this case $\Lambda = \mathrm{diag}(t, -\frac{5}{6} - t)$ is bijective iff $t \in \mathbb{Z}_{\geq 0}$;
$z = -1$: here, $\rho = \upsilon_6 \,\text{Ð}\, \upsilon_8$; its Λ is bijective iff $t \in -\frac{1}{2} + \mathbb{Z}_{\geq 0}$;
$z = \zeta$: here, $\rho = \upsilon_2 \,\text{Ð}\, 1$; its Λ is bijective iff $t \in -\frac{5}{6} + \mathbb{Z}_{\geq 0}$;
$z = -\zeta$: here, $\rho = \upsilon_8 \,\text{Ð}\, \upsilon_6$; its Λ is bijective iff $t \in -\frac{1}{3} + \mathbb{Z}_{\geq 0}$.

These restrictions on t are needed to avoid the Gamma function poles in (67). Nevertheless, the 'missing' values of t (apart from the forbidden $t \equiv \frac{1}{12} \pmod{\frac{1}{2}}$) are all accounted for: For example, the limit $t \to 0$ with $x = t^{-1/2}, 1, t^{-1}$ recovers $S, \mathscr{X}$ etc for $1 \oplus \upsilon_2$, $1 \,\text{Ð}\, \upsilon_2$, $1 \,\text{Ð}\, \upsilon_2$, respectively.

The free generators over m of $\mathcal{M}^{\lambda^{hol}}(\rho)$ for these ρ are now easy to find. Fix $0 \leq \mathrm{Re}\,t < 1$, and note that $\lambda^{hol} = \mathrm{diag}(t, \frac{1}{6} - t)$ when $\mathrm{Re}\,t \leq \frac{1}{6}$, and otherwise $\lambda^{hol} = \mathrm{diag}(t, \frac{7}{6} - t)$ .Consider first the case where ρ is irreducible; then ρ is tight and Proposition 3.3(b) tells us $\dim \mathcal{M}_w^{\lambda^{hol}}(\rho)$ for all $w < 0$. In particular, if $\mathrm{Re}\,t \leq \frac{1}{6}$ then $w^{(1)} = 0$ and $w^{(2)} = 2$, and $\mathbb{X}^{(1)}(\tau)$ is the first column of $\Xi(\tau)$ given above; if instead $\mathrm{Re}\,t > \frac{1}{6}$ then $w^{(i)} = 6, 8$, $\mathbb{X}^{(1)}(\tau)$ is the first column of $\Xi(\tau)$ at $w = 6$ for $\Lambda = \mathrm{diag}(t, \frac{1}{6} - t)$. In both cases, $\mathbb{X}^{(2)}(\tau) = \mathrm{D}\mathbb{X}^{(1)}(\tau)$.

The holomorphic vvmf for the 2-dimensional indecomposable representations was discussed at the end of Sect. 3.7, and we find that for our four such ρ, $\dim \mathcal{M}_w^{\lambda^{hol}}(\upsilon_{2a} \,\text{Ð}\, \upsilon_{2b}) = \dim \mathcal{M}_w^{\lambda^{hol}}(\upsilon_{2a}) + \dim \mathcal{M}_w^{\lambda^{hol}}(\upsilon_{2b})$. The 1-dimensional case was worked out in Sect. 4.1, and we find that in all cases $\{w^{(1)}, w^{(2)}\} = \{2a, 2b\}$ with $\mathbb{X}^{(i)}(\tau)$ given by the appropriate column of the fundamental matrix at $w = 2a$ and $2b$ respectively.

The choice $z = \pm\zeta$, i.e. $t \equiv \frac{1}{12} \pmod{\frac{1}{2}}$, corresponds here to two logarithmic representations. Consider for concreteness $z = \xi_{12}$. A weakly holomorphic vvmf for it is $\eta(\tau)^2 \binom{\tau}{1}$. This generates all of $\mathcal{M}_0^!$, using $\mathbb{C}[J]$ and the differential operators ∇_i; together with

$$q^{1/12}\begin{pmatrix} \pi i q^{-1} - 242\pi i + (-140965\pi i - 55440\tau)q + \cdots \\ -55440q + \cdots \end{pmatrix}$$

it freely generates $\mathcal{M}_0^!$ over $\mathbb{C}[J]$. The five other 2-dimensional logarithmic representations correspond to this one tensored with a $\overline{\Gamma}$ character. The free basis for holomorphic vvmf is $\eta(\tau)^2 \binom{\tau}{1}$ and its derivative.

We can see the Galois action of Sect. 3.6 explicitly here: σ takes $t \mapsto \sigma t$ and $x \mapsto \sigma x$, and it keeps one inside this connected component.

Another class of two-dimensional representations has $\det T = -1$:

$$T = \begin{pmatrix} z & 0 \\ 0 & -z^{-1} \end{pmatrix}, \qquad S = \begin{pmatrix} \frac{-z}{z^2+1} & y \\ \frac{z^4+z^2+1}{y(z^2+1)^2} & \frac{z}{z^2+1} \end{pmatrix}, \tag{68}$$

for arbitrary $y, z \in \mathbb{C}$ provided $y \neq 0$, $z \notin \{0, \pm \mathrm{i}\}$. Irreducibility requires $z \notin \{\pm\zeta, \pm\overline{\zeta}\}$. Irreducible ρ with $zz' = -1$ are isomorphic. Here, $\alpha_1 = \beta_1 = \beta_2 = 1$, so $\mathrm{Tr}\,\Lambda = -\frac{3}{2}$ and $\Lambda = \mathrm{diag}(t, -\frac{3}{2} - t)$ for $z = e^{2\pi \mathrm{i} t}$ and some $t \not\equiv \pm\frac{1}{4}$ (mod 1). Then as before

$$\mathscr{X} = \begin{pmatrix} 24\frac{20t^2+51t+32}{4t+3} & \frac{384x(3t+2)(3t+1)(6t+5)(6t+7)}{(4t+5)(4t+3)^2} \\ \frac{384}{x(4t+1)} & -12\frac{40t^2+18t+1}{4t+3} \end{pmatrix}, \tag{69}$$

$$\Xi(\mathrm{z}) = \mathrm{z}^{1/3} \begin{pmatrix} f(t+\frac{1}{3}; \frac{5}{6}; \mathrm{z}) & \mathscr{X}_{12} f(t+\frac{4}{3}; \frac{5}{6}; \mathrm{z}) \\ \mathscr{X}_{21} f(-\frac{1}{6}-t; \frac{5}{6}; \mathrm{z}) & f(-t-\frac{7}{6}; \frac{5}{6}; \mathrm{z}) \end{pmatrix}, \tag{70}$$

$$y = x\frac{\sqrt{3}}{6912\ 432^{2t}} \frac{\Gamma(2t+\frac{3}{2})^2}{\Gamma(2t+\frac{4}{3})\,\Gamma(2t+\frac{2}{3})}. \tag{71}$$

Again, for irreducible ρ, any possible t yields bijective Λ. The four indecomposable but reducible ρ are:

$z = \zeta$: here, $\rho = \upsilon_2 \mathbin{\text{Đ}} \upsilon_4$ and Λ is bijective iff $t \in -\frac{5}{6} + \mathbb{Z}_{\geq 0}$;
$z = -\zeta$: here $\rho = \upsilon_8 \mathbin{\text{Đ}} \upsilon_{10}$ and Λ is bijective iff $t \in -\frac{1}{3} + \mathbb{Z}_{\geq 0}$;
$z = \overline{\zeta}$: here, $\rho = \upsilon_{10} \mathbin{\text{Đ}} \upsilon_8$ and Λ is bijective iff $t \in \frac{5}{6} + \mathbb{Z}_{\geq 0}$;
$z = -\overline{\zeta}$: here $\rho = \upsilon_4 \mathbin{\text{Đ}} \upsilon_2$ and Λ is bijective iff $t \in \frac{4}{3} + \mathbb{Z}_{\geq 0}$.

Again the 'missing' t correspond to other reducible ρ.

The holomorphic analysis is identical to before. Here $\mathrm{Tr}\,\lambda^{hol} = \frac{1}{2}$ or $\frac{3}{2}$, depending on whether or not $\mathrm{Re}\, t \leq \frac{1}{2}$. In the former case $w^{(i)} = (2,4)$, and in the latter it equals (8,10). $\mathbb{X}^{(i)}(\tau)$ is as before. The indecomposable ρ behave exactly as before.

The final class of two-dimensional representations has $\det T = \overline{\zeta}$:

$$T = \begin{pmatrix} z & 0 \\ 0 & \overline{\zeta} z^{-1} \end{pmatrix}, \qquad S = \begin{pmatrix} \frac{\zeta z}{z^2-\overline{\zeta}} & y \\ \frac{(z^2-1)(z+\overline{\zeta})(z-\overline{\zeta})}{y(z^2-\overline{\zeta})^2} & \frac{-\zeta z}{z^2-\overline{\zeta}} \end{pmatrix}, \tag{72}$$

for arbitrary $y, z \in \mathbb{C}$ provided $y \neq 0$, $z \notin \{0, \pm e^{-\pi \mathrm{i}/6}\}$. Irreducibility requires $z \notin \{\pm 1, \pm\overline{\zeta}\}$; irreducible ρ related by $zz' = \overline{\zeta}$ are equivalent. Here, $\alpha_1 = \beta_0 = \beta_2 = 1$, so $\mathrm{Tr}\,\Lambda = -\frac{7}{6}$ and $\Lambda = \mathrm{diag}(t, -\frac{7}{6} - t)$ for some t satisfying $z = e^{2\pi \mathrm{i} t}$. Then

$$\mathscr{X} = \begin{pmatrix} 24\frac{t(60t+71)}{12t+7} & \frac{10368xt(2t+1)(3t+2)(6t+7)}{(12t+13)(12t+7)^2} \\ \frac{10368}{x(12t+1)} & -4\frac{(6t+7)(60t-1)}{12t+7} \end{pmatrix}, \tag{73}$$

$$\Xi(z) = \begin{pmatrix} f(t;\frac{7}{6};z) & \mathscr{X}_{12} f(t+1;\frac{7}{6};z) \\ \mathscr{X}_{21} f(-\frac{1}{6}-t;\frac{7}{6};z) & f(-t-\frac{7}{6};\frac{7}{6};z) \end{pmatrix}, \tag{74}$$

$$y = x\frac{\sqrt{3}\,2^{1/3}}{10368\;\,432^{2t}}\frac{\Gamma(2t+\frac{7}{6})^2}{\Gamma(2t)\,\Gamma(2t+\frac{4}{3})}. \tag{75}$$

Again, for irreducible ρ, any possible t yields bijective Λ. The four indecomposable but reducible ρ are:

$z = 1$: then $\rho = 1 \ni\!\!\!\!+\, \upsilon_{10}$ and Λ is bijective iff $t \in \mathbb{Z}_{\geq 0}$;
$z = -1$: then $\rho = \upsilon_6 \ni\!\!\!\!+\, \upsilon_4$ and Λ is bijective iff $t \in -\frac{1}{2} + \mathbb{Z}_{\geq 0}$;
$z = \overline{\xi}$: then $\rho = \upsilon_{10} \ni\!\!\!\!+\, 1$ and Λ is bijective iff $t \in \frac{5}{6} + \mathbb{Z}_{\geq 0}$;
$z = -\overline{\xi}$: then $\rho = \upsilon_4 \ni\!\!\!\!+\, \upsilon_6$ and Λ is bijective iff $t \in \frac{4}{3} + \mathbb{Z}_{\geq 0}$.

Again the 'missing' t correspond to the other reducible ρ.

The holomorphic story for irreducible ρ is as before. Here $\mathrm{Tr}\,\lambda^{hol} = \frac{5}{6}$ or $\frac{11}{6}$ depending on whether or not $\mathrm{Re}\,t \leq \frac{5}{6}$. This means $w^{(i)}$ will equal (4,6) or (10,12), respectively. The only new phenomenon here is the indecomposable at $z = \overline{\xi}$: at the end of Sect. 3.7 we learned that $\dim \mathscr{M}_0^{\lambda^{hol}}(\upsilon_{10} \ni\!\!\!\!+\, 1)$ is 0, not 1. We find $w^{(i)} = (4,6)$.

The λ^{hol}-holomorphic two-dimensional theory is also studied in [24], though without quantifying the relation between Fourier coefficients and the matrix S (i.e. his statements are only basis-independent), which as we see involves the Gamma function. Our two-dimensional story can be extended to any triangle group [3].

4.3 *vvmf in Dimensions < 6*

Trivially, the spaces of vvmf for an arbitrary ρ are direct sums of those for its indecomposable summands. Theorem 3.6 reduces understanding the vvmf for an indecomposable ρ, to those of its irreducible constituents. In this section we prove any admissible (ρ, w) is *tight*, provided ρ is irreducible and of dimension < 6 (recall the definition of tight at the end of Sect. 3.4). This means in dimension < 6 we get all kinds of things for free (see Proposition 3.3), including identifying the Hilbert–Poincaré series $H^\lambda(x;\rho)$. These series were first computed in [21], for the special case of exponent $\lambda = \lambda^{hol}$, when T is unitary, and the representation ρ is what Marks calls *T-determined*, which means that any indecomposable ρ' with the same T-matrix is isomorphic to ρ. It turns out that most ρ are T-determined. We will see this hypothesis is unnecessary, and we can recover and generalise his results with much less effort. The key observation is the following, which is of independent interest:

Theorem 4.1. *Let (ρ, w) be admissible and T diagonal. Assume ρ is irreducible and the dimension $d < 6$. Then ρ is tight: an exponent λ is bijective for $(\rho, w+2k)$ iff $Tr\,\lambda = c_{(\rho,w+2k)}$.*

Proof. The case $d = 1$ is trivial, and $d = 2$ is explicit in Sect. 4.2, so it suffices to consider $d = 3, 4, 5$. Without loss of generality (by tensoring with υ_{-w}) we may assume ρ is a true representation of $\overline{\Gamma}$, i.e. that $(\rho, 0)$ is admissible. Let $\alpha_i = \alpha_i(\rho, 0)$, $\beta_j = \beta_j(\rho, 0)$. Let $\mathbb{X}^{(i)}(\tau)$ be the free generators which exist by Theorem 3.4(a), and let $w^{(1)} \leq \cdots \leq w^{(d)}$ be their weights. We will have shown that ρ is tight, if we can show that these $w^{(i)}$ agree with those in the tight Hilbert–Poincaré series (44). This is because this would require all $\dim \mathcal{M}^{\lambda}_{w+2k}(\rho)$ to equal that predicted by $H_{tt}(x)$, as given in Proposition 3.3, and that says λ will be bijective iff λ has the correct trace. In fact it suffices to verify that the values of $w^{(i)} - w^{(1)}$ match the numerator of (44), as the value of $\sum_i w^{(i)}$ would then also fix $w^{(1)}$.

Let n_i be the total number of generators $\mathbb{X}^{(i)}(\tau)$ with weight $w^{(i)} \equiv 2i \pmod{12}$. By Theorem 3.4(b) we know $\sum_i n_i = d$, $n_i + n_{2+i} + n_{4+i} = \alpha_i$, $n_j + n_{3+j} = \beta_j$ for all i, j. These have solutions

$$(n_0, n_1, n_2, n_3, n_4, n_5) = (\alpha_0 - \beta_2 + t - s, \beta_1 - s, \beta_2 - t, s - t + \alpha_1 - \beta_1, s, t) \tag{76}$$

for parameters s, t.

Consider first $d = 3$. The inequalities (8) force $\beta_i = 1$ and $\{\alpha_0, \alpha_1\} = \{1, 2\}$. By Proposition 3.3(a), we can assume without loss of generality (hitting with υ_6 if necessary) that $\alpha_1 = 2$. Then the only nonnegative solutions to (76) are $(n_i) = (0,1,1,1,0,0), (1,1,0,0,0,1), (0,0,0,1,1,1)$. From (26) we see $L := \mathrm{Tr}\,\lambda \in \mathbb{Z}$. Theorem 3.4(b) says $\sum_i w^{(i)} = 12L$. The inequality (42) requires $w^{(1)} = 4L-2$, so the only possibility consistent with the given values of n_i together with $\sum_i w^{(i)} = 12L$ is $(w^{(i)}) = (4L-2, 4L, 4L+2)$, which is the prediction of (44).

Consider next $d = 4$. As before we may assume $(\beta_i) = (2,1,1)$ and $(\alpha_i) = (2,2)$. Then (76) forces $(n_i) = (1,1,1,1,0,0), (1,0,0,1,1,1,1), (2,1,0,0,0,1), (0,0,1,2,1,0)$. Again $L := \mathrm{Tr}\,\lambda \in \mathbb{Z}$ and $\sum_i w^{(i)} = 12L$, and (42) forces $w^{(1)} \in \{3L-3, 3L-1\}$ (if L is odd) or $w^{(1)} \in \{3L-2, 3L\}$ (if L is even).

We claim for each L there is a unique possibility for the $w^{(i)}$ which is compatible with $\sum_i w^{(i)} = 12L$, the listed possibilities for (n_i), the two possible values for $w^{(1)}$ given above, and the absence of a 'gap' in the sense of (46). When L is even this is $(w^{(i)}) = (3L-2, 3L, 3L, 3L+2)$; when L is odd this is $(w^{(i)}) = (3L-3, 3L-1, 3L+1, 3L+3)$. These match (44).

Finally, consider $d = 5$. We may take $(\beta_i) = (1,2,2)$ and $(\alpha_i) = (3,2)$, so

$$\begin{aligned}(n_i) \in \{&(1,2,2,0,0,0), (1,1,1,0,1,1),\\ &(1,0,0,0,2,2), (0,1,2,1,1,0), (0,0,1,1,2,1)\},\end{aligned} \tag{77}$$

$L := \mathrm{Tr}\,\lambda \in \mathbb{Z}$ and $\sum_i w^{(i)} = 12L$.

If $L = 5L'$, then (42) forces $w^{(1)} = 12L' - 4$ or $12L' - 2$. We find the only possible value of $w^{(i)}$ is $(12L' - 4, 12L' - 2, 12L', 12L' + 2, 12L' + 4)$.

If $L = 5L' + 1$, then (42) forces $w^{(1)} = 12L'$. We find the only possible value of $w^{(i)}$ is $(12L', 12L' + 2, 12L' + 2, 12L' + 4, 12L' + 4)$.

If $L = 5L' + 2$, then (42) forces $w^{(1)} = 12L' + 2$ or $12L' + 4$. We find the only possible value of $w^{(i)}$ is $(12L' + 2, 12L' + 4, 12L' + 4, 12L' + 6, 12L' + 8)$.

If $L = 5L' + 3$, then (42) forces $w^{(1)} = 12L' + 4$ or $12L' + 6$. We find the only possible value of $w^{(i)}$ is $(12L' + 4, 12L' + 6, 12L' + 8, 12L' + 8, 12L' + 10)$.

If $L = 5L' + 4$, then (42) forces $w^{(1)} = 12L' + 6$ or $12L' + 8$. We find the only possible value of $w^{(i)}$ is $(12L' + 8, 12L' + 8, 12L' + 10, 12L' + 10, 12L' + 12)$.

All of these agree with (44). □

Proposition 3.3 gives some consequences of tightness.

4.4 Further Remarks

Now let's turn to more general statements. The simplest way to change the weight w has already been alluded to in several places. Namely, suppose we are given any admissible multiplier system (ρ, w) with bijective Λ and fundamental matrix $\Xi(\tau)$. Recall the multiplier υ_w of $\Delta^w(\tau)$. Then for any $w' \in \mathbb{C}$, $(\upsilon_{w'} \otimes \rho, w + 12w')$ is admissible, with bijective and fundamental matrices $\Lambda + w' 1_d$, and $\Delta^{w'}\,\Xi$.

Suppose bijective Λ, Λ' with corresponding fundamental matrices $\Xi(\tau), \Xi'(\tau)$ are known for admissible (ρ, w) and (ρ', w'). Then the dd' columns of the Kronecker matrix product $\Xi(\tau) \otimes \Xi'(\tau)$ will manifestly lie in $\mathscr{M}^!_{w+w'}(\rho \otimes \rho')$, and will generate over $\mathbb{C}[J]$ a full rank submodule of it. By Proposition 3.2 the differential operators ∇_i then generate from that submodule all of $\mathscr{M}^!_{w+w'}(\rho \otimes \rho')$. In that way, bijective exponents and fundamental matrices for tensor products (and their submodules) can be obtained.

The easiest and most important products involve the six one-dimensional $\overline{\Gamma}$-representations υ_{2i}. Here we can be much more explicit. Equivalently, we can describe the effect of changing the weights by even integers but keeping the same representation. For simplicity we restrict to even integer weights.

Proposition 4.1. *Let $(\rho, 0)$ be admissible and T diagonal. Fix a bijective Λ, with corresponding $\Xi(\tau)$. Then for $i = 2, 3, 4, 5$ respectively, the columns for a fundamental matrix for $(\rho, 2i)$ can be obtained as a linear combination over $\mathbb{C}$ of the columns of, respectively,:*

(2) $E_4\Xi = q^\Lambda(1_d + \cdots)$, $\mathrm{D}^2\Xi - E_4\Xi\Lambda(\Lambda - \frac{1}{6}) = q^\Lambda(1728\mathscr{A}_3(\mathscr{A}_3 - \frac{1}{3})q + \cdots)$;
(3) $E_6\Xi = q^\Lambda(1_d + \cdots)$, $E_4\mathrm{D}\Xi - E_6\Xi\Lambda = q^\Lambda(1728\mathscr{A}_2 q + \cdots)$;
(4) $E_4^2\Xi = q^\Lambda(1_d + \cdots)$, $E_6\mathrm{D}\Xi - E_4^2\Xi\Lambda = q^\Lambda(1728\mathscr{A}_3 q + \cdots)$;
(5) $E_4E_6\Xi = q^\Lambda(1_d + \cdots)$, $E_4\left(E_4\mathrm{D}\Xi - E_6\Xi\Lambda\right) = q^\Lambda(1728\mathscr{A}_2 q + \cdots)$, $E_6\left(\mathrm{D}^2\Xi - E_4\Xi\Lambda(\Lambda - \frac{1}{6})\right) = q^\Lambda(1728\mathscr{A}_3(\mathscr{A}_3 - \frac{1}{3})q + \cdots)$.

Proof. Let $\alpha_i = \alpha_i(\rho, w)$ and $\beta_j = \beta_j(\rho, w)$, and write $\Lambda_{(i)}$ for some to-be-determined bijective exponent at weight $w+2i$. Then $\mathrm{Tr}(\Lambda_{(i)}) - \mathrm{Tr}(\Lambda)$ can be read off from (30). Consider first the case $i = 5$. Theorem 3.3(c) says $\mathscr{A}_3(\mathscr{A}_3 - \frac{1}{3})$ has rank β_2, while $\mathscr{A}_2$ has rank α_1. The column spaces of the second and third matrices given above for $i = 5$ have trivial intersection, since any $\mathbb{X}(\tau)$ in their intersection will be a vvmf for $(\rho, w+2)$ which would be divisible by $E_4(\tau)^2 E_6(\tau)$. This would mean $E_{14}(\tau)^{-1}\Delta(\tau)\mathbb{X}(\tau) \in \mathscr{M}^!_w(\rho)$ would lie in the kernel of $\mathscr{P}_\Lambda$, so by bijectivity $\mathbb{X}(\tau)$ must be 0. The desired generators will be linear combinations of α_1 columns of the second matrix with β_1 of the third and $d - \beta_1 - \alpha_1$ of the first. These would define a matrix $\Xi_{(5)}(\tau)$ whose $\Lambda_{(5)}$ has the correct trace, and therefore it must be a fundamental matrix (since its principal part map $\mathscr{P}_{\Lambda_{(5)}}$ is manifestly surjective).

The other cases listed are easier. □

In all these cases $2 \leq i \leq 6$, Proposition 4.1 finds a bijective exponent $\Lambda_{(i)}$ such that $\Lambda_{(i)} - \Lambda$ consists of 0's and 1's (for $i = 6$, $\Lambda_{(6)} = \Lambda + 1_d$ always works).

The case $i = 1$ is slightly more subtle. Note that a fundamental matrix for $w + 2i + 12j$ is $\Delta(\tau)^j$ times that for $w + 2i$. So one way to do $i = 1$ is to first find $i = 4$, then find the $i = 3$ from $w + 8$, then divide by $\Delta(\tau)$. Here is a more direct approach: almost always, the columns of the fundamental matrix for $i = 1$ is a linear combination over $\mathbb{C}$ of the columns of $M_1(\tau) := \mathrm{D}\Xi$, $M_2(\tau) := E_4^2(E_4\mathrm{D}\Xi - E_6\Xi\Lambda)/\Delta = q^\Lambda(1728\mathscr{A}_2 + \cdots)$, and $M_3(\tau) := (E_6^2\mathrm{D}\Xi - E_4^2E_6\Xi\Lambda)/\Delta = q^\Lambda(-1728\mathscr{A}_3 + \cdots)$. Indeed, take any $v \in \mathrm{Null}(\mathscr{A}_2 - \frac{1}{2}) \cap \mathrm{Null}(\mathscr{A}_3 - \frac{1}{3})(\mathscr{A}_3 - \frac{2}{3})$. That intersection has dimension at least $\alpha_1 - \beta_0$, because $\mathrm{Null}(\mathscr{A}_2 - \frac{1}{2})$ has dimension α_1 and $\mathrm{Null}(\mathscr{A}_3 - \frac{i}{3})$ has dimension β_i. Then $M_2 v = q^\Lambda(v/2 + \cdots)$ and $M_3 v = q^\Lambda(jv/3 + \cdots)$ where $j = 1$ or 2, so $2jM_2(\tau)v - 3M_3(\tau)v \in q^(\Lambda + 1_d)\mathbb{C}^d[[q]]$. Generically, this gives $\alpha_1 - \beta_0$ linearly independent vvmf, which together with $d - \alpha_1 + \beta_0$ columns of $M_1(\tau)$ will give the columns of the $i = 1$ fundamental matrix. This method doesn't work for all ρ, e.g. it fails for $\rho = 1$.

Acknowledgements I would like to thank Peter Bantay for a fruitful collaboration, Chris Marks, Geoff Mason, and Arturo Pianzola for conversations, and my students Jitendra Bajpai and Tim Graves. My research is supported in part by NSERC.

References

1. Anderson, G., Moore, G.: Rationality in conformal field theory. Commun. Math. Phys. **117**, 441–450 (1988)
2. Atkin, A.O.L., Swinnerton-Dyer, H.P.F.: Modular forms on noncongruence subgroups. In: Motzkin, T.S. (ed.) Proceedings of Symposia in Pure Mathematics, vol. 19, pp. 1–26. American Mathematical Society, Providence (1971)
3. Bajpai, J., Gannon, T.: In preparation
4. Bántay, P., Gannon, T.: Conformal characters and the modular representation. J. High Energy Physics **0602**, 005 (2006)

5. Bántay, P., Gannon, T.: Vector-valued modular functions for the modular group and the hypergeometric equation. Commun. Number Theory Phys. **1**, 637–666 (2008)
6. Bántay, P., Gannon, T.: In preparation
7. Bolibruch, A.A.: The Riemann-Hilbert problem. Russian Math. Surv. **45**(2), 1–47 (1990)
8. Bolibruch, A.A.: The Riemann-Hilbert problem and Fuchsian differential equations on the Riemann sphere. In: Proceedings of the International Congress of Mathematicians (Zürich, 1994), pp. 1159–1168. Birkhäuser, Basel (1995)
9. Borcherds, R.E.: Automorphic forms with singularities on Grassmannians. Invent. Math. **132**, 491–562 (1998)
10. Borcherds, R.E.: Gross-Kohnen-Zagier theorem in higher dimensions. Duke Math. J. **97**, 219–233 (1999); Correction. Duke Math. J. **105**, 183–184 (2000)
11. Bruinier, J.H., van der Geer, G., Harder, G., Zagier, D.: The 1-2-3 of Modular Forms. Springer, Berlin (2008)
12. Dijkgraaf, R., Maldecena, J., Moore, G., Verlinde, E.: A black hole Farey tale. arXiv: hep-th/0005003 (preprint)
13. Eichler, M., Zagier, D.: The Theory of Jacobi Forms. Progress in Mathematics, vol. 55. Birkhäuser, Boston (1985)
14. Gannon, T.: In preparation
15. Hille, E.: Ordinary Differential Equations in the Complex Domain. Wiley, New York (1976)
16. Kimura, T.: On the Riemann problem on Riemann surfaces. Lecture Notes in Mathematics, vol. 243, pp.218–228. Springer, Berlin (1971)
17. Knopp, M., Mason, G.: Vector-valued modular forms and Poincaré series. Ill. J. Math. **48**, 1345–1366 (2004)
18. Knopp, M., Mason, G.: Logarithmic vector-valued modular forms. Acta Arith. **147**, 261–282 (2011). arXiv:0910.3976.
19. Krauel, M., Mason, G.: Vertex operator algebras and weak Jacobi forms. Int. J. Math. **23**(6), 1250024 (2012)
20. Levelt, A.H.M.: Hypergeometric functions I–IV. Indag. Math. **23**, 361–403 (1961)
21. Marks, C.: Irreducible vector-valued modular forms of dimension less than six. Ill. J. Math. **55**, 1267–1297 (2011). arXiv:1004.3019
22. Marks, C., Mason, G.: Structure of the module of vector-valued modular forms. J. Lond. Math. Soc. **82**, 32–48 (2010). arXiv:0901.4367
23. Mason, G.: Vector-valued modular forms and linear differential operators. Int. J. Number Theory **3**, 377–390 (2007)
24. Mason, G.: 2-Dimensional vector-valued modular forms. Ramanujan J. **17**, 405–427 (2008)
25. Mathur, S.D., Mukhi, S., Sen, A.: Differential equations for correlators and characters in arbitrary rational conformal field theories. Nucl. Phys. **B312**, 15–57 (1989)
26. Milas, A.: Virasoro algebra, Dedekind eta-function and specialized Macdonald's identities. Transform. Groups **9**, 273–288 (2004)
27. Miyaki, T.: Modular Forms. Springer, Berlin/Heidelberg (2006)
28. Miyamoto, M.: A modular invariance on the theta functions defined on vertex operator algebras. Duke Math. J. **101**, 221–236 (2000)
29. Miyamoto, M.: Modular invariance of vertex operator algebras satisfying C_2-cofiniteness. Duke Math. J. **122**, 51–91 (2004)
30. Röhrl, H.: Das Riemann-Hilbertsche Problem der Theorie der linearen Differentialgleichungen. Math. Ann. **133**, 1–25 (1957)
31. Selberg, A.: On the estimation of Fourier coefficients of modular forms. In: Theory of Numbers. Proceedings of Symposia in Pure Mathematics, vol. VIII, pp. 1–15. American Mathematical Society, Providence (1965)
32. Shimura, G.: Introduction to the Arithmetic Theory of Automorphic Functions. Princeton University Press, Princeton (1971)
33. Skoruppa, N.-P.: Über den Zusammenhang zwischen Jacobiformen und Modulformen halbganzen Gewichts. Ph.D thesis, Universität Bonn (1984)

34. Tuba, I., Wenzl, H.: Representations of the braid group B_3 and of $\mathrm{SL}(2, \mathbb{Z})$. Pacific J. Math **97**, 491–510 (2001)
35. Westbury, B.W.: On the character varieties of free products of cyclic groups (2001, preprint)
36. Zhu, Y.: Modular invariance of characters of vertex operator algebras. J. Am. Math. Soc. **9**, 237–302 (1996)

Yangian Characters and Classical $\mathcal{W}$-Algebras

A.I. Molev and E.E. Mukhin

Abstract The Yangian characters (or q-characters) are known to be closely related to the classical $\mathcal{W}$-algebras and to the centers of the affine vertex algebras at the critical level. We make this relationship more explicit by producing families of generators of the $\mathcal{W}$-algebras from the characters of the Kirillov–Reshetikhin modules associated with multiples of the first fundamental weight in types B and D and of the fundamental modules in type C. We also give an independent derivation of the character formulas for these representations in the context of the RTT presentation of the Yangians. In all cases the generators of the $\mathcal{W}$-algebras correspond to the recently constructed elements of the Feigin–Frenkel centers via an affine version of the Harish-Chandra isomorphism.

1 Introduction

1.1 *Affine Harish-Chandra Isomorphism*

Let $\mathfrak{g}$ be a simple Lie algebra over $\mathbb{C}$. Choose a Cartan subalgebra $\mathfrak{h}$ of $\mathfrak{g}$ and a triangular decomposition $\mathfrak{g} = \mathfrak{n}_- \oplus \mathfrak{h} \oplus \mathfrak{n}_+$. Recall that the *Harish-Chandra homomorphism*

$$\mathrm{U}(\mathfrak{g})^{\mathfrak{h}} \to \mathrm{U}(\mathfrak{h}) \tag{1}$$

A.I. Molev (✉)
School of Mathematics and Statistics, University of Sydney, Sydney, NSW 2006, Australia
e-mail: alexander.molev@sydney.edu.au

E.E. Mukhin
Department of Mathematical Sciences, Indiana University – Purdue University Indianapolis, 402 North Blackford St, Indianapolis, IN 46202-3216, USA
e-mail: mukhin@math.iupui.edu

W. Kohnen and R. Weissauer (eds.), *Conformal Field Theory, Automorphic Forms and Related Topics*, Contributions in Mathematical and Computational Sciences 8, DOI 10.1007/978-3-662-43831-2_10,

is the projection of the $\mathfrak{h}$-centralizer $\mathrm{U}(\mathfrak{g})^{\mathfrak{h}}$ in the universal enveloping algebra to $\mathrm{U}(\mathfrak{h})$ whose kernel is the two-sided ideal $\mathrm{U}(\mathfrak{g})^{\mathfrak{h}} \cap \mathrm{U}(\mathfrak{g})\mathfrak{n}_+$. The restriction of the homomorphism (1) to the center $\mathrm{Z}(\mathfrak{g})$ of $\mathrm{U}(\mathfrak{g})$ yields an isomorphism

$$\mathrm{Z}(\mathfrak{g}) \to \mathrm{U}(\mathfrak{h})^W \tag{2}$$

called the *Harish-Chandra isomorphism*, where $\mathrm{U}(\mathfrak{h})^W$ denotes the subalgebra of invariants in $\mathrm{U}(\mathfrak{h})$ with respect to an action of the Weyl group W of $\mathfrak{g}$; see e.g. [7, Sect. 7.4].

In this paper we will be concerned with an affine version of the isomorphism (2). Consider the affine Kac–Moody algebra $\hat{\mathfrak{g}}$ which is the central extension

$$\hat{\mathfrak{g}} = \mathfrak{g}[t, t^{-1}] \oplus \mathbb{C}K,$$

where $\mathfrak{g}[t, t^{-1}]$ is the Lie algebra of Laurent polynomials in t with coefficients in $\mathfrak{g}$. We have a natural analogue of the homomorphism (1),

$$\mathrm{U}\big(t^{-1}\mathfrak{g}[t^{-1}]\big)^{\mathfrak{h}} \to \mathrm{U}\big(t^{-1}\mathfrak{h}[t^{-1}]\big). \tag{3}$$

The vacuum module $V_{-h^\vee}(\mathfrak{g})$ at the critical level over $\hat{\mathfrak{g}}$ is defined as the quotient of the universal enveloping algebra $\mathrm{U}(\hat{\mathfrak{g}})$ by the left ideal generated by $\mathfrak{g}[t]$ and $K + h^\vee$, where $h^\vee$ denotes the dual Coxeter number for $\mathfrak{g}$. The vacuum module $V_{-h^\vee}(\mathfrak{g})$ possesses a vertex algebra structure; see e.g. [11, Chap. 2]. The *center* of this vertex algebra is defined by

$$\mathfrak{z}(\hat{\mathfrak{g}}) = \{S \in V_{-h^\vee}(\mathfrak{g}) \mid \mathfrak{g}[t]\, S = 0\},$$

its elements are called *Segal–Sugawara vectors*. The center is a commutative associative algebra which can be regarded as a subalgebra of $\mathrm{U}\big(t^{-1}\mathfrak{g}[t^{-1}]\big)^{\mathfrak{h}}$. By the results of Feigin and Frenkel [10], $\mathfrak{z}(\hat{\mathfrak{g}})$ is an algebra of polynomials in infinitely many variables, and the restriction of the homomorphism (3) to the subalgebra $\mathfrak{z}(\hat{\mathfrak{g}})$ yields an isomorphism

$$\mathfrak{z}(\hat{\mathfrak{g}}) \to \mathscr{W}({}^L\mathfrak{g}), \tag{4}$$

where $\mathscr{W}({}^L\mathfrak{g})$ is the *classical* $\mathscr{W}$*-algebra* associated with the Langlands dual Lie algebra ${}^L\mathfrak{g}$; see [11] for a detailed exposition of these results. The $\mathscr{W}$-algebra $\mathscr{W}({}^L\mathfrak{g})$ can be defined as a subalgebra of $\mathrm{U}\big(t^{-1}\mathfrak{h}[t^{-1}]\big)$ which consists of the elements annihilated by the *screening operators*; see Sect. 4 below.

Recently, explicit generators of the Feigin–Frenkel center $\mathfrak{z}(\hat{\mathfrak{g}})$ were constructed for the Lie algebras $\mathfrak{g}$ of all classical types A, B, C and D; see [4,5,24]. Our aim in this paper is to describe the Harish-Chandra images of these generators in types B, C and D. The corresponding results in type A are given in [4]; we also reproduce them below in a slightly different form (as in [4], we work with the reductive

Lie algebra $\mathfrak{gl}_N$ rather than the simple Lie algebra $\mathfrak{sl}_N$ of type A). The images of the generators of $\mathfrak{z}(\hat{\mathfrak{g}})$ under the isomorphism (4) turn out to be elements of the $\mathscr{W}$-algebra $\mathscr{W}({}^L\mathfrak{g})$ written in terms of noncommutative analogues of the complete and elementary symmetric functions.

In more detail, for any $X \in \mathfrak{g}$ and $r \in \mathbb{Z}$ introduce the corresponding elements of the loop algebra $\mathfrak{g}[t, t^{-1}]$ by $X[r] = X\,t^r$. The extended Lie algebra $\hat{\mathfrak{g}} \oplus \mathbb{C}\tau$ with $\tau = -d/dt$ is defined by the commutation relations

$$\big[\tau, X[r]\big] = -r\,X[r-1], \qquad \big[\tau, K\big] = 0. \tag{5}$$

Consider the natural extension of (4) to the isomorphism

$$\chi : \mathfrak{z}(\hat{\mathfrak{g}}) \otimes \mathbb{C}[\tau] \to \mathscr{W}({}^L\mathfrak{g}) \otimes \mathbb{C}[\tau], \tag{6}$$

which is identical on $\mathbb{C}[\tau]$; see Sect. 5 for the definition of χ.

1.2 Segal–Sugawara Vectors in Type A

First let $\mathfrak{g} = \mathfrak{gl}_N$ be the general linear Lie algebra with the standard basis elements E_{ij}, $1 \leqslant i, j \leqslant N$. For each $a \in \{1, \dots, m\}$ introduce the element $E[r]_a$ of the algebra

$$\underbrace{\operatorname{End}\mathbb{C}^N \otimes \dots \otimes \operatorname{End}\mathbb{C}^N}_{m} \otimes \mathrm{U} \tag{7}$$

by

$$E[r]_a = \sum_{i,j=1}^{N} 1^{\otimes(a-1)} \otimes e_{ij} \otimes 1^{\otimes(m-a)} \otimes E_{ij}[r], \tag{8}$$

where the e_{ij} are the standard matrix units and U stands for the universal enveloping algebra of $\widehat{\mathfrak{gl}}_N \oplus \mathbb{C}\tau$. Let $H^{(m)}$ and $A^{(m)}$ denote the respective images of the symmetrizer and anti-symmetrizer in the group algebra for the symmetric group $\mathfrak{S}_m$ under its natural action on $(\mathbb{C}^N)^{\otimes m}$; see (26). We will identify $H^{(m)}$ and $A^{(m)}$ with the elements $H^{(m)} \otimes 1$ and $A^{(m)} \otimes 1$ of the algebra (7). Define the elements $\phi_{m\,a}, \psi_{m\,a} \in \mathrm{U}\big(t^{-1}\mathfrak{gl}_N[t^{-1}]\big)$ by the expansions

$$\operatorname{tr} A^{(m)}\big(\tau + E[-1]_1\big)\dots\big(\tau + E[-1]_m\big) = \phi_{m\,0}\,\tau^m + \phi_{m\,1}\,\tau^{m-1} + \dots + \phi_{m\,m}, \tag{9}$$

$$\operatorname{tr} H^{(m)}\big(\tau + E[-1]_1\big)\dots\big(\tau + E[-1]_m\big) = \psi_{m\,0}\,\tau^m + \psi_{m\,1}\,\tau^{m-1} + \dots + \psi_{m\,m}, \tag{10}$$

where the traces are taken over all m copies of $\operatorname{End}\mathbb{C}^N$. The results of [4,5] imply that all elements $\phi_{m\,a}$ and $\psi_{m\,a}$, as well as the coefficients of the polynomials $\operatorname{tr}(\tau + E[-1])^m$ belong to the Feigin–Frenkel center $\mathfrak{z}(\widehat{\mathfrak{gl}}_N)$; see also [25] for a simpler proof. Moreover, each of the families

$$\phi_{11}, \dots, \phi_{NN} \qquad \text{and} \qquad \psi_{11}, \dots, \psi_{NN}$$

is a *complete set of Segal–Sugawara vectors* in the sense that the elements of each family together with their images under all positive powers of the translation operator $T = \operatorname{ad}\tau$ are algebraically independent and generate $\mathfrak{z}(\widehat{\mathfrak{gl}}_N)$.

The elements $\mu_i = E_{ii}$ with $i = 1, \dots, N$ span a Cartan subalgebra of $\mathfrak{gl}_N$. Elements of the classical $\mathscr{W}$-algebra $\mathscr{W}(\mathfrak{gl}_N)$ are regarded as polynomials in the variables $\mu_i[r]$ with $r < 0$. A calculation of the images of the polynomials (9) with $m = N$ and $\operatorname{tr}(\tau + E[-1])^m$ under the isomorphism (6) was given in [4]. The same method applies to all polynomials (9) and (10) to yield the formulas

$$\chi : \operatorname{tr} A^{(m)}\big(\tau + E[-1]_1\big)\dots\big(\tau + E[-1]_m\big) \\ \mapsto e_m\big(\tau + \mu_1[-1], \dots, \tau + \mu_N[-1]\big), \tag{11}$$

$$\chi : \operatorname{tr} H^{(m)}\big(\tau + E[-1]_1\big)\dots\big(\tau + E[-1]_m\big) \\ \mapsto h_m\big(\tau + \mu_1[-1], \dots, \tau + \mu_N[-1]\big), \tag{12}$$

where we use standard noncommutative versions of the complete and elementary symmetric functions in the ordered variables $x_1, \dots, x_p$ defined by the respective formulas

$$h_m(x_1, \dots, x_p) = \sum_{i_1 \leqslant \dots \leqslant i_m} x_{i_1} \dots x_{i_m}, \tag{13}$$

$$e_m(x_1, \dots, x_p) = \sum_{i_1 > \dots > i_m} x_{i_1} \dots x_{i_m}. \tag{14}$$

Relations (11) and (12) can also be derived from the Yangian character formulas as we indicate below; see Sects. 3.1 and 5.

1.3 Main Results

Now turn to the Lie algebras of types B, C and D and let $\mathfrak{g} = \mathfrak{g}_N$ be the orthogonal Lie algebra $\mathfrak{o}_N$ (with $N = 2n$ or $N = 2n + 1$) or the symplectic Lie algebra $\mathfrak{sp}_N$ (with $N = 2n$). We will use the elements $F_{ij}[r]$ of the loop algebra $\mathfrak{g}_N[t, t^{-1}]$, where the F_{ij} are standard generators of $\mathfrak{g}_N$; see Sect. 2.2 for the definitions. For each $a \in \{1, \dots, m\}$ introduce the element $F[r]_a$ of the algebra (7) by

$$F[r]_a = \sum_{i,j=1}^{N} 1^{\otimes(a-1)} \otimes e_{ij} \otimes 1^{\otimes(m-a)} \otimes F_{ij}[r], \tag{15}$$

where U in (7) now stands for the universal enveloping algebra of $\hat{\mathfrak{g}}_N \oplus \mathbb{C}\tau$. We let $S^{(m)}$ denote the element of the algebra (7) which is the image of the symmetrizer of the Brauer algebra $\mathscr{B}_m(\omega)$ under its natural action on $(\mathbb{C}^N)^{\otimes m}$, where the parameter ω should be specialized to N or $-N$ in the orthogonal and symplectic case, respectively. The component of $S^{(m)}$ in U is the identity; see (37) and (38) below for explicit formulas. We will use the notation

$$\gamma_m(\omega) = \frac{\omega + m - 2}{\omega + 2m - 2} \tag{16}$$

and define the elements $\phi_{m\,a} \in \mathrm{U}\big(t^{-1}\mathfrak{g}_N[t^{-1}]\big)$ by the expansion

$$\gamma_m(\omega)\,\mathrm{tr}\, S^{(m)}\big(\tau + F[-1]_1\big)\dots\big(\tau + F[-1]_m\big) \\ = \phi_{m\,0}\,\tau^m + \phi_{m\,1}\,\tau^{m-1} + \dots + \phi_{m\,m}, \tag{17}$$

where the trace is taken over all m copies of $\mathrm{End}\,\mathbb{C}^N$ [we included the constant factor (16) to get a uniform expression in all cases]. By the main result of [24], all coefficients $\phi_{m\,a}$ belong to the Feigin–Frenkel center $\mathfrak{z}(\hat{\mathfrak{g}}_N)$. Note that in the symplectic case $\mathfrak{g}_N = \mathfrak{sp}_{2n}$ the values of m were restricted to $1 \leqslant m \leqslant 2n$, but the result and arguments also extend to $m = 2n + 1$; see [24, Sect. 3.3]. In the even orthogonal case $\mathfrak{g}_N = \mathfrak{o}_{2n}$ there is an additional element $\phi_n' = \mathrm{Pf}\,\tilde{F}[-1]$ of the center defined as the (noncommutative) Pfaffian of the skew-symmetric matrix $\tilde{F}[-1] = [\tilde{F}_{ij}[-1]]$,

$$\mathrm{Pf}\,\tilde{F}[-1] = \frac{1}{2^n n!} \sum_{\sigma \in \mathfrak{S}_{2n}} \mathrm{sgn}\,\sigma \cdot \tilde{F}_{\sigma(1)\,\sigma(2)}[-1] \dots \tilde{F}_{\sigma(2n-1)\,\sigma(2n)}[-1], \tag{18}$$

where $\tilde{F}_{ij}[-1] = F_{ij'}[-1]$ with $i' = 2n - i + 1$. The family $\phi_{2\,2}, \phi_{4\,4}, \dots, \phi_{2n\,2n}$ is a complete set of Segal–Sugawara vectors for the cases $\mathfrak{g}_N = \mathfrak{o}_{2n+1}$ and $\mathfrak{g}_N = \mathfrak{sp}_{2n}$, while $\phi_{2\,2}, \phi_{4\,4}, \dots, \phi_{2n-2\,2n-2}, \phi_n'$ is a complete set of Segal–Sugawara vectors for $\mathfrak{g}_N = \mathfrak{o}_{2n}$.

The Lie algebras $\mathfrak{o}_{2n+1}$ and $\mathfrak{sp}_{2n}$ are Langlands dual to each other, while $\mathfrak{o}_{2n}$ is self-dual. In all the cases we denote by $\mathfrak{h}$ the Cartan subalgebra of $\mathfrak{g}_N$ spanned by the elements $\mu_i = F_{ii}$ with $i = 1, \dots, n$ and identify it with the Cartan subalgebra of ${}^L\mathfrak{g}_N$ spanned by the elements with the same names. We let $\mu_i[r] = \mu_i\, t^r$ with $r < 0$ and $i = 1, \dots, n$ denote the basis elements of the vector space $t^{-1}\mathfrak{h}[t^{-1}]$ so that the elements of the classical $\mathscr{W}$-algebra $\mathscr{W}({}^L\mathfrak{g}_N)$ are regarded as polynomials in the variables $\mu_i[r]$.

Main Theorem. *The image of the polynomial* (17) *under the isomorphism* (6) *is given by the formula*:

type B_n: $$h_m\big(\tau+\mu_1[-1],\dots,\tau+\mu_n[-1],\tau-\mu_n[-1],\dots\tau-\mu_1[-1]\big),$$

type D_n: $$\tfrac{1}{2}\,h_m\big(\tau+\mu_1[-1],\dots,\tau+\mu_{n-1}[-1],\tau-\mu_n[-1],\dots\tau-\mu_1[-1]\big)$$
$$+\tfrac{1}{2}\,h_m\big(\tau+\mu_1[-1],\dots,\tau+\mu_n[-1],\tau-\mu_{n-1}[-1],\dots\tau-\mu_1[-1]\big),$$

type C_n: $$e_m\big(\tau+\mu_1[-1],\dots,\tau+\mu_n[-1],\tau,\tau-\mu_n[-1],\dots\tau-\mu_1[-1]\big).$$

Moreover, the image of the element ϕ_n' *in type* D_n *is given by*

$$\big(\mu_1[-1]-\tau\big)\dots\big(\mu_n[-1]-\tau\big)\,1. \tag{19}$$

In the last relation τ is understood as the differentiation operator so that $\tau\,1=0$.

1.4 Approach and Exposition

The Fourier coefficients of the image of any element of the Feigin–Frenkel center $\mathfrak{z}(\hat{\mathfrak{g}})$ under the state-field correspondence map are well-defined operators (called the *Sugawara operators*) on the Wakimoto modules over $\hat{\mathfrak{g}}$. These operators act by multiplication by scalars which are determined by the Harish-Chandra image under the isomorphism (4); see [11, Chap. 8]. Therefore, the Main Theorem yields explicit formulas for the eigenvalues of a family of the (higher) Sugawara operators in the Wakimoto modules.

Our approach is based on the theory of characters originated in [18] in the Yangian context and in [14] in the context of quantum affine algebras (the latter are commonly known as the q-characters). The theory was further developed in [12] where an algorithm for the calculation of the q-characters was proposed, while conjectures for functional relations satisfied by the q-characters were proved in [15,32]. An extensive review of the role of the q-characters in classical and quantum integrable systems is given in [22]; see also earlier papers [19, 20, 30, 31] where some formulas concerning the representations we dealing with in this paper had been conjectured and studied. In recent work [26, 27] the q-characters have been calculated for a wide class of representations in type B, and associated *extended* T*-systems* have been introduced.

Due to the general results on the connection of the q-characters with the Feigin–Frenkel center and the classical $\mathscr{W}$-algebras described in [14, Sect. 8.5], one could expect that the character formulas would be useful for the calculation of the Harish-Chandra images of the coefficients of the polynomial (17). Indeed, as we demonstrate below, the images in the classical $\mathscr{W}$-algebra are closely related with the top degree components of some linear combinations of the q-characters.

We now briefly describe the contents of the paper. We start by proving the character formulas for some classes of representations of the Yangian $\mathrm{Y}(\mathfrak{g}_N)$ associated with the Lie algebra $\mathfrak{g}_N$ (Sect. 2). To this end we employ realizations of the representations in harmonic tensors and construct special bases of the representation spaces. The main calculation is given in Sect. 3, where we consider particular linear combinations of the Yangian characters and calculate their top degree terms as elements of the associated graded algebra $\mathrm{gr}\,\mathrm{Y}(\mathfrak{g}_N) \cong \mathrm{U}(\mathfrak{g}_N[t])$. In Sect. 4 we recall the definition of the classical $\mathscr{W}$-algebras and write explicit screening operators in all classical types. By translating the results of Sect. 3 to the universal enveloping algebra $\mathrm{U}(t^{-1}\mathfrak{g}_N[t^{-1}])$ we will be able to get them in the form provided by the Main Theorem (Sect. 5). Finally, in Sect. 6 we apply our results to get the Harish-Chandra images of the Casimir elements for the Lie algebras $\mathfrak{g}_N$ arising from the Brauer–Schur–Weyl duality. We show that our formulas are equivalent to those previously found in [16].

2 Characters of Yangian Representations

We will use the RTT-presentation of the Yangians associated with the classical Lie algebras to calculate the characters of certain classes of their representations.

2.1 Yangian for $\mathfrak{gl}_N$

We will let $\mathfrak{h}$ denote the Cartan subalgebra of $\mathfrak{gl}_N$ spanned by the basis elements $E_{11}, \dots, E_{NN}$. The highest weights of representations of $\mathfrak{gl}_N$ will be considered with respect to this basis, and the highest vectors will be assumed to be annihilated by the action of the elements E_{ij} with $1 \leqslant i < j \leqslant N$, unless stated otherwise.

Recall the RTT-presentation of the Yangian associated with the Lie algebra $\mathfrak{gl}_N$; see e.g. [23, Chap. 1]. For $1 \leqslant a < b \leqslant m$ introduce the elements $P_{a\,b}$ of the tensor product algebra

$$\underbrace{\mathrm{End}\,\mathbb{C}^N \otimes \dots \otimes \mathrm{End}\,\mathbb{C}^N}_{m} \tag{20}$$

by

$$P_{a\,b} = \sum_{i,j=1}^{N} 1^{\otimes(a-1)} \otimes e_{ij} \otimes 1^{\otimes(b-a-1)} \otimes e_{ji} \otimes 1^{\otimes(m-b)}. \tag{21}$$

The Yang R-matrix $R_{12}(u)$ is a rational function in a complex parameter u with values in the tensor product algebra $\mathrm{End}\,\mathbb{C}^N \otimes \mathrm{End}\,\mathbb{C}^N$ defined by

$$R_{12}(u) = 1 - \frac{P_{12}}{u}.$$

This function satisfies the Yang–Baxter equation

$$R_{12}(u)\, R_{13}(u+v)\, R_{23}(v) = R_{23}(v)\, R_{13}(u+v)\, R_{12}(u), \tag{22}$$

where the subscripts indicate the copies of the endomorphism algebra $\mathrm{End}\,\mathbb{C}^N$ in (20) with $m = 3$. The *Yangian* $\mathrm{Y}(\mathfrak{gl}_N)$ is an associative algebra with generators $t_{ij}^{(r)}$, where $1 \leqslant i, j \leqslant N$ and $r = 1, 2, \ldots$, satisfying certain quadratic relations. To write them down, introduce the formal series

$$t_{ij}(u) = \delta_{ij} + \sum_{r=1}^{\infty} t_{ij}^{(r)}\, u^{-r} \in \mathrm{Y}(\mathfrak{gl}_N)[[u^{-1}]]$$

and set

$$T(u) = \sum_{i,j=1}^{N} e_{ij} \otimes t_{ij}(u) \in \mathrm{End}\,\mathbb{C}^N \otimes \mathrm{Y}(\mathfrak{gl}_N)[[u^{-1}]].$$

Consider the algebra $\mathrm{End}\,\mathbb{C}^N \otimes \mathrm{End}\,\mathbb{C}^N \otimes \mathrm{Y}(\mathfrak{gl}_N)[[u^{-1}]]$ and introduce its elements $T_1(u)$ and $T_2(u)$ by

$$T_1(u) = \sum_{i,j=1}^{N} e_{ij} \otimes 1 \otimes t_{ij}(u), \qquad T_2(u) = \sum_{i,j=1}^{N} 1 \otimes e_{ij} \otimes t_{ij}(u). \tag{23}$$

The defining relations for the algebra $\mathrm{Y}(\mathfrak{gl}_N)$ can then be written in the form

$$R_{12}(u-v)\, T_1(u)\, T_2(v) = T_2(v)\, T_1(u)\, R_{12}(u-v). \tag{24}$$

We identify the universal enveloping algebra $\mathrm{U}(\mathfrak{gl}_N)$ with a subalgebra of the Yangian $\mathrm{Y}(\mathfrak{gl}_N)$ via the embedding $E_{ij} \mapsto t_{ij}^{(1)}$. Then $\mathrm{Y}(\mathfrak{gl}_N)$ can be regarded as a $\mathfrak{gl}_N$-module with the adjoint action. We will denote by $\mathrm{Y}(\mathfrak{gl}_N)^{\mathfrak{h}}$ the subalgebra of $\mathfrak{h}$-invariants under this action. Consider the left ideal I of the algebra $\mathrm{Y}(\mathfrak{gl}_N)$ generated by all elements $t_{ij}^{(r)}$ with the conditions $1 \leqslant i < j \leqslant N$ and $r \geqslant 1$. By the Poincaré–Birkhoff–Witt theorem for the Yangian [23, Sect. 1.4], the intersection $\mathrm{Y}(\mathfrak{gl}_N)^{\mathfrak{h}} \cap I$ is a two-sided ideal of $\mathrm{Y}(\mathfrak{gl}_N)^{\mathfrak{h}}$. Moreover, the quotient of $\mathrm{Y}(\mathfrak{gl}_N)^{\mathfrak{h}}$ by this ideal is isomorphic to the commutative algebra freely generated by the images of the elements $t_{ii}^{(r)}$ with $i = 1, \ldots, N$ and $r \geqslant 1$ in the quotient. We will use the notation $\lambda_i^{(r)}$ for this image of $t_{ii}^{(r)}$. Thus, we get an analogue of the Harish-Chandra homomorphism (1),

$$\mathrm{Y}(\mathfrak{gl}_N)^{\mathfrak{h}} \to \mathbb{C}[\lambda_i^{(r)} \,|\, i = 1,\dots,N,\ r \geqslant 1]. \tag{25}$$

We combine the elements $\lambda_i^{(r)}$ into the formal series

$$\lambda_i(u) = 1 + \sum_{r=1}^{\infty} \lambda_i^{(r)}\, u^{-r}, \qquad i = 1,\dots,N,$$

which can be viewed as the respective images of the series $t_{ii}(u)$ under the homomorphism (25).

The *symmetrizer* $H^{(m)}$ and *anti-symmetrizer* $A^{(m)}$ in the algebra (20) are the operators in the tensor product space $(\mathbb{C}^N)^{\otimes m}$ associated with the corresponding idempotents in the group algebra of the symmetric group $\mathfrak{S}_m$ via its natural action on the tensor product space $(\mathbb{C}^N)^{\otimes m}$. That is,

$$H^{(m)} = \frac{1}{m!} \sum_{s\in\mathfrak{S}_m} P_s \qquad \text{and} \qquad A^{(m)} = \frac{1}{m!} \sum_{s\in\mathfrak{S}_m} \operatorname{sgn} s \cdot P_s, \tag{26}$$

where P_s is the element of the algebra (20) corresponding to $s \in \mathfrak{S}_m$. Both the symmetrizer and anti-symmetrizer admit multiplicative expressions in terms of the values of the Yang R-matrix,

$$H^{(m)} = \frac{1}{m!} \prod_{1\leqslant a<b\leqslant m} \Big(1 + \frac{P_{a\,b}}{b-a}\Big) \tag{27}$$

and

$$A^{(m)} = \frac{1}{m!} \prod_{1\leqslant a<b\leqslant m} \Big(1 - \frac{P_{a\,b}}{b-a}\Big),$$

where the products are taken in the lexicographic order on the pairs (a,b); see e.g. [23, Sect. 6.4]. The operators $H^{(m)}$ and $A^{(m)}$ project $(\mathbb{C}^N)^{\otimes m}$ to the subspaces of symmetric and skew-symmetric tensors, respectively. Both subspaces carry irreducible representations of the Yangian $\mathrm{Y}(\mathfrak{gl}_N)$. Consider the tensor product algebra

$$\underbrace{\operatorname{End}\mathbb{C}^N \otimes \dots \otimes \operatorname{End}\mathbb{C}^N}_{m} \otimes \mathrm{Y}(\mathfrak{gl}_N)[[u^{-1}]] \tag{28}$$

and extend the notation (23) to elements of (28). All coefficients of the formal series

$$\operatorname{tr} H^{(m)} T_1(u)\, T_2(u+1) \dots T_m(u+m-1) \tag{29}$$

and

$$\operatorname{tr} A^{(m)} T_1(u)\, T_2(u-1) \dots T_m(u-m+1) \tag{30}$$

belong to a commutative subalgebra of the Yangian. This subalgebra is contained in $\mathrm{Y}(\mathfrak{gl}_N)^{\mathfrak{h}}$. The next proposition is well-known and not difficult to prove; see also [3, Sect. 7.4], [13, Sect. 4.5] and [23, Sect. 8.5] for derivations of more general formulas for the characters of the evaluation modules over $\mathrm{Y}(\mathfrak{gl}_N)$. We give a proof of the proposition to stress the similarity of the approaches for all classical types.

Proposition 2.1. *The images of the series* (29) *and* (30) *under the homomorphism* (25) *are given by*

$$\sum_{1 \leqslant i_1 \leqslant \dots \leqslant i_m \leqslant N} \lambda_{i_1}(u)\, \lambda_{i_2}(u+1) \dots \lambda_{i_m}(u+m-1) \tag{31}$$

and

$$\sum_{1 \leqslant i_1 < \dots < i_m \leqslant N} \lambda_{i_1}(u)\, \lambda_{i_2}(u-1) \dots \lambda_{i_m}(u-m+1), \tag{32}$$

respectively.

Proof. By relations (24) and (27) we can write the product occurring in (29) as

$$H^{(m)} T_1(u) \dots T_m(u+m-1) = T_m(u+m-1) \dots T_1(u)\, H^{(m)}. \tag{33}$$

This relation shows that the product on each side can be regarded as an operator on $(\mathbb{C}^N)^{\otimes m}$ with coefficients in the algebra $\mathrm{Y}(\mathfrak{gl}_N)[[u^{-1}]]$ such that the subspace $H^{(m)}(\mathbb{C}^N)^{\otimes m}$ is invariant under this operator. A basis of this subspace is comprised by vectors of the form $v_{i_1,\dots,i_m} = H^{(m)}(e_{i_1} \otimes \dots \otimes e_{i_m})$, where $i_1 \leqslant \dots \leqslant i_m$ and $e_1, \dots, e_N$ denote the canonical basis vectors of $\mathbb{C}^N$. To calculate the trace of the operator, we will find the diagonal matrix elements corresponding to the basis vectors. Applying the operator which occurs on the right hand side of (33) to a basis vector $v_{i_1,\dots,i_m}$ we get

$$T_m(u+m-1) \dots T_1(u)\, H^{(m)} v_{i_1,\dots,i_m} = T_m(u+m-1) \dots T_1(u) v_{i_1,\dots,i_m}.$$

The coefficient of $v_{i_1,\dots,i_m}$ in the expansion of this expression as a linear combination of the basis vectors is determined by the coefficient of the tensor $e_{i_1} \otimes \dots \otimes e_{i_m}$. Hence, a nonzero contribution to the image of the diagonal matrix element corresponding to $v_{i_1,\dots,i_m}$ under the homomorphism (25) only comes from the term $t_{i_m i_m}(u+m-1)\dots t_{i_1 i_1}(u)$. The sum over all basis vectors yields the resulting formula for the image of the element (29).

The calculation of the image of the series (30) is quite similar. It relies on the identity

$$A^{(m)} T_1(u)\dots T_m(u-m+1) = T_m(u-m+1)\dots T_1(u)\, A^{(m)}$$

and a calculation of the diagonal matrix elements of the operator which occurs on the right hand side on the basis vectors $A^{(m)}(e_{i_1}\otimes\cdots\otimes e_{i_m})$, where $i_1<\cdots<i_m$. □

2.2 Yangians for $\mathfrak{o}_N$ and $\mathfrak{sp}_N$

Throughout the paper we use the involution on the set $\{1,\dots,N\}$ defined by $i'=N-i+1$. The Lie subalgebra of $\mathfrak{gl}_N$ spanned by the elements $F_{ij}=E_{ij}-E_{j'i'}$ with $i,j\in\{1,\dots,N\}$ is isomorphic to the orthogonal Lie algebra $\mathfrak{o}_N$. Similarly, the Lie subalgebra of $\mathfrak{gl}_{2n}$ spanned by the elements $F_{ij}=E_{ij}-\varepsilon_i\,\varepsilon_j\,E_{j'i'}$ with $i,j\in\{1,\dots,2n\}$ is isomorphic to the symplectic Lie algebra $\mathfrak{sp}_{2n}$, where $\varepsilon_i=1$ for $i=1,\dots,n$ and $\varepsilon_i=-1$ for $i=n+1,\dots,2n$. We will keep the notation $\mathfrak{g}_N$ for the Lie algebra $\mathfrak{o}_N$ (with $N=2n$ or $N=2n+1$) or $\mathfrak{sp}_N$ (with $N=2n$). Denote by $\mathfrak{h}$ the Cartan subalgebra of $\mathfrak{g}_N$ spanned by the basis elements $F_{11},\dots,F_{nn}$. The highest weights of representations of $\mathfrak{g}_N$ will be considered with respect to this basis, and the highest vectors will be assumed to be annihilated by the action of the elements F_{ij} with $1\leqslant i<j\leqslant N$, unless stated otherwise.

Recall the RTT-presentation of the Yangian associated with the Lie algebra $\mathfrak{g}_N$ following the general approach of [8,35]; see also [1] and [2].

For $1\leqslant a<b\leqslant m$ consider the elements $P_{a\,b}$ of the tensor product algebra (20) defined by (21). Introduce also the elements $Q_{a\,b}$ of (20) which are defined by different formulas in the orthogonal and symplectic cases. In the orthogonal case we set

$$Q_{a\,b}=\sum_{i,j=1}^{N} 1^{\otimes(a-1)}\otimes e_{ij}\otimes 1^{\otimes(b-a-1)}\otimes e_{i'j'}\otimes 1^{\otimes(m-b)},$$

and in the symplectic case

$$Q_{a\,b}=\sum_{i,j=1}^{N} \varepsilon_i\,\varepsilon_j\,1^{\otimes(a-1)}\otimes e_{ij}\otimes 1^{\otimes(b-a-1)}\otimes e_{i'j'}\otimes 1^{\otimes(m-b)}.$$

Set $\kappa = N/2-1$ in the orthogonal case and $\kappa = N/2+1$ in the symplectic case. The R-matrix $R_{12}(u)$ is a rational function in a complex parameter u with values in the tensor product algebra $\operatorname{End}\mathbb{C}^N \otimes \operatorname{End}\mathbb{C}^N$ defined by

$$R_{12}(u) = 1 - \frac{P_{12}}{u} + \frac{Q_{12}}{u-\kappa}.$$

It is well known by [37] that this function satisfies the Yang–Baxter equation (22).

The *Yangian* $\mathrm{Y}(\mathfrak{g}_N)$ is defined as an associative algebra with generators $t_{ij}^{(r)}$, where $1 \leqslant i,j \leqslant N$ and $r = 1,2,\dots$, satisfying certain quadratic relations. Introduce the formal series

$$t_{ij}(u) = \delta_{ij} + \sum_{r=1}^{\infty} t_{ij}^{(r)}\, u^{-r} \in \mathrm{Y}(\mathfrak{g}_N)[[u^{-1}]]$$

and set

$$T(u) = \sum_{i,j=1}^{N} e_{ij} \otimes t_{ij}(u) \in \operatorname{End}\mathbb{C}^N \otimes \mathrm{Y}(\mathfrak{g}_N)[[u^{-1}]].$$

Consider the algebra $\operatorname{End}\mathbb{C}^N \otimes \operatorname{End}\mathbb{C}^N \otimes \mathrm{Y}(\mathfrak{g}_N)[[u^{-1}]]$ and introduce its elements $T_1(u)$ and $T_2(u)$ by the same formulas (23) as in the case of $\mathfrak{gl}_N$. The defining relations for the algebra $\mathrm{Y}(\mathfrak{g}_N)$ can then be written in the form

$$R_{12}(u-v)\, T_1(u)\, T_2(v) = T_2(v)\, T_1(u)\, R_{12}(u-v) \tag{34}$$

together with the relation

$$T^{\,\prime}(u+\kappa)\, T(u) = 1,$$

where the prime denotes the matrix transposition which is defined for an $N \times N$ matrix $A = [A_{ij}]$ by

$$(A')_{ij} = A_{j'i'} \qquad \text{and} \qquad (A')_{ij} = \varepsilon_i\, \varepsilon_j\, A_{j'i'}$$

in the orthogonal and symplectic case, respectively.

We identify the universal enveloping algebra $\mathrm{U}(\mathfrak{g}_N)$ with a subalgebra of the Yangian $\mathrm{Y}(\mathfrak{g}_N)$ via the embedding

$$F_{ij} \mapsto t_{ij}^{(1)}, \qquad i,j = 1,\dots,N.$$

Then $\mathrm{Y}(\mathfrak{g}_N)$ can be regarded as a $\mathfrak{g}_N$-module with the adjoint action. Denote by $\mathrm{Y}(\mathfrak{g}_N)^{\mathfrak{h}}$ the subalgebra of $\mathfrak{h}$-invariants under this action.

Consider the left ideal I of the algebra $\mathrm{Y}(\mathfrak{g}_N)$ generated by all elements $t_{ij}^{(r)}$ with the conditions $1 \leqslant i < j \leqslant N$ and $r \geqslant 1$. It follows from the Poincaré–Birkhoff–Witt theorem for the Yangian [2, Sect. 3] that the intersection $\mathrm{Y}(\mathfrak{g}_N)^{\mathfrak{h}} \cap I$ is a two-sided ideal of $\mathrm{Y}(\mathfrak{g}_N)^{\mathfrak{h}}$. Moreover, the quotient of $\mathrm{Y}(\mathfrak{g}_N)^{\mathfrak{h}}$ by this ideal is isomorphic to the commutative algebra freely generated by the images of the elements $t_{ii}^{(r)}$ with $i = 1, \dots, n$ and $r \geqslant 1$ in the quotient. We will use the notation $\lambda_i^{(r)}$ for this image of $t_{ii}^{(r)}$ and extend this notation to all values $i = 1, \dots, N$. Thus, we get an analogue of the Harish-Chandra homomorphism (1),

$$\mathrm{Y}(\mathfrak{g}_N)^{\mathfrak{h}} \to \mathbb{C}[\lambda_i^{(r)} \,|\, i = 1, \dots, n,\ r \geqslant 1]. \tag{35}$$

We combine the elements $\lambda_i^{(r)}$ into the formal series

$$\lambda_i(u) = 1 + \sum_{r=1}^{\infty} \lambda_i^{(r)} u^{-r}, \qquad i = 1, \dots, N$$

which we will regard as the respective images of the series $t_{ii}(u)$ under the homomorphism (35).

It follows from [2, Prop. 5.2 and 5.14], that the series $\lambda_i(u)$ satisfy the relations

$$\lambda_i(u + \kappa - i)\, \lambda_{i'}(u) = \lambda_{i+1}(u + \kappa - i)\, \lambda_{(i+1)'}(u), \tag{36}$$

for $i = 0, 1, \dots, n-1$ if $\mathfrak{g}_N = \mathfrak{o}_{2n}$ or $\mathfrak{sp}_{2n}$, and for $i = 0, 1, \dots, n$ if $\mathfrak{g}_N = \mathfrak{o}_{2n+1}$, where $\lambda_0(u) = \lambda_{0'}(u) := 1$. Under an appropriate identification, the relations (36) coincide with those for the q-characters, as the $\lambda_i(u)$ correspond to the "single box variables"; see for instance [22, Sect. 7] and [30, Sect. 2]. This coincidence is consistent with the general result which establishes the equivalence of the definitions of q-characters in [14, 18]; see [12, Prop. 2.4] for a proof. The q-characters have been extensively studied; see [13, 14] and [18]. In particular, formulas for the q-characters of some classes of modules were conjectured in [20, 30] and [31] and later proved in [15] and [32]. However, this was done in the context of the new realization of the quantum affine algebras. In what follows we compute some q-characters independently in our setting of the RTT realization of the Yangians.

Introduce the element $S^{(m)}$ of the algebra (20) by setting $S^{(1)} = 1$ and for $m \geqslant 2$ define it by the respective formulas in the orthogonal and symplectic cases:

$$S^{(m)} = \frac{1}{m!} \prod_{1 \leqslant a < b \leqslant m} \Big(1 + \frac{P_{ab}}{b-a} - \frac{Q_{ab}}{N/2 + b - a - 1}\Big) \tag{37}$$

and

$$S^{(m)} = \frac{1}{m!} \prod_{1 \leqslant a < b \leqslant m} \Big(1 - \frac{P_{ab}}{b-a} - \frac{Q_{ab}}{n - b + a + 1}\Big), \tag{38}$$

where the products are taken in the lexicographic order on the pairs (a,b) and the condition $m \leqslant n+1$ is assumed in (38). The elements (37) and (38) are the images of the symmetrizers in the corresponding Brauer algebras $\mathscr{B}_m(N)$ and $\mathscr{B}_m(-N)$ under their actions on the vector space $(\mathbb{C}^N)^{\otimes m}$. In particular, for any $1 \leqslant a < b \leqslant m$ for the operator $S^{(m)}$ we have

$$S^{(m)} Q_{a\,b} = Q_{a\,b} S^{(m)} = 0 \qquad \text{and} \qquad S^{(m)} P_{a\,b} = P_{a\,b} S^{(m)} = \pm S^{(m)} \tag{39}$$

with the plus and minus signs taken in the orthogonal and symplectic case, respectively. The symmetrizer admits a few other equivalent expressions which are reproduced in [24].

In the orthogonal case the operator $S^{(m)}$ projects $(\mathbb{C}^N)^{\otimes m}$ to the irreducible representation of the Lie algebra $\mathfrak{o}_N$ with the highest weight $(m, 0, \dots, 0)$. The dimension of this representation equals

$$\frac{N+2m-2}{N+m-2}\binom{N+m-2}{m}.$$

This representation is extended to the Yangian $\mathrm{Y}(\mathfrak{o}_N)$ and it is one of the Kirillov–Reshetikhin modules. In the symplectic case with $m \leqslant n$ the operator $S^{(m)}$ projects $(\mathbb{C}^{2n})^{\otimes m}$ to the subspace of skew-symmetric harmonic tensors which carries an irreducible representation of $\mathfrak{sp}_{2n}$ with the highest weight $(1, \dots, 1, 0, \dots, 0)$ (with m copies of 1). Its dimension equals

$$\frac{2n-2m+2}{2n-m+2}\binom{2n+1}{m} = \binom{2n}{m} - \binom{2n}{m-2}. \tag{40}$$

This representation is extended to the m-th fundamental representation of the Yangian $\mathrm{Y}(\mathfrak{sp}_{2n})$ which is also a Kirillov–Reshetikhin module. It is well-known that if $m = n+1$ then the subspace of tensors is zero so that $S^{(n+1)} = 0$.

The existence of the Yangian action on the Lie algebra modules here can be explained by the fact that the projections (37) and (38) are the products of the evaluated R-matrices

$$S^{(m)} = \frac{1}{m!} \prod_{1 \leqslant a < b \leqslant m} R_{a\,b}(u_a - u_b), \tag{41}$$

where $u_a = u + a - 1$ and $u_a = u - a + 1$ for $a = 1, \dots, m$ in the orthogonal and symplectic case, respectively; see [17] for a proof in the context of a fusion procedure for the Brauer algebra. The same fact leads to a construction of a commutative subalgebra of the Yangian $\mathrm{Y}(\mathfrak{g}_N)$; see [24]. We will calculate the images of the elements of this subalgebra under the homomorphism (35) and thus reproduce the character formulas for the respective classes of Yangian representations; cf. [22, Sect. 7]. Consider the tensor product algebra

$$\underbrace{\operatorname{End}\mathbb{C}^N\otimes\cdots\otimes\operatorname{End}\mathbb{C}^N}_{m}\otimes\operatorname{Y}(\mathfrak{g}_N)[[u^{-1}]] \tag{42}$$

and extend the notation (23) to elements of (42).

Series B_n

The commutative subalgebra of the Yangian $\operatorname{Y}(\mathfrak{o}_N)$ with $N = 2n + 1$ is generated by the coefficients of the formal series

$$\operatorname{tr} S^{(m)} T_1(u)\, T_2(u+1)\ldots T_m(u+m-1) \tag{43}$$

with the trace taken over all m copies of $\operatorname{End}\mathbb{C}^N$ in (42), where $\mathfrak{g}_N = \mathfrak{o}_N$ and $S^{(m)}$ is defined in (37). It follows easily from the defining relations (34) that all elements of this subalgebra belong to $\operatorname{Y}(\mathfrak{o}_N)^{\mathfrak{h}}$.

Proposition 2.2. *The image of the series* (43) *under the homomorphism* (35) *is given by*

$$\sum_{1\leqslant i_1\leqslant\cdots\leqslant i_m\leqslant N}\lambda_{i_1}(u)\,\lambda_{i_2}(u+1)\ldots\lambda_{i_m}(u+m-1)$$

with the condition that $n+1$ *occurs among the summation indices* $i_1,\ldots,i_m$ *at most once.*

Proof. By [24, Prop. 3.1] the operator $S^{(m)}$ can be given by the formula

$$S^{(m)} = H^{(m)}\sum_{r=0}^{\lfloor m/2\rfloor}\frac{(-1)^r}{2^r\, r!}\binom{N/2+m-2}{r}^{-1}\sum_{a_i<b_i} Q_{a_1b_1}\, Q_{a_2b_2}\cdots Q_{a_rb_r} \tag{44}$$

with the second sum taken over the (unordered) sets of disjoint pairs of indices $\{(a_1,b_1),\ldots,(a_r,b_r)\}$ from $\{1,\ldots,m\}$. Here $H^{(m)}$ is the symmetrization operator defined in (26). Note that for each r the second sum in (44) commutes with any element P_s and hence commutes with $H^{(m)}$.

Recall that the subspace of harmonic tensors in $(\mathbb{C}^N)^{\otimes m}$ is spanned by the tensors v with the property $Q_{a\,b}\, v = 0$ for all $1\leqslant a<b\leqslant m$. By (39) the operator $S^{(m)}$ projects $(\mathbb{C}^N)^{\otimes m}$ to a subspace of symmetric harmonic tensors which we denote by $\mathscr{H}_m$. This subspace carries an irreducible representation of $\mathfrak{o}_N$ with the highest weight $(m,0,\ldots,0)$. Therefore, the trace in (43) can be calculated over the subspace $\mathscr{H}_m$. We will introduce a special basis of this subspace. We identify the image of the symmetrizer $H^{(m)}$ with the space of homogeneous polynomials of degree m in variables $z_1,\ldots,z_N$ via the isomorphism

$$H^{(m)}(e_{i_1}\otimes\cdots\otimes e_{i_m})\mapsto z_{i_1}\ldots z_{i_m}. \tag{45}$$

The subspace $\mathscr{H}_m$ is then identified with the subspace of harmonic homogeneous polynomials of degree m; they belong to the kernel of the Laplace operator

$$\Delta = \sum_{i=1}^{n} \partial_{z_i} \partial_{z_{i'}} + \frac{1}{2}\, \partial^2_{z_{n+1}}.$$

The basis vectors of $\mathscr{H}_m$ will be labeled by the N-tuples $(k_1, \dots, k_n, \delta, l_n, \dots, l_1)$, where the k_i and l_i are arbitrary nonnegative integers, $\delta \in \{0, 1\}$ and the sum of all entries is m. Given such a tuple, the corresponding harmonic polynomial is defined by

$$\sum_{a_1,\dots,a_n} \frac{(-2)^{a_1+\dots+a_n}(a_1 + \dots + a_n)!\, z_{n+1}^{2a_1+\dots+2a_n+\delta}}{a_1! \dots a_n!\, (2a_1 + \dots + 2a_n + \delta)!} \prod_{i=1}^{n} \frac{z_i^{k_i-a_i}\, z_{i'}^{l_i-a_i}}{(k_i - a_i)!\, (l_i - a_i)!}, \tag{46}$$

summed over nonnegative integers a_i satisfying $a_i \leqslant \min\{k_i, l_i\}$. Each polynomial contains a unique monomial (which we call the *leading monomial*) where the variable z_{n+1} occurs with the power not exceeding 1. It is straightforward to see that these polynomials are all harmonic and linearly independent. Furthermore, a simple calculation shows that the number of the polynomials coincides with the dimension of the irreducible representation of $\mathfrak{o}_N$ with the highest weight $(m, 0, \dots, 0)$ and so they form a basis of the subspace $\mathscr{H}_m$.

By relations (34) and (41) we can write the product occurring in (43) as

$$S^{(m)} T_1(u) \dots T_m(u + m - 1) = T_m(u + m - 1) \dots T_1(u)\, S^{(m)}. \tag{47}$$

This relation together with (39) shows that the product on each side can be regarded as an operator on $(\mathbb{C}^N)^{\otimes m}$ with coefficients in the algebra $\mathrm{Y}(\mathfrak{o}_N)[[u^{-1}]]$ such that the subspace $\mathscr{H}_m$ is invariant under this operator. Now fix a basis vector $v \in \mathscr{H}_m$ of the form (46). Denote the operator on the right hand side of (47) by A and consider the coefficient of v in the expansion of $A v$ as a linear combination of the basis vectors. Use the isomorphism (45) to write the vector v as a linear combination of the tensors $e_{j_1} \otimes \dots \otimes e_{j_m}$. We have $S^{(m)} v = v$, while the matrix elements of the remaining product are found from the expansion

$$\begin{aligned} &T_m(u + m - 1) \dots T_1(u)(e_{j_1} \otimes \dots \otimes e_{j_m}) \\ &\qquad = \sum_{i_1,\dots,i_m} t_{i_m j_m}(u + m - 1) \dots t_{i_1 j_1}(u)(e_{i_1} \otimes \dots \otimes e_{i_m}). \end{aligned}$$

The coefficient of v in the expansion of $A v$ is uniquely determined by the coefficient of the tensor $e_{i_1} \otimes \dots \otimes e_{i_m}$ with $i_1 \leqslant \dots \leqslant i_m$ which corresponds to the leading monomial of v under the isomorphism (45). It is clear from formula (46) that if a tensor of the form $e_{j_1} \otimes \dots \otimes e_{j_m}$ corresponds to a non-leading monomial, then the matrix element $t_{i_m j_m}(u+m-1)\dots t_{i_1 j_1}(u)$ vanishes under the homomorphism (35). Therefore, a nonzero contribution to the image of the diagonal matrix element of the operator A corresponding to v under the homomorphism (35) only comes from the term $t_{i_m i_m}(u+m-1)\dots t_{i_1 i_1}(u)$. Taking the sum over all basis vectors (46) yields the resulting formula for the image of the element (43). □

Series D_n

The commutative subalgebra of the Yangian $\mathrm{Y}(\mathfrak{o}_N)$ with $N = 2n$ is generated by the coefficients of the formal series defined by the same formula (43), where the parameter N now takes an even value $2n$.

Proposition 2.3. *The image of the series* (43) *under the homomorphism* (35) *is given by*

$$\sum_{1\leqslant i_1\leqslant\dots\leqslant i_m\leqslant N} \lambda_{i_1}(u)\,\lambda_{i_2}(u+1)\dots\lambda_{i_m}(u+m-1)$$

with the condition that n and n' do not occur simultaneously among the summation indices $i_1,\dots,i_m$.

Proof. As in the proof of Proposition 2.2, we use the formula (44) for the symmetrizer $S^{(m)}$ and its properties (39). Following the argument of that proof we identify the image $S^{(m)}(\mathbb{C}^N)^{\otimes m}$ with the space $\mathcal{H}_m$ of homogeneous harmonic polynomials of degree m in variables $z_1,\dots,z_N$ via the isomorphism (45). This time the harmonic polynomials are annihilated by the Laplace operator of the form

$$\Delta=\sum_{i=1}^{n}\partial_{z_i}\partial_{z_{i'}}.$$

The basis vectors of $\mathcal{H}_m$ will be parameterized by the N-tuples $(k_1,\dots,k_n, l_n,\dots,l_1)$, where the k_i and l_i are arbitrary nonnegative integers, the sum of all entries is m and at least one of k_n and l_n is zero. Given such a tuple, the corresponding harmonic polynomial is now defined by

$$\sum_{a_1,\dots,a_{n-1}}\frac{(-1)^{a_1+\dots+a_{n-1}}(a_1+\dots+a_{n-1})!\,z_n^{a_1+\dots+a_{n-1}+k_n}\,z_{n'}^{a_1+\dots+a_{n-1}+l_n}}{a_1!\dots a_{n-1}!\,(a_1+\dots+a_{n-1}+k_n)!\,(a_1+\dots+a_{n-1}+l_n)!}$$
$$\times\prod_{i=1}^{n-1}\frac{z_i^{k_i-a_i}\,z_{i'}^{l_i-a_i}}{(k_i-a_i)!\,(l_i-a_i)!},\qquad(48)$$

summed over nonnegative integers $a_1,\dots,a_{n-1}$ satisfying $a_i\leqslant\min\{k_i,l_i\}$. A unique *leading monomial* corresponds to the values $a_1=\dots=a_{n-1}=0$.

The argument is now completed in the same way as for Proposition 2.2 by considering the diagonal matrix elements of the operator on right hand side of (47) corresponding to the basis vectors (48). These coefficients are determined by those of the leading monomials and their images under the homomorphism (35) are straightforward to calculate. □

Series C_n

The commutative subalgebra of the Yangian $\mathrm{Y}(\mathfrak{sp}_N)$ with $N = 2n$ is generated by the coefficients of the formal series

$$\operatorname{tr} S^{(m)} T_1(u)\, T_2(u-1) \dots T_m(u-m+1), \tag{49}$$

with the trace taken over all m copies of $\operatorname{End}\mathbb{C}^N$ in (42) with $\mathfrak{g}_N = \mathfrak{sp}_N$ and $S^{(m)}$ defined in (38) with $m \leqslant n$.

Proposition 2.4. *The image of the series* (49) *with* $m \leqslant n$ *under the homomorphism* (35) *is given by*

$$\sum_{1 \leqslant i_1 < \dots < i_m \leqslant 2n} \lambda_{i_1}(u)\, \lambda_{i_2}(u-1) \dots \lambda_{i_m}(u-m+1) \tag{50}$$

with the condition that if for any i *both* i *and* i' *occur among the summation indices as* $i = i_r$ *and* $i' = i_s$ *for some* $1 \leqslant r < s \leqslant m$*, then* $s - r \leqslant n - i$.

Proof. Using again [24, Prop. 3.1] we find that the operator $S^{(m)}$ can be given by the formula

$$S^{(m)} = A^{(m)} \sum_{r=0}^{\lfloor m/2 \rfloor} \frac{1}{2^r\, r!} \binom{-n+m-2}{r}^{-1} \sum_{a_i < b_i} Q_{a_1 b_1} Q_{a_2 b_2} \cdots Q_{a_r b_r} \tag{51}$$

with the second sum taken over the (unordered) sets of disjoint pairs of indices $\{(a_1, b_1), \dots, (a_r, b_r)\}$ from $\{1, \dots, m\}$. Here $A^{(m)}$ is the anti-symmetrization operator defined in (26). For each r the second sum in (51) commutes with any element P_s and hence commutes with $A^{(m)}$.

As with the orthogonal case, the subspace of harmonic tensors in $(\mathbb{C}^N)^{\otimes m}$ is spanned by the tensors v with the property $Q_{ab}\, v = 0$ for all $1 \leqslant a < b \leqslant m$. The operator $S^{(m)}$ projects $(\mathbb{C}^N)^{\otimes m}$ to a subspace of skew-symmetric harmonic tensors which we denote by $\mathscr{H}_m$. Hence, the trace in (49) can be calculated over the subspace $\mathscr{H}_m$. We introduce a special basis of this subspace by identifying the image of the anti-symmetrizer $A^{(m)}$ with the space of homogeneous polynomials of degree m in the anti-commuting variables $\zeta_1, \dots, \zeta_{2n}$ via the isomorphism

$$A^{(m)}(e_{i_1} \otimes \dots \otimes e_{i_m}) \mapsto \zeta_{i_1} \wedge \dots \wedge \zeta_{i_m}. \tag{52}$$

The subspace $\mathscr{H}_m$ is then identified with the subspace of harmonic homogeneous polynomials of degree m; they belong to the kernel of the Laplace operator

$$\Delta = \sum_{i=1}^{n} \partial_i \wedge \partial_{i'},$$

where ∂_i denotes the (left) partial derivative over ζ_i.

The basis vectors of $\mathscr{H}_m$ will be parameterized by the subsets $\{i_1, \ldots, i_m\}$ of the set $\{1, \ldots, 2n\}$ satisfying the condition as stated in the proposition, when the elements $i_1, \ldots, i_m$ are written in the increasing order. We will call such subsets *admissible*. The number of admissible subsets can be shown to be given by the formula (40), which coincides with the dimension of $\mathscr{H}_m$. Consider monomials of the form

$$\zeta_{a_1} \wedge \zeta_{a_1'} \wedge \cdots \wedge \zeta_{a_k} \wedge \zeta_{a_k'} \wedge \zeta_{b_1} \wedge \cdots \wedge \zeta_{b_l} \tag{53}$$

with $1 \leqslant a_1 < \cdots < a_k \leqslant n$ and $1 \leqslant b_1 < \cdots < b_l \leqslant 2n$, associated with subsets $\{a_1, a_1', \ldots, a_k, a_k', b_1, \ldots, b_l\}$ of cardinality $m = 2k + l$ of the set $\{1, \ldots, 2n\}$ such that $b_i \neq b_j'$ for all i and j. We will suppose that the parameters b_i are fixed and label the monomial (53) by the k-tuple $(a_1, \ldots, a_k)$. Furthermore, we order the k-tuples and the corresponding monomials lexicographically.

Now let the subset $\{a_1, a_1', \ldots, a_k, a_k', b_1, \ldots, b_l\}$ be admissible and suppose that the parameters $a_1, \ldots, a_k$ are fixed too. We will call the corresponding monomial (53) *admissible*. Fix $i \in \{1, \ldots, k\}$. Let s be the number of the elements b_j of the subset satisfying $a_i < b_j < a_i'$. By the admissibility condition applied to a_i and a_i', we have the inequality $2(k - i) + s < n - a_i$. Therefore, there exist elements $c_i, \ldots, c_k$ satisfying $a_i < c_i < \cdots < c_k \leqslant n$ so that none of c_j or c_j' with $j = i, \ldots, k$ belongs to the subset $\{a_1, a_1', \ldots, a_k, a_k', b_1, \ldots, b_l\}$. Taking the consecutive values $i = k, k-1, \ldots, 1$ choose the maximum possible element c_i at each step. Thus, we get a family of elements $c_1 < \cdots < c_k$ uniquely determined by the admissible subset. In particular, $c_i > a_i$ for all i.

Note that our condition on the parameters b_i implies that the corresponding monomial $\zeta_{b_1} \wedge \cdots \wedge \zeta_{b_l}$ is annihilated by the operator Δ. We denote this monomial by y and set $x_a = \zeta_a \wedge \zeta_{a'}$ for $a = 1, \ldots, n$. The vector

$$\sum_{p=0}^{k} (-1)^p \sum_{1 \leqslant d_1 < \cdots < d_p \leqslant k} x_{a_1} \wedge \cdots \wedge \hat{x}_{a_{d_1}} \wedge \cdots \wedge \hat{x}_{a_{d_p}} \wedge \cdots \wedge x_{a_k} \wedge x_{c_{d_1}} \wedge \cdots \wedge x_{c_{d_p}} \wedge y,$$

where the hats indicate the factors to be skipped, is easily seen to belong to the kernel of the operator Δ so it is an element of the subspace $\mathscr{H}_m$. Furthermore, these vectors parameterized by all admissible subsets form a basis of $\mathscr{H}_m$. Indeed, the vectors are linearly independent because the linear combination defining each vector is uniquely determined by the admissible monomial $x_{a_1} \wedge \cdots \wedge x_{a_k} \wedge y$ which precedes all the other monomials occurring in the linear combination with respect to the lexicographic order.

Note that apart from the minimal admissible monomial $x_{a_1} \wedge \cdots \wedge x_{a_k} \wedge y$, the linear combination defining a basis vector may contain some other admissible monomials. By eliminating such additional admissible monomials with the use of an obvious induction on the lexicographic order, we can produce another basis of the space $\mathscr{H}_m$ parameterized by all admissible subsets with the property that each basis vector is given by a linear combination of monomials of the same form as above, containing a unique admissible monomial.

By relations (34) and (41) we can write the product occurring in (49) as

$$S^{(m)} T_1(u) \dots T_m(u - m + 1) = T_m(u - m + 1) \dots T_1(u)\, S^{(m)} \tag{54}$$

and complete the argument exactly as in the proof of Proposition 2.2. Indeed, relations (39) and (54) show that the product on each side can be regarded as an operator on $(\mathbb{C}^N)^{\otimes m}$ with coefficients in the algebra $\mathrm{Y}(\mathfrak{sp}_N)[[u^{-1}]]$ such that the subspace $\mathscr{H}_m$ is invariant under this operator. Denote the operator on the right hand side of (54) by A and let v denote the basis vector of $\mathscr{H}_m$ corresponding to an admissible subset $\{i_1, \dots, i_m\}$ with $i_1 < \cdots < i_m$. The properties of the basis vectors imply that a nonzero contribution to the image of the diagonal matrix element of the operator A corresponding to v under the homomorphism (35) only comes from the term $t_{i_m i_m}(u - m + 1) \dots t_{i_1 i_1}(u)$. □

We will need an equivalent formula for the expression (50) from [21, Prop. 2.4]. The argument there is combinatorial and relies only on the identities (36). To state the formula from [21] introduce new parameters $\varkappa_i(u)$ for $i = 1, \dots, 2n + 2$ by

$$\varkappa_i(u) = \lambda_i(u), \qquad \varkappa_{2n-i+3}(u) = \lambda_{2n-i+1}(u) \qquad \text{for} \quad i = 1, \dots, n,$$

and $\varkappa_{n+2}(u) = -\varkappa_{n+1}(u)$, where $\varkappa_{n+1}(u)$ is the formal series in u^{-1} with constant term 1 uniquely determined by

$$\varkappa_{n+1}(u)\, \varkappa_{n+1}(u - 1) = \lambda_n(u)\, \lambda_{n'}(u - 1).$$

Corollary 2.5. *The image of the series* (49) *with* $m \leqslant n$ *under the homomorphism* (35) *can be written as*

$$\sum_{1 \leqslant i_1 < \cdots < i_m \leqslant 2n+2} \varkappa_{i_1}(u)\, \varkappa_{i_2}(u - 1) \dots \varkappa_{i_m}(u - m + 1). \tag{55}$$

Moreover, the expression (55) *is zero for* $m = n + 1$. □

3 Harish-Chandra Images for the Current Algebras

We will use the character formulas obtained in Sect. 2 to calculate the Harish-Chandra images of elements of certain commutative subalgebras of $\mathrm{U}\big(\mathfrak{g}[t]\big)$ for the simple Lie algebras $\mathfrak{g}$ of all classical types. The results in the case of $\mathfrak{gl}_N$ are well-known, the commutative subalgebras were constructed explicitly in [36]; see also [4, 5, 25, 28] and [29].

3.1 Case of $\mathfrak{gl}_N$

Identify the universal enveloping algebra $\mathrm{U}(\mathfrak{gl}_N)$ with a subalgebra of $\mathrm{U}\big(\mathfrak{gl}_N[t]\big)$ via the embedding $E_{ij} \mapsto E_{ij}[0]$. Then $\mathrm{U}\big(\mathfrak{gl}_N[t]\big)$ can be regarded as a $\mathfrak{gl}_N$-module with the adjoint action. Denote by $\mathrm{U}\big(\mathfrak{gl}_N[t]\big)^{\mathfrak{h}}$ the subalgebra of $\mathfrak{h}$-invariants under this action. Consider the left ideal I of the algebra $\mathrm{U}\big(\mathfrak{gl}_N[t]\big)$ generated by all elements $E_{ij}[r]$ with the conditions $1 \leqslant i < j \leqslant N$ and $r \geqslant 0$. By the Poincaré–Birkhoff–Witt theorem, the intersection $\mathrm{U}\big(\mathfrak{gl}_N[t]\big)^{\mathfrak{h}} \cap I$ is a two-sided ideal of $\mathrm{U}\big(\mathfrak{gl}_N[t]\big)^{\mathfrak{h}}$. Moreover, the quotient of $\mathrm{U}\big(\mathfrak{gl}_N[t]\big)^{\mathfrak{h}}$ by this ideal is isomorphic to the commutative algebra freely generated by the images of the elements $E_{ii}[r]$ with $i = 1, \dots, N$ and $r \geqslant 0$ in the quotient. We will denote by $\mu_i[r]$ this image of $E_{ii}[r]$. We get an analogue of the Harish-Chandra homomorphism (1),

$$\mathrm{U}\big(\mathfrak{gl}_N[t]\big)^{\mathfrak{h}} \to \mathbb{C}[\mu_i[r] \,|\, i = 1, \dots, N,\ r \geqslant 0]. \tag{56}$$

Combine the elements $E_{ij}[r]$ and $\mu_i[r]$ into the formal series

$$E_{ij}(u) = \sum_{r=0}^{\infty} E_{ij}[r]\, u^{-r-1} \qquad \text{and} \qquad \mu_i(u) = \sum_{r=0}^{\infty} \mu_i[r]\, u^{-r-1}.$$

Then $\mu_i(u)$ is understood as the image of the series $E_{ii}(u)$ under the homomorphism (56). Consider tensor product algebras

$$\underbrace{\mathrm{End}\,\mathbb{C}^N \otimes \dots \otimes \mathrm{End}\,\mathbb{C}^N}_{m} \otimes \mathrm{U}\big(\mathfrak{gl}_N[t]\big)[[u^{-1}, \partial_u]]$$

and use matrix notation as in (8).

Proposition 3.1. *For the images under the Harish-Chandra homomorphism* (56) *we have*

$$\operatorname{tr} A^{(m)}\big(\partial_u + E_1(u)\big)\dots\big(\partial_u + E_m(u)\big) \mapsto e_m\big(\partial_u + \mu_1(u),\dots,\partial_u + \mu_N(u)\big), \tag{57}$$

$$\operatorname{tr} H^{(m)}\big(\partial_u + E_1(u)\big)\dots\big(\partial_u + E_m(u)\big) \mapsto h_m\big(\partial_u + \mu_1(u),\dots,\partial_u + \mu_N(u)\big). \tag{58}$$

Proof. The argument is essentially the same as in the proof of Proposition 2.1. Both relations are immediate from the cyclic property of trace and the identities

$$\big(\partial_u + E_1(u)\big)\dots\big(\partial_u + E_m(u)\big)A^{(m)} = A^{(m)}\big(\partial_u + E_1(u)\big)\dots\big(\partial_u + E_m(u)\big)A^{(m)},$$

$$H^{(m)}\big(\partial_u + E_1(u)\big)\dots\big(\partial_u + E_m(u)\big) = H^{(m)}\big(\partial_u + E_1(u)\big)\dots\big(\partial_u + E_m(u)\big)H^{(m)},$$

implied by the fact that $\partial_u + E(u)$ is a *left Manin matrix*; see [6, Prop. 18]. □

An alternative (longer) way to proof Proposition 3.1 is to derive it from the character formulas of Proposition 2.1. Indeed, $\partial_u + E(u)$ coincides with the image of the matrix $T(u)e^{\partial_u} - 1$ in the component of degree -1 of the graded algebra associated with the Yangian. Here we extend the filtration on the Yangian to the algebra of formal series $\mathrm{Y}(\mathfrak{gl}_N)[[u^{-1},\partial_u]]$ by setting $\deg u^{-1} = \deg \partial_u = -1$ so that the associated graded algebra is isomorphic to $\mathrm{U}\big(\mathfrak{gl}_N[t]\big)[[u^{-1},\partial_u]]$. Hence, for instance, the element on the left hand side of (57) can be found as the image of the component of degree $-m$ of the expression

$$\operatorname{tr} A^{(m)}\big(T_1(u)e^{\partial_u} - 1\big)\dots\big(T_m(u)e^{\partial_u} - 1\big).$$

The image of this expression under the homomorphism (25) can be found from (32).

There is no known analogue of the argument which we used in the proof of Proposition 3.1 for the B, C and D types. Therefore to prove its counterparts for these types we have to resort to the argument making use of the character formulas of Sect. 2.2.

3.2 Types B, C and D

Recall that $F_{ij}[r] = F_{ij}\,t^r$ with $r \in \mathbb{Z}$ denote elements of the loop algebra $\mathfrak{g}_N[t,t^{-1}]$, where the F_{ij} are standard generators of $\mathfrak{g}_N$; see Sect. 2.

Consider the ascending filtration on the Yangian $\mathrm{Y}(\mathfrak{g}_N)$ defined by

$$\deg\, t_{ij}^{(r)} = r - 1.$$

Denote by $\bar{t}_{ij}^{\,(r)}$ the image of the generator $t_{ij}^{(r)}$ in the $(r-1)$-th component of the associated graded algebra $\operatorname{gr}\mathrm{Y}(\mathfrak{g}_N)$. By [2, Theorem 3.6] the mapping

$$F_{ij}[r] \mapsto \bar{t}_{ij}^{\,(r+1)}, \qquad r \geqslant 0,$$

defines an algebra isomorphism $\mathrm{U}(\mathfrak{g}_N[t]) \to \mathrm{gr}\,\mathrm{Y}(\mathfrak{g}_N)$. Our goal here is to use this isomorphism and Propositions 2.2, 2.3 and 2.4 to calculate the Harish-Chandra images of certain elements of $\mathrm{U}(\mathfrak{g}_N[t])$ defined with the use of the corresponding operators (37) and (38). These elements generate a commutative subalgebra of $\mathrm{U}(\mathfrak{g}_N[t])$ and they can be obtained from the generators (17) of the Feigin–Frenkel center by an application of the vertex algebra structure on the vacuum module $V_{-h^\vee}(\mathfrak{g}_N)$; see [24, Sect. 5].

We identify the universal enveloping algebra $\mathrm{U}(\mathfrak{g}_N)$ with a subalgebra of $\mathrm{U}(\mathfrak{g}_N[t])$ via the embedding $F_{ij} \mapsto F_{ij}[0]$. Then $\mathrm{U}(\mathfrak{g}_N[t])$ can be regarded as a $\mathfrak{g}_N$-module with the adjoint action. Denote by $\mathrm{U}(\mathfrak{g}_N[t])^{\mathfrak{h}}$ the subalgebra of $\mathfrak{h}$-invariants under this action. Consider the left ideal I of the algebra $\mathrm{U}(\mathfrak{g}_N[t])$ generated by all elements $F_{ij}[r]$ with the conditions $1 \leqslant i < j \leqslant N$ and $r \geqslant 0$. By the Poincaré–Birkhoff–Witt theorem, the intersection $\mathrm{U}(\mathfrak{g}_N[t])^{\mathfrak{h}} \cap I$ is a two-sided ideal of $\mathrm{U}(\mathfrak{g}_N[t])^{\mathfrak{h}}$. Moreover, the quotient of $\mathrm{U}(\mathfrak{g}_N[t])^{\mathfrak{h}}$ by this ideal is isomorphic to the commutative algebra freely generated by the images of the elements $F_{ii}[r]$ with $i = 1, \dots, n$ and $r \geqslant 0$ in the quotient. We will write $\mu_i[r]$ for this image of $F_{ii}[r]$ and extend this notation to all values $i = 1, \dots, N$ so that $\mu_{i'}[r] = -\mu_i[r]$ for all i. We get an analogue of the Harish-Chandra homomorphism (1),

$$\mathrm{U}(\mathfrak{g}_N[t])^{\mathfrak{h}} \to \mathbb{C}[\mu_i[r]\,|\,i = 1, \dots, n,\ r \geqslant 0]. \tag{59}$$

We will combine the elements $F_{ij}[r]$ into the formal series

$$F_{ij}(u) = \sum_{r=0}^{\infty} F_{ij}[r]\,u^{-r-1}$$

and write

$$\mu_i(u) = \sum_{r=0}^{\infty} \mu_i[r]\,u^{-r-1}, \qquad i = 1, \dots, N.$$

Then the series $\mu_i(u)$ will be viewed as the image of the formal series $F_{ii}(u)$ under the homomorphism (59).

It is clear from the definitions of the homomorphisms (35) and (59), that the graded version of (35) coincides with (59) in the sense that the following diagram commutes

$$\begin{array}{ccc} \mathrm{U}(\mathfrak{g}_N[t])^{\mathfrak{h}} & \longrightarrow & \mathbb{C}\,[\mu_i[r]] \\ \downarrow & & \downarrow \\ \mathrm{gr}\,\mathrm{Y}(\mathfrak{g}_N)^{\mathfrak{h}} & \longrightarrow & \mathrm{gr}\,\mathbb{C}\,[\lambda_i^{(r+1)}], \end{array} \tag{60}$$

where i ranges over the set $\{1,\dots,n\}$ while $r \geqslant 0$ and the second vertical arrow indicates the isomorphism which takes $\mu_i[r]$ to the image of $\lambda_i^{(r+1)}$ in the graded polynomial algebra with the grading defined by the assignment $\deg \lambda_i^{(r+1)} = r$.

In what follows we extend the filtration on the Yangian to the algebra of formal series $\mathrm{Y}(\mathfrak{g}_N)[[u^{-1},\partial_u]]$ by setting $\deg u^{-1} = \deg \partial_u = -1$. The associated graded algebra will then be isomorphic to $\mathrm{U}\big(\mathfrak{g}_N[t]\big)[[u^{-1},\partial_u]]$. We consider tensor product algebras

$$\underbrace{\operatorname{End}\mathbb{C}^N \otimes \dots \otimes \operatorname{End}\mathbb{C}^N}_{m} \otimes \mathrm{U}\big(\mathfrak{g}_{N'}[t]\big)[[u^{-1},\partial_u]] \tag{61}$$

and use matrix notation as in (15).

Series B_n

Take $\mathfrak{g}_N = \mathfrak{o}_N$ with $N = 2n+1$ and consider the operator $S^{(m)}$ defined in (37). We also use notation (16) with $\omega = N$ and (13). The trace is understood to be taken over all copies of the endomorphism algebra $\operatorname{End}\mathbb{C}^N$ in (61).

Theorem 3.2. *For the image under the Harish-Chandra homomorphism* (59) *we have*

$$\gamma_m(N)\,\operatorname{tr} S^{(m)}\big(\partial_u + F_1(u)\big)\dots\big(\partial_u + F_m(u)\big)$$
$$\mapsto h_m\big(\partial_u + \mu_1(u),\dots,\partial_u + \mu_n(u), \partial_u + \mu_{n'}(u),\dots,\partial_u + \mu_{1'}(u)\big). \tag{62}$$

Proof. The element $\partial_u + F(u)$ coincides with the image of the matrix $T(u)\,e^{\partial_u} - 1$ in the component of degree -1 of the graded algebra associated with the Yangian. Therefore the element on the left hand side of (62) can be found as the image of the component of degree $-m$ of the expression

$$\gamma_m(N)\,\operatorname{tr} S^{(m)}\big(T_1(u)\,e^{\partial_u} - 1\big)\dots\big(T_m(u)\,e^{\partial_u} - 1\big). \tag{63}$$

Hence, the theorem can be proved by making use of the commutative diagram (60) and the Harish-Chandra image of (63) implied by Proposition 2.2. We have

$$\operatorname{tr} S^{(m)}\,\big(T_1(u)\,e^{\partial_u} - 1\big)\dots\big(T_m(u)\,e^{\partial_u} - 1\big)$$
$$= \sum_{k=0}^{m}(-1)^{m-k}\sum_{1\leqslant a_1<\dots<a_k\leqslant m}\operatorname{tr} S^{(m)}\,T_{a_1}(u)\,e^{\partial_u}\dots T_{a_k}(u)\,e^{\partial_u}.$$

Each product $T_{a_1}(u)\,e^{\partial_u}\dots T_{a_k}(u)\,e^{\partial_u}$ can be written as $P\,T_1(u)\,e^{\partial_u}\dots T_k(u)\,e^{\partial_u}\,P^{-1}$, where P is the image in (61) (with the identity component in the last tensor factor) of a permutation $p \in \mathfrak{S}_m$ such that $p(r) = a_r$ for $r = 1,\dots,k$. By the second property in (39) and the cyclic property of trace, we can bring the above expression to the form

$$\sum_{k=0}^{m}(-1)^{m-k}\binom{m}{k}\operatorname{tr} S^{(m)}\, T_1(u)\, e^{\partial_u}\dots T_k(u)\, e^{\partial_u}.$$

Now apply [24, Lemma 4.1] to calculate the partial traces of the symmetrizer $S^{(m)}$ over the copies $k+1,\dots,m$ of the algebra $\operatorname{End}\mathbb{C}^N$ in (20) to get

$$\operatorname{tr}_{k+1,\dots,m} S^{(m)} = \frac{\gamma_k(N)}{\gamma_m(N)}\binom{N+m-2}{m-k}\binom{m}{k}^{-1} S^{(k)}.$$

Thus, by Proposition 2.2, the Harish-Chandra image of the expression (63) is found by

$$\sum_{k=0}^{m}(-1)^{m-k}\,\gamma_k(N)\binom{N+m-2}{m-k}\sum_{1\leqslant i_1\leqslant\dots\leqslant i_k\leqslant N}\lambda_{i_1}(u)\, e^{\partial_u}\dots\lambda_{i_k}(u)\, e^{\partial_u} \tag{64}$$

with the condition that $n+1$ occurs among the summation indices $i_1,\dots,i_k$ at most once. The next step is to express (64) in terms of the new variables

$$\sigma_i(u) = \lambda_i(u)\, e^{\partial_u} - 1, \qquad i = 1,\dots,N. \tag{65}$$

This is done by a combinatorial argument as shown in the following lemma.

Lemma 3.3. *The expression* (64) *multiplied by* $-2\binom{N/2-2}{N+m-2}$ *equals*

$$\sum_{r=0}^{m}\binom{N/2-2}{N+r-3}\sum_{a_1+\dots+a_{1'}=r}\sigma_1(u)^{a_1}\dots\sigma_n(u)^{a_n}\sigma_{n'}(u)^{a_{n'}}\dots\sigma_{1'}(u)^{a_{1'}}$$

$$+\sum_{r=1}^{m}\binom{N/2-2}{N+r-3}\sum_{a_1+\dots+a_{1'}=r-1}\sigma_1(u)^{a_1}\dots\sigma_n(u)^{a_n}$$

$$\times\big(\sigma_{n+1}(u)+2\big)\,\sigma_{n'}(u)^{a_{n'}}\dots\sigma_{1'}(u)^{a_{1'}},$$

where $a_1,\dots,a_{1'}$ *run over nonnegative integers.*

Proof. The statement is verified by substituting (65) into both terms and calculating the coefficients of the sum

$$\sum_{1\leqslant i_1\leqslant\dots\leqslant i_k\leqslant N}\lambda_{i_1}(u)\, e^{\partial_u}\dots\lambda_{i_k}(u)\, e^{\partial_u} \tag{66}$$

for all $0 \leqslant k \leqslant m$, where $n+1$ occurs among the summation indices $i_1, \dots, i_k$ at most once. Note the following expansion formula for the noncommutative complete symmetric functions (13),

$$h_r(x_1 - 1, \dots, x_p - 1) = \sum_{k=0}^{r} (-1)^{r-k} \binom{p+r-1}{r-k} h_k(x_1, \dots, x_p). \tag{67}$$

Take $x_i = \lambda_i(u)\, e^{\partial_u}$ with $i = 1, \dots, n, n' \dots, 1'$ and apply (67) with $p = 2n$ to the first term in the expression of the lemma. Using a similar expansion for the second term we find that the coefficient of the sum (66) in the entire expression will be found as

$$\sum_{r=k}^{m} \binom{N/2-2}{N+r-3} \binom{N+r-3}{r-k} = \binom{N/2-2}{N+k-3} \binom{N/2+m-1}{m-k},$$

which coincides with

$$-2\,(-1)^{m-k}\, \gamma_k(N) \binom{N/2-2}{N+m-2} \binom{N+m-2}{m-k},$$

as claimed. □

Denote the expression in Lemma 3.3 by A_m. Since the degree of the element (63) is $-m$, its Harish-Chandra image (64) and the expression A_m also have degree $-m$. Observe that the terms in the both sums of A_m are independent of m so that $A_{m+1} = A_m + B_{m+1}$, where

$$B_{m+1} = \binom{N/2-2}{N+m-2} \sum_{a_1+\dots+a_{1'}=m+1} \sigma_1(u)^{a_1} \dots \sigma_n(u)^{a_n} \sigma_{n'}(u)^{a_{n'}} \dots \sigma_{1'}(u)^{a_{1'}} +$$

$$\binom{N/2-2}{N+m-2} \sum_{a_1+\dots+a_{1'}=m} \sigma_1(u)^{a_1} \dots \sigma_n(u)^{a_n} \big(\sigma_{n+1}(u)+2\big)\, \sigma_{n'}(u)^{a_{n'}} \dots \sigma_{1'}(u)^{a_{1'}}.$$

Since A_{m+1} has degree $-m-1$, its component of degree $-m$ is zero, and so the sum of the homogeneous components of degree $-m$ of A_m and B_{m+1} is zero. However, each element $\sigma_i(u)$ has degree -1 and its top degree component equals $\partial_u + \mu_i(u)$. This implies that the component of A_m of degree $-m$ equals the component of degree $-m$ of the term

$$-2 \binom{N/2-2}{N+m-2} \sum_{a_1+\dots+a_{1'}=m} \sigma_1(u)^{a_1} \dots \sigma_n(u)^{a_n} \sigma_{n'}(u)^{a_{n'}} \dots \sigma_{1'}(u)^{a_{1'}}.$$

Taking into account the constant factor used in Lemma 3.3, we can conclude that the component in question coincides with the noncommutative complete symmetric function as given in (62). □

As we have seen in the proof of the theorem, all components of the expression in Lemma 3.3 of degrees exceeding $-m$ are equal to zero. Since the summands do not depend on m, we derive the following corollary.

Corollary 3.4. *The series*

$$\sum_{r=0}^{\infty}\binom{N/2-2}{N+r-3}\sum_{a_1+\dots+a_{1'}=r}\sigma_1(u)^{a_1}\dots\sigma_n(u)^{a_n}\sigma_{n'}(u)^{a_{n'}}\dots\sigma_{1'}(u)^{a_{1'}}$$

$$+\sum_{r=1}^{\infty}\binom{N/2-2}{N+r-3}\sum_{a_1+\dots+a_{1'}=r-1}\sigma_1(u)^{a_1}\dots\sigma_n(u)^{a_n}$$

$$\times\big(\sigma_{n+1}(u)+2\big)\,\sigma_{n'}(u)^{a_{n'}}\dots\sigma_{1'}(u)^{a_{1'}}$$

is equal to zero. □

Series D_n

Now take $\mathfrak{g}_N=\mathfrak{o}_N$ with $N=2n$ and consider the operator $S^{(m)}$ defined in (37). We keep using notation (16) with $\omega=N$ and (13).

Theorem 3.5. *For the image under the Harish-Chandra homomorphism* (59) *we have*

$$\gamma_m(N)\,\mathrm{tr}\,S^{(m)}\big(\partial_u+F_1(u)\big)\dots\big(\partial_u+F_m(u)\big)$$

$$\mapsto\tfrac{1}{2}\,h_m\big(\partial_u+\mu_1(u),\dots,\partial_u+\mu_{n-1}(u),\partial_u+\mu_{n'}(u),\dots,\partial_u+\mu_{1'}(u)\big)$$

$$+\tfrac{1}{2}\,h_m\big(\partial_u+\mu_1(u),\dots,\partial_u+\mu_n(u),\partial_u+\mu_{(n-1)'}(u),\dots,\partial_u+\mu_{1'}(u)\big).$$

Proof. We repeat the beginning of the proof of Theorem 3.2 with N now taking the even value $2n$, up to the application of the formula for Yangian characters. This time we apply Proposition 2.3 to conclude that the Harish-Chandra image of the expression (63) is found by

$$\sum_{k=0}^{m}(-1)^{m-k}\,\gamma_k(2n)\binom{2n+m-2}{m-k}\sum_{1\leqslant i_1\leqslant\dots\leqslant i_k\leqslant 2n}\lambda_{i_1}(u)e^{\partial_u}\dots\lambda_{i_k}(u)e^{\partial_u}\qquad(68)$$

with the condition that n and n' do not occur simultaneously among the summation indices $i_1,\dots,i_k$. Introducing new variables by the same formulas (65) we come to the D_n series counterpart of Lemma 3.3, where we use the notation

$$c_r = (-1)^{r-1}\binom{2n+r-2}{n-1}^{-1}.$$

Lemma 3.6. *The expression* (68) *multiplied by* $2c_m$ *equals*

$$\begin{aligned}2c_m &\sum_{\substack{a_1+\dots+a_{1'}=m\\ a_n=a_{n'}=0}} \sigma_1(u)^{a_1}\dots\sigma_{1'}(u)^{a_{1'}}\\
&+ c_m \sum_{\substack{a_1+\dots+a_{1'}=m\\ \text{only one of } a_n \text{ and } a_{n'} \text{ is zero}}} \sigma_1(u)^{a_1}\dots\sigma_{1'}(u)^{a_{1'}}\\
&-\sum_{r=1}^{m}\frac{r\,c_r}{n+r-1}\sum_{\substack{a_1+\dots+a_{1'}=r\\ a_n=a_{n'}=0}} \sigma_1(u)^{a_1}\dots\sigma_{1'}(u)^{a_{1'}}\\
&+\sum_{r=1}^{m}\frac{(n-1)\,c_r}{n+r-1}\sum_{\substack{a_1+\dots+a_{1'}=r\\ \text{only one of } a_n \text{ and } a_{n'} \text{ is zero}}} \sigma_1(u)^{a_1}\dots\sigma_{1'}(u)^{a_{1'}},\end{aligned}$$

where $a_1,\dots,a_{1'}$ *run over nonnegative integers.*

Proof. Substitute (65) into the expression and calculate the coefficients of the sum

$$\sum_{1\leqslant i_1\leqslant\dots\leqslant i_k\leqslant 2n} \lambda_{i_1}(u)\,e^{\partial_u}\dots\lambda_{i_k}(u)\,e^{\partial_u}. \tag{69}$$

The argument splits into two cases, depending on whether neither of n and n' occurs among the summation indices $i_1,\dots,i_k$ in (69) or only one of them occurs. The application of the expansion formula (67) brings this to a straightforward calculation with binomial coefficients in both cases. □

Let A_m denote the four-term expression in Lemma 3.6. This expression equals $2c_m$ times the Harish-Chandra image of (63) and so A_m has degree $-m$. Hence, the component of degree $-m$ of the expression A_{m+1} is zero. On the other hand, each element $\sigma_i(u)$ has degree -1 and its top degree component equals $\partial_u + \mu_i(u)$. This implies that the component of degree $-m$ in the sum of the third and fourth terms in A_m is zero. Therefore, the component of A_m of degree $-m$ equals the component of degree $-m$ in the sum of the first and the second terms. Taking into account the constant factor $2c_m$, we conclude that the component takes the desired form. □

The following corollary is implied by the proof of the theorem.

Corollary 3.7. *The series*

$$-\sum_{r=1}^{\infty} \frac{r\, c_r}{n+r-1} \sum_{\substack{a_1+\dots+a_{1'}=r \\ a_n=a_{n'}=0}} \sigma_1(u)^{a_1} \dots \sigma_{1'}(u)^{a_{1'}}$$

$$+\sum_{r=1}^{\infty} \frac{(n-1)\, c_r}{n+r-1} \sum_{\substack{a_1+\dots+a_{1'}=r \\ \text{only one of } a_n \text{ and } a_{n'} \text{ is zero}}} \sigma_1(u)^{a_1} \dots \sigma_{1'}(u)^{a_{1'}}$$

is equal to zero. □

Series C_n

Now we let $\mathfrak{g}_N = \mathfrak{sp}_N$ with $N = 2n$ and consider the operator $S^{(m)}$ defined in (38). We also use notation (16) with $\omega = -2n$ and (14). Although the operator $S^{(m)}$ is defined only for $m \leqslant n+1$, it is possible to extend the values of expressions of the form (17) and those which are used in the next theorem to all m with $m \leqslant 2n+1$; see [24, Sect. 3.3]. The Harish-Chandra images turn out to be given by the same expression for all these values of m. We postpone the proof to Corollary 5.2 below, and assume first that $m \leqslant n$.

Theorem 3.8. *For all* $1 \leqslant m \leqslant n$ *for the image under the Harish-Chandra homomorphism* (59) *we have*

$$\gamma_m(-2n)\, \mathrm{tr}\, S^{(m)}\big(\partial_u - F_1(u)\big) \dots \big(\partial_u - F_m(u)\big)$$

$$\mapsto e_m\big(\partial_u + \mu_1(u), \dots, \partial_u + \mu_n(u), \partial_u, \partial_u + \mu_{n'}(u), \dots, \partial_u + \mu_{1'}(u)\big). \tag{70}$$

Proof. The element $\partial_u - F(u)$ coincides with the image of the matrix $1 - T(u)e^{-\partial_u}$ in the component of degree -1 of the graded algebra associated with the Yangian. Hence the left hand side of (70) can be found as the image of the component of degree $-m$ of the expression

$$(-1)^m \gamma_m(-2n)\, \mathrm{tr}\, S^{(m)}\big(T_1(u)e^{-\partial_u} - 1\big) \dots \big(T_m(u)e^{-\partial_u} - 1\big). \tag{71}$$

Now we use the commutative diagram (60) and the Harish-Chandra image of (71) implied by Proposition 2.4. We have

$$\mathrm{tr}\, S^{(m)}\, \big(T_1(u)e^{-\partial_u} - 1\big) \dots \big(T_m(u)e^{-\partial_u} - 1\big)$$

$$= \sum_{k=0}^{m} (-1)^{m-k} \sum_{1 \leqslant a_1 < \dots < a_k \leqslant m} \mathrm{tr}\, S^{(m)}\, T_{a_1}(u)e^{-\partial_u} \dots T_{a_k}(u)e^{-\partial_u}. \tag{72}$$

As in the proof of Theorem 3.2, we use the second property in (39) and the cyclic property of trace to bring the expression to the form

$$\sum_{k=0}^{m}(-1)^{m-k}\binom{m}{k}\operatorname{tr} S^{(m)}\, T_1(u)\,e^{-\partial_u}\dots T_k(u)\,e^{-\partial_u}.$$

Further, the partial traces of the symmetrizer $S^{(m)}$ over the copies $k+1,\dots,m$ of the algebra $\operatorname{End}\mathbb{C}^N$ in (20) are found by applying [24, Lemma 4.1] to get

$$\operatorname{tr}_{k+1,\dots,m} S^{(m)} = \frac{\gamma_k(-2n)}{\gamma_m(-2n)}\binom{2n-k+1}{m-k}\binom{m}{k}^{-1} S^{(k)}.$$

By Proposition 2.4 and Corollary 2.5, the Harish-Chandra image of the expression (71) is found by

$$\sum_{k=0}^{m}(-1)^k\,\gamma_k(-2n)\binom{2n-k+1}{m-k}\sum_{1\leqslant i_1<\dots<i_k\leqslant 2n+2}\varkappa_{i_1}(u)\,e^{-\partial_u}\dots\varkappa_{i_k}(u)\,e^{-\partial_u}. \tag{73}$$

Introduce new variables by

$$\sigma_i(u)=\varkappa_i(u)\,e^{-\partial_u}-1, \qquad i=1,\dots,2n+2, \quad i\neq n+2, \tag{74}$$

and $\sigma_{n+2}(u)=\varkappa_{n+2}(u)\,e^{-\partial_u}+1$.

Lemma 3.9. *For $m\leqslant n$ the expression* (73) *multiplied by $2\,(-1)^m\binom{2n-m+1}{n+1}$ equals*

$$\sum_{r=0}^{m}\binom{2n-r+2}{n+1}\sum_{1\leqslant i_1<\dots<i_r\leqslant 2n+2}\sigma_{i_1}(u)\dots\sigma_{i_r}(u)$$
$$-2\sum_{r=0}^{m-1}\binom{2n-r+1}{n+1}\sum_{\substack{1\leqslant i_1<\dots<i_r\leqslant 2n+2\\ i_s\neq n+2}}\sigma_{i_1}(u)\dots\sigma_{i_r}(u),$$

where $n+2$ does not occur among the summation indices in the last sum.

Proof. Substituting (74) into the expression and simplifying gives

$$\sum_{r=0}^{m}\binom{2n-r+2}{n+1}\sum_{1\leqslant i_1<\dots<i_r\leqslant 2n+2}\big(\varkappa_{i_1}(u)\,e^{-\partial_u}-1\big)\dots\big(\varkappa_{i_r}(u)\,e^{-\partial_u}-1\big). \tag{75}$$

Now use the expansion formula for the noncommutative elementary symmetric functions (14),

$$e_r(x_1 - 1, \dots, x_p - 1) = \sum_{k=0}^{r} (-1)^{r-k} \binom{p-k}{r-k} e_k(x_1, \dots, x_p).$$

Taking $x_i = \varkappa_i(u)\, e^{-\partial_u}$ with $i = 1, \dots, 2n+2$, it is straightforward to verify that the coefficient of the sum

$$\sum_{1 \leqslant i_1 < \dots < i_k \leqslant N} \varkappa_{i_1}(u)\, e^{-\partial_u} \dots \varkappa_{i_k}(u)\, e^{-\partial_u}$$

in (75) equals

$$(-1)^{m-k} \binom{n-k}{m-k} \binom{2n-k+2}{n+1}$$

which coincides with

$$2(-1)^{m-k}\, \gamma_k(-2n) \binom{2n-m+1}{n+1} \binom{2n-k+1}{m-k}$$

as claimed. □

For $m \leqslant n$ let A_m denote the expression in Lemma 3.9. Note that A_m coincides with the Harish-Chandra image of (72) multiplied by $\binom{2n-m+2}{n+1}$. The proof of Lemma 3.9 and the second part of Corollary 2.5 show that A_m is also well-defined for the value $m = n+1$ and $A_{n+1} = 0$.

Since the degree of the element (71) is $-m$, for $m \leqslant n$ the expression A_m also has degree $-m$. Hence, the component of degree $-m$ of the expression A_{m+1} is zero; this holds for $m = n$ as well, because $A_{n+1} = 0$. Furthermore, each element $\sigma_i(u)$ has degree -1 and so the component of A_m of degree $-m$ must be equal to the component of degree $-m$ of the expression

$$2 \binom{2n-m+1}{n+1} \sum_{\substack{1 \leqslant i_1 < \dots < i_m \leqslant 2n+2 \\ i_s \neq n+2}} \sigma_{i_1}(u) \dots \sigma_{i_m}(u).$$

The component of $\sigma_i(u)$ of degree -1 equals

$$\begin{cases} -\partial_u + \mu_i(u) & \text{for} \quad i = 1, \dots, n, \\ -\partial_u & \text{for} \quad i = n+1, \\ -\partial_u + \mu_{i-2}(u) & \text{for} \quad i = n+3, \dots, 2n+2. \end{cases}$$

The proof is completed by taking the signs and the constant factor used in Lemma 3.9 into account. □

4 Classical $\mathscr{W}$-Algebras

We define the classical $\mathscr{W}$-algebra $\mathscr{W}(\mathfrak{g})$ associated with a simple Lie algebra $\mathfrak{g}$ following [11, Sect. 8.1], where more details and proofs can be found. We let $\mathfrak{h}$ denote a Cartan subalgebra of $\mathfrak{g}$ and let $\mu_1, \dots, \mu_n$ be a basis of $\mathfrak{h}$. The universal enveloping algebra $\mathrm{U}(t^{-1}\mathfrak{h}[t^{-1}])$ will be identified with the algebra of polynomials in the infinitely many variables $\mu_i[r]$ with $i = 1, \dots, n$ and $r < 0$ and will be denoted by π_0. We will also use the extended algebra with the additional generator τ subject to the relations

$$\big[\tau, \mu_i[r]\big] = -r\, \mu_i[r-1],$$

implied by (5). The extended algebra is isomorphic to $\pi_0 \otimes \mathbb{C}[\tau]$ as a vector space. Furthermore, we will need the operator $T = \operatorname{ad}\tau$ which is the derivation $T : \pi_0 \to \pi_0$ defined on the generators by the relations

$$T\, \mu_i[r] = -r\, \mu_i[r-1].$$

In particular, $T\,1 = 0$. The *classical $\mathscr{W}$-algebra* is defined as the vector subspace $\mathscr{W}(\mathfrak{g}) \subset \pi_0$ spanned by the elements which are annihilated by the *screening operators*

$$V_i : \pi_0 \to \pi_0, \qquad i = 1, \dots, n,$$

which we will write down explicitly for each classical type below,[1]

$$\mathscr{W}(\mathfrak{g}) = \{P \in \pi_0 \mid V_i\, P = 0, \quad i = 1, \dots, n\}.$$

The operators V_i are derivations of π_0 so that $\mathscr{W}(\mathfrak{g})$ is a subalgebra of π_0. The subalgebra $\mathscr{W}(\mathfrak{g})$ is T-invariant. Moreover, there exist elements $B_1, \dots, B_n \in \mathscr{W}(\mathfrak{g})$ such that the family of elements $T^r B_i$ with $i = 1, \dots, n$ and $r \geqslant 0$ is algebraically independent and generates the algebra $\mathscr{W}(\mathfrak{g})$. We will call $B_1, \dots, B_n$ a *complete set of generators* of $\mathscr{W}(\mathfrak{g})$. Examples of such sets in the classical types will be given below. We extend the screening operators to the algebra $\pi_0 \otimes \mathbb{C}[\tau]$ by

$$V_i\big(P \otimes Q(\tau)\big) = V_i(P) \otimes Q(\tau), \qquad P \in \pi_0, \quad Q(\tau) \in \mathbb{C}[\tau].$$

[1]Our V_i essentially coincides with the operator $\overline{V}_i[1]$ in the notation of [11, Sect. 7.3.4], which is associated with the Langlands dual Lie algebra ${}^L\mathfrak{g}$.

4.1 Screening Operators and Generators for $\mathscr{W}(\mathfrak{gl}_N)$

Here π_0 is the algebra of polynomials in the variables $\mu_i[r]$ with $i = 1, \dots, N$ and $r < 0$. The screening operators $V_1, \dots, V_{N-1}$ are defined by

$$V_i = \sum_{r=0}^{\infty} V_{i\,[r]} \Big(\frac{\partial}{\partial \mu_i[-r-1]} - \frac{\partial}{\partial \mu_{i+1}[-r-1]} \Big),$$

where the coefficients $V_{i\,[r]}$ are found from the expansion of a formal generating function in a variable z,

$$\sum_{r=0}^{\infty} V_{i\,[r]}\, z^r = \exp \sum_{m=1}^{\infty} \frac{\mu_i[-m] - \mu_{i+1}[-m]}{m}\, z^m.$$

Define elements $\mathscr{E}_1, \dots, \mathscr{E}_N$ of π_0 by the expansion in $\pi_0 \otimes \mathbb{C}[\tau]$,

$$\big(\tau + \mu_N[-1]\big) \dots \big(\tau + \mu_1[-1]\big) = \tau^N + \mathscr{E}_1\, \tau^{N-1} + \dots + \mathscr{E}_N, \tag{76}$$

known as the *Miura transformation*. Explicitly, using the notation (14) we can write the coefficients as

$$\mathscr{E}_m = e_m\big(T + \mu_1[-1], \dots, T + \mu_N[-1]\big), \tag{77}$$

which follows easily from (76) by induction. The family $\mathscr{E}_1, \dots, \mathscr{E}_N$ is a complete set of generators of $\mathscr{W}(\mathfrak{gl}_N)$. Verifying that all elements $\mathscr{E}_i$ are annihilated by the screening operators is straightforward. This is implied by the relations for the operators on π_0,

$$V_i\, T = \big(T + \mu_i[-1] - \mu_{i+1}[-1]\big)\, V_i, \qquad i = 1, \dots, N-1. \tag{78}$$

They imply the corresponding relations for the operators on $\pi_0 \otimes \mathbb{C}[\tau]$,

$$V_i\, \tau = \big(\tau + \mu_i[-1] - \mu_{i+1}[-1]\big)\, V_i, \qquad i = 1, \dots, N-1, \tag{79}$$

where τ is regarded as the operator of left multiplication by τ. For each i the relation

$$V_i\, \big(\tau + \mu_N[-1]\big) \dots \big(\tau + \mu_1[-1]\big) = 0$$

then follows easily. Indeed, it reduces to the particular case $N = 2$ where we have

$$\begin{aligned}V_1\,\big(\tau+\mu_2[-1]\big)\big(\tau+\mu_1[-1]\big)\\&=\Big(\big(\tau+\mu_1[-1]-\mu_2[-1]\big)\,V_1+\mu_2[-1]\,V_1-1\Big)\big(\tau+\mu_1[-1]\big)\\&=\big(\tau+\mu_1[-1]\big)\,V_1\,\big(\tau+\mu_1[-1]\big)-\big(\tau+\mu_1[-1]\big)=0.\end{aligned}$$

Showing that the elements $T^r\mathscr{E}_i$ are algebraically independent generators requires a comparison of the sizes of graded components of π_0 and $\mathscr{W}(\mathfrak{gl}_N)$.

By the definitions (13) and (14), we have the relations

$$\sum_{k=0}^{m}(-1)^k\,\mathscr{E}_k\,h_{m-k}\big(T+\mu_1[-1],\dots,T+\mu_N[-1]\big)=0 \tag{80}$$

for $m \geqslant 1$, where $\mathscr{E}_0=1$ and $\mathscr{E}_k=0$ for $k>N$. They imply that all elements

$$h_m\big(T+\mu_1[-1],\dots,T+\mu_N[-1]\big),\qquad m\geqslant 1, \tag{81}$$

belong to $\mathscr{W}(\mathfrak{gl}_N)$. Moreover, the family (81) with $m=1,\dots,N$ is a complete set of generators of $\mathscr{W}(\mathfrak{gl}_N)$.

Note that the classical $\mathscr{W}$-algebra $\mathscr{W}(\mathfrak{sl}_N)$ associated with the special linear Lie algebra $\mathfrak{sl}_N$ can be obtained as the quotient of $\mathscr{W}(\mathfrak{gl}_N)$ by the relation $\mathscr{E}_1=0$.

4.2 Screening Operators and Generators for $\mathscr{W}(\mathfrak{o}_N)$ and $\mathscr{W}(\mathfrak{sp}_N)$

Now π_0 is the algebra of polynomials in the variables $\mu_i[r]$ with $i=1,\dots,n$ and $r<0$. The families of generators of the algebras $\mathscr{W}(\mathfrak{o}_N)$ and $\mathscr{W}(\mathfrak{sp}_N)$ reproduced below were constructed in [9, Sect. 8], where equations of the KdV type were introduced for arbitrary simple Lie algebras. The generators are associated with the Miura transformations of the corresponding equations.

Series B_n

The screening operators $V_1,\dots,V_n$ are defined by

$$V_i=\sum_{r=0}^{\infty}V_{i\,[r]}\Big(\frac{\partial}{\partial\mu_i[-r-1]}-\frac{\partial}{\partial\mu_{i+1}[-r-1]}\Big), \tag{82}$$

for $i=1,\dots,n-1$, and

$$V_n = \sum_{r=0}^{\infty} V_{n\,[r]} \frac{\partial}{\partial \mu_n[-r-1]},$$

where the coefficients $V_{i\,[r]}$ are found from the expansions

$$\sum_{r=0}^{\infty} V_{i\,[r]}\, z^r = \exp \sum_{m=1}^{\infty} \frac{\mu_i[-m] - \mu_{i+1}[-m]}{m}\, z^m, \qquad i = 1, \dots, n-1$$

and

$$\sum_{r=0}^{\infty} V_{n\,[r]}\, z^r = \exp \sum_{m=1}^{\infty} \frac{\mu_n[-m]}{m}\, z^m.$$

Define elements $\mathscr{E}_2, \dots, \mathscr{E}_{2n+1}$ of π_0 by the expansion

$$\begin{aligned} &\big(\tau - \mu_1[-1]\big) \dots \big(\tau - \mu_n[-1]\big)\, \tau\, \big(\tau + \mu_n[-1]\big) \dots \big(\tau + \mu_1[-1]\big) \\ &\qquad = \tau^{2n+1} + \mathscr{E}_2\, \tau^{2n-1} + \mathscr{E}_3\, \tau^{2n-2} + \dots + \mathscr{E}_{2n+1}. \end{aligned} \tag{83}$$

All of them belong to $\mathscr{W}(\mathfrak{o}_{2n+1})$. By (77) we have

$$\mathscr{E}_m = e_m\big(T + \mu_1[-1], \dots, T + \mu_n[-1], T, T - \mu_n[-1], \dots, T - \mu_1[-1]\big). \tag{84}$$

The family $\mathscr{E}_2, \mathscr{E}_4, \dots, \mathscr{E}_{2n}$ is a complete set of generators of $\mathscr{W}(\mathfrak{o}_{2n+1})$. The relation

$$V_i \big(\tau - \mu_1[-1]\big) \dots \big(\tau - \mu_n[-1]\big)\, \tau\, \big(\tau + \mu_n[-1]\big) \dots \big(\tau + \mu_1[-1]\big) = 0 \tag{85}$$

is verified for $i = 1, \dots, n-1$ in the same way as for $\mathfrak{gl}_N$ with the use of (79). Furthermore,

$$V_n\, \tau = \big(\tau + \mu_n[-1]\big)\, V_n,$$

so that

$$\begin{aligned} V_n \big(\tau - \mu_n[-1]\big)\, \tau\, \big(\tau + \mu_n[-1]\big) &= \big(\tau\, V_n - 1\big)\, \tau\, \big(\tau + \mu_n[-1]\big) \\ &= \tau\, \big(\tau + \mu_n[-1]\big)\, \big(\tau + 2\mu_n[-1]\big)\, V_n, \end{aligned}$$

which implies that (85) holds for $i = n$ as well.

By (80) all elements

$$h_m\big(T + \mu_1[-1], \dots, T + \mu_n[-1], T, T - \mu_n[-1], \dots, T - \mu_1[-1]\big) \tag{86}$$

belong to $\mathscr{W}(\mathfrak{o}_{2n+1})$. The family of elements (86) with $m = 2, 4, \dots, 2n$ forms another complete set of generators of $\mathscr{W}(\mathfrak{o}_{2n+1})$.

Series C_n

The screening operators $V_1, \dots, V_n$ are defined by (82) for $i = 1, \dots, n-1$, and

$$V_n = \sum_{r=0}^{\infty} V_{n\,[r]} \frac{\partial}{\partial \mu_n[-r-1]},$$

where

$$\sum_{r=0}^{\infty} V_{n\,[r]}\, z^r = \exp \sum_{m=1}^{\infty} \frac{2\mu_n[-m]}{m}\, z^m.$$

Define elements $\mathscr{E}_2, \dots, \mathscr{E}_{2n}$ of π_0 by the expansion

$$\begin{aligned}\big(\tau - \mu_1[-1]\big)\dots\big(\tau - \mu_n[-1]\big)\big(\tau + \mu_n[-1]\big)\dots\big(\tau + \mu_1[-1]\big)&\\ = \tau^{2n} + \mathscr{E}_2\, \tau^{2n-2} + \mathscr{E}_3\, \tau^{2n-3} + \dots + \mathscr{E}_{2n}&.\end{aligned}$$

All of them belong to $\mathscr{W}(\mathfrak{sp}_{2n})$. By (77) we have

$$\mathscr{E}_m = e_m\big(T + \mu_1[-1], \dots, T + \mu_n[-1], T - \mu_n[-1], \dots, T - \mu_1[-1]\big).$$

The family $\mathscr{E}_2, \mathscr{E}_4, \dots, \mathscr{E}_{2n}$ is a complete set of generators of $\mathscr{W}(\mathfrak{sp}_{2n})$. The relation

$$V_i\,\big(\tau - \mu_1[-1]\big)\dots\big(\tau - \mu_n[-1]\big)\big(\tau + \mu_n[-1]\big)\dots\big(\tau + \mu_1[-1]\big) = 0 \qquad (87)$$

is verified for $i = 1, \dots, n-1$ in the same way as for $\mathfrak{gl}_N$ with the use of (79). In the case $i = n$ we have

$$V_n\, \tau = \big(\tau + 2\mu_n[-1]\big)\, V_n,$$

so that

$$\begin{aligned}V_n\,\big(\tau - \mu_n[-1]\big)\big(\tau + \mu_n[-1]\big) &= \Big(\big(\tau + \mu_n[-1]\big)\, V_n - 1\Big)\big(\tau + \mu_n[-1]\big)\\ &= \big(\tau + \mu_n[-1]\big)\big(\tau + 3\mu_n[-1]\big)\, V_n,\end{aligned}$$

and (87) with $i = n$ also follows.

It follows from (80) that the elements

$$h_m\big(T + \mu_1[-1], \dots, T + \mu_n[-1], T - \mu_n[-1], \dots, T - \mu_1[-1]\big)$$

with $m = 2, 4, \dots, 2n$ form another complete set of generators of $\mathscr{W}(\mathfrak{sp}_{2n})$.

Series D_n

The screening operators $V_1, \dots, V_n$ are defined by (82) for $i = 1, \dots, n-1$, and

$$V_n = \sum_{r=0}^{\infty} V_{n\,[r]} \Big(\frac{\partial}{\partial \mu_{n-1}[-r-1]} + \frac{\partial}{\partial \mu_n[-r-1]} \Big)$$

where

$$\sum_{r=0}^{\infty} V_{n\,[r]}\, z^r = \exp \sum_{m=1}^{\infty} \frac{\mu_{n-1}[-m] + \mu_n[-m]}{m}\, z^m.$$

Define elements $\mathscr{E}_2, \mathscr{E}_3, \dots$ of π_0 by the expansion of the *pseudo-differential operator*

$$\big(\tau - \mu_1[-1]\big) \dots \big(\tau - \mu_n[-1]\big)\, \tau^{-1} \big(\tau + \mu_n[-1]\big) \dots \big(\tau + \mu_1[-1]\big)$$
$$= \tau^{2n-1} + \sum_{k=2}^{\infty} \mathscr{E}_k\, \tau^{2n-k-1}.$$

The coefficients $\mathscr{E}_k$ are calculated with the use of the relations

$$\tau^{-1} \mu_i[-r-1] = \sum_{k=0}^{\infty} \frac{(-1)^k\,(r+k)!}{r!}\, \mu_i[-r-k-1]\, \tau^{-k-1}.$$

All the elements $\mathscr{E}_k$ belong to $\mathscr{W}(\mathfrak{o}_{2n})$. Moreover, define $\mathscr{E}_n' \in \pi_0$ by

$$\mathscr{E}_n' = \big(\mu_1[-1] - T\big) \dots \big(\mu_n[-1] - T\big), \tag{88}$$

so that this element coincides with (19). The family $\mathscr{E}_2, \mathscr{E}_4, \dots, \mathscr{E}_{2n-2}, \mathscr{E}_n'$ is a complete set of generators of $\mathscr{W}(\mathfrak{o}_{2n})$. The identity

$$V_i \big(\tau - \mu_1[-1]\big) \dots \big(\tau - \mu_n[-1]\big)\, \tau^{-1} \big(\tau + \mu_n[-1]\big) \dots \big(\tau + \mu_1[-1]\big) = 0 \tag{89}$$

is verified with the use of (79) and the additional relations

$$V_i\, \tau^{-1} = \big(\tau + \mu_i[-1] - \mu_{i+1}[-1]\big)^{-1} V_i, \qquad i = 1, \dots, n-1,$$

and

$$V_n\, \tau^{-1} = \big(\tau + \mu_{n-1}[-1] + \mu_n[-1]\big)^{-1} V_n. \tag{90}$$

In comparison with the types B_n and C_n, an additional calculation is needed for the case $i = n$ in (89). It suffices to take $n = 2$. We have

$$\begin{aligned}
&V_2\,(\tau-\mu_1[-1])(\tau-\mu_2[-1])\,\tau^{-1}\,(\tau+\mu_2[-1])(\tau+\mu_1[-1])\\
&=\Big((\tau+\mu_2[-1])\,V_2-1\Big)(\tau-\mu_2[-1])\,\tau^{-1}\,(\tau+\mu_2[-1])(\tau+\mu_1[-1])\\
&=\Big((\tau+\mu_2[-1])(\tau+\mu_1[-1])\,V_2-2\tau\Big)\,\tau^{-1}\,(\tau+\mu_2[-1])(\tau+\mu_1[-1]).
\end{aligned}$$

Furthermore, applying the operator V_2 we find

$$\begin{aligned}
V_2\,(\tau+\mu_2[-1])(\tau+\mu_1[-1]) &= \Big((\tau+\mu_1[-1]+2\mu_2[-1])\,V_2+1\Big)(\tau+\mu_1[-1])\\
&= 2\,(\tau+\mu_1[-1]+\mu_2[-1])
\end{aligned}$$

and so by (90),

$$V_2\,\tau^{-1}\,(\tau+\mu_2[-1])(\tau+\mu_1[-1]) = 2$$

thus completing the calculation.

The relations

$$V_i\,(\mu_1[-1]-T)\dots(\mu_n[-1]-T) = 0, \qquad i = 1,\dots,n,$$

are verified with the use of (78).

5 Generators of the $\mathscr{W}$-Algebras

Here we prove the Main Theorem stated in the Introduction by deriving it from Theorems 3.2, 3.5 and 3.8.

Choose a basis $X_1,\dots,X_d$ of the simple Lie algebra $\mathfrak{g}$ and write the commutation relations

$$[X_i, X_j] = \sum_{k=1}^{d} c_{ij}^{k}\, X_k$$

with structure constants c_{ij}^k. Consider the Lie algebras $\mathfrak{g}[t]$ and $t^{-1}\mathfrak{g}[t^{-1}]$ and combine their generators into formal series in u^{-1} and u,

$$X_i(u) = \sum_{r=0}^{\infty} X_i[r]\,u^{-r-1} \qquad \text{and} \qquad X_i(u)_+ = \sum_{r=0}^{\infty} X_i[-r-1]\,u^{r}.$$

The commutation relations of these Lie algebras written in terms of the formal series take the form

$$(u-v)\,[X_i(u), X_j(v)] = -\sum_{k=1}^{d} c_{ij}^{\,k}\,\big(X_k(u) - X_k(v)\big),$$

$$(u-v)\,[X_i(u)_+, X_j(v)_+] = \sum_{k=1}^{d} c_{ij}^{\,k}\,\big(X_k(u)_+ - X_k(v)_+\big).$$

Observe that the second family of commutation relations is obtained from the first by replacing $X_i(u)$ with the respective series $-X_i(u)_+$.

On the other hand, in the classical types, the elements of the universal enveloping algebra $\mathrm{U}(\mathfrak{g}[t])$ and their Harish-Chandra images calculated in Proposition 3.1 and Theorems 3.2, 3.5 and 3.8 are all expressed in terms of the series of the form $X_i(u)$. Therefore, the corresponding Harish-Chandra images of the elements of the universal enveloping algebra $\mathrm{U}(t^{-1}\mathfrak{g}[t^{-1}])$ are readily found from those theorems by replacing $X_i(u)$ with the respective series $-X_i(u)_+$.

To be consistent with the definition for the Wakimoto modules in [11], we will write the resulting formulas for the opposite choice of the Borel subalgebra, as compared to the homomorphism (59). To this end, in types B, C and D we consider the automorphism σ of the Lie algebra $t^{-1}\mathfrak{g}_N[t^{-1}]$ defined on the generators by

$$\sigma : F_{ij}[r] \mapsto -F_{ji}[r]. \tag{91}$$

We get the commutative diagram

$$\begin{array}{ccc} \mathrm{U}\big(t^{-1}\mathfrak{g}_N[t^{-1}]\big)^{\mathfrak{h}} & \longrightarrow & \mathbb{C}\,[\mu_i[r]] \\ \sigma\big\downarrow & & \big\downarrow\sigma \\ \mathrm{U}\big(t^{-1}\mathfrak{g}_N[t^{-1}]\big)^{\mathfrak{h}} & \stackrel{\chi}{\longrightarrow} & \mathbb{C}\,[\mu_i[r]], \end{array} \tag{92}$$

where i ranges over the set $\{1,\dots,n\}$ while $r<0$. The top and bottom horizontal arrows indicate the versions of the Harish-Chandra homomorphism defined as in (59), where the left ideal I is now generated by all elements $F_{ij}[r]$ with the conditions $1 \leqslant i < j \leqslant N$ and $r<0$ for the top arrow, and by all elements $F_{ij}[r]$ with the conditions $N \geqslant i > j \geqslant 1$ and $r<0$ for the bottom arrow (which we denote by χ). The second vertical arrow indicates the isomorphism which takes $\mu_i[r]$ to $-\mu_i[r]$.

Note that an automorphism analogous to (91) can be used in the case of the Lie algebra $\mathfrak{gl}_N$ to get the corresponding description of the homomorphism χ and to derive the formulas (11) and (12). However, these formulas follow easily from the observation that $\tau + E[-1]$ is a Manin matrix by the same argument as in the proof of Proposition 3.1.

To state the result in types B, C and D, introduce the formal series

$$\gamma_m(\omega)\,\mathrm{tr}\, S^{(m)}\big(\partial_u + F_1(u)_+\big)\dots\big(\partial_u + F_m(u)_+\big), \tag{93}$$

where we use notation (16) with $\omega = N$ and $\omega = -N$ in the orthogonal and symplectic case, respectively, and

$$F(u)_+ = \sum_{i,j=1}^{N} e_{ij} \otimes F_{ij}(u)_+ \in \mathrm{End}\,\mathbb{C}^N \otimes \mathrm{U}\big(t^{-1}\mathfrak{g}_N[t^{-1}]\big)[[u]].$$

We will assume that in the symplectic case the values of m in (93) are restricted to $1 \leqslant m \leqslant 2n+1$; see [24, Sects. 3.3 and 4.1]. The trace is taken over all m copies $\mathrm{End}\,\mathbb{C}^N$ in the algebra

$$\underbrace{\mathrm{End}\,\mathbb{C}^N \otimes \dots \otimes \mathrm{End}\,\mathbb{C}^N}_{m} \otimes \mathrm{U}\big(t^{-1}\mathfrak{g}_N[t^{-1}]\big)[[u,\partial_u]] \tag{94}$$

and we use matrix notation as in (15). We set

$$\mu_i(u)_+ = \sum_{r=0}^{\infty} \mu_i[-r-1]\,u^r, \qquad i = 1,\dots,n.$$

Proposition 5.1. *The image of the series* (93) *under the homomorphism* χ *is given by the formula*:

type B_n: $\quad h_m\big(\partial_u + \mu_1(u)_+,\dots,\partial_u + \mu_n(u)_+,\partial_u - \mu_n(u)_+,\dots\partial_u - \mu_1(u)_+\big),$

type D_n: $\quad \tfrac{1}{2}\,h_m\big(\partial_u + \mu_1(u)_+,\dots,\partial_u + \mu_{n-1}(u)_+,\partial_u - \mu_n(u)_+,\dots\partial_u - \mu_1(u)_+\big)$

$$+\tfrac{1}{2}\,h_m\big(\partial_u + \mu_1(u)_+,\dots,\partial_u + \mu_n(u)_+,\partial_u - \mu_{n-1}(u)_+,\dots\partial_u - \mu_1(u)_+\big),$$

type C_n: $\quad e_m\big(\partial_u + \mu_1(u)_+,\dots,\partial_u + \mu_n(u)_+,\partial_u,\partial_u - \mu_n(u)_+,\dots\partial_u - \mu_1(u)_+\big).$

Proof. We start with the orthogonal case $\mathfrak{g}_N = \mathfrak{o}_N$. The argument in the beginning of this section shows that the image of the series

$$\gamma_m(N)\,\mathrm{tr}\, S^{(m)}\big(\partial_u - F_1(u)_+\big)\dots\big(\partial_u - F_m(u)_+\big)$$

under the homomorphism given by the top horizontal arrow in (92) is found by Theorems 3.2 and 3.5, where $\mu_i(u)$ should be respectively replaced by $-\mu_i(u)_+$ for $i = 1,\dots,n$. Therefore, using the diagram (92) we find that the image of the series

$$\gamma_m(N)\,\mathrm{tr}\, S^{(m)}\big(\partial_u + F_1^{\,t}(u)_+\big)\dots\big(\partial_u + F_m^{\,t}(u)_+\big) \tag{95}$$

under the homomorphism χ is given by the respective B_n and D_n type formulas in the proposition, where we set $F^t(u)_+ = \sum_{i,j} e_{ij} \otimes F_{ji}(u)_+$. It remains to observe that the series (95) coincides with (93). This follows by applying the simultaneous transpositions $e_{ij} \mapsto e_{ji}$ to all m copies of $\operatorname{End}\mathbb{C}^N$ and taking into account the fact that $S^{(m)}$ stays invariant.

In the symplectic case, we suppose first that $m \leqslant n$. Starting with the Harish-Chandra image provided by Theorem 3.8 and applying the same argument as in the orthogonal case, we conclude that the image of the series

$$\gamma_m(-2n)\, \operatorname{tr} S^{(m)}\big(\partial_u - F_1(u)_+\big) \dots \big(\partial_u - F_m(u)_+\big) \tag{96}$$

under the homomorphism χ agrees with the C_n type formula given by the statement of the proposition. One more step here is to observe that this series coincides with (93). Indeed, this follows by applying the simultaneous transpositions $e_{ij} \mapsto \varepsilon_i \varepsilon_j e_{j'i'}$ to all m copies of $\operatorname{End}\mathbb{C}^N$. On the one hand, this transformation does not affect the trace of any element of (94), while on the other hand, each factor $\partial_u - F_i(u)_+$ is taken to $\partial_u + F_i(u)_+$ and the operator $S^{(m)}$ stays invariant.

Finally, extending the argument of [24, Sect. 3.3] to the case $m = 2n+1$ and using the results of [24, Sect. 5], we find that for all values $1 \leqslant m \leqslant 2n+1$ the coefficients $\Phi^{(s)}_{m\,a}$ in the expansion

$$\gamma_m(-2n)\, \operatorname{tr} S^{(m)}\big(\partial_u + F_1(u)_+\big) \dots \big(\partial_u + F_m(u)_+\big) = \sum_{a=0}^{m} \sum_{s=0}^{\infty} \Phi^{(s)}_{m\,a}\, u^s\, \partial_u^a$$

belong to the Feigin–Frenkel center $\mathfrak{z}(\widehat{\mathfrak{sp}}_{2n})$. The image of the element $\Phi^{(s)}_{m\,a}$ with respect to the isomorphism (4) is a polynomial in the generators $T^r \mathscr{E}_{2k}$ of the classical $\mathscr{W}$-algebra $\mathscr{W}(\mathfrak{o}_{2n+1})$, where $k = 1, \dots, n$ and $r \geqslant 0$; see (84). For a fixed value of m and varying values of n the coefficients of the polynomial are rational functions in n. Therefore, they are uniquely determined by infinitely many values of $n \geqslant m$. This allows us to extend the range of n to all values $n \geqslant (m-1)/2$ for which the expression (93) is defined. □

Corollary 5.2. *Theorem 3.8 holds for all values* $1 \leqslant m \leqslant 2n+1$.

Proof. This follows by reversing the argument of the proof of Proposition 5.1. □

With the exception of the formula (19) for the image of the element ϕ'_n in type D_n, all statements of the Main Theorem now follow from Proposition 5.1. It suffices to note that the coefficients of the polynomial (17) and the differential operator (93) are related via the vertex algebra structure on the vacuum module $V_{-h^\vee}(\mathfrak{g}_N)$. In particular, the evaluation of the coefficients of the differential operator (93) at $u = 0$ reproduces the corresponding coefficients of the polynomial (17). This implies the desired formulas for the Harish-Chandra images in the Main Theorem; see e.g. [11, Chap. 2] for the relevant properties of vertex algebras.

Now consider the element $\mathscr{E}'_n$ of the algebra $\mathscr{W}(\mathfrak{o}_{2n})$ defined in (88) and which coincides with the element (19). To prove that the Harish-Chandra image of the element ϕ'_n introduced by (18) equals $\mathscr{E}'_n$, use the automorphism of the Lie algebra $t^{-1}\mathfrak{o}_{2n}[t^{-1}]$ defined on the generators by

$$F_{k\,l}[r] \mapsto F_{\tilde{k}\,\tilde{l}}[r], \tag{97}$$

where $k \mapsto \tilde{k}$ is the involution on the set $\{1,\dots,2n\}$ such that $n \mapsto n'$, $n' \mapsto n$ and $k \mapsto k$ for all $k \neq n, n'$. Note that $\phi'_n \mapsto -\phi'_n$ under the automorphism (97). Similarly, $\mathscr{E}'_n \mapsto -\mathscr{E}'_n$ with respect to the automorphism of $t^{-1}\mathfrak{h}_{2n}[t^{-1}]$ induced by (97).

As a corollary of the Main Theorem and the results of [24] we obtain from the isomorphism (4) that the elements

$$\mathscr{F}_m = \tfrac{1}{2}\, h_m\big(T + \mu_1[-1], \dots, T + \mu_{n-1}[-1], T - \mu_n[-1], \dots T - \mu_1[-1]\big)$$
$$+ \tfrac{1}{2}\, h_m\big(T + \mu_1[-1], \dots, T + \mu_n[-1], T - \mu_{n-1}[-1], \dots T - \mu_1[-1]\big),$$

with $m = 2, 4, \dots, 2n-2$ together with $\mathscr{E}'_n$ form a complete set of generators of $\mathscr{W}(\mathfrak{o}_{2n})$ (this fact does not rely on the calculation of the image of the Pfaffian). Observe that all elements $T^r \mathscr{F}_{2k}$ with $k = 1, \dots, n-1$ and $r \geqslant 0$ are stable under the automorphism (97). Since the Harish-Chandra image $\chi(\phi'_n)$ is a unique polynomial in the generators of $\mathscr{W}(\mathfrak{o}_{2n})$ and its degree with respect to the variables $\mu_1[-1], \dots, \mu_n[-1]$ does not exceed n, we can conclude that $\chi(\phi'_n)$ must be proportional to $\mathscr{E}'_n$. The coefficient of the product $\mu_1[-1] \dots \mu_n[-1]$ in each of these two polynomials is equal to 1 thus proving that $\chi(\phi'_n) = \mathscr{E}'_n$. This completes the proof of the Main Theorem.

The properties of vertex algebras mentioned above and the relation $\chi(\phi'_n) = \mathscr{E}'_n$ imply the respective formulas for the Harish-Chandra images of the Pfaffians $\operatorname{Pf} \tilde{F}(u)_+$ and $\operatorname{Pf} \tilde{F}(u)$ defined by (18) with the matrix $\tilde{F}[-1]$ replaced by the skew-symmetric matrices $\tilde{F}(u)_+ = [F_{ij'}(u)_+]$ and $\tilde{F}(u) = [F_{ij'}(u)]$, respectively.

Corollary 5.3. *The Harish-Chandra images of the Pfaffians are found by*

$$\chi : \operatorname{Pf} \tilde{F}(u)_+ \mapsto \big(\mu_1(u)_+ - \partial_u\big) \dots \big(\mu_n(u)_+ - \partial_u\big)\, 1,$$
$$\operatorname{Pf} \tilde{F}(u) \mapsto \big(\mu_1(u) - \partial_u\big) \dots \big(\mu_n(u) - \partial_u\big)\, 1,$$

where the second map is defined in (59).

Proof. The first relation follows by the application of the state-field correspondence map to the Segal–Sugawara vector (18) and using its Harish-Chandra image (19). To get the second relation, apply the automorphism (91) to the first relation to calculate the image of $\operatorname{Pf} \tilde{F}(u)_+$ with respect to the homomorphism defined by the top arrow in (92),

$$\operatorname{Pf} \tilde{F}(u)_+ \mapsto \big(\mu_1(u)_+ + \partial_u\big) \dots \big(\mu_n(u)_+ + \partial_u\big)\, 1.$$

Now replace $\tilde{F}(u)_+$ with $-\tilde{F}(u)$ and replace $\mu_i(u)_+$ with the respective series $-\mu_i(u)$ for $i = 1, \dots, n$. □

The isomorphism (4) and the Main Theorem provide complete sets of generators of the classical $\mathcal{W}$-algebras. In types B and C they coincide with those introduced in Sect. 4.2, but different in type D, as pointed out in the above argument.

Corollary 5.4. *The elements $\mathscr{F}_2, \mathscr{F}_4, \dots, \mathscr{F}_{2n-2}, \mathscr{E}_n'$ form a complete set of generators of $\mathcal{W}(\mathfrak{o}_{2n})$.* □

To complete this section, we point out that the application of the state-field correspondence map to the coefficients of the polynomial (17) and to the additional element (18) in type D_n yields Sugawara operators associated with $\hat{\mathfrak{g}}_N$. They act as scalars in the Wakimoto modules at the critical level. The eigenvalues are found from the respective formulas of Proposition 5.1 and Corollary 5.3 as follows from the general theory of Wakimoto modules and their connection with the classical $\mathcal{W}$-algebras; see [11, Chap. 8].

6 Casimir Elements for $\mathfrak{g}_N$

We apply the theorems of Sect. 3 to calculate the Harish-Chandra images of certain Casimir elements for the orthogonal and symplectic Lie algebras previously considered in [16]. Our formulas for the Harish-Chandra images are equivalent to those in [16], but take a different form. We will work with the isomorphism (2), where the Cartan subalgebra $\mathfrak{h}$ of the Lie algebra $\mathfrak{g} = \mathfrak{g}_N$ is defined in the beginning of Sect. 2 and the subalgebra $\mathfrak{n}_+$ is spanned by the elements F_{ij} with $1 \leqslant i < j \leqslant N$. We will use the notation $\mu_i = F_{ii}$ for $i = 1, \dots, N$ so that $\mu_i + \mu_{i'} = 0$ for all i.

Consider the evaluation homomorphism

$$\mathrm{ev} : \mathrm{U}(\mathfrak{g}_N[t]) \to \mathrm{U}(\mathfrak{g}_N), \qquad F_{ij}(u) \mapsto F_{ij} u^{-1},$$

so that $F_{ij}[0] \mapsto F_{ij}$ and $F_{ij}[r] \mapsto 0$ for $r \geqslant 1$. The image of the series $\mu_i(u)$ then coincides with $\mu_i\, u^{-1}$. Applying the evaluation homomorphism to the series involved in Theorems 3.2, 3.5 and 3.8 we get the corresponding Harish-Chandra images of the elements of the center of the universal enveloping algebra $\mathrm{U}(\mathfrak{g}_N)$. The formulas are obtained by replacing $F_{ij}(u)$ with $F_{ij} u^{-1}$ and $\mu_i(u)$ with $\mu_i\, u^{-1}$. Multiply the resulting formulas by u^m from the left. In the case $\mathfrak{g}_N = \mathfrak{o}_{2n+1}$ use the relation

$$u^m\,(\partial_u + F_1 u^{-1}) \dots (\partial_u + F_m u^{-1}) = (u\,\partial_u + F_1 - m + 1) \dots (u\,\partial_u + F_m) \qquad (98)$$

to conclude that the Harish-Chandra image of the polynomial

$$\gamma_m(N)\,\mathrm{tr}\, S^{(m)}(F_1+v-m+1)\dots(F_m+v) \tag{99}$$

with $v = u\,\partial_u$ is found by

$$\sum_{1\leqslant i_1\leqslant\dots\leqslant i_m\leqslant 1'} (\mu_{i_1}+v-m+1)\dots(\mu_{i_m}+v),$$

summed over the multisets $\{i_1,\dots,i_m\}$ with entries from $\{1,\dots,n,n',\dots,1'\}$. By the arguments of [16], the Harish-Chandra image of the polynomial (99) is essentially determined by those for the even values $m=2k$ and a particular value of v.

Corollary 6.1. *For $\mathfrak{g}_N=\mathfrak{o}_{2n+1}$ the image of the Casimir element*

$$\gamma_{2k}(N)\,\mathrm{tr}\, S^{(2k)}\,(F_1-k)\dots(F_{2k}+k-1)$$

under the Harish-Chandra isomorphism is given by

$$\sum_{1\leqslant i_1\leqslant\dots\leqslant i_{2k}\leqslant 1'} (\mu_{i_1}-k)\dots(\mu_{i_{2k}}+k-1), \tag{100}$$

summed over the multisets $\{i_1,\dots,i_{2k}\}$ with entries from $\{1,\dots,n,n',\dots,1'\}$. Moreover, the element (100) *coincides with the factorial complete symmetric function*

$$\sum_{1\leqslant j_1\leqslant\dots\leqslant j_k\leqslant n} \big(l_{j_1}^2-(j_1-1/2)^2\big)\dots\big(l_{j_k}^2-(j_k+k-3/2)^2\big), \tag{101}$$

where $l_i=\mu_i+n-i+1/2$ for $i=1,\dots,n$.

Proof. The coincidence of the elements (100) and (101) is verified by using the characterization theorem for the factorial symmetric functions [34]; see also [16]. Namely, both elements are symmetric polynomials in $l_1^2,\dots,l_n^2$ of degree k, and their top degree components are both equal to the complete symmetric polynomial $h_k(l_1^2,\dots,l_n^2)$. It remains to verify that each of the elements (100) and (101) vanishes when $(\mu_1,\dots,\mu_n)$ is specialized to a partition with $\mu_1+\dots+\mu_n<k$ which is straightforward. □

Similarly, if $\mathfrak{g}_N=\mathfrak{o}_{2n}$ use the same relation (98) to conclude from Theorem 3.5 that the Harish-Chandra image of the polynomial

$$2\,\gamma_m(N)\,\mathrm{tr}\, S^{(m)}(F_1+v-m+1)\dots(F_m+v)$$

is found by

$$\sum_{\substack{1\leqslant i_1\leqslant\cdots\leqslant i_m\leqslant 2n\\ i_s\neq n}} (\mu_{i_1}+v-m+1)\dots(\mu_{i_m}+v)$$
$$+\sum_{\substack{1\leqslant i_1\leqslant\cdots\leqslant i_m\leqslant 2n\\ i_s\neq n'}} (\mu_{i_1}+v-m+1)\dots(\mu_{i_m}+v),$$

where the summation indices in the first sum do not include n and the summation indices in the second sum do not include n'.

Corollary 6.2. *For $\mathfrak{g}_N=\mathfrak{o}_{2n}$ the image of the Casimir element*

$$\gamma_{2k}(N)\,\mathrm{tr}\,S^{(2k)}\,(F_1-k)\dots(F_{2k}+k-1)$$

under the Harish-Chandra isomorphism is given by

$$\tfrac{1}{2}\sum_{\substack{1\leqslant i_1\leqslant\cdots\leqslant i_{2k}\leqslant 2n\\ i_s\neq n}} (\mu_{i_1}-k)\dots(\mu_{i_{2k}}+k-1)+\tfrac{1}{2}\sum_{\substack{1\leqslant i_1\leqslant\cdots\leqslant i_{2k}\leqslant 2n\\ i_s\neq n'}} (\mu_{i_1}-k)\dots(\mu_{i_{2k}}+k-1).$$

Moreover, this element coincides with the factorial complete symmetric function

$$\sum_{1\leqslant j_1\leqslant\cdots\leqslant j_k\leqslant n} \big(l_{j_1}^2-(j_1-1)^2\big)\dots\big(l_{j_k}^2-(j_k+k-2)^2\big),$$

where $l_i=\mu_i+n-i$ for $i=1,\dots,n$.

Proof. The coincidence of the two expressions for the Harish-Chandra image is verified in the same way as for the case of $\mathfrak{o}_{2n+1}$ outlined above. □

Now suppose that $\mathfrak{g}_N=\mathfrak{sp}_{2n}$ and use the relation

$$u^m\,(-\partial_u+F_1\,u^{-1})\dots(-\partial_u+F_m\,u^{-1})=(-u\,\partial_u+F_1+m-1)\dots(-u\,\partial_u+F_m)$$

to conclude from Theorem 3.8 and Corollary 5.2 that the Harish-Chandra image of the polynomial

$$\gamma_m(-2n)\,\mathrm{tr}\,S^{(m)}(F_1+v+m-1)\dots(F_m+v)$$

with $v=-u\,\partial_u$ is found by

$$\sum_{1\leqslant i_1<\cdots<i_m\leqslant 1'} (\mu_{i_1}+v+m-1)\dots(\mu_{i_m}+v),$$

summed over the subsets $\{i_1, \dots, i_m\}$ of the set $\{1, \dots, n, 0, n', \dots, 1'\}$ with the ordering $1 < \dots < n < 0 < n' < \dots < 1'$, where $\mu_0 := 0$. Taking $m = 2k$ and $v = -k + 1$ we get the following.

Corollary 6.3. *For $\mathfrak{g}_N = \mathfrak{sp}_{2n}$ the image of the Casimir element*

$$\gamma_{2k}(-2n) \operatorname{tr} S^{(2k)} (F_1 + k) \dots (F_{2k} - k + 1)$$

under the Harish-Chandra isomorphism is given by

$$\sum_{1 \leqslant i_1 < \dots < i_{2k} \leqslant 1'} (\mu_{i_1} + k) \dots (\mu_{i_{2k}} - k + 1), \tag{102}$$

summed over the subsets $\{i_1, \dots, i_{2k}\} \subset \{1, \dots, n, 0, n', \dots, 1'\}$. Moreover, the element (102) *coincides with the factorial elementary symmetric function*

$$(-1)^k \sum_{1 \leqslant j_1 < \dots < j_k \leqslant n} \left(l_{j_1}^2 - j_1^2\right) \dots \left(l_{j_k}^2 - (j_k - k + 1)^2\right), \tag{103}$$

where $l_i = \mu_i + n - i + 1$ for $i = 1, \dots, n$.

Proof. To verify that the elements (102) and (103) coincide, use again the characterization theorem for the factorial symmetric functions [34]; see also [16]. Both elements are symmetric polynomials in $l_1^2, \dots, l_n^2$ of degree k, and their top degree components are both equal to the elementary symmetric polynomial $(-1)^k e_k(l_1^2, \dots, l_n^2)$. Furthermore, it is easily seen that each of the elements (102) and (103) vanishes when $(\mu_1, \dots, \mu_n)$ is specialized to a partition with $\mu_1 + \dots + \mu_n < k$. □

By using Proposition 3.1 and applying the above arguments, we get explicit formulas for Casimir elements for $\mathfrak{gl}_N$ and their Harish-Chandra images. They essentially coincide with the Capelli-type elements produced by Nazarov [33].

References

1. Arnaudon, D., Avan, J., Crampé, N., Frappat, L., Ragoucy, E.: R-matrix presentation for super-Yangians $Y(\mathrm{osp}(m|2n))$. J. Math. Phys. **44**, 302–308 (2003)
2. Arnaudon, D., Molev, A., Ragoucy, E.: On the R-matrix realization of Yangians and their representations. Annales Henri Poincaré **7**, 1269–1325 (2006)
3. Brundan, J., Kleshchev, A.: Representations of shifted Yangians and finite W-algebras. Mem. Am. Math. Soc. **196**(918) (2008)
4. Chervov, A.V., Molev, A.I.: On higher order Sugawara operators. Int. Math. Res. Not. 2009, no. 9, 1612–1635
5. Chervov, A., Talalaev, D.: Quantum spectral curves, quantum integrable systems and the geometric Langlands correspondence. arXiv:hep-th/0604128

6. Chervov, A., Falqui, G., Rubtsov, V.: Algebraic properties of Manin matrices 1. Adv. Appl. Math. **43**, 239–315 (2009)
7. Dixmier, J.: Algèbres Enveloppantes. Gauthier-Villars, Paris (1974)
8. Drinfeld, V.G.: Quantum groups. In: International Congress of Mathematicians (Berkeley, 1986), pp. 798–820. American Mathematical Society, Providence (1987)
9. Drinfeld, V.G., Sokolov, V.V.: Lie algebras and equations of Korteweg–de Vries type. J. Sov. Math. **30**, 1975–2036 (1985)
10. Feigin, B., Frenkel, E.: Affine Kac–Moody algebras at the critical level and Gelfand–Dikii algebras. Int. J. Mod. Phys. A **7**(suppl. 1A), 197–215 (1992)
11. Frenkel, E.: Langlands Correspondence for Loop Groups. Cambridge Studies in Advanced Mathematics, vol. 103. Cambridge University Press, Cambridge (2007)
12. Frenkel, E., Mukhin, E.: Combinatorics of q-characters of finite-dimensional representations of quantum affine algebras. Commun. Math. Phys. **216**, 23–57 (2001)
13. Frenkel, E., Mukhin, E.: The Hopf algebra Rep $U_q\widehat{\mathfrak{gl}}_\infty$. Selecta Math. **8**, 537–635 (2002)
14. Frenkel, E., Reshetikhin, N.: The q-characters of representations of quantum affine algebras and deformations of W-algebras. In: Recent Developments in Quantum Affine Algebras and Related Topics (Raleigh, 1998). Contemporary Mathematics, vol. 248, pp. 163–205. American Mathematical Society, Providence (1999)
15. Hernandez, D.: The Kirillov–Reshetikhin conjecture and solutions of T-systems. J. Reine Angew. Math. **596**, 63–87 (2006)
16. Iorgov, N., Molev, A.I., Ragoucy, E.: Casimir elements from the Brauer–Schur–Weyl duality. J. Algebra **387**, 144–159 (2013)
17. Isaev, A.P., Molev, A.I., Ogievetsky, O.V.: A new fusion procedure for the Brauer algebra and evaluation homomorphisms. Int. Math. Res. Not. 2012, no. 11, 2571–2606
18. Knight, H.: Spectra of tensor products of finite-dimensional representations of Yangians. J. Algebra **174**, 187–196 (1995)
19. Kuniba, A., Suzuki, J.: Analytic Bethe ansatz for fundamental representations of Yangians. Commun. Math. Phys. **173**, 225–264 (1995)
20. Kuniba, A., Ohta, Y., Suzuki, J.: Quantum Jacobi–Trudi and Giambelli formulae for $U_q(B_r^{(1)})$ from the analytic Bethe ansatz. J. Phys. A **28**, 6211–6226 (1995)
21. Kuniba, A., Okado, M., Suzuki, J., Yamada, Y.: Difference L operators related to q-characters. J. Phys. A **35**, 1415–1435 (2002)
22. Kuniba, A., Nakanishi, T., Suzuki, J.: T-systems and Y-systems in integrable systems. J. Phys. A **44**(10), 103001 (2011)
23. Molev, A.: Yangians and Classical Lie Algebras. Mathematical Surveys and Monographs, vol. 143. American Mathematical Society, Providence (2007)
24. Molev, A.I.: Feigin–Frenkel center in types B, C and D. Invent. Math. **191**, 1–34 (2013)
25. Molev, A.I., Ragoucy, E.: The MacMahon Master Theorem for right quantum superalgebras and higher Sugawara operators for $\widehat{\mathfrak{gl}}_{m|n}$. Moscow Math. J. **14**, 83–119 (2014)
26. Mukhin, E., Young, C.A.S.: Path description of type B q-characters. Adv. Math. **231**, 1119–1150 (2012)
27. Mukhin, E., Young, C.A.S.: Extended T-systems. Selecta Math. (N.S.) **18**, 591–631 (2012)
28. Mukhin, E., Tarasov, V., Varchenko, A.: Bethe eigenvectors of higher transfer matrices. J. Stat. Mech. Theory Exp. 2006, no. 8, P08002, 44 pp
29. Mukhin, E., Tarasov, V., Varchenko, A.: Bethe algebra of Gaudin model, Calogero-Moser space and Cherednik algebra. Int. Math. Res. Not. 2014, no. 5, 1174–1204
30. Nakai, W., Nakanishi, T.: Paths, tableaux and q-characters of quantum affine algebras: the C_n case. J. Phys. A **39**, 2083–2115 (2006)
31. Nakai, W., Nakanishi, T.: Paths and tableaux descriptions of Jacobi–Trudi determinant associated with quantum affine algebra of type D_n. J. Algebraic Combin. **26**, 253–290 (2007)
32. Nakajima, H.: t-analogs of q-characters of Kirillov-Reshetikhin modules of quantum affine algebras. Represent. Theory **7**, 259–274 (2003)

33. Nazarov, M.L.: Quantum Berezinian and the classical Capelli identity. Lett. Math. Phys. **21**, 123–131 (1991)
34. Okounkov, A.: Quantum immanants and higher Capelli identities. Transform. Groups **1**, 99–126 (1996)
35. Reshetikhin, N.Yu., Takhtajan, L.A., Faddeev, L.D.: Quantization of Lie groups and Lie algebras. Leningrad Math. J. **1**, 193–225 (1990)
36. Talalaev, D.V.: The quantum Gaudin system. Funct. Anal. Appl. **40**, 73–77 (2006)
37. Zamolodchikov, A.B., Zamolodchikov, Al.B.: Factorized S-matrices in two dimensions as the exact solutions of certain relativistic quantum field models. Ann. Phys. **120**, 253–291 (1979)

Vertex Operator Algebras, Modular Forms and Moonshine

Geoffrey Mason

Abstract This paper comprises a more-or-less verbatim account of four lectures on monstrous moonshine that I gave in a mini course prior to the main Heidelberg conference.

1 Lecture 1

1.1 The Monster Simple Group

Group theorists conceived the Monster sporadic simple group M in the early 1970s, although it was not officially born until 1982. Many features of M were understood well before that time, however. In particular the complete character table was already known. Here is a small part of it.

	1	$2A$	$2B$
χ_1	1	1	1
χ_2	196883	4371	275
χ_3	21296876	91884	−2324

Let V_i be the M-module that affords the character χ_i. From the character table one can compute branching rules $V_i \otimes V_j = \oplus_k c_{ijk} V_k$. In particular, the tensor square $V_2^{\otimes 2}$ decomposes into the sum of symmetric and exterior squares $S^2(V_2) \oplus \Lambda^2(V_2)$, and the branching rules show that $c_{222} = 1$ with $V_2 \subseteq S^2(V_2)$. So there is a canonical M-invariant surjection $V_2 \otimes V_2 \to V_2$, and it gives rise to a *commutative, nonassociative algebra* structure on V_1 whose automorphism group contains M.

G. Mason (✉)
University of California, Santa Cruz, Santa Cruz, CA 95064, USA
e-mail: gem@ucsc.edu

W. Kohnen and R. Weissauer (eds.), *Conformal Field Theory, Automorphic Forms and Related Topics*, Contributions in Mathematical and Computational Sciences 8,
DOI 10.1007/978-3-662-43831-2_11,

We can formally add an identity element 1 to obtain a unital, commutative algebra

$$B = V_1 \oplus V_2 \tag{1}$$

($V_1 = \mathbb{C}1$) with $M \subseteq \mathrm{Aut}(B)$.

1.2 J and $V^\natural$

Up to an undetermined constant, there is a unique modular function of weight 0 on the full modular group $\Gamma := SL_2(\mathbb{Z})$ with a simple pole of residue 1 at ∞. Such functions can be represented as quotients of holomorphic modular forms of equal weight. For example we have

$$J + 744 = \frac{\theta_{E_8}(\tau)^3}{\Delta(\tau)} = q^{-1} + 744 + 196884q + 21493760q^2 + \ldots \tag{2}$$

$$J + 24 = \frac{\theta_\Lambda(\tau)}{\Delta(\tau)} = q^{-1} + 24 + 196884q + 21493760q^2 + \ldots. \tag{3}$$

Here,

$$J = q^{-1} + 196884q + 21493760q^2 + \ldots \tag{4}$$

is the modular function with constant 0, $\Delta(\tau)$ is the discriminant

$$\Delta(\tau) = q\prod_{n=1}^{\infty}(1-q^n)^{24}, \tag{5}$$

and

$$\theta_L(\tau) = \sum_{\alpha\in L} q^{(\alpha,\alpha)/2} \tag{6}$$

is the theta function of an even lattice L; E_8 and Λ denote the E_8-root lattice and Leech lattice respectively.

John McKay noticed that the first few Fourier coefficients in (4) are simple linear combinations of dimensions of the irreducible M-modules V_i with nonnegative coefficients. This suggests that we replace the Fourier coefficients by the putative M-modules that correspond to them—a sort of 'categorification'. From (3)–(6), the coefficients of J are all nonnegative, so at least they correspond to linear spaces. Shifting the grading by 1 for later convenience, we obtain a $\mathbb{Z}$-graded linear space

$$V^\natural := V_0^\natural \oplus V_2^\natural \oplus V_3^\natural \oplus \ldots \tag{7}$$

with

$$\begin{aligned} V_0^\natural &= V_1 \\ V_2^\natural &= V_1 \oplus V_2 \\ V_3^\natural &= V_1 \oplus V_2 \oplus V_3 \\ &\dots \end{aligned} \tag{8}$$

and with dim $V_n^\natural$ = coefficient of q^{n-1} in (4). McKay's observation was promoted to the conjecture that each $V_n^\natural$ carries a 'natural' action of M. Note that $V_2^\natural$ is identified with the algebra B.

1.3 Monstrous Moonshine

With the conjectured $\mathbb{Z}$-graded M-module $V^\natural$ in hand, for each $g \in M$ we can take the graded trace of g and obtain another q-series

$$Z_g = Z_g(q) := q^{-1} \sum_{n=0}^{\infty} \mathrm{Tr}_{V_n^\natural}(g) q^n. \tag{9}$$

It was John Thompson who first asked what one can say about these additional q-expansions. (There are 174 of them, one for each conjugacy class of M). We have $Z_{1A}(1, q) = J$ by construction, and from the character table and (8) we see that

$$\begin{aligned} Z_{2A}(q) &= q^{-1} + 4372q + 96256q^2 + \dots \\ Z_{2B}(q) &= q^{-1} + 276q - 2048q^2 + \dots . \end{aligned}$$

In a celebrated paper, John Conway and Simon Norton resoundingly answered Thompson's question. They gave overwhelming evidence for the conjecture that each of the trace functions (9) was a *hauptmodul* for a subgroup of $SL_2(\mathbb{Q})$ commensurable with $SL_2(\mathbb{Z})$. This means that for each g we have a subgroup $\Gamma_g \subseteq SL_2(\mathbb{Q})$ with $|\Gamma_g : \Gamma_g \cap \Gamma|, |\Gamma : \Gamma_g \cap \Gamma| < \infty$ such that the following hold:

(i) Each Z_g is the q-expansion of a modular function of weight zero on Γ_g. (10)

(ii) If $\mathbb{H}$ is the complex upper half-plane, the compact Riemann surface $\Gamma_g \setminus \mathbb{H}^*$ is a Riemann sphere whose function field is $\mathbb{C}(Z_g)$.

If $g = 1A$ then of course $Z_1 = J$ and $\Gamma_1 = \Gamma$. Conway–Norton proposed formulae for each Z_g. For example,

$$Z_{2B}(q) = \frac{\eta(\tau)^{24}}{\eta(2\tau)^{24}} + 24, \tag{11}$$

where $\eta(\tau)$ is the Dedekind eta-function (a 24th root of the discriminant (5))

$$\eta(\tau) = q^{1/24} \prod_{n=1}^{\infty} (1 - q^n). \tag{12}$$

$Z_{2B}(q)$ is a hauptmodul for the index 3 subgroup $\Gamma_0(2) \subseteq \Gamma$.

1.4 Vertex Algebras

The problem is now to define a natural action of the Monster M on $V^\natural$ so that the graded traces Z_g satisfy the Conway–Norton moonshine conjectures (10). Borcherds' radical proposed solution involved the idea of a *vertex algebra*, which may be defined as follows. It is a pair $(V, \mathbf{1})$ consisting of a nonzero $\mathbb{C}$-linear space V and a distinguished vector $\mathbf{1} \neq 0$. Moreover, V is equipped with bilinear products

$$\mu_n : V \otimes V \to V \quad (n \in \mathbb{Z}),$$
$$u \otimes v \mapsto u(n)v \ \ (u, v \in V),$$

satisfying the following axioms for all $u, v, w \in V$:

There is $n_0 = n_0(u, v) \in \mathbb{Z}$ such that $u(n)v = 0$ for $n \geq n_0$, (13)

$v(n)\mathbf{1} = 0$ $(n \geq 0)$ and $v(-1)\mathbf{1} = v$, (14)

For all $p, q, r \in \mathbb{Z}$ we have

$$\sum_{i=0}^{\infty} \binom{p}{i} \{u(r+i)v\}(p+q-i)w = \tag{15}$$

$$\sum_{i=0}^{\infty} (-1)^i \binom{r}{i} \{u(p+r-i)v(q+i)w - (-1)^r v(q+r-i)u(p+i)w\}.$$

Thanks to (13), both sums in (15) are finite, so that (15) is sensible.

At this point the reader may well be asking, where did these identities come from, what are they good for, and what do they have to do with Monstrous Moonshine? The point of these lectures is to address these questions.

We begin by specializing (15) in various ways. It is convenient to consider $u(n) \in \mathrm{End}(V)$ to be the linear operator $v \mapsto u(n)v$ $(v \in V)$. Taking $r = 0$, the binomial $\binom{r}{i}$ vanishes unless $i = 0$ and (15) reduces to the operator identity

$$[u(p), v(q)] = \sum_{i=0}^{\infty} \binom{p}{i} \{u(i)v\}(p+q-i), \tag{16}$$

called the *commutator formula*. Similarly, taking $p = 0$ yields the *associativity formula*

$$\{u(r)v\}(q) = \sum_{i=0}^{\infty}(-1)^i \binom{r}{i} \{u(r-i)v(q+i) - (-1)^r v(q+r-i)u(i)\}. \quad (17)$$

With $n_0(u,v)$ as in (13), we obtain

$$\sum_{i=0}^{\infty}(-1)^i \binom{r}{i} \{u(p+r-i)v(q+i) - (-1)^r v(q+r-i)u(p+i)\} = 0 \quad (18)$$

whenever $r \geq n_0$, which is sometimes referred to as *commutativity*.

Assuming (13), it is not too hard to show that (15) is a consequence of the commutator and associativity formulas, and thus is *equivalent* to them. There are other equivalent ways to reformulate (15) that are useful. We explain one of them (cf. (24)) in the next section.

1.5 Locality and Quantum Fields

In the succeeding two sections we will explain how the idea of a vertex algebra corresponds to the physicist's 2-*dimensional conformal field theory*.

The important idea of a *vertex operator*, or *quantum field*, or simply *field*, defined on an arbitrary linear space V is as follows. It is a formal series

$$a(z) := \sum_{n\in\mathbb{Z}} a_n z^{-n-1} \in \mathrm{End}(V)[[z, z^{-1}]]$$

of operators a_n on V such that if $v \in V$ then $a_n v = 0$ for all large enough n. Set

$$\mathfrak{F}(V) = \{a(z) \in \mathrm{End}(V)[[z, z^{-1}]] \mid a(z) \text{ is a field}\}.$$

$\mathfrak{F}(V)$ is a linear subspace of $\mathrm{End}(V)[[z, z^{-1}]]$.

If $(V, \mathbf{1})$ is a vertex algebra, we set

$$Y(u,z) := \sum_{n\in\mathbb{Z}} u(n) z^{-n-1} \quad (u \in V),$$

where we are using notation introduced in the previous section. By construction,

$$\{Y(u,z) \mid u \in V\} \subseteq \mathfrak{F}(V), \quad (19)$$

and we can think of Y as a linear map

$$Y : V \to \mathfrak{F}(V),\ u \mapsto Y(u,z). \tag{20}$$

We use obvious notation when manipulating fields, e.g.,

$$Y(u,z)v := \sum_n \{u(n)v\}z^{-n-1} \in V[[z]][z,z^{-1}].$$

In this language, (14) reads

$$Y(u,z)\mathbf{1} = u + \sum_{n\leq -2} \{u(n)\mathbf{1}\}z^{-n-1}.$$

In particular, it follows that the Y map (20) is *injective.*

A pair of fields $a(z), b(z) \in \mathfrak{F}(V)$ are called *mutually local* if

$$(z_1 - z_2)^k [a(z_1), b(z_2)] = 0 \ \text{(some integer } k \geq 0). \tag{21}$$

This means that the (operator) coefficients of each monomial $z_1^p z_2^q$ in the following identity coincide:

$$(z_1 - z_2)^k a(z_1)b(z_2) - (z_1 - z_2)^k b(z_2)a(z_1) = 0. \tag{22}$$

Indeed,

$$\begin{aligned}
&(z_1 - z_2)^k a(z_1)b(z_2)\\
&= \sum_{i=0}^{k} (-1)^i \binom{k}{i} z_1^{k-i} z_2^i \sum_m a_m z_1^{-m-1} \sum_n b_n z_2^{-n-1}\\
&= \sum_p \sum_q \left\{ \sum_{k-i-m=-p} \ \sum_{i-n=-q} (-1)^i \binom{k}{i} a_m b_n \right\} z_1^{-p-1} z_2^{-q-1}\\
&= \sum_p \sum_q \left\{ \sum_{i=0}^{k} (-1)^i \binom{k}{i} a_{p+k-i} b_{q+i} \right\} z_1^{-p-1} z_2^{-q-1}.
\end{aligned}$$

Therefore also

$$\begin{aligned}
&(z_1 - z_2)^k b(z_2)a(z_1) = (-1)^k (z_2 - z_1)^k b(z_2)a(z_1)\\
&= (-1)^k \sum_p \sum_q \left\{ \sum_{i=0}^{k} (-1)^i \binom{k}{i} b_{q+k-i} a_{p+i} \right\} z_1^{-p-1} z_2^{-q-1},
\end{aligned}$$

whence locality (21), (22) holds if, and only if, for all integers p, q, and some nonnegative integer k we have

$$\sum_{i=0}^{k}(-1)^i \binom{k}{i} \{a_{p+k-i}b_{q+i} - (-1)^k b_{q+k-i}a_{p+i}\} = 0.$$

The last display is identical with the commutativity formula (18) if we take $r = k$. Because (18) holds for all $r \geq n_0$, it certainly holds for some positive integer k in place of r. Combining this with (19), we have established

$$\text{If } (V, \mathbf{1}) \text{ is a vertex algebra then any two vertex operators}$$
$$Y(u, z_1), Y(v, z_2)\ (u, v \in V) \text{ are mutually local fields.} \tag{23}$$

In a similar vein, let $\delta(z) := \sum_{n\in\mathbb{Z}} z^n$ be the formal delta-function, and consider the identity

$$z_0^{-1}\delta\left(\frac{z_1 - z_2}{z_0}\right) Y(u, z_1)Y(v, z_2) - z_0^{-1}\delta\left(\frac{z_2 - z_1}{-z_0}\right) Y(v, z_2)Y(u, z_1)$$
$$= z_2^{-1}\delta\left(\frac{z_1 - z_0}{z_2}\right) Y(Y(u, z_0)v, z_2). \tag{24}$$

Here, the delta-functions are expanded as power series in the *second variable* in the numerator, e.g.,

$$\delta\left(\frac{z_1 - z_2}{z_0}\right) = \sum_{n\in\mathbb{Z}} z_0^{-n}(z_1 - z_2)^n$$
$$= \sum_{n=0}^{\infty}(z_1/z_0)^n(1 - z_2/z_1)^n + \sum_{n>0}(z_0/z_1)^n \left(\sum_{i\geq 0}(z_2/z_1)^i\right)^n.$$

With this convention, identifying the operator coefficients for each monomial $z_0^p z_1^q z_2^r$ on the lhs and rhs of (24) yields exactly the identity (15).

1.6 CFT Axioms

Equation (23) is the 'main' axiom for (2-dimensional) *conformal field theory* (CFT). We now discuss the other axioms. Let $(V, \mathbf{1})$ be a vertex algebra, and introduce the endomorphism

$$D : V \to V,\ u \mapsto u(-2)\mathbf{1}. \tag{25}$$

Using (14) and associativity (17) with $q = 2$, we have

$$\begin{aligned}\{u(n)v\}(-2)\mathbf{1} &= \sum_{i=0}^{\infty}(-1)^i\binom{n}{i}\{u(n-i)v(-2+i)\mathbf{1}\}\\ &= u(n)v(-2)\mathbf{1} - nu(n-1)v(-1)\mathbf{1}\\ &= u(n)v(-2)\mathbf{1} - nu(n-1)v. \end{aligned}\tag{26}$$

Therefore,

$$\begin{aligned}[D, Y(u,z)]v &= \sum_n \{Du(n)v - u(n)Dv\}z^{-n-1}\\ &= \sum_n \{(u(n)v)(-2)\mathbf{1} - u(n)v(-2)\mathbf{1}\}z^{-n-1}\\ &= \sum_n \{-nu(n-1)v\}z^{-n-1}\\ &= \sum_n \{(-n-1)u(n)v\}z^{-n-2}\\ &= \frac{d}{dz}Y(u,z)v,\end{aligned}$$

where d/dz is the formal derivative. Hence, we obtain

$$[D, Y(u,z)] = \frac{d}{dz}Y(u,z).$$

If we take $u = v = w = \mathbf{1}$ and $p = q = r = -1$ in (15) we find that $\mathbf{1}(-2)\mathbf{1} = \mathbf{1}(-2)\mathbf{1} + \mathbf{1}(-2)\mathbf{1}$. Thus $\mathbf{1}(-2)\mathbf{1} = 0$, that is $D\mathbf{1} = 0$.

We have arrived at the following set-up: a quadruple $(V, Y, \mathbf{1}, D)$ consisting of a linear space V, a distinguished nonzero vector $\mathbf{1} \in V$, an endomorphism $D : V \to V$ with $D\mathbf{1} = 0$, and a linear injection $Y : V \mapsto \mathfrak{F}(V)$, satisfying the following for all $u, v \in V$:

$$\begin{aligned}&\underline{\text{Locality:}}\ Y(u,z_1), Y(v,z_2) \text{ are mutually local fields,}\\ &\underline{\text{Creativity:}}\ Y(u,z)\mathbf{1} = u + O(z),\\ &\underline{\text{Translation covariance:}}\ [D, Y(u,z)] = d/dzY(u,z).\end{aligned}\tag{27}$$

The axioms (27) amount to a mathematical formulation of 2-dimensional CFT, and we have shown that a vertex algebra $(V, \mathbf{1})$ naturally defines a CFT. Conversely if $(V, Y, \mathbf{1}, D)$ is a CFT then it can be shown that $(V, \mathbf{1})$ is a vertex algebra. Basically, this means that the full strength of (15) can be recovered (27).

The nomenclature in (27) is fairly standard in the physical literature, and we use it in what follows. In addition, $\mathbf{1}$ is the *vacuum vector*, V is a *Fock space*, elements in V are *states*, Y is the *state-field correspondence*, $u(n)$ is the nth *mode* of $Y(u,z)$. Creativity is interpreted to mean that the state u is created from the vacuum by the field $Y(u,z)$ corresponding to u.

There are several other useful identities that follow without difficulty from our axiomatic set-up. Among them we mention the following.

$$Y(\mathbf{1},z) = \mathrm{Id}_V, \tag{28}$$

$$Y(u,z)\mathbf{1} = e^{zD}u = \sum_{n=0}^{\infty} \frac{D^n u}{n!} z^n,$$

$$u(n)v = (-1)^{n+1} \sum_{i=0}^{\infty} \frac{(-D)^i}{i!} v(n+i)u. \tag{29}$$

(29) is called *skew-symmetry*.

2 Lecture 2

2.1 Lie Algebras and Local Fields

Certain infinite-dimensional Lie algebras naturally give rise to mutually local fields. In this section we discuss some important examples that illustrate some of the ideas developed so far.

1. *Affine algebras.*

Let L be a (complex) Lie algebra with bracket $[a,b]$ $(a,b \in L)$, equipped with a *symmetric, invariant, bilinear form* $\langle\ ,\ \rangle : L \otimes L \to \mathbb{C}$. (Invariant means that $\langle [a,b],c\rangle = \langle a,[b,c]\rangle$ for $a,b,c \in L$). The associated affine Lie algebra is $\hat{L} := L \otimes \mathbb{C}[t,t^{-1}] \oplus \mathbb{C}K$ with central element K and bracket

$$[a \otimes t^m, b \otimes t^n] = [a,b] \otimes t^{m+n} + m\delta_{m+n,0}\langle a,b\rangle K.$$

There is a *triangular decomposition*

$$\hat{L} = \hat{L}^- \oplus \hat{L}^0 \oplus \hat{L}^+$$

with

$$\hat{L}^- := \{a \otimes t^m \mid m < 0\}, \hat{L}^+ := \{a \otimes t^m \mid m > 0\}, \hat{L}^0 := \{a \otimes t^0\} \oplus \mathbb{C}K.$$

Let W be a (left) L-module. Extend W to a $\hat{L}^+ \oplus \hat{L}^0$-module by letting $\hat{L}^+$ *annihilate* W; K act as a scalar l called the *level*. The *induced module*

$$V = V(l, W) := \mathrm{Ind}_{\hat{L}^+ \oplus \hat{L}^0}^{\hat{L}}(W) \cong S(\hat{L}^-) \otimes W \tag{30}$$

is a left $\hat{L}$-module affording the representation π, say. (The linear isomorphism in (30) comes from the Poincaré–Birkhoff–Witt theorem). A typical vector in V is a sum of vectors that look like

$$(b_1 \otimes t^{n_1}) \dots (b_k \otimes t^{n_k}) \otimes w \ (b_i \in L, w \in W, n_1 \leq \dots \leq n_k \leq -1),$$

and

$$\begin{aligned} \pi(a \otimes t^n)\{(b_1 \otimes t^{n_1}) \dots (b_k \otimes t^{n_k}) \otimes w\} = \\ (a \otimes t^n)(b_1 \otimes t^{n_1}) \dots (b_k \otimes t^{n_k}) \otimes w \end{aligned} \tag{31}$$

where the product on the left is in the universal enveloping algebra of $\hat{L}$.

Set

$$Y(a, z) := \sum_{n \in \mathbb{Z}} \pi(a \otimes t^n) z^{-n-1} \ (a \in L). \tag{32}$$

It is easy to see that if $n + \sum_i n_i > 0$ then (31) reduces to 0. In particular, $Y(a, z) \in \mathfrak{F}(V)$. The following calculation, showing that the fields $Y(a, z)$ $(a \in L)$ are mutually local of order 2 (i.e. we may take $k = 2$ in (21)), gives a first insight into how locality comes into play. Thus

$$\begin{aligned} &(z_1 - z_2)^2 [Y(a, z_1), Y(b, z_2)] \\ &= (z_1 - z_2)^2 \sum_{m,n \in \mathbb{Z}} [\pi(a \otimes t^m), \pi(b \otimes t^n)] z_1^{-m-1} z_2^{-n-1} \\ &= (z_1 - z_2)^2 \sum_{m,n \in \mathbb{Z}} \pi([(a \otimes t^m), (b \otimes t^n)]) z_1^{-m-1} z_2^{-n-1} \\ &= (z_1 - z_2)^2 \sum_{m,n \in \mathbb{Z}} \{\pi([a, b] \otimes t^{m+n}) + m\delta_{m+n,0} \langle a, b \rangle \pi(K)\} z_1^{-m-1} z_2^{-n-1} \\ &= (z_1 - z_2)^2 \left\{ \sum_{p \in \mathbb{Z}} \pi([a, b] \otimes t^p) \sum_{m \in \mathbb{Z}} z_1^{-m-1} z_2^{m-p-1} \right. \\ &\quad \left. + \langle a, b \rangle \pi(K) \sum_{m \in \mathbb{Z}} m z_1^{-m-1} z_2^{m-1} \right\} \end{aligned}$$

$$= z_2^{-p-2}(z_1 - z_2)^2 \sum_{p\in\mathbb{Z}} \pi([a,b]\otimes t^p) \sum_{m\in\mathbb{Z}} z_1^{-m-1} z_2^{m+1} +$$

$$z_2^{-2}(z_1 - z_2)^2 \langle a,b\rangle \pi(K) \sum_{m\in\mathbb{Z}} m z_1^{-m-1} z_2^{m+1}$$

$$= z_2^{-p}\left(\frac{z_1}{z_2} - 1\right)^2 \sum_{p\in\mathbb{Z}} \pi([a,b]\otimes t^p)\delta\left(\frac{z_1}{z_2}\right) - \left(\frac{z_1}{z_2} - 1\right)^2 \langle a,b\rangle \pi(K)\delta'\left(\frac{z_1}{z_2}\right). \tag{33}$$

Here $\delta(z)$ is as in Sect. 1.4 (cf. comments preceding (24)), and $\delta'(z) := \sum_{n\in\mathbb{Z}} nz^{n-1}$. Now check that $(z-1)^k\delta(z) = 0$ for $k \geq 1$, $(z-1)^k\delta'(z) = 0$ for $k \geq 2$. In particular, (33) vanishes and $(z_1 - z_2)^2[Y(a,z_1), Y(b,z_2)] = 0$, as asserted.

When $W = \mathbb{C}v_0$ is the trivial 1-dimensional L-module we can go further, and see the beginnings of a CFT. Here,

$$\begin{aligned} V &= V(l, \mathbb{C}v_0) \cong S(\hat{L}^{-1}) \otimes \mathbb{C}v_0 \\ &= S(\oplus_{m=1}^{\infty} L\otimes t^{-m}) \otimes \mathbb{C}v_0 \\ &= \mathbb{C}(1\otimes v_0) \oplus (L\otimes t^{-1}) \otimes \mathbb{C}v_0 \\ &\qquad \oplus \left(L\otimes t^{-2} \oplus S^2(L\otimes t^{-1})\right) \otimes \mathbb{C}v_0 \oplus \dots \\ &\cong \mathbb{C}\mathbf{1} \oplus L \oplus (L \oplus S^2(L)) \oplus \dots \end{aligned} \tag{34}$$

where we have used the natural identification $L \xrightarrow{\cong} L\otimes t^{-1}$, $a \mapsto a\otimes t^{-1}$, set $1\otimes v_0 = \mathbf{1}$, and dropped v_0 from the notation for convenience.

In this way, the field $Y(a,z)$ (32) is associated with the state $a \in V$. $Y(a,z)$ is creative (cf. (27)) because

$$\begin{aligned} Y(a,z)\mathbf{1} &= \sum_{n\in\mathbb{Z}} \{\pi(a\otimes t^n)(1\otimes v_0)\} z^{-n-1} \\ &= \sum_{n=0}^{\infty} \{1\otimes (a\otimes t^n)v_0\} z^{-n-1} + \sum_{n=-1}^{-\infty} \{(a\otimes t^n)\otimes v_0\} z^{-n-1} \\ &= a\otimes t^{-1}\otimes v_0 + \sum_{n=-2}^{-\infty} \{(a\otimes t^n)\otimes v_0\} z^{-n-1} \\ &= a + O(z). \end{aligned}$$

(Because $\mathbb{C}v_0$ is the trivial L-module then $(a\otimes t^n)v_0 = 0$ for $n \geq 0$). $Y(a,z)$ is also translation covariant (loc. cit.): if $m \geq 1$ then

$$[d/dt, Y(a,z)](b \otimes t^{-m})$$

$$= \frac{d}{dt}\sum_{n\in\mathbb{Z}}\{a \otimes t^n.b \otimes t^{-m} \otimes v_0\}z^{-n-1} + m\sum_{n\in\mathbb{Z}}\{a \otimes t^n.b \otimes t^{-m-1} \otimes v_0\}z^{-n-1}$$

$$= \frac{d}{dt}\sum_{n<0}\{a \otimes t^n.b \otimes t^{-m} \otimes v_0\}z^{-n-1} +$$

$$\frac{d}{dt}\sum_{n\geq 0}\{[a,b] \otimes t^{n-m} \otimes v_0 + n\delta_{n,m}\langle a,b\rangle K \otimes v_0\}z^{-n-1} +$$

$$m\sum_{n\in\mathbb{Z}}\{a \otimes t^n.b \otimes t^{-m-1} \otimes v_0\}z^{-n-1}$$

$$= \sum_{n<0}\{na \otimes t^{n-1}.b \otimes t^{-m} \otimes v_0 - ma \otimes t^n.b \otimes t^{-m-1} \otimes v_0\}z^{-n-1} +$$

$$\sum_{n\geq 0}\{(n-m)[a,b] \otimes t^{n-m-1} \otimes v_0\}z^{-n-1} + m\sum_{n\in\mathbb{Z}}\{a \otimes t^n.b \otimes t^{-m-1} \otimes v_0\}z^{-n-1}$$

$$= \sum_{n<0}\{na \otimes t^{n-1}.b \otimes t^{-m} \otimes v_0\}z^{-n-1} + m\sum_{n\geq 0}\{a \otimes t^n.b \otimes t^{-m-1} \otimes v_0\}z^{-n-1} +$$

$$\sum_{n\geq 0}\{(n-m)[a,b] \otimes t^{n-m-1} \otimes v_0\}z^{-n-1}$$

$$= \sum_{n<0}\{na \otimes t^{n-1}.b \otimes t^{-m} \otimes v_0\}z^{-n-1} + m\sum_{n\geq 0}\{[a,b] \otimes t^{n-m-1} \otimes v_0\}z^{-n-1} +$$

$$m(m+1)\langle a,b\rangle\{K \otimes v_0\}z^{-m-2} + \sum_{n\geq 0}\{(n-m)[a,b] \otimes t^{n-m-1} \otimes v_0\}z^{-n-1}$$

$$= \sum_{n<0}\{na \otimes t^{n-1}.b \otimes t^{-m} \otimes v_0\}z^{-n-1} + \sum_{n\geq 0} n\{[a,b] \otimes t^{n-m-1} \otimes v_0\}z^{-n-1} +$$

$$m(m+1)\langle a,b\rangle\{K \otimes v_0\}z^{-m-2}$$

$$= -\frac{d}{dz}\left\{\sum_{n<0}\{a \otimes t^{n-1}.b \otimes t^{-m} \otimes v_0\}z^{-n} + \sum_{n\geq 0}\{[a,b] \otimes t^{n-m-1} \otimes v_0\}z^{-n} + \right.$$

$$\left. m\langle a,b\rangle\{K \otimes v_0\}z^{-m-1}\right\}$$

$$= -\frac{d}{dz}\left\{\sum_{n\in\mathbb{Z}}\{a \otimes t^{n-1}.b \otimes t^{-m} \otimes v_0\}z^{-n}\right\}$$

$$= -\frac{d}{dz}Y(a,z)b \otimes t^{-m}.$$

This shows that $[D, Y(a,z)] = d/dzY(a,z)$ where $D = -d/dt$, and because $\mathbf{1}$ is independent of t then $D\mathbf{1} = 0$. It should come as no surprise that in fact $(V(l, \mathbb{C}v_0), Y, 1 \otimes v_0, -d/dt)$ is a vertex algebra/CFT. Indeed, based on what we already know, the result follows from the following general result.

$$\begin{aligned}&V \text{ is a linear space with } 0 \neq \mathbf{1} \in V,\ D \in \mathrm{End}(V), \text{ and mutually local,}\\ &\text{translation covariant, creative fields } y(u,z) \in \mathfrak{F}(V)\ (u \in S \subseteq V).\\ &\text{If } V \text{ is spanned by states } u_1(n_1)\ldots u_k(n_k)\mathbf{1}\ (u_i \in S, n_i \in \mathbb{Z}) \text{ then}\\ &\text{there is a vertex algebra } (V, Y, \mathbf{1}, D) \text{ with } Y(u,z) := y(u,z)\ (u \in S).\end{aligned} \tag{35}$$

In this situation, we say that S *generates* V. Thus $(V(l, \mathbb{C}v_0), Y, 1 \otimes v_0, -d/dt)$ is a vertex algebra generated by $L = L \otimes t^{-1}$. We will denote this vertex algebra by $V(L, l)$.

2. *Virasoro algebra.* (Several aspects of this case are similar to the previous one, so we give less detail).

The Virasoro algebra is the Lie algebra with underlying linear space $Vir := \oplus_{n \in \mathbb{Z}} \mathbb{C}L_n \oplus \mathbb{C}K$ with central element K and bracket

$$[L_m, L_n] = (m-n)L_{m+n} + \frac{m^3 - m}{12}\delta_{m+n,0}K. \tag{36}$$

(The denominator 12 is conventional here; it can be removed by rescaling). There is a triangular decomposition

$$Vir = Vir^+ \oplus Vir^0 \oplus Vir^-$$

with

$$Vir^+ := \oplus_{n>0}\mathbb{C}L_n,\ Vir^0 := \mathbb{C}L_0 \oplus \mathbb{C}K,\ Vir^- = \oplus_{n<0}\mathbb{C}L_n.$$

Let $W = \mathbb{C}v_0$ be the 1-dimensional Vir^0-module such that $L_0 v_0 = hv_0$, $Kv_0 = cv_0$, extend to a $Vir^+ \oplus Vir^0$-module by letting Vir^+ *annihilate* v_0, and form the induced module

$$\begin{aligned} V = V(c,h) &= \mathrm{Ind}_{Vir^+ \oplus Vir^0}^{Vir} W \\ &\cong S(Vir^-) \otimes \mathbb{C}v_0 \\ &= S(\oplus_{n<0}\mathbb{C}L_n) \otimes \mathbb{C}v_0 \\ &\cong \mathbb{C}\mathbf{1} \oplus \mathbb{C}L_{-1} \oplus \ldots \end{aligned}$$

where $\mathbf{1} := 1 \otimes v_0$. h and c are called the *conformal weight* and *central charge* respectively. Introduce

$$Y(\omega, z) := \sum_{n\in\mathbb{Z}} L_n z^{-n-2}. \tag{37}$$

One sees easily that $Y(\omega, z) \in \mathfrak{F}(V)$. Note the slight change in convention regarding powers of z in (37), which is standard. The reader may enjoy proving that $Y(\omega, z)$ is a (self-) local field. Indeed, we have

$$(z_1 - z_2)^4[Y(\omega, z_1), Y(\omega, z_2)] = 0. \tag{38}$$

Note that

$$Y(\omega, z)\mathbf{1} = \sum_{n\in\mathbb{Z}}\{L_n\mathbf{1}\}z^{-n-2} = h\mathbf{1}z^{-2} + L_{-1}\mathbf{1}z^{-1} + L_{-2}\mathbf{1} + \dots. \tag{39}$$

So there is no chance that $Y(\omega, z)$ is creative, because $L_{-1}\mathbf{1}$ is nonzero by construction. Furthermore, as it stands ω is just an abstract symbol, not a state in V. We do not deal systematically with these issues here, but move on to the definition of *vertex operator algebra*, where in some sense they get resolved.

2.2 Vertex Operator Algebras

A *vertex operator algebra* (VOA) is a vertex algebra with additional structure that arises from a special Virasoro field of the type discussed in Sect. 2.1. Specifically, a VOA is a vertex algebra/CFT $(V, Y, \mathbf{1}, D)$ together with a distinguished state $\omega \in V$ (called the *conformal* or *Virasoro* vector) such that the following hold:

(1) $Y(\omega, z) = \sum_{n\in\mathbb{Z}} L(n)z^{-n-2}$ and the modes $L(n)$ generate an action of the Virasoro algebra *Vir* (36) in which K acts on V as a scalar c, called the *central charge* of V.

(2) $L(0)$ is a *semisimple* operator on V. Its eigenvalues lie in $\mathbb{Z}$, are *bounded below*, and have *finite-dimensional* eigenspaces.

(3) $D = L(-1)$.

This definition requires some discussion. Because $(V, Y, \mathbf{1}, D)$ is a vertex algebra, the fields $Y(u, z)$ $(u \in V)$ are required to be mutually local and creative. In particular, $Y(\omega, z)$ is necessarily self-local—a condition that can be independently verified (38). Furthermore, comparison with (39) shows that in the present situation we must have $L(0)\mathbf{1} = L(-1)\mathbf{1} = 0$ (otherwise $Y(\omega, z)$ is not creative) and $\omega = L(-2)\mathbf{1}$ (because ω is created from the vacuum by the field which corresponds to it). Note that $L(n) = \omega(n+1)$.

The associativity formula (17) yields

$$(L(-1)u)(n) = (\omega(0)u)(n) = \omega(0)u(n) - u(n)\omega(0) = [L(-1), u(n)].$$

Thanks to (3) and the last display, translation covariance may then be written

$$d/dzY(u,z) = [L(-1), Y(u,z)] = Y(L(-1)u, z).$$

In particular, $Du = L(-1)u = u(-2)\mathbf{1}$, and (3) is consistent with (25).

For $n \in \mathbb{Z}$ we let V_n be the $L(0)$-eigenspace with eigenvalue n. According to (2), we have the fundamental *spectral decomposition* (into finite-dimensional graded pieces)

$$V = \bigoplus_{n=n_0}^{\infty} V_n \tag{40}$$

where n_0 is the smallest eigenvalue of $L(0)$. Because $L(0)\mathbf{1} = 0$ then $\mathbf{1} \in V_0$.

We usually denote a VOA by the quadruple $(V, Y, \mathbf{1}, \omega)$. It is a model for the creation and annihilation of *bosons* (particles of integer spin).

The vertex algebra $V(L, l)$ can sometimes be given the structure of a VOA—we just have to find the right conformal vector. We describe two important cases where this can be achieved.

1. *Heisenberg algebra, or free bosonic theories.*

Here, the Lie algebra L is *abelian* (i.e. $[a,b] = 0$ $(a, b \in L)$) of dimension l, equipped with the (unique) nondegenerate symmetric bilinear form $\langle\ ,\ \rangle$ (which is automatically invariant). K acts as the identity. The conformal vector is $\omega := 1/2\sum_{i=1}^{l} v_i(-1)v_i$ where $\{v_i\}$ is an orthonormal basis of L, and it transpires that the central charge is $c = l$. The grading by $L(0)$-eigenvalues (40) coincides with the natural tensor product grading in which $L \otimes t^{-m}$ has degree m (cf. (34)). This is the *rank l Heisenberg* VOA. It models l free (noninteracting) bosons. The special case when $l = 24$ underlies the *bosonic string*.

2. *Kac–Moody theories, or WZW models.*

In this case, L is a *finite-dimensional simple* Lie algebra, and $\langle\ ,\ \rangle$ is the *Killing form* (which is unique up to an overall scalar). The conformal vector is similar to the last case, namely $\omega = 1/2\sum_{i=1}^{\dim L} v_i(-1)v_i$ for an orthonormal basis $\{v_i\}$ of L. The central charge is $c = l \dim L/(l + h^\vee)$, and we obtain a VOA as long as $l + h^\vee \neq 0$ ($h^\vee$ is the *dual Coxeter number* of the root system associated to L).

2.3 Super Vertex Algebras

Physically realistic theories incorporate both bosons and fermions. Axiomatically, this corresponds to *super vertex (operator) algebras* (SV(O)A). We limit ourselves here to the basic definitions.

The Fock space for a SVA is a linear *superspace*, i.e. a linear space V equipped with a $\mathbb{Z}_2$-grading $V = V^0 \oplus V^1$, and a nonzero vacuum vector $\mathbf{1} \in V^0$. Here and below, superscripts will always lie in $\{0, 1\}$ regarded as the two elements of $\mathbb{Z}/2\mathbb{Z}$. We write $|u| = p$ if $u \in V^p$. V^0 and V^1 are called the *even* and *odd* parts of V respectively.

There is a correspondence $u \mapsto Y(u, z)$ between states $u \in V$ and mutually local, creative fields $Y(u, z) := \sum_{n\in\mathbb{Z}} u(n)z^{-n-1}$, and we have

$$u(n) : V^p \to V^{p+|u|}.$$

Finally, we require the super version of the basic identity (15), namely

$$\sum_{i=0}^{\infty} \binom{p}{i} \{u(r+i)v\}(p+q-i)w =$$
$$\sum_{i=0}^{\infty}(-1)^i \binom{r}{i} \{u(p+r-i)v(q+i)w - (-1)^{r+|u||v|}v(q+r-i)u(p+i)w\}.$$

The delta-function version of this (cf. (24)) reads

$$z_0^{-1}\delta\left(\frac{z_1-z_2}{z_0}\right) Y(u,z_1)Y(v,z_2) - (-1)^{|u||v|}z_0^{-1}\delta\left(\frac{z_2-z_1}{-z_0}\right) Y(v,z_2)Y(u,z_1)$$
$$= z_2^{-1}\delta\left(\frac{z_1-z_0}{z_2}\right) Y(Y(u,z_0)v,z_2). \tag{41}$$

Note that the substructure $(V^0, Y, \mathbf{1})$ is a vertex algebra. As in the case of vertex algebras, these axioms are equivalent to a SCFT for which super locality, the super analog of (22), is as follows:

$$(z_1 - z_2)^k[Y(u,z_1), Y(v,z_2)] = (-1)^{|u||v|}(z_1 - z_2)^k[Y(v,z_2), Y(u,z_1)]$$

A SVOA is a quadruple $(V, Y, \mathbf{1}, \omega)$ such that analogs of (1)–(3) of Sect. 2.2 hold. The only change is that eigenvalues of $L(0)$ are allowed to lie in $1/2\mathbb{Z}$. $L(0)$ leaves V^0 invariant, and on this subspace the eigenvalues lie in $\mathbb{Z}$. Thus $(V^0, Y, \mathbf{1}, \omega)$ is a VOA.

There are further variations on this theme, where it is assumed that additional special states and fields exist. These lead to so-called $N = 1$ SCFT, $N = 2$ SCFT, etc. They play a rôle in certain geometric and physical applications, although we will not discuss them here.

3 Lecture 3

3.1 Modules Over a VOA

Suppose that $(V, Y, \mathbf{1})$ is a VA. A module over this structure, i.e. a V-module, is a linear space W and a linear map $Y_W : V \to \mathfrak{F}(W)$, $v \mapsto Y_W(v, z) = \sum_{n\in\mathbb{Z}} v_W(n) z^{-n-1}$ such that $Y_W(\mathbf{1}, z) = Id_W$ and the analog of (15) holds, i.e. for all $u, v \in V, w \in W$ we have

$$\sum_{i=0}^{\infty} \binom{p}{i} \{u(r+i)v\}_W(p+q-i)w = \tag{42}$$

$$\sum_{i=0}^{\infty} (-1)^i \binom{r}{i} \{u(p+r-i)_W v(q+i)_W w - (-1)^r v(q+r-i)_W u(p+i)_W w\}.$$

As before, there are a number of auxiliary consequences of this identity. We mention only that the fields $Y_W(u, z)$ $(u \in V)$ are *mutually local*. Generally W has no analog of the vacuum vector, so creativity has no meaning for the fields $Y_W(u, z)$.

Suppose that $(V, Y, \mathbf{1}, \omega)$ is a VOA. A module over this structure is a V-module (in the previous sense) such that $L_W(0)$ is a semisimple operator (on W) with finite-dimensional eigenspaces. The eigenvalues of $L_W(0)$ are truncated below in the following sense: given an eigenvalue λ, there are only *finitely many* eigenvalues of the form $\lambda - n$ $(n \in \mathbb{N})$. In particular, we have a spectral decomposition of W analogous to (40).

W is called *irreducible*, or *simple*, if the only subspaces invariant under all modes $u_W(q)$ $(u \in V, q \in \mathbb{Z})$ are 0 and W. It is easy to see that $\sum_{n\in\mathbb{Z}} W_{\lambda+n}$ is always an invariant subspace. Hence, if W is a simple V-module then the spectral decomposition takes the form

$$W = \bigoplus_{n=0}^{\infty} W_{h+n} \tag{43}$$

for some uniquely determined $h = h_W \in \mathbb{C}$ called the *conformal weight* of W.

We give a few examples of V-modules.

1. If $(V, Y, \mathbf{1}, \omega)$ is a VOA then V is itself a V-module, called the *adjoint* module.
2. Suppose that V is a VOA and $W \subseteq V$ satisfies $u(n)w \in W$ $(u \in V, w \in W)$. Then W is a V-module. W is called an *ideal* of V, because we also have $w(n)u \in W$ (use skew-symmetry (29)). It follows that V/W is the Fock space of a VOA (the *quotient VOA of* V) in which the mode $(u + W)(q)$ of $Y(u + W, z)$ is the operator induced on V/W by $u(q)$. We call V *simple* if the only ideals are the trivial ones V and 0. For example, any Heisenberg VOA $V(l, \mathbb{C}v_0)$ is simple.

3. If $(V, Y, \mathbf{1}, \omega)$ is a SVOA (cf. Sect. 2.3) the odd part V^1 is a module over the even part V^0. In this case the conformal weight of V^1 lies in $1/2\mathbb{Z}$.
4. Recall the rank l Heisenberg VOA $V(l, \mathbb{C}v_0)$ (cf. Sect. 2.2) generated by a rank l abelian Lie algebra L. For an L-module W we constructed (Sect. 2.1) a space $V(l, W)$ and mutually local fields $Y(a, z) \in \mathfrak{F}(V(l, W))$ $(a \in L)$. It is not hard to see that $V(l, W)$ is a $V(l, \mathbb{C}v_0)$-module, and it is simple whenever $\dim W = 1$.

3.2 Lattice Theories

For our purposes, an *integral lattice* L is a finitely generated free abelian group equipped with a positive-definite, symmetric, $\mathbb{Z}$-valued bilinear form $\langle\, ,\, \rangle : L \times L \to \mathbb{Z}$. Let $l := \mathrm{rk} L$. Set $H := \mathbb{C} \otimes_{\mathbb{Z}} L$ and let $\langle\, ,\, \rangle$ also denote the linear extension to H. We regard H as an abelian Lie algebra equipped with a symmetric invariant bilinear form. As such, there is the associated Heisenberg VOA $V(l, \mathbb{C}v_0)$ (cf. Sects. 2.1 and 2.2).

Fix $\beta \in L$. Let $\mathbb{C}e^\beta$ be the 1-dimensional linear space spanned by e^β, regarded as an H-module through the action

$$\alpha.e^\beta := \langle \alpha, \beta \rangle e^\beta \quad (\alpha \in H). \tag{44}$$

Associated to this L-module is the simple $V(l, \mathbb{C}v_0)$-module $V(l, \mathbb{C}e^\beta)$. Note that $\mathbb{C}e^0$ is the trivial L-module, so that it can be identified with $\mathbb{C}v_0$. Also, we have a linear isomorphism $V(l, \mathbb{C}e^\beta) \cong S(\hat{H}^-) \otimes \mathbb{C}e^\beta$ (cf. (30)). Form the Fock space

$$\begin{aligned} V_L &:= \bigoplus_{\beta \in L} V(l, \mathbb{C}e^\beta) \\ &\cong S(\hat{H}^-) \otimes \bigoplus_{\beta \in L} \mathbb{C}e^\beta \\ &= S(\hat{H}^-) \otimes \mathbb{C}[L]. \end{aligned} \tag{45}$$

(It is convenient to identify the *group algebra* $\mathbb{C}[L]$ of L with $\oplus_\beta \mathbb{C}e^\beta$). We discuss the following result:

V_L carries the structure of a SVOA; if L is an *even* lattice (i.e. $\langle \beta, \beta \rangle \in 2\mathbb{Z}$ for $\beta \in L$), then V_L is a VOA.

$S(\hat{H}^-)$ is naturally identified with the Heisenberg VOA itself, and in particular it is generated (cf. (35)) by the fields $Y(\alpha, z)$ $(\alpha \in H)$. Because each $V(l, \mathbb{C}e^\beta)$ is a Heisenberg module, the $Y(\alpha, z)$ naturally extend to (mutually local) fields on V_L.

To get a generating set of fields for V_L we would need to extend the set of $Y(\alpha, z)$ to a larger set of mutually (super) local fields by defining fields $Y(1\otimes e^{\beta}, z)$ ($\beta \in L$) directly. We will skip the details here. Recall (cf. Sect. 2.2) that the conformal vector for the Heisenberg VOA is $\omega := 1/2\sum_{i=1}^{l} v_i(-1)v_i$ for an orthonormal basis $\{v_i\}$ of H. This state is also taken as the conformal vector of V_L. In particular, the central charge of V_L is the rank l of L. The field $Y(\omega, z) = \sum_n L(n)z^{-n-2}$ determined by ω is defined in the natural way, i.e. on $V(l, \mathbb{C}e^{\beta})$ it acts as $Y_{V(l,\mathbb{C}e^{\beta})}(\omega, z)$. Since each summand in (45) is a Heisenberg module, $L(0)$ acts semisimply on each of them, and therefore on V_L. We consider the eigenvalues and eigenspaces of $L(0)$ in the next section. Finally, we note that V_L is a simple VOA if L is even.

3.3 *Partition Functions*

Suppose that $(V, Y, \mathbf{1}, \omega)$ is a VOA of central charge c (cf. Sect. 2.2, axiom 1)), and spectral decomposition (40) into $L(0)$-eigenspaces. The *partition function* of V is the formal q-series

$$Z(q) = Z_V(q) := q^{-c/24}\sum_{n=n_0}^{\infty} \dim V_n q^n. \tag{46}$$

(This is the first place that c has played a rôle in the proceedings). Generally, for a simple V-module W with spectral decomposition (43), the corresponding partition function is

$$Z(q) = Z_W(q) := q^{h-c/24}\sum_{n=0}^{\infty} \dim V_n q^n. \tag{47}$$

These expressions make sense because $L(0)$-eigenspaces in both cases are finite-dimensional. Indeed, it will be convenient to define the partition function for any graded space in the same way, as long as it too makes sense. One can often check the VOA axioms regarding the conformal vector (Sect. 2.2, axiom 2)) by directly computing the corresponding partition function. We will carry this out in the case of the Fock spaces for the Heisenberg VOA and the lattice theory V_L.

For the rank l Heisenberg theory $V = V(l, \mathbb{C}v_0)$ we saw (34) that V has a tensor decomposition $S(\oplus_{m=1}^{\infty} L\otimes t^{-m})\otimes \mathbb{C}v_0$ (L is the abelian Lie algebra of rank l). It is not hard to see that the $L(0)$-grading respects this decomposition, and that $L\otimes t^{-m}$ is an eigenspace with eigenvalue m. Since symmetric powers are multiplicative over direct sums, we obtain

$$Z_{V(l,\mathbb{C}v_0)}(q) = q^{-l/24}\prod_{m=1}^{\infty} (\text{partition function of } S(L\otimes t^{-m}))$$

$$= q^{-l/24} \prod_{m=1}^{\infty} (1 + q^m + q^{2m} + \ldots)^l$$

$$= q^{-l/24} \prod_{m=1}^{\infty} (1 - q^m)^{-l} = \eta(q)^{-l},$$

$\eta(q)$ being the eta function (12).

We turn to the lattice theory V_L. The partition function for V_L is the product of those for the two factors $S(\hat{H}^-)$ and $\mathbb{C}[L]$ in (45). Moreover, the first of these is just the partition function for the Heisenberg theory that we just computed. As for the second factor, using the module version of associativity (17) we have

$$\begin{aligned} L(0).1 \otimes e^\beta &= 1/2 \sum_{i=1}^{l} (v_i(-1)v_i)(1).1 \otimes e^\beta \\ &= 1/2 \sum_{i=1}^{l} \sum_{j=0}^{\infty} \{(v_i(-1-j)v_i)(1+j) + v_i(-j)v_i(j)\} 1 \otimes e^\beta \\ &= 1/2 \sum_{i=1}^{l} \{v_i(0)v_i(0)\} 1 \otimes e^\beta \\ &= 1/2 \sum_{i=1}^{l} \langle v_i, \beta \rangle^2 1 \otimes e^\beta = 1/2 \langle \beta, \beta \rangle 1 \otimes e^\beta. \end{aligned}$$

(Here, we used that $v_i(j)$ moves across the tensor sign if $j \geq 0$ and annihilates e^β if $j \geq 1$, as well as (44). The last equality holds because $\{v_i\}$ is an orthonormal basis of H). The upshot is that $1 \otimes e^\beta$ is an eigenvector for $L(0)$ with eigenvalue $1/2\langle \beta, \beta \rangle$. We therefore see that

$$\text{partition function of } \mathbb{C}[L] = \sum_{\beta \in L} q^{1/2\langle \beta, \beta \rangle} = \theta_L(q)$$

is the theta function of L (6). Altogether then, we obtain

$$Z_{V_L}(q) = \frac{\theta_L(q)}{\eta(q)^l}, \tag{48}$$

and in particular the $L(0)$-eigenspaces are indeed finite-dimensional.

Let $L_0 \subseteq L$ consist of those $\beta \in L$ such that $\langle \beta, \beta \rangle \in 2\mathbb{Z}$. Because L is an integral lattice, L_0 is a sublattice of L with $|L : L_0| \leq 2$. If $L = L_0$ then L_0 is an even lattice and V_L is a VOA. If $|L : L_0| = 2$, choose $\gamma \in L \setminus L_0$. Then, with an obvious notation, there is a decomposition

$$V_L = S(\hat{H}^-) \otimes \mathbb{C}[L_0] \oplus S(\hat{H}^-) \otimes \mathbb{C}[L_0 + \gamma],$$

where $S(\hat{H}^-) \otimes \mathbb{C}[L_0], S(\hat{H}^-) \otimes \mathbb{C}[L_0 + \gamma]$ are the parts of V_L graded by $\mathbb{Z}$ and $1/2 + \mathbb{Z}$ respectively. In this case, V_L is a SVOA and $S(\hat{H}^-) \otimes \mathbb{C}[L_0]$ and $S(\hat{H}^-) \otimes \mathbb{C}[L_0 + \gamma]$ are the even and odd parts.

With this discussion, we have at last made contact with the ideas of Sect. 1.2. For if we take L to be the *Leech lattice* Λ (a self-dual, even lattice of rank 24), then according to (48) we have

$$Z_{V_\Lambda}(q) = \frac{\theta_\Lambda(q)}{\Delta(q)}, \tag{49}$$

and (using (5)) this is the partition function (3). Similar comments apply to (2), which is now seen to be the partition function for V_{3E_8}.

Thanks to (48) and known transformation properties of θ- and η-functions, it follows that the partition function $Z_{V_L}(q)$ of a lattice theory is a modular function of weight zero on a congruence subgroup of the modular group. We derived this result only after explicitly computing the partition function, but in fact there is a large class of VOAs for which *a priori* results about the partition function and its transformation properties can be proved without explicitly knowing what the partition function is. This is the class of *regular* VOAs.

One point that we will not pursue but that deserves mention is this: the partition function of a VOA is a *formal* q-expansion, with no *a priori* convergence properties. On the other hand, at least for a regular VOA, the partition function turns out to be *holomorphic* in the complex upper half-plane $\mathfrak{H}$ when we think of it as a function $Z_V(\tau)$ with $q = e^{2\pi i \tau}$, $\tau \in \mathfrak{H}$. For this reason, we now write partition functions as functions of τ rather than q.

Although there will be no time to develop the general theory of regular VOAs in these lectures, we can illustrate some of the ideas using the lattice theory V_L. If V is an arbitrary VOA, the set of modules over V are the objects of a category V-Mod. A *morphism* $f : W_1 \to W_2$ between two V-modules W_1, W_2 is a linear map such that

$$f(u(n)w) = u(n)(f(w)) \quad (u \in V, n \in \mathbb{Z}, w \in W_1).$$

In terms of fields, this reads $fY_{W_1}(u,z) = Y_{W_2}(u,z)f$. Roughly speaking, V is called *rational* if V-Mod is *semisimple*, i.e. every V-module is a direct sum of simple V-modules. (In fact, one has to include additional types of modules that we did not discuss in Sect. 3.1). It can be shown that a rational VOA has only *finitely many* (isomorphism classes of) simple V-modules. A VOA is regular if it is both rational in the above sense and satisfies an additional condition that we will not discuss here.

If L is an even lattice as before then V_L is indeed a regular VOA. It therefore has only finitely many inequivalent simple modules, and in fact they are enumerated by the *quotient group* L^0/L where L^0 is the *dual lattice* of L. If we set $E := \mathbb{R} \otimes_{\mathbb{Z}} L$ then the *dual lattice* is

$$L^0 := \{\alpha \in E \mid \langle \alpha, \beta \rangle \in \mathbb{Z} \, (\beta \in L)\}.$$

Because L is integral and positive-definite then $L \subseteq L^0$ is a subgroup of finite index. The simple V_L-modules have a structure that is parallel to V_L itself. The Fock spaces are

$$\begin{aligned} V_{L+\gamma} &:= \bigoplus_{\beta \in L+\gamma} V(l, \mathbb{C}e^{\beta}) \\ &\cong S(\hat{H}^-) \otimes \bigoplus_{\beta \in L+\gamma} \mathbb{C}e^{\beta} \\ &= S(\hat{H}^-) \otimes \mathbb{C}[L+\gamma], \end{aligned} \tag{50}$$

(compare with (45)), where $L + \gamma \in L^0/L$. The partition function is

$$Z_{V_{L+\gamma}}(\tau) = \frac{\theta_{L+\gamma}(\tau)}{\eta(\tau)^l},$$

which is once again a modular function of weight zero on a congruence subgroup of the modular group. Indeed, one knows that the linear space

$$P := \langle \theta_{L+\gamma}(\tau)/\eta(\tau)^l \mid L + \gamma \in L^0/L \rangle$$

spanned by these partition functions furnishes a representation of the modular group (through the usual action $\tau \mapsto \frac{a\tau+b}{c\tau+d}$). This set-up is conjectured to hold for all rational VOAs V; that is, if P is the span of the partition functions of the (finitely many) simple V-modules then P affords a representation of the modular group that factors through a principal congruence subgroup. This phenomenon is often called *modular-invariance* of rational VOAs.

An important special case of these ideas arises when the VOA V is not only rational, but has (up to isomorphism) a *unique* simple module, namely the adjoint module V. We call such a V *holomorphic*. Then our discussion of modular-invariance shows that the partition function $Z_V(\tau)$ of a holomorphic VOA is a modular function on the full modular group (perhaps with a character). For example, since the simple V_L-modules are indexed by the cosets of L in L^0, it follows that V_L is holomorphic if, and only if, $L = L^0$ is *self-dual*. The Leech lattice Λ and orthogonal sums of the E_8 root lattice are examples of self-dual lattices, and indeed their partition functions (2), (3) are modular functions on the full modular group.

4 Lecture 4

4.1 The Lie Algebra on V_1

We have seen that a regular VOA that is holomorphic (i.e. has a unique simple module) has a partition function that is a modular function of weight 0 on the full modular group. A case in point is the Leech lattice theory V_Λ, which has central charge 24 ($=\mathrm{rk}\Lambda$) and partition function $Z_{V_\Lambda}(\tau) = J + 24 = q^{-1} + 24 + 196,884q + \ldots$. Our goal now is to construct a holomorphic VOA $V^\natural$, also of central charge 24, whose partition function is J (4), which has constant term 0. This is the[1] *Moonshine module.*

Although the VOAs V_Λ and $V^\natural$ have partition functions differing only in their constant term, many of their algebraic properties are quite different. Indeed, these properties are to a large extent governed by the constant term. For this reason, we begin with a general discussion of this point. We restrict attention to VOAs of *CFT-type*, which means that in the spectral decomposition (40) the pieces V_n vanish for $n < 0$ and $V_0 = \mathbb{C}\mathbf{1}$. (Recall that we always have $\mathbf{1} \in V_0$). There are many interesting VOAs that are *not* of CFT-type, nevertheless CFT-type theories are natural from a physical standpoint because they arise from 'unitarity' assumptions. Be that as it may, our basic assumption here is that the spectral decomposition of V has the shape

$$V = \mathbb{C}\mathbf{1} \oplus V_1 \oplus V_2 \oplus \ldots.$$

For states $u, v \in V_1$ we define a bracket [ ] by setting $[uv] := u(0)v$. Now

$$L(0)u(0)v = [L(0), u(0)]v + u(0)L(0)v = u(0)v.$$

($[L(0), u(0)] = 0$ by translation covariance and $L(0)v = v$ because $v \in V_1$). This shows that $u(0)v \in V_1$, so that we have a bilinear product $[\ ] : V_1 \times V_1 \to V_1$. One can check that this makes V_1 into a Lie algebra. (Use (29) for skew-symmetry $[uv] = -[vu]$ and the associativity formula (17) for the Jacobi identity $[[uv]w] + [[wu]v] + [[vw]u] = 0$).

For a *VOA V* of CFT-type and central charge $c = 24$, the partition function has the general shape $Z_V(\tau) = q^{-1} + \dim V_1 + \ldots$. So for such theories, the constant term is the *dimension* of the Lie algebra on V_1.

If L is an even lattice of rank l, the nature of the partition function (48) of the lattice theory shows that

$$Z_{V_L}(\tau) = q^{-l/24}(1 + (l + |L_2|)q + \ldots), \tag{51}$$

[1]It is expected that there is a *unique* VOA with partition function J, but this remains open.

where $L_2 = \{\alpha \in L \mid \langle \alpha, \alpha \rangle = 2\}$ are the *roots* of L. In particular, V_L is of CFT-type. The Lie algebra on $(V_L)_1$ is *reductive*, being a direct sum $\mathfrak{a} \oplus \mathfrak{g}$ where $\mathfrak{a}$ is abelian and $\mathfrak{g}$ is semisimple with root system L_2. (The set of roots in an even lattice always carries the structure of a semisimple root system embedded in the ambient Euclidean space $E = \mathbb{R} \otimes L$). For example, if $L = 3E_8$ then the Lie algebra on $(V_L)_1$ is semisimple, being the sum of three copies of the E_8 Lie algebra. (Note that $\dim E_8 = 248$, so that $\dim(V_L)_1 = 744$, in agreement with (2)). Similarly, the Leech lattice Λ has *no* roots, whence $(V_\Lambda)_1$ is abelian of rank $l = 24$.

Because J has no constant term, a VOA $V^\natural$ with partition function J and central charge $c = 24$ necessarily has *no* corresponding Lie algebra. In particular, $V^\natural$ cannot be a lattice theory, because the weight one piece never vanishes for a lattice theory (cf. (51)).

4.2 Automorphisms

Let V be a (S)VOA. An *automorphism* of V is an invertible linear map $g : V \to V$ such that $g(\omega) = \omega$ and $gv(q)g^{-1} = g(v)(q)$ for all v, q, i.e.

$$gY(v,z)g^{-1} = Y(g(v),z) \quad (v \in V). \tag{52}$$

We give some basic examples of automorphisms.

1. One checks (use induction and (16) or (17)) that for $n \geq 0$,

$$(u(0)^n v)(q) = \sum_{i=0}^{n} (-1)^i \binom{n}{i} u(0)^{n-i} v(q) u(0)^i \quad (u, v \in V,\ q \in \mathbb{Z}).$$

   Therefore,

$$\begin{aligned} \left(e^{u(0)}.v\right)(q) &= \sum_{n=0}^{\infty} \frac{1}{n!} (u(0)^n v)(q) \\ &= \sum_{n=0}^{\infty} \sum_{i=0}^{n} \frac{(-1)^i}{i!(n-i)!} u(0)^{n-i} v(q) u(0)^i \\ &= e^{u(0)} v(q) e^{-u(0)}, \end{aligned}$$

   showing that (52) holds with $g = e^{u(0)}$. Furthermore, if V is of CFT-type (cf. Sect. 3.3) and $u \in V_1$ then

$$\begin{aligned} u(0)\omega &= u(0)L(-2)\mathbf{1} \\ &= [u(0), L(-2)]\mathbf{1} + L(-2)u(0)\mathbf{1} \end{aligned}$$

$$= [u(0), \omega(-1)]\mathbf{1}$$
$$= -[\omega(-1), u(0)]\mathbf{1}$$
$$= -\sum_{i=0}^{\infty}(-1)^i(\omega(i)u)(-1-i)\mathbf{1}$$

$$= -\{(L(-1)u)(-1) - (L(0)u)(-2) + (L(1)u)(-3)\}\mathbf{1} = 0.$$

(For the last two equalities use translation covariance, $L(0)u = u$ (because $u \in V_1$), $L(1)u \in V_0 = \mathbb{C}\mathbf{1}$, $L(n)u \in V_{1-n} = 0$ for $n \geq 2$, and $\mathbf{1}(q) = \delta_{q+1,0}\mathrm{Id}_V$ (cf. (28)).

It follows from this calculation that if V is a VOA of CFT-type then $\{e^{u(0)} \mid u \in V_1\}$ is a set of automorphisms of V. In the previous section we learned that V_1 carries the structure of a Lie algebra with bracket $[uv] = u(0)v$. Now we see that the usual action of the associated Lie group $\mathfrak{G}$ generated by exponentials e^{ad_u} extends to an action of $\mathfrak{G}$ as automorphisms of V.

2. Suppose that V is a SVOA. Then there is a canonical involutorial automorphism which acts as $+1$ on the even part of V and -1 on the odd part.
3. A related example (and the one we will need later) is an involutorial automorphism t of a lattice VOA V_L, defined to be a lifting of the -1 automorphism of the lattice L. t also acts as -1 on the abelian Lie algebra $\mathbb{C} \otimes L$ and then acts as naturally on the associated Heisenberg VOA (cf. (34)—where L is the Lie algebra, not the lattice!) and on V_L, where

$$t(u \otimes e^{\beta}) = t(u) \otimes e^{-\beta} \quad (u \in S(\hat{H}^-)) \tag{53}$$

(cf. (45)).

If g is an automorphism of V then $gY(\omega, z)g^{-1} = Y(g(\omega), z) = Y(\omega, z)$, in particular $gL(0)g^{-1} = L(0)$. Thus g acts on the eigenspaces of V, i.e. the homogeneous pieces V_n. We may therefore define additional partition functions

$$Z_V(g, \tau) := q^{-c/24} \sum_{n=n_0}^{\infty} (\mathrm{Tr}_{V_n} g)q^n.$$

Let's compute this trace function for the automorphism t of V_L. It is clear from (53) that the only contributions to the trace arise from states $u \otimes e^0$, i.e. from states in the Heisenberg VOA Fock space $S(\hat{H}^-)$. Therefore by (34),

$$Z_{V_L}(t, \tau) = \text{Trace } t \text{ on } q^{-l/24} S(\oplus_{m>0} H \otimes t^{-m})$$
$$= \text{Trace } t \text{ on } q^{-l/24} \bigotimes_{m>0} S(\mathbb{C}u \otimes t^{-m})^l$$

$$
\begin{aligned}
&= q^{-l/24} \prod_{m>0} (1 - q^m + q^{2m} - \ldots)^l \\
&= q^{-l/24} \prod_{m>0} (1 + q^m)^{-l} \\
&= \left(\frac{\eta(\tau)}{\eta(2\tau)} \right)^l . \qquad (54)
\end{aligned}
$$

This is a modular function of weight 0. If $l = 24$ it is almost equal to (11)!

4.3 Twisted Sectors

Let (V, Y) be a VOA and g an automorphism of V of finite order R. A *g-twisted V-module*, or *g-twisted sector*, is a generalization of V-module (to which it reduces if $g = 1$). Precisely, it is a pair (W_g, Y_g) consisting of a Fock space W_g and a Y-map $Y_g : V \to \mathfrak{F}(W_g),\ u \mapsto Y_g(u, z)$ where

$$
Y_g(u, z) := \sum_{n \in r/R + \mathbb{Z}} u(n) z^{-n-1} \in \mathrm{End}(W)[[z^{1/R}, z^{-1/R}]]
$$

whenever $g(u) = e^{-2\pi i r/R}$ $(r \in \mathbb{Z})$, and $Y_g(\mathbf{1}, z) = \mathrm{Id}_{W_g}$. The *twisted vertex operators* $Y_g(u, z)$ are required to satisfy twisted analogs of the basic identity (15). In the delta-function formulation (cf. (24)) this reads

$$
\begin{aligned}
&z_0^{-1} \delta\left(\frac{z_1 - z_2}{z_0} \right) Y_g(u, z_1) Y_g(v, z_2) - z_0^{-1} \delta\left(\frac{z_2 - z_1}{-z_0} \right) Y_g(v, z_2) Y_g(u, z_1) \\
&\quad = z_2^{-1} \left(\frac{z_1 - z_0}{z_2} \right)^{-r/R} \delta\left(\frac{z_1 - z_2}{z_0} \right) Y_g(Y(u, z_0)v, z_2). \qquad (55)
\end{aligned}
$$

Finally, the operator $L_g(0)$ (the zero mode of $Y_g(\omega, z)$) is required to be semisimple with finite-dimensional eigenspaces. The eigenvalues satisfy a truncation condition analogous to that for V-modules (cf. the discussion in Sect. 3.1 preceding display (43)). There is an obvious notion of irreducible (or simple) g-twisted module, and as in the untwisted case (cf. (43)) the spectral decomposition of a simple g-twisted module takes the form

$$
W_g = \bigoplus_{n=0}^{\infty} (W_g)_{h_g + n/R} \qquad (56)
$$

for a scalar h_g (the conformal weight). Needless to say, the twisted sector has an associated partition function

$$Z_{W_g}(\tau) := q^{-c/24+h_g} \sum_{n=0}^{\infty} \dim(W_g)_n q^{n/R}.$$

Let us specialize to the case of the (even, self-dual) Leech lattice Λ with its associated VOA V_Λ and canonical involution t (cf. Sect. 4.2). In this case there is (up to isomorphism) a *unique* simple t-twisted module, denoted by $V_\Lambda(t,\tau)$. The following transformation law can be proved:

$$Z_{V_\Lambda}(t, -1/\tau) = Z_{V_\Lambda(t)}(\tau).$$

Using (54), the partition function of the t-twisted sector must be

$$\begin{aligned} Z_{V_\Lambda(t)}(\tau) &= \left(\frac{\eta(-1/\tau)}{\eta(-2/\tau)}\right)^{24} \\ &= 2^{12}\left(\frac{\eta(\tau)}{\eta(\tau/2)}\right)^{24} \\ &= 2^{12} q^{1/2} \prod_{n=1}^{\infty} (1 + q^{n/2})^{24}, \end{aligned} \tag{57}$$

(using the transformation law $\eta(-1/\tau) = (\sqrt{\tau}/i)\eta(\tau)$). Because the central charge is $c = 24$, it follows that the conformal weight of $V_\Lambda(t,\tau)$ is $3/2$.

Similarly to the rank 24 Heisenberg VOA, the product term in (57) is the partition function of a symmetric algebra $S(\oplus_{n>0} H \otimes t^{-n/2})$ (cf. (34)). This suggests how one might try to construct the twisted sector, though we must skip the details here. (The curious factor 2^{12} turns out to correspond to a Clifford algebra. Cf. Sect. 4.5 for further comment).

4.4 *The Moonshine Module*

Retaining the notation of the previous section, consider

$$V_\Lambda \oplus V_\Lambda(t). \tag{58}$$

The involution t acts naturally on the twisted sector: in the 'usual way' on $S(\oplus_{n>0} H \otimes t^{-n/2})$ and as -1 on the 2^{12} constant part. The Moonshine Module is then defined to be the space of t-invariants

$$V^\natural := V_\Lambda^+ \oplus V_\Lambda(t)^+. \tag{59}$$

Now every state $u \otimes e^{\beta} \in V_{\Lambda}$ $(\beta \neq 0)$ produces a t-invariant $u \otimes e^{\beta} + t(u) \otimes e^{-\beta}$. On the other hand, the partition function of the Heisenberg VOA (consisting of states $u \otimes e^{0}$) is $1/\Delta(\tau)$ and the graded trace of t is $\Delta(\tau)/\Delta(2\tau)$ (the case $l = 24$ of (54)). It follows that

$$\begin{aligned} & Z_{V_{\Lambda}^{+}}(\tau) \\ &= (Z_{V_{\Lambda}}(\tau) - 1/\Delta(\tau))/2 + (1/\Delta(\tau) + \Delta(\tau)/\Delta(2\tau))/2 \\ &= (Z_{V_{\Lambda}}(\tau) + \Delta(\tau)/\Delta(2\tau))/2 \\ &= ((q^{-1} + 24 + 196884q + \ldots) + q^{-1}(1 - 24q + 276q^{2} + \ldots))/2 \\ &= q^{-1} + 98580q + \ldots. \end{aligned}$$

On the other hand, a similar calculation using (54) and the nature of the twisted sector as a symmetric algebra shows that

$$\begin{aligned} Z_{V_{\Lambda}(t)^{+}}(\tau) &= 2^{12}(\Delta(\tau)/\Delta(\tau/2) - q\Delta(\tau/2)/\Delta(\tau))/2 \\ &= 2^{11}q^{1/2}\left(\prod_{n=1}^{\infty}(1 + q^{n/2})^{24} - \prod_{n=1}^{\infty}(1 - q^{n-1/2})^{24}\right) \\ &= 98304q + \ldots. \end{aligned}$$

Altogether, we have

$$\begin{aligned} Z_{V^{\natural}}(\tau) &= Z_{V_{\Lambda}^{+}}(\tau) + Z_{V_{\Lambda}(t)^{+}}(\tau) \\ &= (q^{-1} + 98580q + \ldots) + (98304q + \ldots) \\ &= q^{-1} + 196884q + \ldots. \end{aligned} \tag{60}$$

It is clear from the above that $Z_{V^{\natural}}(\tau)$ is a modular function of weight 0 and level at most 2, and it is easy to check that in fact it is invariant under the full modular group. Thus from the q-expansion we arrive at the identity

$$Z_{V^{\natural}}(\tau) = J.$$

The space (58) has the structure of an *abelian intertwining algebra*, a generalization of VOA and SVOA. The main missing ingredient, which we cannot go into here, is the definition of fields $Y_{V_{\Lambda} \oplus V_{\Lambda}(t)}(u, z)$ for states u in the twisted sector satisfying an appropriate variation of the basic identity (15), (24). Once this is done, t is seen to be an automorphism of this larger structure. Then it is easy to see that the t-invariant subspace $V^{\natural}$, together with the restriction of the fields to this subspace, defines the structure of a VOA on $V^{\natural}$ with central charge 24. Furthermore, $V_{\Lambda}^{+} \oplus V_{\Lambda}(t)^{-}$ is a SVOA with even part V_{Λ}^{+}. (Indeed, it is an $N = 1$ superconformal field theory, a term we alluded to but did not define in Sect. 2.3).

Consider a VOA V of CFT-type (cf. Sect. 4.1) with *trivial* Lie algebra V_1:

$$V = V_0 \oplus V_2 \oplus \dots$$

($V^\natural$ satisfies these conditions, as follows from (60)). If $u, v \in V_2$, define $u.v := u(1)v$. It is easy to check that $u(1)v \in V_2$, so that we have a (nonassociative) bilinear product on V_2. By skew-symmetry (29), $v(1)u = u(1)v - L(-1)u(2)v + L(-1)^2u(3)v/2 - \dots$. But $u(2)v \in V_1 = 0, u(3)v \in \mathbb{C}\mathbf{1}$ and $L(-1)\mathbf{1} = 0$, and all other $u(q)v$ ($q \geq 4$) lie in V_n with $n < 0$ and hence also vanish. The upshot is that $u(1)v = v(1)u$, so that V_2 has the structure of a commutative, nonassociative algebra. In the case of $V^\natural$, this is precisely the algebra B that we discussed in Sect. 1.1.

4.5 Aut $V^\natural$

Consider the CFT

$$V_\Lambda = \mathbb{C}\mathbf{1} \oplus (V_\Lambda)_1 \oplus \dots$$

where Λ is, as before, the Leech lattice. Because Λ has no roots, it follows from (51) that $\dim(V_\Lambda)_1 = 24$, and the Lie algebra on $(V_\Lambda)_1$ is *abelian*. So the automorphisms $e^{u(0)}$ ($u \in (V_\Lambda)_1$) generate a 24-dimensional complex torus T. Additional automorphisms of V_Λ arise from the automorphism group $Co_0 := \mathrm{Aut}(\Lambda)$ of the Leech lattice, and there is a (nonsplit) short exact sequence

$$1 \to T \to \mathrm{Aut}V_\Lambda \to Co_0 \to 1.$$

The automorphism t of Λ (or of V_Λ) is a central involution of Co_0, and the quotient $Co_1 := Co_0/\langle t \rangle$ is the largest sporadic (simple) Conway group of order $2^{21} \dots$.

Because t acts as -1 on T, its only fixed elements are those of order at most 2. So the *centralizer* $C(t)$ of t in AutV_Λ (the elements that commute with t) is described by another short exact sequence (also nonsplit)

$$1 \to 2^{24} \to C(t) \to Co_0 \to 1.$$

($2^{24} = \mathbb{Z}_2^{24}$ consists of the elements in T of order at most 2). Note that $|C(t)| = 2^{46} \dots$.

As regards the Monster, the relevance of $C(t)$ is that it *preserves* the decomposition (59). This is a bit subtle: t acts trivially by definition, but the action of $C(t)/\langle t \rangle$ is *projective* on $V_\Lambda(t)^+$. When it is linearized, we obtain a third group $\hat{C}$ occurring as the middle term of a short exact sequence

$$1 \to 2^{1+24} \to \hat{C} \to Co_1 \to 1,$$

where now 2^{1+24} is the (nonabelian) linearization of the projective action of the 2^{24}-group. 2^{1+24} is a so-called *extra-special* group. It is familiar in physics (24×24 Pauli matrices) and the theory of theta-functions. It has a unique faithful irreducible representation, realizable on the 2^{12}-dimensional Clifford algebra that we identified at the end of Sect. 4.3. We have $|\hat{C}| = 2^{46} \ldots$ and

$$\hat{C} \subseteq \mathrm{Aut} V^{\natural}$$

It turns out that the decomposition (59) breaks the symmetry of $V^{\natural}$ in the sense that there are further automorphisms that do *not* preserve (59) and hence do not lie in $\hat{C}$. The Monster M is the full automorphism group of $V^{\natural}$ and also of the algebra B, and

$$|M| = 2^{46}3^{20}5^{9}7^{6}11^{2}13^{3}17.19.23.31.41.47.59.71$$

These results are not easily obtained, and we say no more about them here.

The graded traces $Z_{V^{\natural}}(g, \tau)$ for $g \in M$ turn out to be hauptmoduln as described in Sect. 1.1. This result is also difficult. We end these Notes with the computation for a single automorphism g of order 2 that acts trivially on V_{Λ}^{+} and as -1 on $V_{\Lambda}(t)^{+}$. A previous calculation shows that its graded trace is a modular function of weight 0 and level 2. Specifically,

$$\begin{aligned} Z_{V^{\natural}}(g, \tau) &= Z_{V_{\Lambda}^{+}}(\tau) - Z_{V_{\Lambda}(t)^{+}}(\tau) \\ &= (q^{-1} + 98556q + \ldots) - (98304q + \ldots) \\ &= q^{-1} + 276q + \ldots \end{aligned}$$

is the hauptmodul for the Monster element $2B$ (11).

Further Reading

The textbooks and monographs listed below cover the material in these Notes and much more.

1. Conway, J., et al.: ATLAS of Finite Groups. Clarendon Press/Oxford University Press, Oxford/New York (1985)
2. Dong, C., Lepowsky, J.: Generalized Vertex Algebras and Relative Vertex Operators. Birkhäuser, Boston (1993)
3. Dong, C., Mason, G. (eds.) Moonshine, the Monster, and Related Topics. Contemporary mathematics. vol. 193. American Mathematical Society, Providence (1996)
4. Frenkel, I., Lepowsky, J., Meurman, A.: Vertex Operator Algebras and the Monster. Academic, Boston (1988)
5. Gannon, T.: Moonshine Beyond the Monster: The Bridge Connecting Algebra, Modular Forms, and Physics. Cambridge University Press, Cambridge (2006)

6. Kac, V.: Vertex Algebras for Beginners, 2nd edn. University Lecture Series, vol. 10. American Mathematical Society, Providence (1998)
7. Lepowsky, J., Li, H.: Introduction to Vertex Operator Algebras and Their Representations. Birkäuser, Boston (2004)
8. Lepowsky, J., McKay, J., Tuite, M. (eds.) Moonshine: The First Quarter Century and Beyond. London Mathematical Society Lecture Note Series, vol. 372. Cambridge University Press, Cambridge (2010)
9. Mason, G., Tuite, M.: Vertex operators and modular forms. In: A Window into Zeta and Modular Physics. Mathematical Sciences Research Institute Publications, vol. 57. Cambridge University Press, Cambridge (2010)
10. Matsuo, A., Nagatomo, K.: Axioms for a vertex algebra and the locality of quantum fields. Math. Soc. Jpn. Mem. **4** (1999)

MIX
Papier aus verantwortungsvollen Quellen
Paper from responsible sources
FSC® C105338

If you have any concerns about our products,
you can contact us on
ProductSafety@springernature.com

In case Publisher is established outside the EU,
the EU authorized representative is:
Springer Nature Customer Service Center GmbH
Europaplatz 3, 69115 Heidelberg, Germany

Printed by Libri Plureos GmbH
in Hamburg, Germany